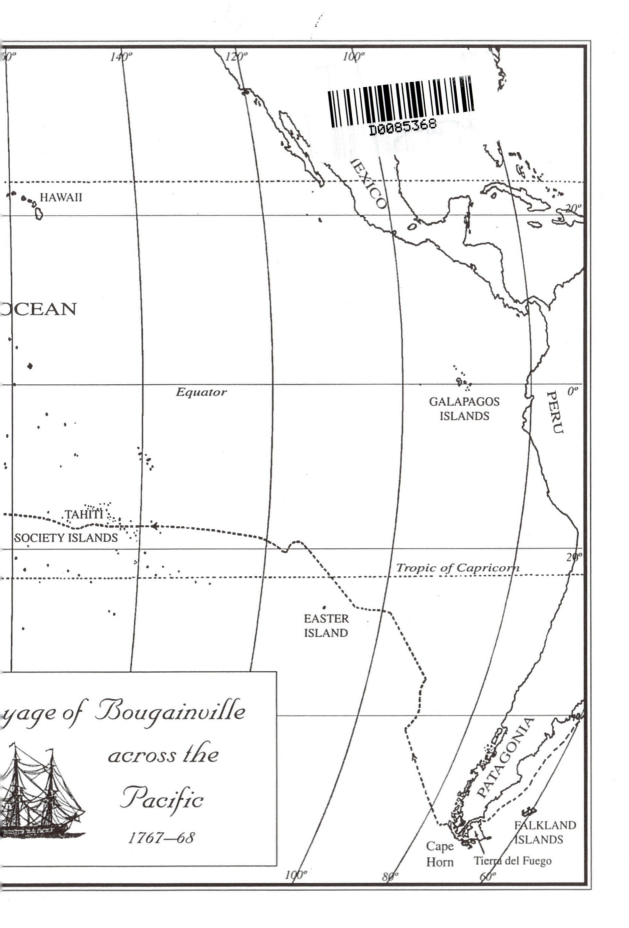

140° 120° 100°

MEXICO

HAWAII

20°

OCEAN

Equator

GALAPAGOS
ISLANDS

PERU

0°

TAHITI
SOCIETY ISLANDS

Tropic of Capricorn

20°

EASTER
ISLAND

Voyage of Bougainville

across the

Pacific

1767—68

PATAGONIA

FALKLAND
ISLANDS

Cape
Horn

Tierra del Fuego

100° 80° 60°

WORKS ISSUED BY
THE HAKLUYT SOCIETY

———

THE PACIFIC JOURNAL OF
LOUIS-ANTOINE DE BOUGAINVILLE
1767-1768

THIRD SERIES
NO. 9

THE
PACIFIC JOURNAL
OF
LOUIS-ANTOINE DE BOUGAINVILLE
1767–1768

Translated and edited by
JOHN DUNMORE

THE HAKLUYT SOCIETY
LONDON
2002

Published by The Hakluyt Society
c/o Map Library
British Library, 96 Euston Road,
London NW1 2DB

SERIES EDITORS
W. F. RYAN
ROBIN LAW

ISBN 0 904180 78 6
ISSN 0072 9396

British Library Cataloguing-in-Publication Data
A catalogue record for this book is
available from the British Library

Typeset by Waveney Typesetters, Wymondham, Norfolk
Printed in Great Britain at
the University Press, Cambridge

CONTENTS

List of Illustrations and Maps   viii
Preface   ix

EDITORIAL NOTE
    The Problem of Equivalence   xi
    Naval titles   xii
    Measurements   xii
    Currency   xiii
    Navigational Practices   xiv
    Textual Note   xv
    Abbreviations Used   xvi

INTRODUCTION
    The Background   xix
    The Participants   xxiv
    The Ships   xlii
    The Instructions   xliii
    The Voyage   xlix
    The Achievements   lxx

THE JOURNAL – 14 November 1767 to 12 December 1768   I

APPENDICES
    1. The Muster Roll   183
    2. The Journal of Caro   199
    3. The Journal of Vivez   223
    4. The Journal of Fesche   249
    5. The Journal of Nassau–Siegen   281
    6. The Commerson Documents   296

BIBLIOGRAPHY   306

INDEXES   315

ILLUSTRATIONS AND MAPS

Illustrations

1. Journal of Bougainville, first page. *Reprinted by permission of the Archives de la Marine (Colonies), Paris.*   xvii

2. View of the French settlement, Malouines, in 1763-1764, by Dom Pernetty, *Histoire abrégée d'un voyage aux îles Malouines.*   xxi

3. New Cythera, coloured drawing believed to be by Charles de Romainville. *Reprinted by permission of the Bibliothèque Nationale, Paris, Service des Cartes et Plans*   56

4. Pastel drawing by an unknown artist, representing Bougainville and his officers with islanders in Tahiti. *Held in the Rex Nan Kivell collection, Canberra, reprinted with the permission of the National Library of Australia, Canberra.*   65

5. Document taking possession of the Louisiades Archipelago on behalf of the King of France. *Reprinted by permission of the Archives de la Marine (Colonies), Paris.*   111

Maps

1. Voyage of Bougainville across the Pacific, 1767–68.   Inside cover

2. Tuamotu Archipelago to Tahiti, 22 March – 18 April 1768.   49

3. Tahiti, 4 April – 15 April 1768.   58

4. Samoa, 3 May – 5 May 1768.   80

5. Vanuatu, 22 May – 28 May 1768.   90

6. Vanuatu to New Guinea, 28 May – 8 August 1768.   98

PREFACE

The journal of Louis-Antoine de Bougainville's Pacific voyage of 1767-8 lay unpublished in the Archives Nationales in Paris until 1977 when Étienne Taillemite, then the Archives' Chief Conservator, brought out his impressive two-volume *Bougainville et ses compagnons autour du monde 1766–1769*, as the first in the series 'Voyages et Découvertes' published and printed by the Imprimerie Nationale, Paris, in association with the Publications de la Sorbonne. Taillemite had previously drawn attention to the little-known journal of Jean-Louis Caro, and the decision was taken to include the journals of those who had accompanied Bougainville on his circumnavigation. A more rounded picture of the first major French voyage of exploration could thus be presented to the French public.

The expedition had been known to the public not long after its return to France, as the result of Bougainville's own narrative, *Voyage autour du monde*, published in France in 1771, re-issued in two volumes in 1772 and available in that same year in an English translation by J.R. Forster. The *Voyage* remained available in one form or another and in several languages over the next two hundred years, but the original shipboard journal stayed buried in the vaults of the Archives Nationales. As Patrick O'Reilly, the secretary of the *Société des océanistes*, pointed out in a prefatory note to the Taillemite edition, there had been no scholarly and authentic account of the voyage available to French or other readers.

In providing an English translation and edition of Bougainville's original journal of his Pacific crossing, together with a selection from the journals of those who sailed with him, one must therefore pay a tribute to the painstaking labours of Mr Taillemite. Footnotes have been supplied in this present publication to take into account the interests and requirements of English readers, but where a note has been taken in its entirety from the French edition it is given between single quotes with the acknowledgement 'ET'.

Thanks must also be expressed to a number of people who have helped to elucidate some of the questions that remained unanswered or unclear: Professor G. G. Aymonin, of the Laboratoire de Phanérogamie at the Musée national d'Histoire naturelle, Paris, Emeritus Professor Kenneth Quinn, Dr Edward Duyker, Melbourne, librarians at Massey University, Palmerston North and Albany campuses, Dr W. R. Sykes, Landcare Research, Sandy Bartle, Te Papa Museum, Wellington, Professor Glynnis M. Cropp, Dr Ly Tio-Fane, Mauritius, Mme Jeannine Monnier, Maurice Recq, Brest, Mrs Fay Clayton, and the late Dr Norman Austin, with particular thanks to Captain R. J. Campbell for guidance on the interpretation of technical nautical matters.

My thanks must also be expressed to the Council of the Hakluyt Society, its trustees and honorary series editors, Professor Robin Law and Professor W. F. Ryan, for their assistance and their confidence over a period of years that have made it possible for the journals of the three leading French eighteenth-century Pacific explorers, Surville, La Pérouse and now Bougainville, to take their place in the Society's publications. Special thanks are due to Professor Law for his painstaking assistance during preparation of the present volume.

EDITORIAL NOTE

The Problem of Equivalence

A number of problems arise when translating French journals and documents into English owing to differences in the naval hierarchies of the two countries in the eighteenth century, as well as in the varying uses of weights and measures. Currencies, of course, fluctuated then as they do now. A balance has to be found between translating the word by its dictionary equivalent, which can create a misleading impression, and leaving all such words in French, highlighting them by the use of italics, which can be tiresome and in the end unhelpful. A compromise has been sought by in general using the English word, even though a purist would object that a true equivalence does not exist, and leaving the words in French and in italics in cases where an English approximation would leave a false impression. Thus, a league or a mile will be found in the text as preferable to *lieue* or *mille*, and lieutenant or clerk instead of *lieutenant* or *écrivain*.

Similar problems arise with the use of place names in the manuscripts. In the translated text, the English equivalent of a place name has been used in accordance with general usage. There are occasions when a literal translation has been judged more informative in the particular context or in the case of a descriptive name, and consequently preferable. For instance, the French seldom if ever used the name Falkland Islands; in the translation this has been found to be an appropriate name when the island group is referred to in a general way, but the Malouine Islands or Malouines has been used when references to the French settlement or French activities have made this preferable. In the case of the Straits of Magellan, the French tended (and still tend) to use the singular, in the sense of a single suitable channel, and this preference has been reflected in the translation. In their references to the island or land of Espiritu Santo, the French did not use the Spanish term but the French equivalent, and this has been reflected in the translation as the 'Land of the Holy Spirit'. In the case of present-day Mauritius, the eighteenth-century name of the Isle de France has been used throughout.

The term 'Savage' in reference to islanders and other native people is retained throughout the translation. Lest it should give offence, it may be necessary to draw attention to the capital letter – used in the journals and reports – and to point out that in the context of the times it referred to a person living in a state close to nature, the word being derived from the late Latin *salvaticus*, or forest-dweller. During the eighteenth century the question of the 'Noble Savage', a man living in an ideal non-urbanized state, uncorrupted by property rights, political structures and inherited social hierarchies, was widely discussed, especially after the publication of Jean-Jacques Rousseau's essay on the foundations of inequality in human societies.

Naval titles

Capitaine ('captain') was a term describing the person in charge of a ship. The lowest rank in this hierarchy was a *capitaine de flûte*, or storeship captain. The *flûte* being a transport vessel, normally accompanying a warship, its captain was often a man who had served in merchant ships, a professional sailor and one who had been accepted in the navy either on a permanent basis or with a temporary commission, i.e. an officer of the class known as 'blue' from the predominant colour of his uniform.

A *capitaine de brûlot* was an officer in charge of a fireship. This vessel was used in naval engagements, loaded with explosive materials, set on fire and directed at enemy ships. The command of such a vessel required experience, daring and skill. The appointment implied acceptance in the naval service, as a fighting officer, as against the storeship captain whose main function was the conveying of stores; however, many fireship captains were 'officers of the blue'.

The rank of *capitaine de frégate* was gradually falling into disuse in the eighteenth century, the rank of *capitaine de vaisseau*, usually senior to it, being in the process of replacing it. However, it was found useful during the Seven Years War in order to provide rank and status to experienced officers from the 'blue' class. Bougainville, although captain of a frigate, was given a commission as *capitaine de vaisseau*, the senior grade, applying to a captain in charge of a warship of 20 guns or more, roughly the equivalent of a post-captain. In the modern French navy, the rank of *capitaine de frégate* has been restored, being the intermediate rank between *capitaine de corvette* and *capitaine de vaisseau*.

A *chef d'escadre* was a senior officer in charge of several ships and accordingly of superior rank to a *capitaine de vaisseau*. The term 'commodore' is a fair approximation.

Entry into the naval service for sons of the nobility was normally through the *Gardes de la Marine*, which it would be misleading to translate as 'guard' unless the context made it clear that it was not intended to refer to a sentry or a person watching over people or property. Young men could also join a ship as *volontaires*. These volunteers were usually youths with links to the merchant service, such as the two sons of Duclos, and with some experience of the sea – Fesche had previously served as a junior pilot – who planned to pursue a career in the merchant service and, in some cases, in the royal navy. They were regarded as junior officers.

In the case of warrant officers and other lower ranks, the equivalent British grades are given, but with some reservations.

Measurements

In the eighteenth century, measurements had not been standardized and varied from country to country and often from one district to another. Attempts to standardize units of measure throughout France had been made at frequent intervals; indeed, one of the requests made in the list of complaints known as the *Cahiers de Doléances* in 1788 was for 'One king, one law, one weight, one measure'. There was one major reform in 1737 when the *lieue* or league was standardized, but even so there was one league applicable to the transport of grain, one for the postal service and one for general purposes, in addition to the *lieue marine* used by sailors. Major reforms began with the French Revolution, the metric system being gradually introduced from 1793-5.

It is possible to translate some French terms into English: a *brasse*, for instance, is normally translated as a fathom, a *lieue* as a league, and an *encablure* as a cablelength, but these are forms of usage and are not precise equivalents.

Measures of length in use at the time of Bougainville's voyage (with their approximate English equivalent in brackets) were:

Lieue (league): 2000 *toises* for the transport of grain, 2200 *toises* for postal services, 2400 *toises* for general road travel.

Lieue marine: one twentieth of a degree (3 nautical miles), or 5.557 km.

Mille marin (nautical mile): 1852 m.

Toise: 6 *pieds* (feet): 1.949 m.

Pied de roi: 12 *pouces* (inches): 32.5 cm.

Pouce: 12 *lignes:* 2.7 cm.

Ligne: 12 *points*: 2.26 mm.

Point: equivalent to 0.188 mm.

Brasse (fathom) or 5 *pieds:* 1.624 m. The English fathom is 6 feet or 1.82 m.

Encablure (cablelength): equivalent to 185.2 m (one-tenth of a nautical mile).

Weights and measures of capacity also require caution in estimating equivalents:

Barique (barrel, cask): 230 litres.

Livre (pound): 489 gm.

Pot de Paris: 1.90 litres.

Quintal (hundredweight): 100 *livres* or 48.95 kg.

Tonneau de mer (barrel, tun): 2,000 *livres* or 979 kg.

The term *quart* currently refers to a quarter of a litre, but it was also used to refer to a quarter of a *pot*.

Longitudes given by Bougainville were based on the Paris meridian, 2°20′14″ East of Greenwich.

Currency

Units of currency similarly present some difficulty. The terms *livre, sol, denier* may strike a chord with those who recall the pre-decimal *l. s. d.*, but it would be misleading to equate French and English currencies. The term *franc* goes back to the fourteenth century; the *livre* with which it was synonymous changed in value over the centuries as its gold content fluctuated, usually downwards, with the royal fortunes; while the *louis* made its appearance in the mid-seventeenth century. The following values were current at the *time* when Bougainville was writing:

Louis: gold coin worth 24 *livres*.

Écu: silver coin worth 6 *livres*.

Livre: silver coin worth 20 *sols*.

Sol: copper coin worth 12 *deniers*.

Liard: copper coin worth 3 *deniers*.

Denier: copper coin.

It is difficult to estimate the value of France's currency in British terms. Following the disastrous Seven Years War, France's financial situation was unenviable. Even comparing average wages is hazardous, as conditions varied from one country or one region to another. One estimate, based on a comparison of the annual income of a Lyons silk weaver, recorded in Voltaire's 1754 *Dictionnaire philosophique* (in the article entitled 'Feasts') as being approximately 639 *livres*, gives when compared with similar incomes of English skilled workers a value for a French *livre* of just under one shilling or five modern pence. The *Boudeuse*'s muster roll gives the monthly pay of sailors and others, applicable to the years 1765-8, and is useful for appreciating income relativities. A master gunner, master carpenter and master sailmaker earned between 480 and 600 *livres*. Sailors were paid between 144 and 216 *livres*. In addition, they were provided with some form of accommodation and food of varying quality; this was balanced by their lengthy absences from home.

Navigational Practices

The position of a ship, once out of the sight of land or away from a traditional trade route, was determined by calculating the point of intersection between two imaginary lines, the latitude, or the distance north or south from the Equator, and the longitude, or distance east or west from a known point, such as the port of departure or more generally the capital of the home country, London (Greenwich) for British sailors, Paris for French.

The latitude was reasonably easy to determine by observing the altitude of the noonday sun and using an almanac. The astrolabe and later the octant and the sextant were used for these observations, two mirrors bringing the image of the sun in coincidence with the horizon. Reasonably clear weather allowed quite accurate observations.

Longitudes, however, were more difficult to determine, requiring a comparison between shipboard time and the time at the point of departure or in Paris. Measuring the speed of a ship was difficult and unreliable: the log line (*ligne de loch)*, a rope of some 150 fathoms in length, marked at intervals by means of knots and with a steadying board at the end, was thrown overboard. The number of knots passing the rail during a specified interval was then counted, usually aloud, and provided, at least in theory, the information required – in fact, it depended on the sailors operating it not shortening what was a boring task by upturning the sandglass before it had emptied. Allowance also needed to be made for the ship's progress being affected by currents. The naval regulations provided for a 30-second sandglass, and a spacing between knots of $\frac{1}{120}$ of a degree or just over 47 *pieds*; ship's pilots however often shortened this distance to 41 *pieds*, considering this to be both safer and more accurate, as it theoretically allowed for drag..

The French used as a reference book a national almanac, the *Connaissance des temps*, first published in 1679 and regularly updated, which contained tables for the determination of longitudes from observations of the eclipses of Jupiter's satellites or of the moon (lunar distances). Little accuracy could be attained, however, until the true time on board could be calculated. Pendulum clocks were of little use. A great improvement came aboard when the Englishman John Harrison invented the chronometer; this, however was not available until after 1766, by which time Bougainville had already left. Bougainville had on board with him the latest model of the octant, perfected in England in the 1730s; several of the officers also had one and comparisons were frequently established, often with the help of the astronomer Véron, enabling an average to be determined. On fine days, the degree of accuracy was fairly satisfactory; on others, guesswork tended to dominate. Once on land, however, checks could be carried out with other instruments, mostly by Véron, and the ship's position recalculated.

Hence, on numerous occasions, the word 'Estimated' (or 'Estd') appears in the journal. It corresponds to the English expressions 'Dead Reckoning', 'By Reckoning' or 'By Account'. The term 'Estimated' (for *Estimé*) has been retained in the translation as being, if anything clearer to the reader. The method used to determine the position of the *Boudeuse* is recorded in the journals, so that the reader knows whether clouds or bad weather interfered and made accurate observations of the elevation of the sun or the angle of the moon impossible. Bougainville often gives the observed and the estimated latitude when a change in the cloud cover makes both possible, thus enabling a check to be made on the accuracy of the estimates. Additional information about the methods used is also often supplied, frequently in naval or scientific terminology. Thus the terms 'ortive' and 'occiduous' are found in the text: broadly, these signify eastern and western or, in reference to a planet, its rising or setting (e.g. 'the rising amplitude' of a planet). Compass variations were calculated by observing the magnetic bearing of the sun at its rising (ortive amplitude) or setting (occiduous amplitude) and comparing it with the sun's true bearing.

Bougainville's log also gives the distance covered by the ship – or '*courses*' (translated as 'run') – during the previous 24 hours (i.e. from midday to midday). This is stated in minutes of a degree, corresponding to a nautical mile (e.g. for the first entry in the journal, the route is given as $70'40''$, equivalent to 70.65 miles or 131 km). In almost every case, however, these measurements need to be treated with caution, on account of the combined effects of drift, winds and currents. Bougainville often adds 'Estimated' to the distances he records.

The direction in which the vessel travelled can be calculated according to the new estimated or true position. The French (and indeed the Spanish) method of recording the direction differed from the one current in England. In this edition, the French style has been retained. The corresponding English compass directions are as follows:

N¼NE: NbyE	NE¼N: NEbyN	NE¼E: NEbyE	E¼NE:EbyN
E¼SE: EbyS	SE¼E: SEbyE	SE¼S: SEbyS	S¼SE: SbyE
S¼SW: SbyW	SW¼S: SWbyS	SW¼W: SWbyW	W¼SW:WbyS
W¼NW:WbyN	NW¼W: NWbyN	NW¼W: NWbyW	NW¼N: NWbyN
N¼NW: NbyW			

Greater precision can be achieved by the addition of degrees and minutes.

Textual Note

Bougainville's shipboard journal is held at the Archives Nationales, Paris, under the reference Marine 4 JJ 142, N° 17. A second copy is in existence, also held at the Archives Nationales, but it is the work of an unknown copyist some years later and contains a number of errors.

1. Journal of Bougainville, first page.
   Reprinted by permission of the Archives de la Marine (Colonies), Paris.

INTRODUCTION

The Background

The Seven Years War, lasting from 1756 to 1763, exhausted France economically and ended the dominant position she had held in European politics under the Old Regime. British naval superiority and the French government's policy of concentrating her war effort on the continent had brought about the end of French colonial developments in Canada and India. To some, the loss of distant settlements and outposts was of little consequence, but to many and especially to those who had been involved in France's overseas campaigns the new situation appeared both humiliating and dangerous. Their British rivals were no longer laying the foundations for an imposing overseas empire: they were consolidating it. The future loomed ominous, with Britain apparently set on building up a global colonial empire, controlling sea routes and dominating world trade.

This fear would come to the forefront during the Revolutionary and Napoleonic wars. In the interim, the uppermost thought of French politicians and patriots was that the peace of 1763 was only a temporary truce, forced on the combatants by sheer exhaustion, and that the struggle would soon be resumed. Every effort should therefore be directed towards preparing for the next stage. One obvious move was the occupation of unsettled territories at key points on sea routes, in order to forestall some of Britain's likely colonizing moves, and to provide new bases to make up for those that had been lost. It was a policy rendered even more desirable by the closing of Canada to French immigration and the expulsion of considerable numbers of French settlers from Acadia, part of Nova Scotia, and the destruction of their farms at the beginning of the war. This *Grand Dérangement*, as it was called, an early form of ethnic cleansing, was deeply resented by men like Bougainville. Although most of the Acadians had been sent to Louisiana and to various British settlements, a number had found their way to France, and Bougainville saw in them a humanitarian cause, a useful lever by which to move opinion in favour of his plans, as well as a good source of potential recruits.

The fall of Quebec and the consequent collapse of French Canada had led to his return to France, at first as a paroled prisoner. Allowed to serve in Germany, he was wounded and had to return to Paris in 1761. He would play no further part in the war and he was therefore free to draw up his plans well before the Treaty of Paris in 1763 put a final end to the hostilities. His years in Parisian and London scientific circles in the mid-1750s had familiarized him with issues of exploration and in particular with Britain's interest in the Pacific Ocean. He knew Charles de Brosses, whose highly influential *Histoire des découvertes aux terres australes* had appeared in 1756. By the end of 1761 his thoughts about Pacific exploration and the need for French bases on sea routes had crystallized sufficiently to enable him to approach the Minister of Marine, the Duc de Choiseul, with a firm proposal.[1]

His plans were well received, but refined over the following year. The French government was in no position to promote or finance a voyage of exploration into the Pacific, so the first step would be the colonization of the uninhabited Falkland Islands, largely by Acadian settlers who 'would be quite willing to go there as the climate where it is proposed to establish their settlement is about the same as that of Acadia, their former homeland.'[2] Bougainville did not conceal the fact that strategic considerations drove him to propose this and made it relatively easy for the government to support the plan: 'seeing that the North was closed to us, I thought of giving to my country in the Southern Hemisphere what she no longer possesses in the northern one. I searched and found the Malouine Island.'[3] If the deserted Falklands could be held by France, they would bar the way to British ambitions, just as the Isle de France could threaten British shipping lanes in the Indian Ocean.

The possibility of establishing a base on the islands went back almost half a century. Laurens Olivier, in the *Comte de Lamoignon*, having sighted three islands in their neighbourhood, had suggested a more thorough survey by a French expedition, as they appeared suitable for a settlement and as a port of call.[4] Others, notably Doublet and Duquesnel, with the Spanish South American trade in mind, also pressed for a French settlement: 'It would be easy and not too costly to settle there, and in a short while to provide [the islands] with all kinds of plants and grain crops and vegetables, as well as to populate them with cattle and poultry. There is fish in abundance, and this discovery deserves attention.'[5]

Two ships were obtained, the *Aigle* and the *Sphinx*. Bougainville and his relatives provided the funds, and in September 1763 the small expedition – France's first colonial venture since the war – was under way. The first settlement, Port Saint-Louis, was established in February 1764 and a few months later Bougainville returned to France to report on his success. He was well received and the Duc de Choiseul had little difficulty in obtaining Louis XV's approval for the formal taking over of a Malouines colony. Bougainville drew up further proposals for the full recognition of what was now known as the Saint Malo Company, including some official financial support for the next stages of the project. There were, Bougainville pointed out, over 300 Acadian families living around St Malo, totalling 1,300 people, who could be transported to the islands; this would require, in addition to the *Aigle*, two storeships and two

---

[1] In his published narrative, Bougainville seems to disclaim that the establishment of a base at the Falkland Islands was his own idea: 'At the beginning of 1763, the French court decided to form a settlement on these islands. I offered to the Minister to establish it at my own expense:' *Voyage*, p. 48. However, there is strong evidence that informal discussions preceded this decision and that Bougainville played a major role at the earliest stages.

[2] Memoir in AN, Colonies, F²A 20.

[3] Note dated 4 July 1764 in BN, NAF 9407-11. The name Malouine is derived from St Malo and commemorated the discoveries and sightings made by ships from this port earlier in the century. In the context of this expedition the term Falklands and Malouines are therefore interchangeable. The Spanish equivalent, Malvinas, is regularly used by Argentinians.

[4] 'Copie du journal du Sieur Laurens Olivier', 1714, BN, NAF 9438.

[5] 'Relation de la nouvelle découverte des isles Cébaldes', 1711, BN, NAF 9438.

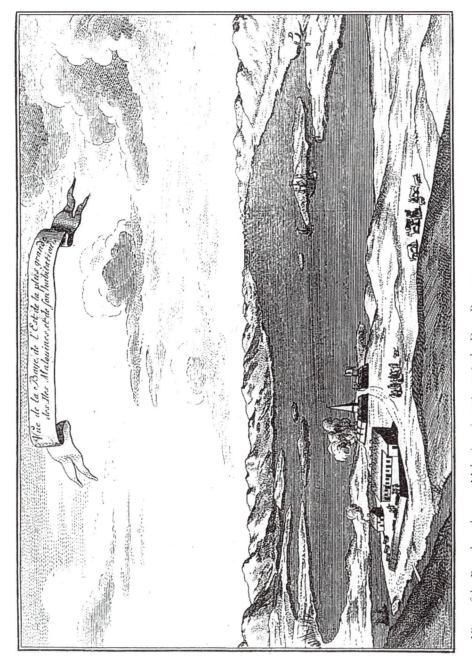

2. View of the French settlement, Malouines, in 1763–1764, by Dom Pernetty, *Histoire abrégée d'un voyage aux îles Malouines.*

sloops, and adequate supplies for subsistence, seeds and agricultural implements. The Minister unequivocally endorsed the proposal: 'Approved in full', he wrote in the margin.

There were, however, some warnings signs about. The Spanish ambassador had already made preliminary enquiries about French intentions. The newspapers at Amsterdam and The Hague published brief reports on disquiet being expressed in London, and discussed whether the Dutch, through Sebald de Weert in 1600, might not have a claim to the islands on the basis of prior discovery, preceding the English by some ninety years, so that 'the French were merely the third in line.'[1]

By then Bougainville was on his way back to Port Saint-Louis in the *Aigle*. He arrived at the beginning of 1765 and found the small settlement in good heart; the winter had not been too harsh, the cattle had survived in the open air and in fact done quite well, the crops were promising, with the exception of wheat. The chaplain had performed one wedding and two christenings. But there was a shortage of timber, both for building and for firewood.[2] Bougainville therefore set out for the Straits of Magellan to get some. He was successful in this mission, surveyed the first narrows and the bays beyond and met some of the local inhabitants. More unexpected was the presence of three ships, which he identified as British: they were Byron's *Dolphin, Tamar* and *Florida* on their way to the Pacific Ocean. Byron had already been to the Falklands without seeing the French settlement; conversely the French had been unaware of the English presence. Bougainville and Byron did not meet in the Straits: they merely sighted each other and went on their way, but it was an ominous encounter.

Back at Port Saint-Louis at the end of March, Bougainville got ready to sail back to France. He sailed on 28 April 1765 and dropped anchor at St Malo on 13 August. He then hastened to Paris and discovered that his colony was doomed. The British had become fully aware of his plans and of the danger this presented to their shipping route. Byron had in fact taken formal possession of the Falkland Islands, laying out a sketchy vegetable garden at a commodious harbour he named Port Egmont, thus bringing Britain firmly into the equation.[3] Byron's third ship, the *Florida*, had been sent back to London with news of the sighting of Bougainville in the Straits, removing any doubts about French activities. The Admiralty took immediate steps to ensure that Britain's claims were respected and within three months the *Jason*, captain Macbride, was on her way to the Falklands with clear instructions to inform 'any subjects of a foreign Power' found there that 'the Islands belonged to Great Britain, and

---

[1] There were a number of periodicals in French published in the Netherlands at this time; this enabled journalists to avoid the censorship procedures in France and to comment more freely on contemporary issues. They also enabled other governments, especially the British, to present their case on contentious matters. The main ones were the *Gazette de la Haye* and the *Gazette de Hollande*, also known as the *Gazette d'Amsterdam*: for the controversy over priority of claims to the Malouines as among the French, English and Dutch, see the *Gazette de la Haye*, 13 July & 14 August 1764, *Gazette de Hollande*, 13 August 1764. The complex arguments which arose out of Bougainville's actions are analysed in minute detail by Martin-Allanic, *Bougainville*, I, pp. 167-419. For early French language periodicals, see Levi, *Guide to French Literature: Beginnings to 1789*.

[2] See Pernetty, *Histoire d'un voyage aux isles Malouines fait en 1763 et 1764*, a useful and reasonably reliable source.

[3] See Harlow, *The Founding of the Second British Empire*, I, pp. 25-7.

that, since his Majesty had given orders for the settlement thereof … any such persons were to be offered transport on his Majesty's ships to some port in the Dominions of the Power to which they belonged.'[1]

The Spanish for their part, after some tactful enquiries through diplomatic channels, had been expressing their increasing concern to Versailles. The islands were geographically part of South America, and thus within the Spanish sphere of colonial influence. They were allied to France through the Family Compact and did not at first oppose Bougainville's activities, but they were uneasy about claims of sovereignty. Once the French government began to transform Port Saint-Louis from a commercial enterprise into a colonial settlement, with a member of Bougainville's family as governor holding a royal warrant, mild protests turned into a firmly worded request for the French to desist from further encroachment into Spanish territory.

At first, the French were unwilling to sacrifice a settlement that held such appealing strategic possibilities. But to hold on to the Falklands would mean alienating Spain – and now possibly lead to a new war with England. The political disadvantages far outweighed any likely gains the settlement might bring. Louis XV consequently agreed to ask Bougainville to abandon the settlement and recognize the Spanish claim. The Bourbon ranks would then be closed, and the British would be forced to claim right of ownership by prior discovery – which was not a principle the British were anxious to establish in international law.

Bougainville went to Madrid in April 1766 and again in September to discuss terms with the Spanish Government and France's representative in Madrid. The King of Spain had no wish to impose any special hardship on the pleasant Frenchman, and he agreed to refund the Saint Malo Company's investments plus interest at five per cent, and to meet any costs to be incurred by Bougainville up to April 1767 in connection with the cession of the colony, its buildings and equipment. The total involved exceeded 600,000 *livres*.[2] Spain hoped in addition that France, in recognizing Spanish prior claims, would condemn the British for establishing Port Egmont, but Louis XV was not willing to venture into a confrontation with Britain at this point. He had agreed to give up what had become the French colony of the 'Malouines': the question of whether England had a legal right to its new settlement on the 'Falklands', or whether Spain had a better claim to its 'Malvinas', could be left to later – and indeed it was still a matter for bitter disputes in the late twentieth century.

Compelled to accept the inevitable, Bougainville revived his original plans. The settlement on the Malouines had been planned merely as the first stage of a voyage into the Pacific Ocean, where new discoveries might be made and further French outposts might be established. The Duc de Choiseul had endorsed the proposal back in 1762 and it had been officially approved the following year. There was even more

---

[1] Ibid. The instructions are reprinted in full in Corney, *The Quest and Occupation of Tahiti*, II, pp. 441-5.
[2] The letter of agreement signed by Bougainville on 4 October 1766 is quoted in Martin-Allanic, *Bougainville*, I, pp. 377-8. Although it included an admission that his settlement was 'illegal', Bougainville continued to regard the Falklands as having been an unclaimed archipelago to which the first settlers were entitled. A case in point was Mauritius, discovered by the Portuguese, settled by the Dutch in 1698, left deserted in 1710, renamed Isle de France by the French in 1715, and turned into a prosperous French colony until 1814, when Britain took it over under the terms of the Treaty of Paris.

reason to proceed now on account of recent British activities in the Pacific. Information coming from London made it clear that further Pacific voyages were being envisaged. Byron had already sailed; soon there would be Wallis and Carteret; scientists on both sides of the Channel were pressing for more expeditions, drawing attention to the coming transits of Venus which would see James Cook take his *Endeavour* around the world and France send out its own savants to the Indian Ocean and elsewhere. Science may have been a respectable pretext, but other voices had also been raised across the Channel for settlements in the vast ocean, notably that of James Callander, who had translated De Brosses's influential book and adapted it so as to present a case for English, as against French, exploration of the South Seas.

The time was now ripe for Bougainville's great undertaking. Within twenty-four hours of his return from Spain with copies of all the agreements that had been signed in Madrid, Bougainville was hard at work making final arrangements for his voyage around the world. Practically everything had been prepared, with the full assistance of the Minister of Marine. There was no reason to delay. France was ready to send its first officially-sponsored expedition into the Pacific Ocean. A month after his return to Paris, Bougainville sailed in the *Boudeuse* from the Loire, impatient to set out on the great undertaking which had been his dream for so many years.

The Participants
1. The *Boudeuse*

Louis-Antoine de Bougainville was born in Paris on 12 November 1729 in a family of respected lawyers. His father Pierre-Yves was a notary at the nearby Châtelet, and in 1741 he was appointed *échevin* or councillor of the city of Paris. This led to his ennoblement, becoming a member of the *noblesse de robe*, 'nobility of the gown' as against the *noblesse d'épée*, or 'nobility of the sword', to which most, but by no means all the leading families of the kingdom belonged. This ennoblement opened many doors to Bougainville who soon proved himself to be a talented and highly personable young man about town. In addition, he made a rapid transition from the world of gowns and law books to the world of swordsmen by joining the army. After some time with the musketeers in 1750, he served in the Picardy Regiment – the Bougainville family had originated from that province – rising to aide-de-camp in 1754.

In October of the same year he went to the French Embassy in London as secretary to the Duc de Mirepoix. He met a number of leading figures in English society, including Lord George Anson who had completed a circumnavigation in 1744. He also met scientists with whom he could discuss the works and ideas of the French savants under whom he had studied or with whom he had talked at length for, although he had the traditional solid grounding in the classics which was the basis of all education and culture in his day, he also had a sound knowledge of science and mathematics. In 1755 he published his *Traité du calcul intégral* which helped him to establish a reputation in learned circles; in January 1756 he was elected a member of the Royal Society.

The international situation, however, had been rapidly worsening and Europe was sliding into what would become the Seven Years War. Bougainville returned to army service. Appointed aide-de-camp to Montcalm, he sailed for Canada in March 1756.

He was wounded in June 1758 and returned to France. He was back in Canada in the following year, in time to witness the fall of Quebec. He continued to defend what remained of French Canada against superior forces, but was taken prisoner near Montreal in September 1760. Freed on parole, he was sent back to France and soon after allowed to take part in military operations in Germany. He was wounded in July 1761 and sent back to Paris; for him, the war was now over.[1]

He now turned his mind to his plans to explore the Pacific Ocean, ensuring that France would have a part to play there in years to come. He had acquired a considerable knowledge of the sciences of navigation and astronomy from each of his crossings of the Atlantic and had established firm friendships with men like Duclos-Guyot. Although he would depend on their help during his circumnavigation, he was by no means a dilettante, an idle army officer playing at sea captain. He had been an able student and had gained practical experience by closely observing and assisting the officers whenever he was at sea; and indeed his ability as a commander and as a navigator remained unquestioned during the voyage. He had in addition a very thorough knowledge, acquired through study and discussions with scientists and naval officers, of earlier voyages to the Pacific. The influence of his older brother, Jean-Pierre, should not be overlooked. Jean-Pierre was a lawyer, in poor health (he died in 1763), a man of considerable learning, well connected in Paris circles, who had planned for his brother a more colourful career than his own state of health would ever allow him.

Bougainville's first duties however concerned the establishment of a colony on the Falkland Islands. This was his own idea, carefully thought out and linked with his desire to set up a French base that would serve later French expeditions to the Pacific while assisting in the resettlement of French Canadian refugees. He sailed twice to the Falklands, in September 1763 and again in July 1764. On the second occasion, he went to the Straits of Magellan to obtain supplies of timber. Pressure from Spain and indirectly from Britain forced the abandonment of the settlement, which Bougainville handed over to the Spanish authorities in April 1767. He then continued on his way with the *Boudeuse* and the *Étoile*, passing through the Straits and into the Pacific Ocean, returning to France in March 1769.

His subsequent career was mostly associated with the navy, to which he was officially transferred on March 1770, and with administrative and political matters. He considered leading an expedition to the North Pole and drew up careful plans, which the Minister considered for a time, but the proposal was shelved and Bougainville busied himself with naval problems. His position in the navy was still ambiguous: not only had he transferred from the army, but strictly he could be counted among the

---

[1] On Bougainville's youth and the Canadian campaigns, see de Kerallain, *Les Français au Canada: la jeunesse de Bougainville et la Guerre de Sept ans*; and La Roncière, *Bougainville*. Kerallain is a descendant of Bougainville. La Roncière's book has a readable and extensive section on the Canadian campaigns, pp. 15–93, compared with pp. 119–71 for the voyage and pp. 172–80 for the aftermath – this may be due to this biography being published during the German Occupation when Britain was often depicted as an enemy of France and this biography appeared in a collection directed by Abel Herman, a prominent collaborator. La Roncière provides a valuable list of documents for the early period of Bougainville's life.

'officers of the blue', men who had often a long experience of the sea but gained in the merchant service, and who were accepted on naval ships only during wartime – after which, to use Bougainville's own words, these men 'born in the honest middle class' seldom found permanent employment in the navy. Traditionally, a career in the naval service was available only to those who came from the *noblesse d'épée*, the 'reds' as they were known.[1] He had however powerful protectors at court, and a well-established reputation as an able and enterprising man. War anyhow was approaching once more. The American colonies began their struggle to free themselves from English rule and gradually France was drawn into the fight.

Bougainville joined the *Terpsichore* as second in command in May 1775 and obtained his own command, that of the *Bien-Aimé*, in January 1777. He fought in several naval engagements for the next six years, commanding the *Guerrier* from April 1778 and the *Languedoc* from June to September 1780. A brief interlude ashore allowed him to marry, on 27 January 1781, Marie-Joséphine de Longchamps-Montendre, of an old naval family. They would have four sons, the eldest of whom, Hyacinthe, would one day head his own expedition to the Pacific.

Bougainville was back at sea at the beginning of March 1781, in the *Auguste* bound for the West Indies and North America. In September he took part in the Battle of the Chesapeake, and in the following January in the capture of St Kitts, but the French fleet was defeated at the Battle of the Saintes near the island of Guadeloupe in April, and Bougainville returned to France. In 1784 he was elected a member of the *Académie de Marine* and inducted in the Order of Cincinnatus by the grateful United States. He spent the next few years helping to plan major French expeditions to the Pacific, such as La Pérouse's and D'Entrecasteaux's. The French political climate was rapidly darkening. At first, he was able to play a part in attempting to reform the navy, became commander of the Brest Squadron in October 1790 and was promoted vice-admiral in January 1792. But the situation was becoming intolerable. Unable to establish order among the naval personnel, he resigned and retired to his property in Normandy. This did not save him from being arrested as a suspected royalist and imprisoned for two months in 1794.

Freed, he soon returned to his influential role in Paris. He was appointed to the committee planning Bonaparte's campaign in Egypt, and helped with the preparations for Baudin's voyage to Australia, on which his son Hyacinthe would embark in 1800. He was appointed a Senator in 1799 and Rear-Admiral in 1802 when finally he retired from the navy. He became Grand Officer of the Legion of Honour in 1804 and a Count of the Empire four years later. His final official function was to chair the court of enquiry that followed the naval defeat at Trafalgar. He died on 20 August 1811, aged eighty-one.

His second-in-command, *Nicolas-Pierre Duclos-Guyot*, undeniably an officer of the blue since he came from a family of artisans, was born at St Malo on 14 December

---

[1] The distinction and its ramifications, and the privileges enjoyed by young members of the nobility, increasingly contested as the eighteenth century progressed, have been analysed in Aman, *Les Officiers bleus dans la marine française au XVIIIe siècle*. Most of the officers who sailed with Bougainville on his voyage around the world were 'blues'.

1722 and first sailed at the age of twelve on one of the French India Company's vessels. Promoted *enseigne* in 1742 and lieutenant a year later, he transferred to privateers during the War of the Austrian Succession, resuming his service with the company in 1749 and making a six-year voyage to Peru. On his return he qualified for the rank of captain, but soon resumed service on privateers when the War of American Independence broke out. He was remarkably successful as captain of the privateer *Victoire*, taking no fewer than seven prizes in the 1757. His ship was transferred to the royal service and he sailed to Canada with the rank of *lieutenant de frégate*. In 1759, commanding the *Chézine*, he crossed to Canada with a passenger on board, Bougainville, with whom he established a close friendship. His time in Canada was marked by frequent actions and some notable successes, and in 1764 he was raised to the rank of fireship captain.

He sailed to the Malouines in 1763 in the *Aigle*, then stayed in France helping to prepare for the planned circumnavigation. However, when he returned from the circumnavigation, he found that as an officer of the blue there was no opening for him in the navy, and he returned to merchant ships, sailing to India and China between 1771 and 1775. He served briefly in the *Belle-Poule* in 1776–7 and was then appointed port captain at the Isle de France. He left in 1781 in the storeship *Osterley*, sailing to Buenos Aires. He retired on health grounds in 1784 with the rank of *lieutenant de vaisseau* and a government pension. However, the Revolution called him back. Given the rank of *capitaine de vaisseau*, he took charge of the *America* in May 1792 and sailed to Santo Domingo, but a mutiny on board in July 1793 forced him to return to France. He died near St Malo on 10 March 1794.

He was highly esteemed by Bougainville, who had been a talented pupil of his back in early Canadian days. 'This officer is assuredly one of the best in Europe and few could claim such extensive periods of service both in peace and in wartime,' wrote Bougainville, an opinion endorsed by the naval administrators who noted in his record: 'His success in everything he has undertaken makes him worthy of recommendation.'[1]

Five young officers of the red, four of them almost contemporaries from the *Gardes de la Marine*, sailed in the *Boudeuse*. The oldest, *Alexandre de Lamotte-Baracé de Bournand*, was born in Anjou on 17 September 1736, joining the *Gardes* at Brest in 1754 and sailing to Canada the following year. He was taken prisoner when his ship was captured off Newfoundland even before war had been formally declared, but was freed on parole in January 1757. He then served with the Knights of Malta, but did not return to sea until April 1763. Just prior to his appointment to the *Boudeuse* with the rank of *enseigne*, he had sailed to Guyana and Santo Domingo. He was promoted lieutenant during the circumnavigation, but found his health so affected by the hardships he had to endure that he eventually withdrew from the naval service with a modest pension. The Revolution had a serious effect on his financial situation and there is a record of a special invalid's pension being granted to him in 1802. He died

---

[1] Duclos-Guyot files in AN, C¹ 174, p. 1965, quoted in Taillemite, *Bougainville*, I, pp. 64–5. There is another file on him at AN, Marine C7:92; and a personal file at the Service Historique de la Marine, CC7.

some time after 1816 when his name disappears from the records, by which time he would have been in his early eighties.[1]

Henri de Fulques d'Oraison was born in Provence on 16 January 1739. He served from May 1757 to February 1758 when he too was taken prisoner. Repatriated, he sailed once more, to be captured for the second time in August 1759. Freed in April 1760, he returned to active service until the end of the war. He then accompanied the Chevalier de Ternay on a mission to various European countries, from England to Russia. On his return, he was promoted to *enseigne* and shortly after was appointed to the *Boudeuse*. Like Bournand, he found his health seriously affected by the voyage, possibly developing tuberculosis. He was given a modest pension and transferred to the army. Campaigning on land proved much more wholesome for him than on sea and he began a highly successful career that saw him rise to the rank of *maréchal de camp* in 1790. This progress was stalled for some time on account of his aristocratic origins, but by October 1795 the revolutionary extremists had been removed and D'Oraison was appointed commander of the Brest forces and later to a similar position at Besançon. He finally retired with the rank of general in December 1813 and died in Paris on 22 May 1819. He too therefore, in spite of suffering ill-health from the hardships of the circumnavigation, reached his eighties.[2]

The third *enseigne*, Jean-Jacques-Pierre de Gratet du Bouchage, however, did not survive the voyage. After joining the *Gardes* at Toulon at the same time as D'Oraison, he served in a succession of ships, mostly in Mediterranean waters, being promoted *enseigne* just before joining the *Boudeuse*. He apparently caught dysentery at Batavia and died at Port-Louis on 26 November 1768, aged twenty-nine.[3]

Two slightly younger officers were promoted *enseignes* during the voyage. *Jean-Baptiste-François de Suzannet* had joined the *Gardes* at Rochefort in August 1757 and was soon involved in the campaigns of the Seven Years War. He had already sailed to Guyana and to Canada. On his return from the circumnavigation, he commanded several small transport vessels, then with the coming of the War of American Independence began a career which saw him obtain several commands and rise to the rank of *lieutenant de vaisseau* in 1778. He took part in a number of naval combats, being finally captured off Puerto Rico in April 1782. Promoted *capitaine de vaisseau* in May 1786, he took over command of the *Capricieuse* and then of the *Andromaque*, but he refused to join the republican side when the Revolution broke out, was allowed to retire in 1792 and emigrated. He joined the force of émigrés attempting with British support to land in Brittany. He was killed during the landing of June 1795 at Quiberon.

His career was paralleled by the other young *garde*, the Breton *Jacques-Marie de Cramezel de Kerhué* (found in some of the journals as *Kerué* and *Kervê*). Born on 25 March 1741, he had joined the navy at the age of sixteen and almost immediately

---

[1] Bournand's personal file is held at AN, Marine C7:231.
[2] Reference to D'Oraison are found in AN, Marine C7:231 and B4:114, f° 201. His mission to northern Europe is mentioned in the biography of Ternay by Linyer de la Barbée, *Le Chevalier de Ternay*, I, p. 174. A biographical note also appears in Six, *Dictionnaire biographique des généraux et amiraux français de la Révolution et de l'Empire*, II, p. 267.
[3] On Du Bouchage, see Taillemite, *Bougainville*, I, p. 67.

went to sea, in the *Magnifique* bound for Louisbourg. He remained in Canada until the following year and returned to France to take part in a number of campaigns until the end of the war. He was given little respite after this, sailing in the *Sceptre* in 1764 and the *Étourdie* in 1765 bound for St Pierre and Miquelon, the remaining French possessions at the mouth of the St Lawrence. It was then that he met another young cadet, Suzannet, and both soon afterwards received instructions to report to Bougainville. Appointed *enseigne* during the voyage, but later than Suzannet, a delay he considered a hurtful slight, he completed the circumnavigation without any impairment to his health and soon after his return was appointed to the *Rossignol* for a voyage to Santo Domingo. A lieutenant by 1778, he took part in several campaigns during the War of American Independence, including the Battle of Ushant in July 1778. He was later taken prisoner. Freed after a time, he returned to active service, obtained his own command, that of the *Barbeau* in 1783, and pursued his successful career until the Revolution. He was promoted to *capitaine de vaisseau*, but increasingly rejected the republican philosophy, resigned and joined the émigrés abroad. Like Suzannet, he took part in the Quiberon landings of June 1795 and was killed at about the same time.[1]

The last officer to join the *Boudeuse*, in October 1766, was *Josselin Le Corre*, born in Brittany on 29 August 1727, an officer of the blue who had served in merchant ships since 1742, working his way up from seaman and leading seaman to boatswain, eventually qualifying as a ship's master. He had seen service in the navy briefly in 1750, and returned to war service during the Seven Years War, sailing to Canada in 1753-4 and taking part in anti-corsair activities; taken prisoner in 1758, he disappears from naval records until 1763. In January 1766, he married Augustine, daughter of Alexandre Guyot, a relative of Duclos-Guyot who recommended him to Bougainville for the circumnavigation. He proved to be a most reliable and able officer during the entire voyage, and Bougainville sought his transfer to the navy on his return, recommending him for the rank of storeship captain, but his background made him a weak competitor against better connected officers. The Brest *Intendant* pointed out that, although Le Corre had to his credit four periods of service on royal ships, he had completed eleven on merchant vessels which had brought him their own benefits: 'the pecuniary advantages these kinds of navigation produce counterbalance the favours that should in preference be given to those who are only employed in the King's service.'[2] Le Corre consequently returned to the merchant service and made his way to the Isle de France some time in 1770. He is in all probability the Le Corre (or Le Cor) who sailed in the *Marquis-de-Castries* on the Marion Dufresne expedition planned to return Ahutoru, the Tahitian who had accompanied Bougainville to France, to Tahiti. He survived this voyage and returned to France. He returned to the Isle de France in January 1777 in command of the *Confiance* and made several trading voyages in the Indian Ocean until 1780, subsequently returning to France and retirement. He died at St Servan on 23 August 1785.[3]

---

[1] The personal files of Suzannet and Kerhué are held at AN, Marine C7:314 and C7:77.
[2] Quoted in Duyker, *An Officer of the Blue*, p. 118.
[3] See on him Duyker, 'Josselin and Alexander Le Corre', pp. 9-13; Duyker, *An Officer of the Blue*, pp. 118-19, 197; *Dictionnaire de biographie mauricienne*, pp. 1490-91. On the family, see Duyker, *Of the Star and the Key*, p. 12.

There were three *volontaires* under Bougainville's direct command, young men who hoped to pursue a career in the navy. They were seldom youths who lacked experience at sea, often had connections that could be of use in furthering their plans, but usually lacked the aristocratic background that would have enabled them to join the navy through the normal channel of the *Gardes de la Marine*. Two were related to the *Boudeuse*'s second-in-command; they were in fact the two sons of Duclos-Guyot, *Alexandre* and *Pierre*. Both had sailed with their father while in their early teens, joining him in the *Aigle* in 1763 for the first voyage to the Falkland Islands. The former, but not apparently Pierre, went on the second voyage of 1764-5. They then both sailed from France in the *Boudeuse*, but once the Malouines settlement had been handed over to the Spanish, Bougainville or their father felt it preferable to separate them. Accordingly, *Pierre* transferred to the *Étoile* while the ships were in port at Rio de Janeiro in July 1767. He was clearly a pleasant young man and able to get on well with the often irascible Philibert Commerson, helping the naturalist with the writing up of his shipboard journal. His later career is not known, and it can be presumed that he subsequently served in trading vessels.

Alexandre Duclos-Guyot, probably named after his uncle Alexandre, a source of occasional confusion, had a more prominent career in the royal service, gradually overcoming the disadvantages of being an officer of the blue. After his return to France, he served in merchant ships, but the War of American Independence opened new avenues for him. He was already highly regarded: a report dated 1774 by the naval commissioner at St Malo states that he was 'a steady individual, very diligent, very knowledgeable and in every respect the worthy son of an equally worthy father.'[1] He joined his father in the *Flamand* in January 1778, sailing to the Isle de France. He was captured a year later, but exchanged against other prisoners a year later. He then served under Bougainville as *lieutenant de frégate* in the *Languedoc*, obtaining his first command, the corvette *Mortemart* in October 1780. Other commands followed; he ended his war service with the temporary rank of fireship captain, made a voyage to China in command of a storeship, and became assistant port lieutenant at the Isle de France, remaining there from 1785 to 1788. The Revolution provided him with new opportunities and in 1791 he sailed to the Isle de France in command of the *Philippe d'Orléans*. Appointed *lieutenant de frégate* at the beginning of January 1792, he was about to return to France when he was asked to sail to Buenos Aires to obtain supplies for the island. He reached that port in February 1793, but the international situation had worsened, his ship was impounded and he was imprisoned until 1802. Back in France, he was pensioned off, and after a time decided to travel to Spain to seek some indemnity from the authorities for his ill-treatment. Negotiations were protracted and complicated by the continually changing political situation in that country. He is reported as having disappeared during the withdrawal of French troops from Spain in 1813 and he is believed to have been killed by guerilla fighters in Andalusia.

The third *volontaire* to sail in the *Boudeuse* was *Charles-Félix-Pierre Fesche*, the best known of the junior officers on account of the informative journal he wrote during

---

[1] Quoted in Taillemite, *Bougainville*, I, p. 75. His file is in AN, Marine C7: 376. An article on him is included in the *Dictionnaire de biographie mauricienne*, no. 27, p. 801.

the voyage. He was apparently descended from a Sardinian family and claimed to be distantly related to the Bonaparte family through the emperor's mother Letizia, whose half-brother was Giuseppe Fesch, eventually Grand Almoner of France and Cardinal-Archbishop of Lyons. He was born in Paris on 10 October 1745, was introduced to Bougainville at an early stage and sailed with him in the *Aigle* as a ship's apprentice on the first and second voyages to the Falklands in 1763 and 1764. He returned to the islands a third time, joining the *Étoile* in October 1765, but was transferred to the *Boudeuse* a year later. He remained at the Isle de France when Bougainville sailed from Port-Louis in November 1768, seemingly at the request of the administrator Pierre Poivre, but this request must have been rather broadly formulated since Fesche found no opening available to him in the colony. Disappointed, he returned to France in December 1769, only to find there was no position vacant for him in the French navy. He then transferred to the army, serving for some years with a cavalry regiment. However, the sea called him back and he commanded several privateers during the American war, being twice taken prisoner, but succeeding in making his escape back to France. He returned to the army, serving for a time in the cavalry. The Revolutionary Wars offered the chance of further promotions and we find him serving in Germany and in France, in time as lieutenant and quartermaster. He was however getting on in years, his health was becoming affected by his continual campaigning, and some of his later postings were to various corps of veterans. He entered the veterans' home at the Invalides in 1802 and died, presumably in Paris, on 27 January 1810 in his sixty-fifth year.[1]

The *Boudeuse*'s clerk – the French term is *écrivain* or 'scribe' – seems to have been ill-suited for a long voyage around the world. *Louis-Antoine Starot de Saint-Germain* did not belong to the naval administrative corps. Born in the Dauphiné region on 27 September 1731, he had studied law, then obtained a position as administrator of stores in Guyana. He held this post from 1763 to 1766 when, back in France, he applied for a similar position in the French possessions in the Indian Ocean. He was instead appointed to the *Boudeuse*. Recently married and no seaman, certainly no intrepid explorer, he bitterly regretted having had to embark on a lengthy circumnavigation. His journal, a valuable record in many respects, is peppered with acid comments. 'What is the use of such a voyage?' he commented, criticizing with some pertinence but little understanding of the problems faced by Bougainville the lack of close surveys of the coasts the ship sailed along and the inadequate information obtained about the islands encountered on the way. He was left behind at the Isle de France in a poor state of health and made his way back to France later. After some time spent working in the naval offices at Rochefort he was sent to Santo Domingo. His career suffered in the 1770s from accusations of shady dealings from which an enquiry exonerated him in 1776. Nevertheless, a number of rumours about him were still circulating, and his father among others refused to speak to him and threatened to disinherit him. He sought the Minister's assistance and was granted a warrant as commissioner in the naval administrative service, which helped to rehabilitate him. However, he decided to retire in September 1779, on health grounds, with a substantial

---

[1] A file on him is held in AN, Marine C7:106. Notes on the early part of the voyage are contained in a file at BN, NAF 11651 and his *Journal de navigation* is kept at the Musée national d'Histoire naturelle, Paris.

pension. His later years were spent in his home district, writing poetry and a play. He died on 11 December 1823, having at the ripe age of ninety-two outlived practically all the members of the Bougainville expedition.[1]

The surgeon *Louis-Claude Laporte* belonged to a family associated with the sea and with medicine. His father was a 'master barber' living in Brest at a time when a skilled barber provided paramedical or indeed medical services, since bleedings – to rid the body of what were believed to be harmful 'humours' – were a major part of medical science. Louis-Claude engaged in more advanced studies, qualifying through medical examinations for a masterate in surgery and medicine, and joining the navy as a surgeon. His evident ability and devotion on the circumnavigation earned him wide recognition. He served in several ships after his return, maintained his reputation by his treatment of the wounded and his energy in preventing infection and sickness during the War of American Independence, and in 1779 was appointed assistant medical officer for the port of Brest.[2]

The chaplain *Jean-Baptiste Lavaisse* (or *Lavaysse,* found also as *Lavoys*) was a Franciscan Minorite, born near Autun in central France but attached to the Paris house. Bougainville had originally been approached by Commerson on behalf of his brother-in-law who was a priest – a suggestion which might cause some surprise as relations between the latter two men were somewhat strained, possibly as the result of Commerson's rumoured liaison with Jeanne Baret. There was insufficient time to discuss the possibility of his joining the expedition, as he lived some distance from Paris, and Bougainville decided to waste no more time on the matter and to approach the Archbishop of Paris. Father Lavaisse had no ties, indeed few resources of any kind, and all he asked was that 'on his return to France, he could be sure of some adequate means of subsistence.'[3] He would, following the death of the *Étoile*'s chaplain, assume some additional functions during the voyage, but there are few references to him in the various journals. When he reached the Isle de France, he was asked to remain behind 'to carry out the tasks related to his ministry, of which the island is in the greatest need.' However, he did not stay long, as we find him sailing back to France in February 1769. His health had been impaired, as is evidenced by a medical certificate of January 1770 declaring him unfit for any further naval chaplaincy. He retired to the Autun district, 'carrying out such functions as his health still allows.' A request for financial assistance, dated July 1787, reminds the authorities that he had joined the Bougainville expedition on the understanding that his needs would be adequately met on his return, but that nothing had so far been done. The Minister of Marine then went only as far as to write to the bishop, recommending Lavaisse to his good offices, but nothing more is known of the ailing and possibly by now ageing priest. The coming storm of the Revolution would, if anything, worsen the position of such men.

The *Boudeuse* also included one passenger whose role on board soon became more

---

[1] A file on Saint-Germain is held at AN, Marine C7:293 and 296. His notebooks formed the basis for the publication by La Roncière of 'Routier inédit d'un compagnon de Bougainville: Louis-Antoine Starot de Saint-Germain, écrivain de la *Boudeuse*'.

[2] 'We know little about [him]', wrote Taillemite, *Bougainville*, I, p. 72, giving as his only sources AN, Marine B³ 570 f. 267 and B³ 664 f. 99.

[3] AN, Marine C7: 171, quoted in Martin-Allanic, *Bougainville*, I, p. 477.

than that of a mere observer. On a number of occasions, *Charles-Nicolas-Othon d'Orange et de Nassau-Siegen* (Prince of Nassau) helped Bougainville and defused some tense situations that had begun to develop with islanders. Born on 9 January 1745 and well connected in France as in Germany, he had joined the army in 1761 during the closing years of the Seven Years War. The end of this conflict in 1763 saw him back in Paris and subject to all the temptations the society of his time could offer to an outwardly wealthy and idle young man. He had some flaws of character, not unusual in a youth of his background; he was proud, impetuous and inclined to fight duels. His wealth was only relative and he began to accumulate debts; in addition, he was rumoured to be involved with several young ladies, one being the celebrated young actress Sophie Arnould, while another claimed he was the father of her child. His family were becoming concerned about his future. He was related to the former Secretary of State for the Navy, the Comte de Maurepas, who invited Bougainville to dinner and introduced him to the young Nassau-Siegen. The two had a long conversation, rapidly established a firm friendship, and Bougainville readily accepted the suggestion that the young man might join him on the expedition. He would sail at his own expense, but his family were delighted by the arrangement since it removed him from the frivolities of French society and the possibility of incurring further debts. Bougainville had few occasions to regret his presence on board: he was far more helpful and companionable than demanding. One is tempted to make a comparison between this passenger and Joseph Banks in Cook's *Endeavour* and the strain between the latter two brought about by Banks's demands, which led to his refusal, to Cook's considerable relief, to sail on the second voyage.[1] Back in France, his health unimpaired and his enthusiasm for adventure undimmed, Nassau-Siegen returned to the army, but sought an appointment in the navy with the senior rank of *capitaine de vaisseau*. This caused a storm of protests among other officers and the proposal was shelved. He then took part in operations connected with the War of American Independence, drew up a plan for the invasion of Jersey in the Channel Islands and even contemplated the possibility of setting up his own kingdom in West Africa. His debts were again mounting up, but a marriage to a Russian princess restored his finances. This led to his moving to Russia, where he became a protégé of Catherine II and eventually achieved his ambition to attain high rank in the navy by becoming a Russian admiral. He was buried in the Ukraine.[2]

2. The *Étoile*

A couple of years older than Bougainville, *François Chesnard de la Giraudais*, born at St Malo in 1727, had been at sea longer than any other officer on the voyage. He had first gone to sea at the age of five with his father on voyages connected with commerce and fisheries. During the War of the Austrian Succession, he served in the

---

[1] By then, in the words of Cook's biographer, 'Mr Banks, we must conclude, had come by an unusually swelled head:' Beaglehole, *Life of Captain James Cook*, p. 293.

[2] A biography, including chapters on his later life, was written by the Marquis d'Aragon under the title *Un Paladin au XVIIIe siècle*. Comments on his life in France appear in memoirs of the time, such as Bachaumont's *Mémoires secrets*; Métra's *Correspondance secrète politique et littéraire*, X, p. 175; and the Duc des Cars' *Mémoires*, I, pp. 174 and 268-70. A file on him is held in AN, Marine, C7:224.

Saint-Laurent, still under his father, and took part in his first sea fight at the age of seventeen. Transferred to the *Lys* later in that same year, 1744, still only seventeen, he held the rank of lieutenant; on this occasion he was wounded. The following year, in yet another sea fight, he was taken prisoner. Freed a year later, he served in the privateer *Gloire* which was to claim a total of thirty-six prizes before being herself captured in 1747. When peace was restored, La Giraudais returned to trading and fishing vessels, sailing regularly to Newfoundland. He was taken prisoner in 1755 during the British pre-emptive arrest of sailors and ships, prior to the formal outbreak of the Seven Years War. He was freed six months later, qualified for his master's certificate in September 1757 and took over command of a privateer, which earned him yet another period of captivity after a running battle that lasted ten hours. In 1759 he led a convoy of sixteen ships bound for Canada and was involved in yet another sea fight near Quebec. In the following year, he returned to Canada with a convoy of six merchantmen, captured seven vessels on the way, but was finally cornered by five enemy ships; after a cat-and-mouse battle lasting more than a fortnight, realizing he could not get away, he burned his ships and made his way to Montreal. Asking to take final despatches from the French commander to Paris, he slipped away on a schooner around the north of Newfoundland. Commissioned *lieutenant de frégate* for the remainder of the war, he was put in charge of five gunboats. But the navy had nothing for him in peacetime. He gladly accepted Bougainville's offer to command the *Sphinx* on the first voyage to the Falklands. He then took over the *Étoile* for the second expedition to the French settlement. He was given a commission as fireship captain for the duration of the circumnavigation; in August 1770 this warrant was turned into a permanent commission. However, there was little active service offering in the navy and financial difficulties forced him back to the merchant service and private trade. He sailed to the Isle de France and became involved in questionable transactions in Africa and South America. In March 1776 he was sentenced *in absentia* to three months imprisonment, but he had already died in Zanzibar in November 1775 at the age of forty-eight.[1]

His second-in-command was an experienced officer from the French India Company. *Jean-Louis Caro* had served with the company since his early youth and proceeded through all its grades, from junior apprentice to captain. He had obtained his first command, with the rank of *lieutenant de frégate* on the snow *Fine* bound for the Isle de France. By 1765, as second lieutenant, he was sailing to China in the *Villevault*. He was back in July 1766 and had been promoted first lieutenant when Bougainville offered him the position of second-in-command of the *Étoile*, for which he received a temporary commission in the royal navy as *lieutenant de frégate*. On his return, on the recommendation of Bougainville who had valued his skill and his determination, he was sent to China as second officer of the *Duc-de-Duras*. He returned to China in 1773–4 and to sailed to India in 1775–6. He died in 1787.

Jean-Louis Caro wrote a careful journal during the circumnavigation, which is typical of the man. It was down-to-earth, giving the image of a practical man who

---

[1] La Giraudais' personal file is held at AN, Marine under reference C7:163. He appears in the *Dictionnaire de biographie mauricienne*, no. 23, p. 689; see also Martin-Allanic, *Bougainville*, I, pp. 89–90.

had little time for shipboard squabbles – which allowed him to ignore Commerson's gibes – and little interest in the philosophical theories of his day, expressing his views, favourable or otherwise, on the people, and being more given to rely on a simple Breton faith.[1]

Four other officers of the blue[2] sailed in the *Étoile*. Of these two, Landais and La Fontaine bear the addition 'deceased', inserted at a later date, but this should not be taken as an indication of longevity in the case of the other two, Donat and Lavarye-Leroy, because although we know very little about Donat, we do know that Lavarye-Leroy died of wounds in 1781, aged fifty-five.

Joseph Donat was a relative of Chesnard de la Giraudais and the brother of Henry Donat de Lagarde who had sailed in the *Sphinx* to the Falklands. Henry had been quite prominent in coastal reconnaissance in the islands, but he did not sail on the circumnavigation.[3] There are a few mentions of Joseph in connection with boats sent to attempt to survey islands in the neighbourhood of New Ireland and the Moluccas, but few details. His service on board was apparently quite routine and satisfactory, and he completed the voyage back to France in the *Étoile*, thereafter presumably returning to the merchant service.

Details about *La Fontaine-Villaubrun* are equally sketchy. He was a fairly late appointment to the *Étoile*, replacing René Hercouet who had sailed on a previous voyage to the Falklands but was unavailable to take up the position of second lieutenant previously offered him. La Fontaine is reported as befriending Ahutoru on his early visits and looking for spices with Commerson on an island in the Moluccas. He was one of those left behind at the Isle de France at the request of the local authorities and some time later was commanding a merchant vessel in Indian seas.

Much more is known about *Pierre Landais*, from St Malo, born on 20 February 1734. He first went to sea at the age of eleven, sailing in a merchant vessel. He underwent his baptism of fire a year later, but was taken prisoner after that. The War of the Austrian Succession over, he returned to the merchant service, rising to lieutenant in 1752, but when the Seven Years War broke out he joined a privateer and was taken prisoner in 1756. Freed, he returned to privateering in 1760, but later joined the navy as an officer of the blue with the rank of lieutenant. While serving in the *Zéphyr* in September 1762 he was wounded and for a third time taken prisoner. Peace was signed a year later, and he served, still as officer of the blue, in the storeships *Garonne* and *Barbue* sailing to Newfoundland. This period of service ended in December 1764, and he was subsequently appointed to the *Étoile* with the rank of *lieutenant de frégate*. He was put in charge of boats surveying various islands later in the voyage, but criticized for firing on islanders without adequate justification. On a final occasion, he had failed to report to the *Boudeuse* for instructions and was reprimanded for it, leading to a violent argument with the Prince of Nassau who, if Commerson is to be believed, was tempted to draw his pistol and shoot him. After this, Donat was usually

---

[1] Personal file in AN, Marine C7:53, with his Journal under reference 4JJ 1. His later career is mentioned in AN, Colonies C² 289 f° 23.

[2] They are listed as 'merchant officers' in Vivez's journal, Rochefort MS, title page.

[3] Martin-Allanic, *Bougainville*, I, p. 108; Taillemite, *Bougainville*, I, p. 80 stresses that little is known about him and Lafontaine.

put in charge. In spite of a difficult nature, he was highly regarded as an experienced sailor and a particularly brave one. His appointment in the navy was confirmed, and he rose to *lieutenant de frégate* in January 1773 and fireship captain three months later. However, an appointment as port lieutenant of Rochefort in December 1775 displeased him and he retired at the beginning of 1776. He then joined the naval forces of the young United States, took command of the *Flamand* bound for New Hampshire and in 1778 was given command of USS *Alliance*, sailing to northern waters under John Paul Jones – with whom he soon fell out. He had taken American nationality, but left the States in 1789, travelled to Turkey, Germany and England before returning to France in 1791, rejoining the French navy in July 1792 with the rank of *capitaine de vaisseau*. He was given command of the *Patriote* for various operations in the Mediterranean, but once again fell out with his superiors. He was nevertheless promoted to rear-admiral from 1 January 1793 and took over the *Côte d'Or* at Brest. The anarchy that plagued the navy during the more difficult time of the Revolution caused him to resign and he returned for a time to America. He attempted to obtain a French naval pension after the end of the revolutionary period but it was declined and he sailed in 1802, apparently for good, to the United States. He died some time after 1815 in his eighties. A crotchety individual right to the end, he had not surprisingly fallen out with the equally difficult Commerson during the circumnavigation.[1]

Pierre-Marie Lavarye-Leroy,[2] like many on the expedition a native of St Malo, was born in the year 1726, joined the merchant service at the age of eleven and eventually rose to the rank of merchant captain. His record prior to his joining the *Aigle* in 1763 was most impressive: he had completed twelve trading voyages to the West Indies, Africa, North America and the Indian Ocean, eight on privateers and three on ships of the royal navy. Bougainville considered him an able and reliable officer, senior in experience and rank to most of the others and a suitable person to take over Port Saint-Louis in the Malouines. He was appointed port captain of Port Saint-Louis and later given a special bonus of 300 *livres* in recognition of this service. While in the islands, he had a small sloop constructed, *La Croisade*, which enabled him to survey some of the lesser known coasts and inlets of the archipelago. When the colony had been handed over to the Spanish authorities, he sailed to Montevideo with Bougainville who appointed him to the *Étoile*. He was back in the Indian Ocean in 1774, assisting the French consul at Surat, and on his return to France in 1778 applied to join the navy. Bougainville continued to support him, writing a testimonial that secured his entry into the service with the rank of *lieutenant de frégate* and a special bonus of 800 *livres*. In August 1780 he joined Bougainville in the *Languedoc*, subsequently transferred to the *Saint-Esprit* and finally to the *Héros*. He was badly wounded at the Battle of La Praya on 16 April 1781 and died four days later.[3]

Two *volontaires* were appointed to the *Étoile*. The first of these, *Jean-Robert-Suzanne*

---

[1] Landais' file is at AN, Marine CC7 and a biographical note appears under his name in the *Dictionnaire biographique des généraux et amiraux français…*, X, p. 51. See also Taillemite, *Bougainville*, I, pp. 80-82.
[2] As is the case with a number of the other officers and members of the expedition, his name is spelt in a variety of ways. He is sometimes referred to as Leroy, Leroi, Le Roy, or Le-Roy, sometimes as Lavarie, Lavari or Lavary.
[3] See AN, Marine C7:174, 182.

Lemoyne de Montchevry, born at Cayenne, Guyana, on 8 March 1750, was the son of the Lemoyne, commissary general at Rochefort, who had supervised the equipping of the *Étoile* and later travelled to Nantes to supervise work on the *Boudeuse*. The youth had already been at sea as a *volontaire* in 1765 and his father, whose efficiency and energy had impressed Bougainville as it had his own superiors, had little difficulty in getting Jean-Robert accepted for the *Étoile*. He would transfer to the *Boudeuse* while at Rio de Janeiro in July 1767 in exchange for Pierre Duclos-Guyot and be promoted *garde de la marine* in August. However, he had a weak chest and died at the Isle de France on 15 November 1768.[1] The second, *Alexandre-Joseph Riouffe* was born at Cannes in southern France, probably also in 1750. He too had sailed for the first time in 1765, in the *Garonne* bound for Cayenne. The ship returned to Rochefort in June 1766 and in early 1767 Riouffe was appointed to the *Étoile*. He completed the circumnavigation and was back in France on 24 April 1769. He held a minor administrative position from 1770 to 1772, then sailed in a transport vessel along the French coast. He served in the storeship *Bricole* in 1774, going to the West Indies, and from then on rose steadily through the ranks, taking part in a number of major engagements during the American War. He obtained his first command in 1784, that of the frigate *Flore*, and was appointed assistant administrator of the port of Rochefort in 1786. The Revolution, whose ideals he shared, gave new opportunities. He sailed to Santo Domingo in command of the frigate *Inconstante* in December 1791, and was promoted to *capitaine de vaisseau* in January 1792. On 29 November 1793, he was seriously wounded in an engagement against two English vessels while on escort duties and was taken to Jamaica. He died at Kingston ten days later.[2]

The clerk *Michau* (also *Michaud, Michaux*) was a land-based official whose only time at sea appears to have been his voyage in the *Étoile*. He had first worked in the artillery section of the Rochefort military college, and after serving there for ten years was transferred to the administrative side. He was probably in his mid to late twenties when he was appointed to the *Boudeuse*, but soon after transferred to the *Étoile*. He needed, according to a report, to save money, which may be why he sought this position in which he would be housed and fed, however modestly. He seems to have fulfilled his duties satisfactorily, writing reports to the Minister on damage suffered by the *Étoile* when a Spanish vessel ran foul of the storeship in Montevideo. His name appears as a co-signatory to the various Acts of Possession. Apart from these few mentions, his name does not appear in the journals, with one exception, a brief note in Pierre Duclos-Guyot's journal: 'Mr Michau, clerk, was placed under arrest on the 15th inst. in the afternoon.'[3] This was possibly the outcome of an altercation between one of the officers and the clerk who was unused to shipboard discipline. He completed the circumnavigation in the *Étoile* and returned to his desk job at Rochefort, rising gradually to the position of chief clerk in the general stores department, a grade he

---

[1] On Lemoyne, see his file at AN, Marine C7:180 and a report on his death in C¹ 183, f. 883.

[2] Riouffe's personal file is held AN, Marine C¹ 206. There is an account of his final days in Troudé, *Batailles navales de la France*, II, p. 313.

[3] Commerson/Duclos-Guyot Journal, entry of 15–16 May 1768. At the time the expedition was approaching the New Hebrides. His file is held at AN, Marine C7:207.

reached in 1786. His name appears in the staff lists until 1792, as the revolutionary storm gathered strength, after which it vanishes.

The surgeon *François Vivez* or *Vivès* is far better known, not merely on account of the journal he kept in the *Étoile* but because of his outstanding career in the naval service, a milieu in which he was born and which he never left. He was a man of severe principles, devoted to his duties and expecting the same standard from others. A careful observer of human nature, although with little sympathy towards some of the islanders he encountered, especially in Melanesia, he was also highly conscientious and eager to find ways to alleviate the sufferings of those who became his patients.[1] He was born in Rochefort on 14 September 1744, the son of a naval surgeon. He first went to sea at the age of seven in the *Formidable* with his father, enrolled as a pupil-surgeon just before his twelfth birthday and qualified as assistant surgeon in May 1760. He worked from 1759 to 1762 at the hospital for the wounded at Rochefort, still under his father, then moved to Aix in 1763. He served as second surgeon in the storeships *Coulisse* and *Garonne* in 1764 and 1765, and joined the *Étoile* as medical officer in 1766, qualified as medical officer and ship's surgeon after his return to France and served as such in various ships from 1772 to 1782. This period included no fewer than seven major naval battles and involved his remaining at sea for lengthy cruises. Vivez disembarked at Boston in December 1782, suffering from scurvy, his hearing badly affected as a consequence of attending to the wounded next to firing guns. He returned to France and in 1786 asked to be allowed to retire on a pension. This was declined, and instead he was appointed to the medical school at Rochefort. He gradually rose through the various grades to medical officer first class and in June 1801 was placed in charge of the convalescent wing of the convict hospital. He finally retired in December 1811. He had married the sister of a naval surgeon whose first husband had also been a surgeon in the services; she died in 1787 and he later married the daughter of an exchange dealer who had three brothers in the navy. He died at Rochefort on 3 September 1828.

The chaplain, Fr *François-Nicolas Buet*, was a native of Quimper. His involvement in the expedition was tragically brief. On 17 June 1767, while the *Étoile* was at anchor in Rio de Janeiro, he went ashore to say Mass in the cathedral and visit other churches in the town; at some time between 6 and 7 p.m. he was about to make his way back to the ship, together with the young apprentice Constantin, when an argument developed with some locals. A hostile crowd gathered that attacked both of them. Constantin was hit but survived by shamming death; Fr Buet was beaten with oars and held down in the water until drowned. His body was found the next morning washed up on the shore, and he was buried in the parish cemetery on the 19th.[2]

---

[1] Taillemite, *Bougainville*, I, pp. 83-5, devotes several paragraphs to an analysis of his personality, and quotes comments on Vivez by the Inspector General of Health, J. R. Quoy, giving a fair but favourable picture of the surgeon, under whom he had studied at Rochefort: *Notice sur Mr Vivès*, MS in La Rochelle library, 1834, and 'Mémoires inédits de Jean-René Quoy (1790-1869).'

[2] A report on this incident is included in a letter by Michau to the Minister dated 12 September 1767, AN, Marine B4 110, and further details are provided by Commerson in his journal, entry of 17 June 1767. There is little doubt that the chaplain had visited local churches as he had already supplied a report on them for Commerson's journal. The latter, however, does add a note of criticism to his account of the killing: 'One should not conceal the fact that the unfortunate abbé brought this catastrophe on his own

The engineer-cartographer (*ingénieur-cartographe*) *Charles Routier de Romainville* was born in Paris in 1742 and joined the army in 1756. He was wounded in August 1762 at the Battle of Johannesberg. He was appointed to serve in the Malouines settlement as engineer-geographer. He had sailed in the *Aigle* on 6 October 1764 and was promoted lieutenant in November. On arrival in the Malouines, he drew up a plan for a new settlement, to be known as Ville Dubuc, and remained in the islands until the end of April 1767 when he embarked in the *Boudeuse* for Rio de Janeiro. The Spanish offered him a post with the rank of captain, but he refused, preferring to join the complement of the *Étoile* as cartographer and draughtsman. During the voyage, he drew a number of maps and some 'charming pen sketches', many of which have been lost.[1] He left the expedition at the Isle de France, where he remained until the end of 1769, subsequently being appointed to the Pondicherry regiment. Back in the Isle de France in 1772, he was sent to take possession of the Seychelles. He was promoted to the rank of captain in August 1779 and to chief engineer (*ingénieur en chef*) on the island of Bourbon in May 1781. He resigned at the end of that year, and there are few records of his subsequent career. He applied in 1792 for the Cross of St Louis, but it was not a propitious time – the Order was abolished by the republican government that very year. A relative enquired as to his whereabouts in 1818, but the Ministry had no knowledge of him at that time: if he was still alive, he would by then have been seventy-six.

The astronomer *Pierre-Antoine Véron* was born in a poor working class family in Normandy in 1736, but in spite of living in an age when birth and class greatly mattered, he earned the respect and admiration of all who met him. He began life as a gardener and had little chance to further his studies until he was twenty; then, thanks to an uncle in Rouen who recognized his special talents, he was able to enrol at the school of hydrography and from there was sent to Paris Royal College where the mathematician Lalande took him under his protection. Lalande used his influence to get him accepted as assistant pilot in the *Diadème* in 1762 and in the *Sceptre* in 1763. He met on the latter vessel a similarly gifted young man, Charles-François de Charnières, with whom he sailed in the *Malicieuse* in the following year and with whom he collaborated on astronomical observations and on perfecting an astronomic glass called a megameter. Lalande then recommended him to the Minister of Marine for the Bougainville expedition. 'From what you told me', replied the Minister, 'I had no hesitation in issuing the required orders for Mr Véron to sail in the storeship *Étoile*.'[2] Bougainville soon realized his ability and relied on him as a matter of course for verifying the ships' positions. When the expedition reached the Isle de France, the *Intendant*, Pierre Poivre, asked Bougainville to let him have Véron's services for scientific work on the island and in particular for calculations related to the 1769 Transit of Venus. Poivre sent him as astronomer on an expedition to the Moluccas and the

---

head. He was a Breton, was excited by the wine he had drunk and behaved in a way that would have caused him to be knocked out as much in our French ports as in Rio: 'Journal', 17 June 1767.

[1] Martin-Allanic, *Bougainville*, I, p. 633, mentions a 'charming view' of a Strait of Magellan bay , in BN, NAF 9407, f. 38, some maps of Tahiti in private ownership, ibid., I, p. 691, one of the waterfall in New Ireland, also in private ownership, ibid, I, p. 765. The reference to his 'charming drawings … so light and fluid', is from ibid., II, p. 975.

[2] AN, Colonies B 125 f. 609, quoted in Taillemite, *Bougainville*, I, p. 90.

Philippines in the corvette *Vigilant* to fetch spice plants for the island. On the return journey, Véron fell ill during a call at Timor and he died soon after reaching Port-Louis on 1 June 1770. Bougainville remembered him as a 'gentle person', devoted to his work, mild and rather shy. Commerson, who fell out with a number of his fellow voyagers, had always remained on friendly terms with him, and on his death he named a flower after him, the *Veronia tristiflora*, 'a star-shaped flower that shows itself for only a few hours, and which against a dark background is spattered with tears.'[1] The French government, wishing to repay the devotion of 'so rare a man' who had died 'victim of his zeal', agreed to grant a pension to each of his five brothers and sisters.[2]

Probably the best known personality in the *Étoile* was the naturalist *Philibert Commerson*. He was the son of a small-town lawyer, born at Châtillon-sur-Chalaronne on 18 November 1727. His father grudgingly agreed to humour his early passion for botany and sent him to study medicine at Montpellier. He travelled throughout the district, collecting plants, and after graduating continued his travels, collecting in central France and the Alps, and turned for a while to the study of fishes, drawing up a list of rare types from the Mediterranean, earning thanks and praise for this from the Queen of Sweden. He wrote a *Martyrologe de la botanique*, a series of biographical notes in praise of those who had lost their life in the service of the natural sciences.[3] He became a resident doctor at Toulon-sur-Arroux, in the Charolais district in 1760, having married the daughter of a nearby lawyer, 'a sensitive plant', as he wrote to a friend, 'whom I propose to place, not in my herbarium but in the nuptial bed.' She was seven years older than he, 'but intelligent, well read' and with some claims to a fortune of 40,000 *livres*. She died a couple of years later and Commerson moved to Paris in 1764, leaving their infant son in the care of his brother-in-law, the Abbé Beau. He met a number of leading botanists in Paris, and in time was advised to join the Bougainville expedition to the Falklands, which he discovered would eventually include a circumnavigation. Bougainville met him, found him a tense, but appealing and undeniably enthusiastic savant, and both agreed on his appointment as 'Royal Botanist and Naturalist' – medical duties were mentioned but were relegated to the background since the *Étoile* had her own surgeon – with the quite substantial annual salary of 2,000 *livres*. In addition, Bougainville agreed to his being accompanied on the voyage by a valet.

Commerson proved to be a keen collector of botanical specimens, but he was given few opportunities to go ashore. Like all landsmen in ships, he was annoyed by the indifference of the naval men, most of them ignorant of the natural sciences and unwilling or unable to remain at anchor for lengthy periods and provide a botanist with the time and leisure to gather plants or other items. Commerson was more than an enthusiast; he was often carried away by his own passion and impatient towards others. The *Étoile* he described as 'that hellish den where hatred, insubordination, bad faith, brigandage, cruelty and all kinds of disorders reign.'[4] His notes contain a number

---

[1] Quoted in Cap, *Philibert Commerson*, p. 37.
[2] AN, Marine C7: 344, file on Véron; Taillemite, *Bougainville*, I, pp. 90–92.
[3] This title, somewhat pompous, as indeed was Commerson's style, was adapted by one of his biographers, Montessus, who wrote a *Martyrologe et biographie de Commerson* in 1859.
[4] Lefranc, *Bougainville et ses compagnons*, p. 93.

of critical comments directed at those who sailed with him; Caro is described as stupid, Landais as a persecutor, Vivez as an incompetent surgeon who tried to poison him. His notes would have produced an interesting, if biased travel story, but he wrote up little more than a few pages; his scientific notes and collections were similarly not written up, and delay in getting them to France resulted in their being neglected until recent times. Commerson's standing on board was not helped when it was discovered that his valet was a woman, a clear breach of naval regulations and no doubt a subject for some hilarity among the officers and crew. It may have been the reason why Commerson did not complete the circumnavigation, but landed with his valet at the Isle de France, at the request of the *Intendant* but possibly also as the result of Bougainville's suggestion to Pierre Poivre. The naturalist's time on the island was quite pleasant at the early stage, and he sailed to Madagascar and Reunion in 1770-71; he continued his botanical work throughout 1772, putting together a substantial collection, but his situation had become less easy after the replacement of Poivre by a new *Intendant* and his health had deteriorated. He died on 13 March 1773, aged forty-five.[1]

Coupled with the name of Commerson is that of his valet *Jeanne Baret*. She became the first woman to complete a voyage around the world. Born on 27 July 1740 at La Comelle, near Autun, the daughter of a labourer, she was orphaned at a relatively early age and worked at a servant, eventually moving to Toulon-sur-Arroux, possibly in 1760 or a little earlier. She had acquired some education, being able to read and write, a rare accomplishment among servants. Her claim to have been ruined and forced into service as the result of losing a lawsuit seems doubtful, as is her statement that she had previously worked for a Genevan in Paris. Whether she was employed by Mrs Commerson upon the latter's marriage is uncertain; she was certainly working as 'governess' after the birth of the Commerson baby and Mrs Commerson's death in April 1762. In August 1764, she was five months pregnant, officially registering this pregnancy with a lawyer in a town thirty kilometres away. Shortly after this, Jeanne and Philibert left together to settle in Paris. There is clear evidence that relations between Commerson and his brother-in-law, the Abbé Beau, were strained, in all probability as the result of the affair, and Commerson was never again able to see his young son. Jeanne gave birth in Paris, but the baby died shortly afterwards, and in December 1766 the couple left Paris for Rochefort and embarked in the *Étoile*. Although she did her best to conceal her sex, sharing her master's cabin to avoid close contact with other servants or the crew, rumours soon began to circulate in the ship, but whether Bougainville, in the *Boudeuse*, discovered the situation at an early stage — which could explain a reference to Commerson's being placed under some form of arrest while in South America — is not certain. Baret was certainly a strong, tough

---

[1] An increasing number of studies are being written on Commerson as well as on Baret. A colloquium has been held on him and his work at the Centre universitaire de la Réunion in 1973, articles on him have appeared in publications of the Musée national d'Histoire naturelle de Paris, of the Montpellier university alumni organization, in local histories, in the *Dictionnaire de biographie mauricienne* and in connection with specialist or local exhibitions. A major collection of essays, *Philibert Commerson: le découvreur du* bougainvillier, was published in 1993 by the Association Saint-Guignefort at Châtillon-sur-Chalaronne. A Rue Philibert Commerson commemorates his memory in his native town of Châtillon-sur-Chalaronne.

woman, 'my beast of burden' as Commerson called her, helping him on arduous collecting expeditions ashore. But in time, the truth came out and Bougainville was officially advised. Commerson was then reprimanded, but nothing further could be done.[1] It may, however, have decided Bougainville to land Commerson and his servant at the Isle de France, thus avoiding the need for an embarrassing report and possible court of inquiry on his arrival at St Malo. The Tahitian Ahutoru expressed early interest in Baret, whose feminine features appealed to him, but he expressed the same interest towards another youthful crew member. When Baret landed on botanizing expedition, a group of Tahitians set upon her, apparently intent on raping her. This episode led to some speculation about the possible existence of a sixth sense among Polynesian people, enabling them to identify a female more easily than the European sailors among whom Baret had lived for so long. It appears more likely that Ahutoru identified her as a *mahu* or transvestite, a class in no way condemned or ostracized in Tahitian society,[2] and that the incident ashore was not a sexual attack but an attempt to verify her gender. Once at the Isle de France, she continued to assist her master in his work, although it is not clear whether she went with him to Madagascar. After his death, now dressed in ordinary female attire, she opened a tavern, as is indicated by her being fined in December 1773 for having served drinks during Mass time. On 27 May 1774 she married Jean Dubernat, a former army sergeant and returned to France with him, probably a year later. She thus completed her circumnavigation. She then lived in relative obscurity, although still regarded as 'an extraordinary woman' and was given a pension of 200 *livres* in 1785.[3] She died on 5 August 1807.[4] Commerson had named a plant after her, the *baretia*, now known as *quivisia*.

The Ships

The two ships which were made available for the expedition were a frigate, the *Boudeuse*, and the storeship *Étoile,* which had already proved her worth on a voyage to the Falkland Islands. The frigate was brand-new and naturally held greater status than the storeship, but Bougainville was about to undertake a voyage which involved close inshore navigation among little known or, hopefully, quite unknown islands, reef-strewn and dangerous. As James Cook was to prove with his stolid but reliable and manoeuvrable *Endeavour*, appearance and status were not factors of great value on a voyage of exploration. However, one should bear in mind the first stage of the expedition, the handing over of the Malouines settlement to the Spanish authorities following an agreement reached by the kings of France and of Spain, for which a

---

[1] The royal ordinance prohibiting women on board ships was dated 15 April 1689: 'His Majesty forbids the officers of his ships and the men of their crews from taking women on board to spend the night there and for any longer than the time of a short visit on penalty of one month's suspension for officers and fifteen days in irons for the crew'. Interestingly, Commerson's suspension or arrest in South America was for one month. A 1765 ordinance had added that the prohibition applied 'even under pretext of public feasts or rejoicings' and specified that the month's loss of pay also applied to petty officers.

[2] See on the *mahu*, Oliver, *Ancient Tahitian Society*, pp. 371–4.

[3] Document in AN, Marine C7 17.

[4] Henriette Dussourd gives the date as 1816: *Jeanne Baret*, p. 79. However, Jeannine Gerbe, in 'La Famille bressane de Philibert Commerson', gives a reproduction of the death certificate of 'Jeanne Baret, âgée de soixante-sept ans'. A story about her, 'Breaking the Bonds', was included in a collection by Shirley Fenton Huie, *Tiger Lilies: Women Adventurers in the South Pacific* pp. 14–21.

suitably impressive ship was required. And impressive she looked, with her hull painted black, the upper works red, the wales yellow, all the sails and rigging brand-new.

The *Boudeuse* was built at the Indret shipyards near Nantes, work starting in June 1765 with the official launching in March 1766. She was 125 feet (40.60 metres) long, 32 feet (10.50 m) wide, of 960 tons, carrying 26 eight-pounders. The crew were accommodated in the between-deck, in a space 1.60 m high which they had to share in the earlier part of the voyage with cattle, sheep and pigs; cages for the poultry were provided along the quarter-deck. Below, by the store rooms, accommodation was available for petty officers, the chief gunner, the chaplain and the surgeon, the latter two having the advantage of a private cabin approximately eight feet long and seven wide. The officers each had a small private cabin, although this privacy was ensured by no more than partitions made of sailcloth. The captain's quarters were part of the main council room, used for meals, meetings and work on the charts. In the hold, space had to be found for food supplies, water casks and minor items such as nails, hatches and glass beads to be used for bartering with native people encountered during the voyage. These cramped quarters were to be shared for almost three years by some 220 men.

The fact that the *Boudeuse* was new was considered by some officials to be a distinct advantage, almost flattering to Bougainville, but this view was not shared by most naval personnel who realized that the usual preliminary trials would have to be carried out during the voyage itself. Work had to be done at an early stage to give the ship a better balance, and for much of the crossing to South America and the Falklands further minor adjustments had to been made. This led Bougainville to consider sending the *Boudeuse* back to France, and sailing to the Pacific in the more stolid and well tried *Étoile*. However, he gradually changed his opinion as the frigate got over her initial problems, and before long his complaints were directed at the slower and at times wallowing *Étoile*.

The *Étoile* was a storeship of 480 tons, built as a trading vessel back in 1759 and bought by the navy in 1762. She was 104 feet (33.80 m) long, 27 feet (9 m) wide, but carried a complement of only 120 men. She had been used to take emigrants to Guyana in 1764 and had then sailed to the Malouines in 1765. She needed some refitting for the planned circumnavigation, but by and large was a reliable ship which met the demands that were placed on her, and proved her reliability on subsequent expeditions to the West Indies and the Indian Ocean, being finally decommissioned in 1778 after almost twenty years of service. The *Boudeuse* would also prove her reliability later in her career, taking part in the War of American Independence and the early Revolutionary Wars, until her final decommissioning in 1800.

The Instructions

Bougainville had discussed his plans with various senior officials since the early 1760s and over the years he refined his proposal in line with political realities and suggestions that were made to him by both the Minister of Marine and the scientific world. What remained from his original programme was the settlement on the Falkland

Islands. He had contemplated the possibility of another settlement in northern California and a search for the southern lands. When he was forced to give up his Malouines colony, he revised what he had envisaged in 1761-2, discarded the idea of further colonial settlements and concentrated on his planned search for lands in the southern hemisphere.

At first, he had been tempted to seek these austral lands somewhere to the south-east of the Falklands, where Bouvet de Lozier had found, back in 1739, land which he named Cap de la Circoncision, believing it to be a headland of the fabled southern continent and not, as it in fact was, an isolated island.[1] French interest in a possible continent somewhere in the south-eastern Indian Ocean or even further east went back to the early sixteenth century, when the Frenchman Binot Paulmyer de Gonneville returned from what was believed to be an attractive country south-east of the Cape of Good Hope. It would require the voyages of Yves de Kerguelen and James Cook to dispel the rumours that had accumulated over the years about this region. It was evident from Bougainville's closer study of the history of exploration, from discussions with scientists and historians and from the activities of the British that the Pacific Ocean itself held far more promise. All this confirmed what he had been told during his stay in London, and led him to give up any idea of continuing the work of Bouvet, as he also did his plan for a French settlement in California: the trouble his colony on the Falklands had caused with the Spanish made it clear that any other attempt to encroach on what was nominally part of the Spanish colonial empire would be politically unwise.

He consequently drafted out a new proposal, soon after his return from Madrid on 16 October 1766. The first stage of the voyage would be the handing over of the Malouines settlement to the Spanish, after which he would sail into the Pacific, either through the Straits of Magellan or by Cape Horn, and would survey 'as much of and as best he will be able to, the lands lying between the Indies and the western coast of America, various parts of which have been sighted by navigators and named Diemen's Land, New Holland, Carpentaria, Land of the Holy Ghost, New Guinea, etc'. After this, he would sail to China, with a possible call at the Philippines, and endeavour to find some island within reach of the China coast that could enable the French India Company to trade, probably illegally, with China.

This was an ambitious programme, not altogether realistic, but Bougainville was casting his net wide, leaving it to the French government to decide which areas should be given preference, unless – which would be preferable – he was given *carte blanche*. Ministry officials promptly studied the proposal and revised it, after which it was approved, restructured in the form of instructions, and signed by the Minister and by Louis XV. Consideration took little time because, although Bougainville's plan is undated, the instructions are dated 26 October 1766. Revision was in fact minimal: apart from changes to the wording amending his suggestions into instructions and adding a few details, Bougainville's proposal was accepted in its entirety, including the crossing to China, which the lack of food and water and the worsening health of his crews would compel him to abandon:

---

[1] On Bouvet, see Dunmore, *French Explorers*, I, pp. 196-202.

Memoir from the King to serve as instructions to Mr de Bougainville, infantry colonel and capitaine de vaisseau *for the duration of the campaign, concerning the operations he will be undertaking*[1]

H.M. having had the frigate *Boudeuse* fitted out at the port of Nantes and the storeship
Étoile at the port of Rochefort in order to sail to the Malouines, has entrusted to Mr de
Bougainville the command of the frigate *Boudeuse* which will be the first to leave.
H.M.'s intention is that after leaving Nantes, Mr de Bougainville will go to the River
Plate to join two frigates H. Catholic M.[2] has sent from his European ports, which are to
wait for him there.

After meeting up with the two Spanish frigates, he will take them to the Malouine
Islands and he will hand over these islands to the officers of H. Catholic M. in accordance with the orders which have been addressed to the French commander of the said
islands, of which a copy shall be handed to Mr de Bougainville.[3]

During his stay in the Malouine Islands, Mr de Bougainville shall be in command of
the French nationals.

After handing the Malouine Islands over to the Spanish and after the storeship *Étoile*
shall have joined him and placed herself under his orders, he will leave with his two vessels and sail for China by way of the South Sea. He shall be free to cross the Strait of
Magellan or round Cape Horn according to what the time of year and the winds dic-
tate.

During his crossing to China, he will examine in the Pacific Ocean as many as possi-
ble and as best he can the lands lying between the Indies and the western shores of
America of which several were sighted by navigators and called Diemen Land, New
Holland, Carpentaria, Land of the Holy Spirit, New Guinea etc.

Knowledge of these islands or continent being very slight, it is of interest to improve
it. Furthermore, as no European nation has any establishment or claim over these lands,
it can only be in France's interest to survey them and take possession of them should
they offer items useful to her trade and her navigation.

With this in mind, the area that Mr de Bougainville must concentrate on examining
is especially the one stretching from the fortieth degrees of southern latitude towards
the north, surveying what may lie between the two tropics.

It is in those climates that one finds rich metals and spices. Mr de Bougainville will
examine the soils, trees and main productions; he will bring back samples and drawings
of everything he may consider worthy of attention. He will note as far as is possible all
the places that could serve as ports of call for ships and everything relating to navigation.
In this connection, H.M. leaves it to the zeal Mr de Bougainville has already shown to
obtain all possible information.

As soon as Mr de Bougainville lands in unknown places, he will see to it that posts
bearing the arms of France are erected and draw up Acts of Possession in the name of
His Majesty, without however leaving anyone behind to establish a settlement and he
will bring back the documents he draws up.

He will organize his navigation so as to be able to leave China at the end of January

---

[1] In AN, Colonies A 10, f. 144 (original) with a copy with the word 'approved' in the King's handwriting in AN, Colonies F²A, f. 116.

[2] I.e. the King of Spain.

[3] The commander in charge of the Malouines settlement was Bougainville's own cousin,
Bougainville de Nerville, a lawyer formerly resident in Paris.

1768 at the latest, the monsoon not allowing a departure after that date. Mr de Bougainville will be allowed to call at the Philippines should his navigation require it, he will take his two vessels there as well as to China if he considers it appropriate or he may send the *Étoile* by a different route if he considers this necessary when he leaves the lands situated between the Equator and the Tropic of Cancer.

H.M. requests Mr de Bougainville to see whether there is an island within reach of the China coast that could serve as a depot for the India Company to trade with China.

Since the length of Mr de Bougainville's voyage should not exceed two years and as events during his campaign may require him to make straight for the Isle de France, H.M. gives him the option of landing or not landing on the China coast. He will be guided in this matter by the circumstances in which he finds himself.

In any case, Mr de Bougainville will do everything in his power to call at the Isle de France in order to find out whether the state of peace has been maintained or whether we are at war with some nation and whether H.M. has despatched any instructions concerning some particular destination for the two vessels.

Should Mr de Bougainville consider it absolutely necessary for his return to France to separate his two vessels, H.M. authorizes him to do so, and to burn one of the two vessels if, in unknown seas, the deteriorating health of his crews or some accident made it impossible for him to sail with the two ships, and if he finds himself in some inhabited port, H.M. authorizes him to sell the vessel which he would not be in a position to bring back to France. Whichever course Mr de Bougainville adopts in one or other circumstance, he will get a report drawn up on the need that arose and the decision made. This report shall be signed by the officers and clerks of the two vessels.

H.M. in addition authorizes Mr de Bougainville to pay during his voyage bonuses either in cash, or in clothes and other items taken on the said vessels to petty officers and sailors who deserve them and to increase their ration if he considers it necessary and to effect such changes as may be required by the health of the crews and the clerk of each vessel shall see to these on Mr de Bougainville's written instructions.

Such are the general instructions H.M. has sent in connection with the aims He has in mind and the advantages that may derive from the campaign Mr de Bougainville is to undertake. H.M. reserves the right to grant upon the return of the two vessels the rewards the officers and crews may have earned during the campaign following the accounts Mr de Bougainville shall give of their zeal and their services.

Drawn up at Versailles on 26 October 1766.

A few days later, a second set of instructions was sent to Bougainville concerning the running of the ships and the discipline policies to be adopted towards the crew:

Royal Memoir to Serve as Instructions to Mr de Bougainville

H.M. having handed over to Mr de Bougainville, commanding the frigate *Boudeuse*, fitted out in the Nantes river for a special mission, an instruction on what he is to carry out during his campaign for which he has sent him a warrant of *capitaine de vaisseau*, he will now explain the steps he has to take concerning the running and discipline to be maintained both in the said frigate and in the storeship *Étoile* which is to be equipped at Rochefort under the command of Mr La Giraudais, fireship captain, which after joining Mr de Bougainville at the place that shall be advised to him is to sail under his orders.

Mr de Bougainville shall pay and ensure that others pay the greatest attention to the preservation of the rigging and equipment including spares taken on the frigate

Boudeuse, he will avoid any being used beyond ordinary requirements except in case of absolute necessity evidenced by reports drawn up in due form and he will take the same care in respect of the food ensuring that during the first months of his campaign those victuals that keep less well are consumed.

He will provide Mr de Saint-Germain, appointed by order of H.M. to carry out the functions of clerk in the said frigate, the assistance and facilities he will need for the task he has been given.

He will see that all the expenditure of whatever nature that may be incurred for the service of the said frigate in the way of purchases of supplies and munitions or of food and refreshments are set out in statements made out by the said Mr de Saint-Germain and certified by him as due to be reported as evidence relating to of the accounts that will have to be drawn up for all the expenses relating to his expedition.

H.M. urges him to see that all sections relating to discipline and orderliness as well as the good order to be maintained in the said frigate are scrupulously observed, in accordance with the ordinances, regulations and decision issued in consequence.

He will keep a precise journal of his navigation of which upon his return one copy shall be sent to the Secretary of State in charge of the Department of Marine and will see that the officers and pilots going with him keep theirs to be handed upon their return to the commander of the port where they shall disembark in accordance with the requirements of the Naval Ordinance of 25 March 1765.

He will report on his return on those who have shown the greatest application and will advise them that the surest way to deserve their promotion is to live within the bounds of the most rigid discipline and to obtain all the knowledge that is necessary to the navy.

As he may meet during his navigation ships belonging to various princes and states, H.M. considers it appropriate to advise him of the steps to be taken in connection with salutes.

He will salute all the ships of crowned heads flying the colours of an admiral, vice-admiral and rear-admiral. In the category of crowned heads are included the King of Spain, the King of the Two Sicilies, the Grand Lord, the King of England, the King of Sardinia, the King of Sweden, the King of Portugal, the King of Denmark and the Republic of Venice.

However, he will not salute English vessels wherever he meets them and whatever flag they are flying and will not request any salute from them, he will act similarly for places that are under the domination of the King of England if he is forced to enter any.

In respect of Spanish warships, he will act in respect of salutes to be returned in accordance with the agreement made in 1734 with H. Catholic M., that is to say the ships of H.M. and those of the King of Spain flying colours of a similar status will not require or return any salute, but in respect of other colours the one who is saluted first shall reply with two shots fewer if the one who saluted is inferior to him by only one degree, with four shots fewer if the flag that gave the salute is inferior to him by two degrees, and so on, so that a flag of vice-admiral shall reply by two shots fewer to a rear-admiral, by four shots fewer to a commodore and by six shots fewer to the commanding officer of one or more ships. The salute shall be returned in the same ratio by rear-admirals and commodores.

As for fortified places and the main fortresses, they will reply with an equal number of shots to the salute of an admiral and vice-admiral's ensign, two shots fewer to a rear-admiral, four fewer to a commodore and six fewer to the commander of one or more vessels. Spanish naval posts and main fortresses that must be given a salute are those shown on the attached list.

As for the Dutch, he will act in accordance with the Ordinance of 1689.

He will similarly follow the requirements of the same ordinance in respect of salutes to be given to the various princes and states.

Prior to his departure, he will send to Mr La Giraudais commanding the storeship *Étoile* the signals of recognition to be used when the said storeship joins the frigate *Boudeuse*.

When on their way back the two ships have reached the latitude of the Azores, Mr de Bougainville shall instruct Mr La Giraudais to make for Rochefort where the storeship *Étoile* is to be decommissioned and he shall continue alone on his way to Brest, where upon arrival he will receive instructions from the commander of the said port for her laying up which he is to see to with orderliness and diligence and will see that everything taken off is placed in the stores after the survey has been completed in accordance with what H.M. has instructed in this respect.[1]

After he had sailed from Nantes, damage to his ship forced Bougainville to put into Brest, at which time he received complementary instructions from the Minister, dated 1 December 1766. They reflect Bougainville's concern about the suitability of the still untried *Boudeuse* for a lengthy voyage around the world and the permission he had sought to sail in the *Étoile* should circumstances warrant it:

As for the decision you are envisaging, to leave [when at the Falkland Islands] the command of the *Boudeuse* to take over that of the *Étoile*, sending back to France the former vessel with a crew of 160 men which you consider sufficient, the King leaves you the option of acting in accordance with what you consider at the time is necessary for the good of His service and the success of your mission ... I am enclosing the letter the King is writing to you to authorize you to take command of the *Étoile* and give to Mr de Bournand that of the *Boudeuse*...

The Minister added a note in his own handwriting:

You will be the master, Sir, of the decision whether to continue your expedition in accordance with what has been planned with the two ships the King has entrusted to your command or to send back the frigate *Boudeuse* if you do not consider she is suitable for the purpose intended. H.M. leaves it to your prudence in this matter and leaves you free to continue on your way with both vessels or simply with one of them by choosing whichever will suit you best.[2]

La Giraudais's instructions were sent to him on 12 January 1767, requiring him to sail without delay for the Falklands and place himself under Bougainville's orders. He was given, however, the option of calling at Cape Verde Islands should he need to take in some urgently needed refreshments, without delaying there unduly. A further letter followed on the 16th confirming that he was to follow any course decided by Bougainville in respect of the *Étoile* – which in the light of Bougainville's uncertainty about the long-term seaworthiness of the frigate could have included his taking over

---

[1] AN, Marine B² 382, ff. 483 and 484. 'Orders similar to these were at the time being handed over to all officers commanding a ship that was on the point of sailing' (ET).

[2] AN, Marine B² 382, ff. 85, 492; Taillemite, *Bougainville*, I, p. 25.

the *Étoile* and sending back the *Boudeuse* under the command of Bournand. La Giraudais was also given a copy of the general instructions concerning the running of his ship and the various salutes to be given to vessels and naval fortresses of other countries.

The instructions therefore allowed Bougainville considerable latitude, with the exception of the handing over of the Malouines which was the primary motivation for the expedition. Once this was done, he was free to choose which ship he sailed with, should he consider that one was unsatisfactory for a lengthy voyage, and to alter his route across the Pacific as circumstances warranted. Any evaluation of his achievements should be measured against the almost free hand he was given to use his judgment and the actual difficulties he had to cope with, including inadequate charts and rapidly diminishing food supplies.

The Voyage

Bougainville arrived at Nantes on 1 November 1766, accompanied by the Prince of Nassau. The frigate *Boudeuse* was still being readied for the long voyage up the Loire River at Mindin. He spent several days meeting his officers and on the 5th went up to Paimboeuf and Mindin, where he formally took over his ship. Final arrangements were made and everything was ready by the 13th. He was able to sail on the afternoon of the 15th. The *Étoile*, which was still being outfitted at Rochefort, would join him later, in South America.

Misfortune struck almost immediately. A storm broke the mainmast and caused other damage. Bougainville had no alternative but to put into Brest for repairs. He seized the occasion to decrease the heights of the masts and to exchange his twelve-pounder guns for lighter eight-pounders. He wrote to the Minister, suggesting that he would complete the transfer of the Falklands to the Spanish authorities in the *Boudeuse* and then proceed with his voyage of exploration in the *Étoile*. He added that, being already behind in the schedule planned for the voyage, he thought it no longer possible to go to China. The furthest he might be able to go would be the Philippines and 'landing on some of the Moluccas'. The Minister had no objections to raise: Bougainville was given freedom to choose whichever course suited the circumstances.[1] Consequently, his failure to cross the Pacific to China – which, had he done so by crossing the Pacific in a north-easterly direction from the Tuamotus, would have not produced much, if anything, in the way of new discoveries – was not brought about by a neglect of his instructions, nor even by the lack of food and supplies in the later part of the voyage, but was an early change of plan whch the authorities had earlier approved.

The *Boudeuse* was finally able to sail on 5 December. Some minor changes had been made to the complement, some men being left behind sick at the hospital or paid off for other reasons, a few deserting, others being taken on as replacements, the total complement dropping from 222 to 216. The first stage of the voyage took the French to Montevideo; the crossing was uneventful, the *Boudeuse* dropping anchor in the estuary of the River Plate on 31 January 1767. Almost at the same moment, back

---

[1] Correspondence in AN, Colonies F² A 20, and AN, Marine B² 382.

in Rochefort, the *Étoile* was getting ready to sail after a series of irritating delays which would mean that the two ships would not be able to join each other at Montevideo as had been originally planned. Bougainville went to Buenos Aires to finalize the agreement covering the handing over of the French settlement in the Falklands. There were no problems and a pleasant three weeks were spent socializing and observing the life of the Spanish colonists and administrators. On 1 March the *Boudeuse* was finally able to sail for the Falklands, accompanied by two Spanish frigates, the *Liebre* and the *Esmeralda*, and the tartan[1] *Nuestra Señora de los Remedios*. The *Boudeuse* reached the Malouines colony on the 22nd, the Spanish frigates arriving soon after.

The settlers had been somewhat anxious when they had first sighted sails appearing on the horizon, fearing they might have been English vessels that had come to drive them off. They told Bougainville of their surprise when, in the previous December, a ship flying English colours had appeared at the entrance to the bay. It was Macbride's *Jason*, sent by the Admiralty to warn off any subject of a foreign nation. In accordance with his instructions, the *Jason* despatched a letter to Nerville, the governor of the Malouines who was also Bougainville's cousin, pointing out that the islands had been discovered by 'a subject of the Crown of England, sent out by the Government for that purpose, and of right belonging to His Majesty', and asking 'upon what authority you have erected a settlement on the said Islands'.[2] Nerville was given one hour in which to reply. To some extent, this made it easier to stall for time: he replied that no one on the Malouines spoke or read English, and in turn asked Macbride what he was doing in the islands. Nerville's envoy, Desperriers, was kept on board the *Jason*, invited to say for dinner, during which Macbride stressed that he had no hostile intentions, but that the French had to leave what was British territory. Desperriers returned to the Malouines settlement at one in the morning, still uneasy. Nerville had taken what few defensive steps he could, and the French waited anxiously for the morning. Macbride, however, was as good as his word. There were a few more, relatively friendly, exchanges of views, and, the winds having turned, the *Jason* sailed away, warning the French that they or some other British ship would return in a month or two. Nerville and his settlers were consequently relieved to find that the vessels now approaching belonged to friendly nations, and equally relieved that the handing over of the colony would put an end, as far as they were concerned, to British intervention.

The formal handing over took place on 1 April. The settlers were given the option of remaining on the islands, under Spanish rule. Thirty-seven accepted. 'I obtained for the French who are staying here the most advantageous conditions I could. They all

---

[1] A tartan (or tartane) was a small single-masted vessel with a large lateen sail and a foresail, common in the Mediterranean.

[2] Nerville drew up a detailed report on this incident, with copies of Macbride's letter and his own reply, which he forwarded for Versailles. It was copied there and forwarded to Spain. A third copy, in Spanish, later reached London. They are held respectively in Affaires Etrangères, Espagne, vol. 548, ff. 337–8; AN, Colonies, F²A 20; and British Museum, Add. 32.603 f. 38. The text of Nerville's report is given in Taillemite, *Bougainville*, I, pp. 486–7. See also Goebel, *The Struggle for the Falkland Islands*, pp. 234–5. Fesche in his report on the effect the presence of the *Jason* had on the colonists, spells Macbride's name as 'M. Abbergh': 'Journal', entry of 24 March 1767.

plan to return to France within a few years', wrote Bougainville.[1] Some fifteen of them were taken on by Bougainville to replace some of his sailors who were sick, or who had deserted, Montevideo being notorious for the number of sailors who jumped ship. Nerville came as a passenger, travelling as far as Montevideo, intending to sail back to France. The Spanish frigates left on 27 April, Bougainville waiting for the *Étoile* which, according to the original plan, was to have joined the *Boudeuse* at the Malouines. He waited through May, making a few excursions around the islands for which he had harboured such hopes and which he was about to leave for ever, and loading such supplies as he could buy from the Spanish. Then on 1 June he weighed anchor.[2] The *Étoile* meanwhile had completed a difficult crossing to Montevideo where she had arrived at the beginning of May. Eventually receiving orders, through the intermediary of the two Spanish frigates, to proceed to Rio de Janeiro, in Portuguese Brazil, the *Étoile* sailed, arriving at Rio on 13 June. The two ships finally came together there a week later.

Bougainville was now eager to set out on his circumnavigation. He had now decided that he would go with both vessels, the *Boudeuse* having settled down after her initial setbacks, and the *Étoile* likely to be a useful consort with additional food supplies on board and some promising members of the expedition such as Commerson, La Giraudais and Vivez. A lengthy stay at Rio had never been envisaged, and relations between the French and the local authorities were cool at best, especially after the *Étoile*'s chaplain was murdered on shore, no one being arrested for the crime. On the credit side, Commerson had been able to set out on botanizing expeditions in the surroundings, and it seems that it was on this occasion that he collected samples of the bougainvillaea plant he was to name after the commander.[3] Even so, Bougainville reported that, like Véron who was attempting to carry out some observations, 'Mr de Commerçon, a famous naturalist ... is encountering similar obstacles'.[4] On 15 July, the French sailed for Montevideo to complete their preparations.

The Spanish were far more friendly towards the French than the Portuguese had been. Even then, there were setbacks and delays. The *Étoile* collided with a Spanish vessel, and the damage she sustained compelled Bougainville to send her to La Encenada for repairs. He himself went to Buenos Aires for news and supplies. The Jesuits were being expelled from South America, which meant that the unfortunate French colonists could not find room on any ships bound for Europe. There was little he could do for them in the meantime beyond sympathizing. When ships were found for them, conditions were so crowded that Bougainville was forced to make an official protest. He returned to Rio, leaving Du Bouchage behind to look after the expedition's interests and purchase further supplies. Days went by. Work on the *Étoile* was

---

[1] AN, Colonies, F²A 20. The return of the colonists to France in the shorter and the longer period was far less straightforward than Bougainville anticipated. A number of them were in fact landed in Spain. The King of Spain rejected claims for compensation by these unfortunate settlers, and finally in August 1774 under Louis XVI the French government met their demands. See Taillemite, *Bougainville*, I, p. 219-20n.

[2] The little English colony established at Port Egmont struggled on until 1774; the Spanish abandoned their Malvinas settlement in 1811, leaving the archipelago once more uninhabited. Argentina continued her efforts to claim the islands, but there were no colonists as such until the 1830s; Britain formally established the Falklands as a colony in 1833-4.

[3] Monnier et al., *Philibert Commerson: le découvreur du bougainvillier*, p. 3.

[4] 'Journal', entry of 5 July 1767.

making little progress, food was hard to obtain, the weather was often stormy and the authorities had their own problems in finding transport for the Jesuits and in keeping an eye on the Portuguese of Brazil, with whom relations were often at breaking point. Bougainville had to be patient and satisfy himself with chronicling the events unfolding around him.

'We learnt this morning that the Jesuits in Chile were placed under arrest on 18 August.' 'I went today with Mr de Commerçon to see some bones we were told came from giants. They are not human bones.' 'We also went to see a former officer from this country they say is 114 years old. He no longer sees, can hear well, can still walk. His face is only that of a nonagenarian.' 'E to ESE storm, rain, frightful weather. A schooner on her way from Montevideo has been lost on the northern coast. She had 20 people on board ... only one officer and 2 sailors survived.' 'They have been waiting daily for the arrival of the *Situao* or silver convoy from Potosi, which is slow in arrival ... the *situao* arrived this morning 24ᵗʰ [October]. It is carrying no more than 500,000 piastres of which 270,000 are for the royal treasury. This sum would be hardly enough for 6 months' pay to the soldiers who are owed for 17 months'.[1] A schooner used for taking supplies to the *Étoile* had to struggle against the tide and the bad weather: 28 October: 'the schooner of Isidoro is loaded and I am going'; 29 October: 'the schooner has grounded in the Rio Chuelo, one has to wait'; 30 October: 'the schooner being still grounded, I am going to La Encenada in the sloop from the *Esmeralda*'. Finally, on 31 October, Bougainville was able to take the *Etoile* back and return to his own *Boudeuse*. On 4 November, he was able to write in his journal, 'I have had everything we had at Montevideo cleared up and have arranged everything for our departure'. He was over-optimistic. For the next week, he could only make a few brief entries in his journal, reporting on the bad weather and the unfavourable winds. Finally, on the 13ᵗʰ the winds altered in his favour and departure became possible.

On 14 November 1767, a year almost to the day since the *Boudeuse* had sailed from Nantes, the expedition finally left the Atlantic coast and set out on the major part of the voyage. Putting an end to the Malouines venture and getting the ships ready for the crossing of the Pacific Ocean had been a long and frustrating affair, but they had finished their work and could now claim their reward.

Their high spirits were dashed within a few days. Expressions such as 'deplorable weather', 'heavy seas', 'unbearable rolling' peppered the journals. 'Nevertheless, this is the beginning of summer in this cursed climate', commented Bougainville. The storm caused the loss of much of the livestock and resulted in damage to the *Étoile*'s rigging and sails. But soon the sun returned, the sea became smoother and the opening to the Straits of Magellan came into sight. Even so, the ships spent four days struggling to enter, and even after a full week they had only reached as far as the second narrows. But the French were able to meet once again the Patagonians and Fuegans they had come upon on their previous visit to the Straits. Relations were cordial, gifts and embraces being shared with exclamations of joy – Commerson was offered, or understood he was being offered, one of the men's daughters in exchange for an old blue coat.

---

[1] Sundry journal entries, 25 September to 23 October 1767.

The legend of the Patagonian giants was fully exploded. The French had seen no giants on their first visit, and they saw none now. The inhabitants' maximum height was five foot nine inches, the average was five six – 'they were my size', wrote Bougainville, 'and through God's grace I am only five foot six'.[1] The ships struggled on, anchoring on 18 December in Bougainville Bay, on Brunswick Peninsula, south of Cape Isidoro and not far from present-day Puntas Arenas. This was where the *Aigle* had dropped anchor in 1765, when fetching timber for the Malouine colony. The weather was harsh, even though this was the southern hemisphere summer. Flurries of snow combined with rain and bitter wind to make their twelve-day stay an unpleasant experience. Nevertheless, the Prince of Nassau went botanizing with Commerson, and later went with Bougainville to Tierra del Fuego, the shore opposite the anchorage, where they carried out soundings and worked to correct their unsatisfactory charts.

The next anchorage, Fortescue Bay, reached on New Year's Day 1768, was equally unattractive. In the morning, the decks were covered with four inches of snow. It was not easy to leave, with sudden storms and unfavourable winds. 'Frightful weather', commented Bougainville on 4 January, 'I do believe this is the worst climate in the world.' The French nevertheless went ashore, climbed steep hills and rowed into small coves. Their spirits were further depressed when a small boy died, having swallowed some glass he had been given. In the morning, the Patagonians were nowhere to be seen. 'They flee from nefarious strangers whom they believe to have come only to destroy them.' The expedition remained in the bay until the 16th, when the ships struggled out and made their way westward. In these Straits, with islands and rocks scattered along the sinuous channels, the larger *Boudeuse* struggled more than the storeship. 'On her own, the *Étoile* would have cleared the strait long ago', commented Bougainville, but he deleted these despondent jottings from his published narrative, having later changed his mind and not wishing to sound ungrateful towards his well-meaning protectors.

Finally, the wind abated and even veered in their favour. On 26 January the weary French saw the open sea, and the nightmare that had lasted almost two months was at an end. When they sailed into the Pacific, they celebrated their deliverance with a *Te Deum*.[2] A steady south-easterly breeze now drove them forward, enabling Bougainville to omit a call at Juan Fernandez Islands. Instead, he began a search for 'David's Land'.

Davis Land

One of the most enduring myths of the Pacific, Davis Land owed its supposed existence to the activities of buccaneers, mostly English, who infested the western ocean in the late seventeenth century. One of these was Edward Davis, captain of the *Batchelor's Delight*, who was sailing from the Galapagos Islands in 1687 when he came upon an island with, to the west of it, a stretch of 'pretty high land'. This caught the imagination of numerous navigators because, apart from being a new discovery, it might

---

[1] Bougainville, *Journal*, I, p. 98.
[2] Caro, 'Journal', 27 January 1768.

indicate the presence of the fabulous Terra Australis. The interest of this prospect was enhanced by the possibility, or in some minds the likelihood, that it contained gold and silver.

The main report came from the *Batchelor's Delight*'s surgeon, Lionel Wafer, in 1699:

> We steer'd South and by East, half Easterly, until we came to the Latitude of 27 deg. 20 min. S when about 2 Hours before Day, we fell in with a small, low, sandy island, and heard a great roaring Noise, like that of the Sea beating upon the Shore, right a-head of the Ship ... So we plied off till Day, and then Stood in again with the Land; which proved to be a small flat Island, without the Guard of any rocks ... To the Westward, about 12 leagues by Judgment, we saw a range of high Land which we took to be Islands, for there were several Partitions in the Prospect. This Land seem'd to reach about 14 or 16 Leagues in a Range , and there came thence great Flocks of Fowls ... The small island bears from Copayapo almost due East 500 Leagues; and from the Gallapago's under the Line, 600 Leagues.[1]

A concluding sentence in Dampier's *A New Voyage round the World* linked this discovery to the endless search for the great south land: 'This might probably be the coast of Terra Australis.' Over the centuries, Terra Australis had become endowed with fabulous riches and wealthy inhabitants with whom it might be possible to trade. The reference to 'Copayapo' – or Copiapo in Chile – reinforced the growing hope that Edward Davis's land might produce the kind of mineral wealth the Spanish had found in South America. When Surville's expedition set out in 1769 to seek, among other places, Davis Land, this hope was turned into a belief: 'it was a natural thing to believe that it was much richer than the other countries because it was situated ... by the southern latitude of 27 to 28 degrees, which is that of Copiago [sic], where the Spanish obtain gold in enormous quantities.'[2]

Not surprisingly, there were inaccuracies in Wafer's report, one of which was the reference to the land as lying east of Chile. Navigators had little difficulty in recognizing this as an error for west. Less often queried was the distance of 500 leagues, or 1,500 nautical miles, which is excessive for a ship sailing from the Galapagos on a south-easterly course, even allowing for incorrect estimates of the actual route. Carteret argued that Davis would have reached a position some 200 leagues west of Copiapo.[3] The French astronomer Pingré suggested that a copyist had been at fault and that 'the small island is not 500 leagues distant, but only 150'.[4]

Roggeveen, sailing in 1722, had looked for Davis Land without success, but was compensated by his discovery of Easter Island. Carteret similarly found no sign of land in the area suggested by Wafer and concluded that Davis had merely encountered the islands of San Felix and San Ambrosio.[5] Wallis had also kept a look-out for Davis Land with an equal lack of success. Bougainville was not aware of the routes

---

[1] Joyce, *A New Voyage and Description of the Isthmus of America*, p. 125, reprinted in Dampier, *A Collection of Voyages*, III, p. 393.

[2] Pierre Monneron, 'Extrait du journal d'un voïage fait sur le Vau. Le St Jean-Baptiste', AN, Marine 4JJ 143:23, p. 2.

[3] Wallis, *Carteret's Voyage*, I, p. 52.

[4] Pingré, *Mémoire sur les découvertes faites dans la Mer du Sud*, p. 69.

[5] Wallis, *Carteret's Voyage*, II, pp. 495–6. Carteret called them 'St Paul and St Ambrose'.

followed by Wallis and Carteret, and of their failure to discover any sign of Davis Land, but he did not have any great faith in the fabled land – which in common with most French writers he referred to as 'David's Land'. He sailed through the waters where some believed it would be found and, as he reported in a brief sentence, merely 'passed over David's Land without sighting it'. The nearest he came to any land in this area was the small island of Sala-y-Gomez, north-west of Easter Island; he sailed north of it in 17–18 February without seeing it.

Unhappily for him, Surville knew nothing of recent searches for Davis Land when he set on a voyage of exploration and trade in 1769. Hearing garbled reports of the discovery of Tahiti by Wallis, establishing consciously or otherwise a phonological link between Wallis and Davis/David, he hastily set sail, hoping to forestall any attempt by England to lay claim to Davis Land, and spent days sailing through empty seas, wasting time and lives on this fruitless search.[1] His endeavours played an important part in eliminating Davis Land from the charts. Cook on his second voyage and La Pérouse in 1786 completed the work and the myth was finally laid to rest. As Pingré was to write of Edward Davis and his fellow buccaneers, they were 'better at piracy than at determining the positions of the places where they found themselves.'[2]

The Tuamotus
The French were now sailing roughly west-north-west across the vast emptiness of the southern Pacific. Bougainville started up at the beginning of March the distilling machine he had been given to provide drinking water for the men of the *Boudeuse*; it produced on average a barrelful of water a day. Then, on 22 March, the first small islands were seen, mostly low atolls covered in coconut trees. He named them The Four Facardins (Les Quatre Facardins), after a popular story of the day. These and the succession of low islands that followed, were so small and so isolated that at first he thought they must be uninhabited. He was so surprised to find men along the shore that he took them at first for European castaways, but more appeared, carrying long pikes which they waved menacingly at the ship. He called the atoll the Island of Lancers (Isle des Lanciers), the next one he named Harp Island (Isle de la Harpe) on account of its bow-like shape. More hove into sight, all defended by coral reefs, almost impossible to approach. 'Bad country, dangerous archipelago', he commented, and 'The Dangerous Archipelago [Archipel Dangereux]' was the name he used in his Act of Possession. He was not alone in his judgement. Roggeveen had already named Takapoto 'The Disastrous Island' and Arutua 'More Trouble'; Byron gave Pukapuka the name of 'Islands of Danger'.

There was lasting puzzlement over the presence of inhabitants on these small islands. The French had no knowledge of the Polynesians' skills in long-distance navigation, and wondered 'Who the devil went and placed them on a small sandbank like this one, and as far from the continent as they are? And surely not many ships would have passed this way, [because] since Quiros passed here in 1603, we are the first in 165 years.'[3] Bougainville altered course to west-south-west on 27 March to get out of this

---

[1] See Dunmore, *The Expedition of the St Jean-Baptiste to the Pacific*, pp. 25–9.
[2] Pingré, *Mémoire sur les découvertes*, p. 69.
[3] Caro, 'Journal', 21–22 March 1768.

labyrinth of islands which offered little hope of making a landing and obtaining water or food supplies. Before leaving the area, he recorded his discoveries in the first of a series of formal Acts of Possession, claiming the islands on behalf of the King of France. The dangers that surrounded the islands prevented him, however, from following the usual practice of burying a copy ashore.

The weather moreover was deteriorating. Soon it was back to 'the most horrible weather in the world.' It was damp everywhere in the ships, fresh food was lacking and scurvy began to affect the crew. Bougainville changed his course to south-south-east and soon the sun began to shine and, better still, a couple of high islands appeared along the northern horizon. The first of these was Mehetia, which Bougainville named Le Boudoir, previously sighted by Wallis and named by him Osnabrug Island.[1] More important was the land to the north-west, 'a larger stretch of land' that promised relief at last. This was Tahiti, not a new discovery as the French thought, since Wallis had called there the year before, but an island that would acquire considerable significance over the years.

New Cythera

At first, the winds were unhelpful. It took two days to come close enough to examine the land. It seemed at first to consist of two islands separated by a channel, until it was realized that 'it was only a wide bay with these low lands behind it'. Bougainville was anxious to anchor as soon as he could and thus did not go far enough along the northern coast to find Matavai Bay, where Wallis had brought the *Dolphin* and where Cook would bring his *Endeavour* in 1769. The entire stretch of coast appeared to be dangerous, defended by a long line of breakers with few openings, and Bougainville was happy enough to sail through the first safe opening he found and drop anchor near the shore. The *Boudeuse* and the *Étoile* had entered Hitiaa Lagoon. It was to prove a 'detestable anchorage' and cost the French a total of six anchors.

On the credit side, they found the islanders friendly and helpful. They soon surrounded the ships in their canoes, bringing fruit and other gifts. The women were beautiful and extremely welcoming – to the extent that the French, still under the influence of European restraint in sexual practices, were at times embarrassed. This was a new society, happy and uninhibited, enchanting to the younger officers and to men like Commerson who, believers in Jean-Jacques Rousseau's theories of the evolution of societies, believed they had finally come upon 'the state of natural man, born essentially good, free from all preconceptions, and following, without suspicion and without remorse, the gentle impulse of an instinct that is always sure because it has not yet degenerated into reason.'[2] Wallis had encountered armed opposition to his landings, which he had quelled with gunfire. The Tahitians had learnt two things from that episode: that European weapons were dangerously effective, and that white visitors left after a relatively short time, being easily placated with food and women. They applied this lesson to the newly arrived French, so that the stay was marred only by a

---

[1] 'Its about three Leags Round, and the whole one continuous Mountain which may easily been seen fifteen or sixteen Leags distance:' Carrington, *The Discovery of Tahiti*, p. 131.

[2] Commerson, 'Post-scriptum sur l'île de la Nouvelle-Cythère', BN, NAF 9407/146-7, published in the *Mercure de France* of November 1769.

couple of incidents, for which the French quickly apologized and made amends. When Bougainville decided to make an example by drawing lots among four soldiers found guilty of a murder and hang one of the malefactors, the Tahitians intervened, successfully pleading for clemency.

The stay at Tahiti acquired a significant influence in European thinking. Bougainville highlighted it in his published narrative, giving the island a classical link by naming it New Cythera (Nouvelle Cythère), after the Greek island where the goddess Aphrodite was said to have risen from the sea. He reinforced this by rearranging early encounters with Tahitian women into a Boticellian scene that has become famous:

> In spite of all our precautions, one young woman came aboard onto the poop, and stood by one of the hatches above the capstan. This hatch was open to give some air to those who were working. The young girl negligently allowed her loincloth to fall to the ground, and appeared to all eyes as Venus showed herself to the Phrygian shepherd. She had the Goddess's celestial form.[1]

He makes no mention of this in his journal, but the Prince of Nassau does record that a Tahitian man came on board with his wife when the *Boudeuse* had anchored, and displayed her charms to the bemused Frenchmen; none, however, took up her challenge and the woman left, looking somewhat displeased. But the printed *Voyage autour du monde* had a wide readership and this passage and other reports on Tahiti, including Commerson's eulogium, were discussed in salons and learned academies. They appealed to some as important comments on sexual freedom and the position of women in society, but to many they could be related to the theory of the Noble Savage. Rousseau's essay on the origin of inequality among men had been written in 1754 and was therefore quite fresh in people's mind, but the search for an ideal primitive society still uncorrupted by modern ideas of property and social rank had preceded it. The myth of a golden age, of an uncomplicated pastoral world, was much older, and all civilizations have their legend of a 'dreamtime' or a lost paradise. As societies became more complex and more rigid, the yearning became more marked – and by the eighteenth century much of Europe was suffering under rigid social structures that caused a great deal of unhappiness. Hopes that North America might prove to be the home of a primitive society of Noble Savages, or 'Good Hurons', disappeared as explorers advanced into the interior or became better acquainted with Amerindian societies. Isolated islands in the vast Pacific, preserved by distance from the influence of 'progress', seemed more promising. By the 1760s, Rousseau's ideas, often simplified and misunderstood, were all the rage. Bougainville's elegant prose, buoyed by classical allusions, Commerson's enthusiastic reports to friends at home, comments by Nassau-Siegen and a number of the younger officers, helped to create the legend of Tahiti, a South Sea island paradise where the backbreaking labour of the peasants back home was unnecessary, where the sun shone over sparkling sands, and where men and women were free from the complex and irksome restrictions morality and property-owning imposed on sexual relations. The legend did not survive in this form for very long, but enough of it was left, even after the end of the Romantic

---

[1] Bougainville, *Voyage*, p. 190.

era, for the very name of Tahiti, admittedly a beautiful island, to retain its aura of magic and idleness.

Ahutoru

Bougainville was able to temper his original enthusiasm for Tahitian society by observing the 'Cytheran' he took with him as passenger and talking with him over a lengthy period. Ahutoru, the brother of the Hitiaa chief Ereti (or Reti), had gone on board the *Étoile* even before she had dropped anchor. He was reluctant to leave, had dinner and stayed overnight, going ashore in the morning to be greeted by a crowd of several hundred Tahitians. It is likely that the information he then gave his people on the French visitors, the kindly manner they had treated him on board and what seemed to be their intentions in putting into Hitiaa Lagoon, reassured the Tahitians and encouraged them to make the French welcome. Ahutoru was among those who greeted Bougainville when he landed from the *Boudeuse* in the morning. He later returned to the *Étoile*, becoming a fairly regular visitor. When the French were about to depart, he went with his brother to Bougainville's ship and asked to be accepted as a passenger. After some hesitation, Bougainville agreed and ensured that he was well looked after. He named him Louis de Cythère, but Ahutoru, following the custom of taking his protector's name, soon changed it to Poutavéri or Boutavéri (i.e. Bougainville); later, when he came under the protection of Marion Dufresne who organized the expedition that was to return him to Tahiti, he changed it to Mayoa.

He was intelligent and shrewd, with a gift for mimicry, although he never found it easy to acquire a satisfactory knowledge of the French language. Bougainville considered him somewhat lazy, timid, but pleasant and good-humoured. His interest in women and his regret at the French sailing past islands where female company might be found were evident. At an early stage during his stay on the *Étoile*, he identified Jeanne Baret as someone who might be a woman – as he did another youthful-looking member of the crew – but easily accepted that she was out of bounds to him and probably a *mahu* or transvestite; after this, he took pleasure in being combed and powdered in the French style by her, and she does not appear to have objected to this.[1] He is not reported as having been a particularly handsome man, probably aged around thirty. He was eager to visit other Polynesian islands, but expressed some contempt towards the darker-skinned Melanesians. He was nervous at first at the thought of meeting other Europeans, and Bougainville reports that he was reluctant to land at Buru and be forced to display his knock knees. However, once in France, he had no difficulty in making himself at home in the salons of Versailles and Paris, attending the opera and other functions. He was introduced to the king and to numerous members of the nobility, and became the protégé of the Duchesse de Choiseul who later helped to finance his voyage back to the Isle de France. He was interviewed by various scientists, including the mathematician and traveller La Condamine and Pereire, a specialist on languages. He walked alone in Paris, apparently never losing his way. He

---

[1] As indicated by Vivez in a lengthy report on the Baret affair. Vivez adds that other young crew members 'whose face looked effeminate were constantly being tormented by the Savages who followed them everywhere and would seize them if they did not produce the proofs of their gender'.

visited public gardens and became the subject of a poem by Jacques Delille, in which he clasps his arms around a tree 'he had known since his earliest days … he bathes it with his tears, covers it with kisses.'[1] Certain *philosophes*, however, criticized Bougainville for having taken Ahutoru from his home to display him in Europe, forcing Bougainville to defend himself by pointing out that he had simply given in to pressure from the Tahitian. He met all Ahutoru's living expenses in Paris, and organized and helped to finance his return to Tahiti. Marion Dufresne, himself an enthusiastic Rousseauist, volunteered to take him from the Isle de France to Tahiti. Ahutoru sailed from France in March 1770, reached the Isle de France in October, stayed a year and left in the *Mascarin* with Marion on 18 October 1771. Unhappily, he had contracted smallpox, and died on 6 November.[2]

The French sailed from Tahiti on 15 April with a good supply of fruit, pigs and poultry. Before leaving, while not omitting to bury a formal Act of Possession near the shore, Bougainville had a garden prepared in which wheat, maize, peas and other vegetables were sown. He also gave the Tahitians turkeys and geese, in the hope that they would allow them to breed and in time help to supplement the local diet. Ahutoru endeavoured to get him to call at some of the nearby islands, where as he pointed out there were some attractive women. But Bougainville knew that his supplies did not allow a leisurely voyage among the islands, and he was not sure that Ahutoru's so-called islands were anything more than atolls similar to those he had seen in the Tuamotus. This belief was reinforced when, soon after the departure from Hitiaa, Tetiaroa atoll came into sight.

Language problems made it difficult to obtain clear information from Ahutoru. The Tahitian referred to Tetiaroa as Oumatia, which may have been a reference to the high island of Moorea further off. Other names were equally confusing: 'Aimeo, Maoroua, Aca, Oumatia, Tapoua, Massou, Papara, Aiatea, …' Aimeo was a former name for Moorea,[3] Maoroua being its later form; Aca remains unidentifiable; Tapoua Massou was probably Tubuai Manu; Papara was a clan district in Tahiti itself; Aiatea is Raiatea. Other names given were districts of Tahiti and of Raiatea. Bougainville suspected that days could be wasted on searching for these places in the hope of constructing an adequate chart. It was more important to press on towards the west.

No further land was seen until 3 May at dawn, when a high island appeared to the north-west. It was soon evident that there were three islands altogether. Canoes came out towards the ships, but Ahutoru could not understand the language of the islanders manning them. The people were not related to the Tahitians and lacked the charm of the New Cytherans, for 'a woman who had come in one of the canoes was hideous.'[4]

---

[1] Delille, *Les Jardins*, II. The Abbé Delille was a mediocre but popular poet, whose *Les Jardins*, with the subtitle *ou l'art d'embellir les paysages*, was first published in 1782. His complete works were edited and published by Lefèvre in 1844, as *Oeuvres*, the Ahutoru poem appearing in Lefèvre's vol. I, p. 68. Delille added a note to the effect that he had not invented the scene, but merely transposed it to the Royal Gardens. See Martin-Allanic, *Bougainville*, II, p. 992.

[2] It was the worst epidemic of smallpox the island had suffered for seventeen years: Dunmore, *French Explorers*, I, p. 170, quoting Freycinet, *Voyage autour du monde*, I, p. 386n. On Marion Dufresne, see Duyker, *An Officer of the Blue*; and Dunmore, *French Explorers*, I, pp. 166-95.

[3] Teuira Henry, *Tahiti*, p. 97.

[4] Bougainville, 'Journal', 4-5 May 1768.

The French had reached the Samoan group to which, on account of the number of canoes they saw, they gave them the name of Navigators' Islands (Archipel des Navigateurs). The first three islands form the Manua group, Tau with the two smaller islands of Olosega and Ofu. On the morning of the 5th, the French sighted a long and high island to the north-west. This was Tutuila, and as shown by Bougainville's charts they had a distant view of the southern coast of Upolu in the evening. The weather then became overcast, causing Bougainville to exclaim 'How much patience, O God, this navigation requires!'[1] The lack of charts, the possibility that a reef might suddenly appear at night, the rapidly dwindling supplies, gave Bougainville a feeling of growing insecurity. The ships had been almost continually at sea for six months since their departure from Montevideo, and it was unwise to tarry in these little-known seas. On the 11th at dawn a fairly high land was seen to the west-south-west, a lonely island he named La Solitaire, then L'Enfant Perdu. The 'Lost Child' soon divided into two distinct islands; they were the Hoorn Islands, Futuna and Alofi, discovered by Jacob Le Maire in May 1616. The passage of the French is commemorated in the name of the 400 m (1,310 ft) high Alofi peak, which is now known as Mt Bougainville.

Venereal Disease

Scurvy was now once gain making its appearance. 'Several officers have a somewhat inflamed mouth and traces of scurvy in their gums', wrote Bougainville on 14 May. There was not only scurvy on board, but a disease which, to the sailors, could only be what was known as 'the venereal distemper'. Saint-Germain reported its appearance in his journal a couple of days later: 'We have begun to notice that some sailors, two in number, have caught an illness, it is believed at the island of Cythera; it has manifested itself by chancres [venereal ulcers].'[2] On the following day, he reported that six soldiers were suffering from the same trouble, and on the 22nd mentioned that several cases had occurred in the *Étoile*, adding 'in praise of the continence or good fortune of the officers, none has been affected in this way'. Vivez gave the eventual number as approximately twenty in the *Boudeuse* and twelve in the *Étoile*. The sickness has usually been referred to as syphilis, but Bougainville noted 'Several venereal diseases ... have recently appeared on board both ships. They are of all the types known in Europe.'[3] The Tahitian passenger, he noted, was 'ridden with them' and he was made to understand that 'his compatriots are not greatly concerned about this sickness.' As he had no clear information about the arrival of Wallis's *Dolphin* the previous year, he speculated about how the disease could have reached the islands, and even considered that spontaneous generation might well be a possibility.

Ahutoru's comment suggests that the disease was endemic in the islands; if it was not, the question that arose was which navigator was responsible for its introduction.

---

[1] 'Journal', 9–10 May 1768. Bougainville, according to a comment by his son Hyacinthe, was not noted for his patience, BN, NAF 9407–57. However, others included similar comments in their journal, e.g. Caro, who wrote on 9–10 May that the weather was 'très pitoyable'.

[2] In La Roncière, 'Routier inédit', p. 236.

[3] 'Journal', 18–19 May 1768.

The easiest answer would be that the first shipload of sailors and marines to enjoy the favours of the island women had introduced, in spite of their captain and officers' precautions, a sickness with which seamen especially were familiar. This would give that dubious honour to the men of the *Dolphin*. Indeed, the Tahitians were to call it *Apa na peritane*, the British illness. Wallis had up to twenty men affected by 'a disorder too common amongst unthinking seamen' when the ship sailed from Plymouth. All were reported as cured before the *Dolphin* reached Tahiti. The surgeon 'Affirmed upon his Honour that no man onb[oar]d was affected by any sort of disorder, that they could communicate to the Natives.'[1] How sensitive the matter became is evidenced by the addition in red ink and by a different hand of the words 'No Venereal'. Wallis would reject any suggestion that his men were responsible in the account of the voyage written by Hawkesworth: 'The reproach ... must be due either to him [Bougainville] or to me, to England or to France; I think myself happy to be able to exculpate myself and my country beyond the possibility of doubt.'[2]

Not surprisingly, Bougainville reacted against the accusation that his men were guilty of bringing the disease to the islands – and thereby indirectly infecting James Cook's crew. In the second edition of his *Voyage*, he replied, 'It is with little justification that they accuse us of having brought to the unfortunate Tahitians the illness that we could suspect more justly was communicated to them by M. Wallas's [sic] crew.'[3] The dispute, reflecting the age-old tension between the English and the French, dragged on, James Cook querying whether any significance could be attached to the description *apa na peritane*, as the Tahitians would have been hard put to distinguish between the two nationalities when the *Dolphin* and the *Boudeuse* called at the island. There was also the possibility of some Spanish galleon reaching Polynesia, as well as some early contacts between American Indians and the people of Oceania.[4] Yet another suggestion is that Ahutoru was referring to yaws and not syphilis. Yaws is a tropical disease, endemic in the Pacific, caused by a virus very similar to the causative organism of syphilis.[5] The symptoms are very similar in both cases, and yaws acts, to some extent as an immunizing agent against syphilis. It is significant that in the Pacific, where yaws remains a problem syphilis is uncommon. Conversely, where yaws has been largely eliminated, as in present-day Tahiti, syphilis is on the increase. This raises the possibility of both Wallis's men and Bougainville's being exonerated. One cannot attach too much importance to the Englishman's claim that, by the time his expedition reached Tahiti, his sailors were free from disease. A venereal infection rendered sailors liable to a loss of pay – a practice that did not encourage them to report to the surgeon – and would have prevented them from going ashore. Contact with yaws-infected women, somewhat immunized against venereal infection, would not have caused a serious recurrence of the disease among the sailors, but yaws-free

---

[1] Carrington, *The Discovery of Tahiti*, p. 186.

[2] Hawkesworth, *An Account of the Voyages*, I, p. 489.

[3] *Voyage* (2nd ed., 1771), II, pp. 115-16.

[4] On these possible contacts, see Hornell, 'Was there pre-Columbian contact between the people of Oceania and America?'; Heyerdahl, *American Indians in the Pacific*; Langdon, *The Lost Caravel*.

[5] The micro-organism in yaws is *Spirocaeta pallidula*; in syphilis, it is *Spirochaeta pallida*. The one may indeed be a mutation of the other.

women may have become infected, so that it had become established in the island and known as 'the British disease' by the time Bougainville and Cook arrived.

The accusations and counter-claims eventually faded as more serious accusations were levelled against European navigators for introducing other illnesses, many of them fatal to the islanders, and bringing about in the nineteenth century a serious fall in the island populations. The old Franco-British rivalry subsided in time, but for Bougainville and other navigators of his day the argument had been a bitter embarrassment.

Espiritu Santo

Eleven days after sailing from the Hoorn Islands, the French sighted land ahead. Two islands appeared, stretching roughly north-west, with a pass between. The northernmost, Bougainville named Aurora and the southernmost Pentecost (Isle de la Pentecoste), it being the Feast of the Pentecost, or Whit Sunday. The ships had now reached the New Hebrides, present-day Vanuatu. Aurora is Maéwo Island, but Pentecost retained its name, as well as being known by its native name, Raga. Bougainville rounded the northern cape of Aurora, sighting far to the north a single peak much like Mehetia, which he named Pic de l'Étoile, just as he had first called Mehetia Pic de la Boudeuse. It was the small island of Méré Lava in the Banks group. Now more islands hove into sight to the south-west and the west. The archipelago received the name of the Great Cyclades – once again a derivation from the Greek. The feeling grew, however, that these islands had been visited before, and that this was the land that Quiros had discovered in 1606 and called La Austrialia del Espíritu Santo (The Southern Land of the Holy Ghost). If this were so, the French should find the bay in which Quiros had anchored, St Philip and St James Bay (Bahía San Felipe y Santiago), but more important still, they might have reached the edge of the fabled southern continent. Charles de Brosses had reported the discovery in his *Histoire des navigations* and quoted from Quiros's enthusiastic report to the King of Spain: 'the extent of the lands recently discovered is equal to the whole of Europe and Asia minor as far as the Caspian Sea.'[1] Bougainville read this to his officers and 'Bournand, not moving a muscle, does not know what to make of his leader's evident emotion.'[2]

The search for this 'Land of the Holy Spirit' formed part of Bougainville's instructions, and it was still linked in the minds of many, as it had been in those of Quiros and his backers a century and a half earlier, with the existence of a southern continent. A belief in this southern land, this *terra australis*, had plagued cartographers and explorers since the days of the Renaissance. Belief in a flat world had gradually waned during the Middle Ages, with the argument shifting towards the possibility that the antipodes might be inhabited. Then the discussion had moved further, towards the likely size of the continent and the wealth of its people. Geophysicists developed the theory that a land mass was necessary to counterbalance the continents of the northern hemisphere – for rocks and soil are heavier than seawater, and if there was no continent down under the globe would tip over. Navigators, however, gradually whittled away

[1] De Brosses, *Histoire*, II, p. 334.
[2] Martin-Allanic, *Bougainville*, I, p. 714.

at its estimated size. Magellan's voyage cut a swathe across the blank spaces of the maps, and others, Mendaña, Quiros himself, Schouten and Le Maire, Roggeveen, Dampier, Tasman, drove it back into the south-west corner of the charts. The great and little-known Australian continent itself was being eroded from the west and north. If a *terra australis* existed, it could only be in the ill-defined area where Mendaña and Quiros had discovered their Solomons and their Espiritu Santo.

Bougainville was rather more down-to-earth than the apparent emotion and the enthusiasm of some of his young officers suggest. His belief in a large extent of land was held in check as he sailed through the islands of the New Hebrides. But he knew from his reckonings that he could not be far from Quiros's landfall of St Philip and St James, the commodious bay where the Spanish navigator had established his ill-fated New Jerusalem (Nueva Jerusalen). Unfortunately he missed it, because he was approaching Espiritu Santo from the east, whereas Quiros had sailed down from the north. Consequently the Spanish put into St Philip and St James on the north coast, while Bougainville rounded the island along its south coast, hauled along offshore islands and sailed out through what became known as Bougainville Strait. This does not mean that sailing into and through the Great Cyclades represents any kind of failure: there was more to his mission that finding Quiros's landfall. The nibbling away at the possible limits of the southern continent was a negative achievement, to which Bougainville made a contribution then and in the weeks to come. What is of greater value is the charting of the great archipelago through which he sailed, and the descriptions he was able to give of the islands and their people. To achieve this, he needed to visit an island or two, without delaying unduly or endangering the lives of his men.

On the morning of 23 May, Bougainville went ashore with two boats, accompanied by one armed boat from the *Étoile*. Firewood was obtained, as well as bananas and coconuts, in exchange for a few gifts. The Prince of Nassau went with him, and the presence of this awe-inspiring aristocrat was helpful, for when the islanders began to display some hostility, he advanced towards them and quelled them by his bearing and his obvious courage. The customary Act of Possession was buried at the foot of a tree, allowing France to lay a claim to the island group, for the Spanish had displayed no interest in it over the years. It was little more than a gesture – but Bougainville could hardly suspect that one day, under the name of New Hebrides bestowed on them by Cook, they would be jointly administered by France and her traditional rival, England.

In the late afternoon, just as the party was about to leave, the islanders suddenly attacked with stones and arrows. A few musket shots were needed to disperse them. Communication had not been easy, as Ahutoru did not speak their language, and their appearance was unsavoury. 'Many of them were covered with running scabs; others had horrible sores covering part of the body.'[1] The French named the place the Island of Lepers (Isle des Lépreux). The symptoms denoted leprosy, yaws and tropical ulcers. Yet the natives of Aoba, to give it its local name, are today as clean and as healthy as those of neighbouring islands. It is obvious that Bougainville had landed among an afflicted section of the population, and possibly one that had only recently been engaged in a war with neighbouring tribes.

[1] La Roncière, 'Routier inédit', p. 240.

From Aoba, the ships sailed west, passing between the islands of Malakula and Malo, entering the vast blank area between Vanuatu and Australia. There was still a possibility that the New Hebrides formed an offshore extension of the southern continent, but this seemed unlikely. Bougainville was sailing outside the area Quiros was reported to have explored – even though Spanish longitudes and latitudes were uncertain, 'Where then is his great land?', Bougainville wondered, adding '*Davus sum, non Oedipus*' or 'I am here as to serve my country, not answer riddles.'[1]

At least one other riddle was being solved at the time. Bougainville went over to the *Étoile* to discuss the Jeanne Baret affair. His attention had been drawn to it after the episode in Tahiti, in which she was set upon by a group of island men, although it is likely that he had been aware of it much earlier. But the presence of a woman on board a naval vessel was such a breach of the regulations that he could no longer ignore it and had to go over to the *Étoile* to deal with it. However, there was little that could be done, beyond making arrangements to contain any scandal, and he returned to his ship in a fairly philosophical mood. 'I admire her determination', he wrote', 'I have taken steps to ensure that she suffers no unpleasantness. The Court will, I think, forgive her for this infraction to the ordinances. Her example will hardly be contagious.'[2]

Of greater moment was their destination. The ships were heading straight for what is now known as Queensland, and for the Great Barrier Reef which was to bring Cook's *Endeavour* so close to destruction a couple of years later. Fate, once again, was kind to Bougainville, issuing a warning on the evening of 4 June in the shape of a low island with sandbanks and rocks. He named it Diane Reef (Bâture de Diane). A further warning appeared on the 6th, breakers covering a wide area; more were sighted a little later. 'A third warning that I should not persist in seeking land along this parallel.' The reef is now called Bougainville Reef.

He set a course for north-north-east, thus turning away from the coast of Australia, which some of his sailors claimed to have seen, 'a low land SW of the breakers'. All agreed that it was unwise in the extreme to go any closer to this dangerous coast. Caro, in the *Étoile*, summed up the general feeling in his ship when he wrote: 'It would appear that this part of New Holland is as full of shoals and reefs as the other.'[3] Bougainville defended his decision in his journal:

> I had planned to approach land along the parallel of 15 to 16°, not that I had any doubts that the southern land of the Holy Ghost was anything other than the great Cyclades, and that Quiros falsified either his discoveries or his narrative … The encounter with this succession of breakers does not allow me to seek Quiros's southern continent here … Other reasons peculiar to our situation urge me to sail towards known countries. I have only enough bread for three months at most, wood and water for hardly a month, we have been 7 months at sea … Furthermore I have no anchors left to sacrifice to save the ships in such dangerous waters.[4]

---

[1] 'Journal', entry of 28–29 May 1768.
[2] Ibid. As Jeanne Baret was officially part of the expedition, as Commerson's valet, a biographical note on her is included in the section of this Introduction on 'The Participants'.
[3] 'Journal', 6–7 June 1768.
[4] Ibid.

He sailed away, believing that the nearby land would offer no opening for a colony that would be of use to the metropolis. This stretch of the Queensland coast is indeed difficult for ships to approach, it has a semi-tropical climate and would not have appealed to French farming settlers. It might have been possible to make for Carpentaria – indeed, Bougainville's instructions included a mention of it – but this would have meant sailing through Torres Strait, the existence of which was still unconfirmed. The charts were vague, and if there was no navigable channel, Bougainville could find himself imprisoned in a wide gulf, from which he would emerge only after much perilous battling against the prevailing winds. It was a risk conditions on board did not allow him to take.

Sailing roughly north, the expedition came upon the south coast of New Guinea. At first, Bougainville found himself caught in a bay, named the Cul-de-Sac de l'Orangerie, which did not appear safe enough to risk a landing. Struggling on an easterly course in order to round some cape and sail north and west for the Dutch settlements, he was foiled for several days by contrary winds and by the fast currents that flow towards Torres Strait. A succession of reefs and islands required wearying tacking and constant watchfulness. He named this the Louisiade Archipelago (Archipel de Louisade), after the ageing Louis XV. Finally, after a fortnight, he reached the end of the chain, hailing it as Cap de la Délivrance. Time was pressing. The rations had once again to be reduced, and Saint-Germain recorded wryly that he had shared a rat with the Prince of Nassau and hoped that the others would not acquire a taste for such a delicacy. The journals all reflect the growing weariness. 'O, Bellin, how much you are costing us!' Bougainville exclaimed. Caro wrote that he would be a fool if ever he were to be persuaded into another such expedition. 'Of what use is this voyage to the nation?' wondered Saint-Germain, and he went on to list what could have been achieved if only better planning and greater good fortune had made it possible.

Rediscovering the Solomons

It is somewhat ironical that as Saint-Germain was penning his series of complaints, the expedition was on the verge of new and important discoveries. The charting of the Louisiade Archipelago was of course a significant contribution to the knowledge of the western Pacific, but soon more land appeared as the French sailed north to avoid the dangers they knew to be lurking around north-western New Guinea. 'What is this land?' pondered Bougainville. It could not be New Britain or the neighbourhood of Dampier Strait.

It was the southern islands of the Solomon Islands, Rannonga and Vella Lavella, soon followed by the Treasury Islands and a great land to the north, which Bougainville named Choiseul Bay, on what was to become Choiseul Island, after the Minister of Marine, his friend and protector. He tried to land, but darkness and tide were against it, and the dark-skinned Melanesians less friendly than the Polynesians of Tahiti. He veered to the north-west, along another large island, which now bears his name, as does the strait that divides it from Choiseul. He sailed along the northern coast of Bougainville Island, past the smaller island of Buka – so named because the islanders called out 'boca, boca' a number of times at the passing French, inviting them to go ashore. Then the whole island group fell behind while the French puzzled over their inadequate charts and stared ahead for what might come next.

They had rediscovered the Solomon Islands, two centuries almost to the day since Alvaro de Mendaña had discovered a great island he named Santa Ysabel, followed by an entire archipelago. They became known as the Islas de Salomon, but he died in 1595 while trying to make his way back to them and Quiros, ten years later, also failed to reach them. In their days, determining a latitude was relatively easy, but calculating a longitude was far more difficult. Errors of latitude were usually of a minor nature, errors of longitude could easily add up to several degrees. Mendaña, anxious to obtain official support for a return voyage, had no doubt understated the distance between Peru and the Solomons, and at the same time waxed lyrical over the riches his new discoveries could bring. Hernan Gallego, the expedition's pilot, underestimated the distance by over 2,000 miles. Disbelief about the extent of the Pacific Ocean tended to support the view that the Solomon Islands were not unreasonably distant from South America. When Mendaña was appointed 'Governor of the Western Isles', the title supported the view that they were the western islands of Spanish South America. This did not necessarily contradict the theory that they were near New Guinea: the Pacific had simply been shrunk by popular opinion.

Not surprisingly, when later expeditions struggled across the ocean and reached the Santa Cruz and New Hebrides groups, they tended to give up their search and veered north. Seventeenth-century mapmakers mostly placed the Solomons close to New Guinea, often leaving their western and southern coastline blank, since they suspected they might be attached to the famous Terra Australis. Others felt that they actually were the New Hebrides, where Mendaña and Quiros had ended up in their attempts to return to the Solomons. A Dutch map of 1622, Hessel Gerritsz's 'Great Chart of the South Sea' boldly labels the entire group 'Islas de Salomon'. And if they were the New Hebrides, then they might not exist as a separate island group at all: Peter Heylen's *Cosmography in Four Books* of 1657, a hundred years after Mendaña's voyage, lumps the Solomons with 'Utopia, New Atlantis, Faierie Land, the Painter's Wives Island and the Land of Chivalry'.[1] If doubts were cast on their existence and they could not be fitted where geographers thought they might be, cartographers were tempted to leave them out altogether or move them to some other, less crowded part of the Pacific. The cartographer D. R. Dudley changed his mind about them between 1647, when he placed them close to New Guinea in his *Arcano del Mare*, and the second edition of 1661 when they were shown near the Marquesas Island, 4,000 miles away.[2] The Islands of Solomon were well on their way to becoming geography's most mobile land masses. On the eve of their rediscovery, doubts about their existence were gaining ground. Both Bougainville and Carteret, who sighted them, sailed from Europe doubting their existence or confused about their possible location. A comprehensive chart of the Pacific, drawn in 1756 by Robert de Vaugondy for De Brosses's *Histoire des navigations aux terres australes* shows 'land seen by Gallegos' to the south of Easter Island, and the Solomons much further east than they really are, while

---

[1] Maggs Bros Catalogue 491, *Australia and the South Seas*, London, 1927, quoted in Jack-Hinton, *The Search for the Islands of Solomon*, p. 200. Jack-Hinton's is a most detailed study of the various theories held about the islands and their eventual rediscovery.

[2] Jack-Hinton, ibid., p. 228.

Espiritu Santo is attached to the Australian continent, with Guadalcanal labelled 'Santa Cruz'. Confusion was total. Not surprisingly, Callander, who translated and adapted De Brosses's book, suggested that the Solomons were probably a fiction.[1] Bougainville was consequently not in a frame of mind that could lead him to identify the islands of Choiseul and Bougainville as part of the Solomons. It would take a careful analysis by European geographers and the detailed work of Buache and above all Claret de Fleurieu to piece together the contributions made by Bougainville, Carteret, Surville and Shortland, and finally solve the problem that had bedevilled Pacific cartographers some more than two centuries.

New Ireland to the East Indies

The expedition reached New Ireland on 6 July 1768, and at last found a haven in Port Praslin. They were unaware of the discovery of St George's Channel and consequently believed they were in New Britain. Carteret had preceded them and, as they soon found out when they explored further along the shore, had anchored close by: a sailor, looking for cockles along the beach, found a lead plaque with the remains of an inscription in English. Looking for further traces, Bougainville came upon signs of an English camp, some three miles to the north-west, dating back, he estimated, some four months. He was wrong – Carteret had put into what he named Gower's Harbour at the end of August 1767.[2] It was not a bad place for Bougainville to rest his men after the strains of the previous months. There was a good beach, four brooks and a picturesque waterfall, there was plenty of firewood, and no inhabitants. The sick could walk along the shore in safety, the others could go fishing or wash their clothes. Commerson went botanizing, the Prince of Nassau went hunting, Véron observed an eclipse of the sun, the officers supervised the cleaning of the ships and the restowing of the stores. But there was very little in the way of fruit in the dense forest that covered the land, and the rain poured down almost incessantly. 'Shall I always have to write bad weather, strong gale, storm, constant rain?' sighed Bougainville. 'Our crews are worn out, this land is providing them with nothing but an insalubrious air and extra labour.'

The weather delayed their departure and hid from them the pass that Carteret had discovered and which would have provided a quicker route to New Guinea and the Dutch East Indies. Instead, Bougainville rounded the southern tip of New Ireland and proceeded north along the eastern coast, sailing past a succession of islands from which, from time to time, canoes came out to examine the strange vessels. Attacks and skirmishes were not uncommon, for not only were the French arousing the islanders' fears, they were entering an area where inter-tribal warfare was almost endemic. Bougainville named the islands after his officers, respecting their order of seniority. Duclos-Guyot already had his small island in St George Bay; it was now the turn of Bournand, D'Oraison, Du Bouchage and Suzannet. Most of these names have been replaced by the original native ones, Feni, Tanga, Lihir, Tabar.

By the beginning of August, the ships had come to the northernmost island of the

---

[1] Callander, *Terra Australis Cognita*, III, p. 711.
[2] Wallis, *Carteret's Voyage*, I, pp. 179–80.

New Ireland chain and were making almost due west towards the East Indies. They sailed north of the Admiralty Islands until they were close to the Equator. It was Commerson's turn to see his name given to an island – and although it is more frequently known as Sae, the French name is still found in use today. One island, sighted on 8 August and part of the Kaniet group, appeared to be well populated, to judge from the number engaged in fishing around it, and as none stopped work to approach the ships or even to pay attention to them, Bougainville named it Anchorites Island (Isle des Anchorètes).

On the following morning, a line of low islands appeared. They were well wooded but protected by fierce-looking reefs: 'A disastrous encounter.' Bougainville named this the Echiquier, the Chessboard. It was the Ninigo group, west-north-west of Manus Island.

On the 11ᵗʰ, the French sighted the coast of northern New Guinea. It was wise to sail further north, which meant almost following the line of the Equator, in stifling heat. Not surprisingly, scurvy was spreading rapidly in both ships; by late August the surgeons counted forty-five cases. The *Étoile* was forcing the *Boudeuse* to take in sail, slowing progress even further; the storeship was wallowing and creaking abominably. 'There have been many arguments over where Hell is situated, truly we have found it,' wrote Bougainville.[1] Eventually, the long struggle around New Guinea came to an end, and he reached the island of Seram, the first of the Moluccas. Hoping to get a good reception, he hoisted the Dutch colours, only to find the place deserted: the locals, afraid of the Dutch, had fled to the hills. There was nothing for it but to sail on to the next island, Buru, where an important trading post of the Dutch East India Company was situated. Strictly speaking, it was out of bounds to all except Dutch ships, but Bougainville used tact and courtesy, and his written request for help was accepted, and during the six days he spent there he took on board rice, oxen, sheep and other supplies – highly priced but badly needed.

The *Boudeuse* and the *Étoile* sailed from Buru on 7 September 178, but without the help of a Dutch pilot. A labyrinth of islands lay ahead of them and their charts were imprecise, but it was the policy of the Dutch authorities to keep out all intruders and to defend their monopoly by publishing as little information as possible about East Indian waters and to encourage the spread of rumours about the dangers of inter-island navigation. The Dutch officials at Buru were not prepared to risk incurring the wrath of their superiors by going out of their way to assist the French. Fortunately, the dangers were exaggerated, although Bougainville could not know this as he began to thread his way through the seas and straits, but he had the good fortune to have on board a French sailor with experience in the Molucca Sea, a man 'who has been sailing in these waters for the last six or seven years and has made three or four crossings from Batavia to Amboyna, Buton and Buru.'[2] A local man met on the way also proved helpful. The French proceeded with growing confidence towards Java, buying supplies from the natives as they went.

On 28 September, they saluted the fort of Batavia with a twenty-one gun salute,

---

[1] 'Journal', 16–17 August 1768.
[2] Caro, 'Journal, 12–13 September 1768.

which was returned in full. On the following day, they went ashore, finding that as the result of their east-to-west crossing, the date was actually the 30th. The Dutch received them 'in the best manner possible'. The twenty-eight sick on board the two ships were cared for at the local hospital. Supplies were purchased and loaded on board for the return home, but after a week illness appeared among the crew – this time, dysentery, a frequent scourge in Batavia and the cause of many deaths during the eighteenth century. To escape the pestilential climate of the city, Bougainville went to anchor outside the harbour and, warned about the impending monsoonal change, decided to cut short his stay. He did not alter the dating of his shipboard journal until he had reached the Isle de France; consequently, according to his reckoning, he left Batavia on 17 October 1768.

On 7 November, the *Boudeuse* reached the Isle de France; the storeship would follow the next day. It is ironical that having sailed this far with no major mishap, the *Boudeuse* ran aground at the entrance to Port-Louis while under the guidance of the local pilot. Bougainville took over and extricated himself without damage and landed at last on French soil. A few days later, Du Bouchage died of dysentery. Others left the expedition at this point: Commerson with his valet, Véron making arrangement to travel to Pondicherry where he hoped to observe the Transit of Venus, a number of others at the request of the island's Intendant, Pierre Poivre. Some soldiers joined the local military forces, a number of sailors were transferred to the Port-Louis hospital, a few remaining behind on the island or transferring to other ships. Bougainville did not wish to tarry longer than was necessary, and he left in the *Boudeuse* on 12 December, leaving the *Étoile* to be careened, cleaned and repaired. Since the expedition was now virtually over, there was no further need for the two ships to travel together. From this point on, Bougainville made only brief, perfunctory notes in his journal.

There was one strange episode during the final stage of the voyage home. In the Atlantic, Bougainville caught up with a slow-moving British ship. It was Carteret's *Swallow*, returning home after her circumnavigation at a pace which belied her name. Bougainville already knew that Carteret was not far ahead of him, for when he called at the island of Ascension on 4 February, 'they brought me from a cave a bottle in which were recorded the names of two English vessels, the first being *The King of Portugal* which passed here in August 1765, the second the *Swallow* of Mr Carteret returning from a voyage around the world, arrived here on 31 January and departed 1 February 1769. He had sailed from the Cape on 6 January. I sent the bottle back after adding the name of the *Boudeuse*.'[1] Carteret refused Bougainville's offer of help, but handed to the young Pierre Duclos-Guyot, who spoke English and went across to the *Swallow*, some letters destined for France which he had collected at Cape Town. He also sent Bougainville an arrow 'he had found on his voyage round the world, a voyage that he was far from suspecting we had likewise made.'[2] A little later the two ships parted, the *Boudeuse* leaving the *Swallow* wallowing behind. 'His ship was very

---

[1] Bougainville, 'Journal', 4 February 1769.
[2] Bougainville stopped writing up his journal with the entry of 14-15 February. The account of his meeting with Carteret comes from his published narrative, *Voyage*, pp. 386-7. See also on the meeting, Wallis, *Carteret's Voyage*, I, pp. 266-73, II, pp. 440-43.

small', commented Bougainville, 'went very badly, and when we took leave of her, she remained as it were at anchor. How he must have suffered in so bad a vessel may well be conceived.'

The *Boudeuse* dropped anchor at St Malo on 16 March 1769. She had been away for two years and four months, during which only seven men had been lost. The *Étoile*, which reached France a month later, had lost only two men as the result of illness. Bougainville could not resist a final Latin tag to mark the end of the voyage and the joy the officers and the sailors felt at being home at last: *Puppibus et laeti nautae imposuere coronas*. 'And the sailors, rejoicing, covered the poop with crowns'.[1]

The Achievements

The voyage of Bougainville made a greater impression on the general public, in France and elsewhere, than it did on the world of science. Étienne Taillemite has described the muted reactions of the geographers and astronomers of the time. Although Bougainville's friends, de Brosses, Buffon, Lalande, Poissonnier, La Condamine, all eagerly listened to his verbal reports, others showed much less interest. Thus, no mention of the publication of Bougainville's narrative can be found in the pages of the *Journal des Savants* during the year 1771, when the *Voyage autour du monde* appeared, and it is only referred to, fairly briefly, in November 1772 when the second edition was published.[2] Nor was much attention paid by the *Académie des Sciences* or the *Académie de Marine*.

This relative lack of interest on the part of the scientific world was due to a number of factors. Firstly, the expedition had not been conceived as a scientific enterprise, backed by scientific bodies or provided, as was La Pérouse's fifteen years later, with lists of tasks, geographical and other, to be undertaken as the voyage proceeded.[3] There was consequently nothing the various scientific bodies were waiting for when Bougainville returned.

Secondly, Bougainville chose to write his own account of the voyage with a literary skill that ensured its immediate success among a wide range of educated readers. This militated against its being accepted by the world of science as a serious work embodying the results of exploration and scientific endeavour. The lack of navigational details, of precise latitudes and longitudes, which guaranteed that his narrative

---

[1] Virgil, *Aeneid*, IV, 418.

[2] 'The *Journal* did condescend to advise its readers of the publication of the second edition but restricts itself to the following brief comment: "This second edition of the *Voyage* of Mr de Bougainville had long been necessary and the author has made it even more interesting than the first." This laconism contrasts with the attention paid by writers in this journal to Cook's voyage, the account of which translated by Fréville received a ten-page review in the December 1772 issue:' Taillemite, *Bougainville*, I, p. 111, quoting the *Journal des Savants* for November 1772, p. 766, and December 1772, pp. 771–31.

[3] The instructions given to La Pérouse in 1785 included a detailed section on work to be done on astronomy, geography, navigation, physics and 'the various branches of Natural History', as well as detailed memoirs from the *Académie royale des Sciences* and the *Société royale de Médecine*. See Dunmore, *The Journal of Jean-François de Galaup de la Pérouse 1785–1788*, I, pp. cx–cl. Other French expeditions to the Pacific after Bougainville were Surville's in 1769–70 and Marion Dufresne's in 1771–3, but these were essentially commercial undertakings. Kerguelen in 1773 received a total of fourteen memoirs on work to be carried out in the fields of geology, botany and zoology, but he did not actually enter the Pacific Ocean.

would flow and captivate the reader, was criticized by the savants, and even seen by some as evidence that his claim to have made new discoveries in the South Seas was suspect.[1] The jealousy of a number of his contemporaries was compounded by political rivalries. Bougainville's patrons had been the Choiseuls – the Duc de Choiseul-Praslin and the Duc de Choiseul-Stainville, his cousin – and he had been the protégé of the Marquise de Pompadour. But she had died in 1764 and the Choiseuls were dismissed and exiled to their country estates in 1770. With them went a number of politicians and court officials, to be replaced by supporters of the rival groups. Bougainville, who had named the island of Choiseul and Port Praslin after the former minister, found himself deprived of influential protectors.

However, the *Année littéraire* of September 1771 did pay him the tribute of a lengthy review.[2] It was fair and concluded: 'The account of this voyage around the world can be most useful to those who intend travelling through the seas he mentions. They will find when they read it various nautical or geographical notices that correct dangerous accepted errors.'[3] Then in 1772 it reviewed, quite briefly, the second edition of Bougainville's *Voyage*, and at this point one realizes the drawback under which Bougainville was struggling, for it reviewed shortly afterwards the French translation of James Cook's first voyage, hailing the work of the naturalists Joseph Banks and Daniel Solander, both 'scientists enjoying throughout Europe a well deserved reputation.'[4] Philibert Commerson's work meanwhile was still unknown in France, his botanical collections and his notes languishing unsorted back in Mauritius, but Commerson had preceded Banks and Solander in the Pacific and had his work been published, or even substantially outlined in Bougainville's *Voyage*, it would have ensured a distinguished place both for him and for the whole expedition in the annals of science.[5] The world of savants would have then had to pay greater attention to the scientific aspects of the voyage and would have been less overawed by the achievements of the British.

Instead, Commerson gained notoriety more than fame by his report on Tahiti, which was published well before Bougainville's own narrative, and to a considerable extent predisposed the educated public to expect a readable and somewhat exciting, possibly even sensational, account of the expedition. Commerson had written a letter to an influential friend, the astronomer Lalande, in which he praised New Cythera as 'the only part of the world in which men live without vices, without preconceptions, without needs, without dissension.' This tropical paradise, where private ownership

---

[1] 'Mr de Bougainville takes good care to conceal the latitude, the longitude and anything that might reveal the position of the island he claims to have discovered:' Bachaumont, in *Mémoires secrets*, XIX, p. 66. In two earlier references, Bachaumont shed doubt on the discovery of New Cythera and 'other marvels' supposed to have been encountered during the voyage.

[2] Written by Élie Fréron, a critic and journalist, not a scientist. His review was published in volume VI, pp. 49–63.

[3] Quoted in Taillemite, *Bougainville*, I, p. 120.

[4] *Année littéraire*, issues of 30 July 1772 (for Bougainville) and 24 August 1772 (for Cook).

[5] See Hume, 'Seamen and Savants', pp. 21–6. Hume points out that 'two large collections of plants shipped back to France from Brazil and the La Plata region were never seen again', and that his collection of fish was discovered in Buffon's attic after the latter's death. He also quotes George Cuvier's nineteenth century judgment on Commerson: 'Had his work been published, he would have been in the foremost ranks of naturalists.'

of land was unknown and consequently where legislation was not needed to protect property, where there were no boundaries to signal exclusive ownership, no exploitation and no poaching, above all where free and uninhibited love was the rule, caught the public imagination. For true believers, this was the home of Jean-Jacques Rousseau's primitive, happy man, the Noble Savage successive generations dream about.

The 'Post-Scriptum sur l'île de la Nouvelle-Cythère ou Taïty, par M. Commerson, docteur en médecine' was written in April 1768,[1] but reached a much wider readership when it was published in the *Mercure de France* in November 1769. At the time, Bougainville was still working on his narrative and had only just presented his personal report to the king. Commerson had written in a similar vein to his friend Crassous and to Doctor Durnolin, of Cluny.[2] His views had therefore been well publicized. His paean in praise of Tahiti, set against the background of ongoing arguments about Rousseau's theories, was further highlighted by the presence in Paris of Ahutoru. Bougainville's exotic passenger was an amiable young man, not really fluent in French but interested in all he saw about him and eager to please. He reinforced Commerson's report on his people, especially as, to all accounts, he displayed a keen interest in women.[3] While Bougainville dealt with all the probems of the voyage's aftermath, discussed with officials the future of the Falkland Islands and future French policies towards the Pacific, and worked on his own narrative, the salons and by extension the influential world of the *philosophes* gossiped about sexual mores in New Cythera.

Meanwhile, an account of the earlier expedition to the Falklands had been published by Dom Pernetty, a Benedictine monk and a talented naturalist who had sailed with Bougainville as chaplain and botanist in the *Aigle* in 1763. The *Journal historique d'un voyage fait aux îles Malouines en 1763 et 1764, pour les connaître et y former un établissement* appeared in Berlin in 1769. It drew substantially from his shipboard journal and, from the literary point of view, it was not very readable. Nevertheless, it was well received, but Pernetty decided to seek a wider readership by rewriting it. This more popular version came out in Paris in 1770.[4] Although this added to the attention being paid to Bougainville, it did have the effect of focussing attention on his attempt to colonize the Falklands, which was still a topic engaging the attention of diplomats and commentators. To some extent, therefore, Pernetty's book added emphasis to the preliminary stages of the voyage, and may have contributed to the feeling that the crossing of the Pacific was a secondary issue, the travels of an army officer, not a naval explorer, of a literary man, a courtier, something of a dilettante, and not in the same league as the voyages of Anson, Byron and Cook whose narratives were more 'serious' and more scientifically presented.

By now, French translations of major English publications were becoming available. Anson's *Voyage autour du monde* had come out in Amsterdam and Leipzig in 1749,

---

[1] The MS copy is held at BN, NAF 9407/146–7.

[2] The letter to Crassous, with additional details, was published in the *Décade philosophique* for June 1797, pp. 133–42; the communication to Durnolin was published in the *Annales* of the Mâcon Academy in 1857, vol. II, pp. 329 ff. See Martin-Allanic, *Bougainville*, II, pp. 992–3.

[3] 'His great passion is for women, which he indulges in indiscriminately:' Bachaumont, *Mémoires secrets*, IV, p. 310.

[4] *Histoire d'un voyage aux îles Malouines fait en 1763 et 1764, avec des observations sur le détroit de Magellan et sur les Patagons.*.

Byron's similarly titled narrative had been published in Paris in 1767. A translation of an account of Cook's first voyage became available in 1772,[1] and Hawkesworth's longer work, which included the voyages of Byron, Carteret, Wallis and Cook, which appealed to many readers in spite of its shortcomings, was being translated and would soon be on sale to readers, both savants and the wider public.[2] And men like Dalrymple were in regular touch with De Brosses and other French correspondents. Once again, the British had claimed the high ground, and French geographers and historians would have to fight back to establish a French claim to having played a significant role in the exploration of the Pacific, in which Bougainville had been a major figure.[3]

The original published narrative, while it ensured Bougainville's reputation among a wide public over the centuries and helped to ensure his place among France's major historical figures, did not establish quite so firmly his renown among geographers and scientists. For his achievements to be realized fully one needed easy access to all the original documents, something which did not occur until 1977. A fuller and fairer reassessment of his achievements can now be attempted.

The French government did not promote the Bougainville expedition as part of a pre-planned voyage of exploration from which commercial, scientific or politico-strategic gains might be made as, it can be argued, the British did with Byron, Carteret or Cook. It was an undertaking which Bougainville suggested, to follow his handing over to the Spanish of his French Falkland Islands settlement. In the aftermath of the disastrous Seven Years War, the combined transfer and circumnavigation suited the plans of the Duc de Choiseul, as it did the broad policy of Louis XV. But it would be unfair to view the voyage as an appendix to the Falklands transfer, or as a consolation prize to make up for Bougainville's disappointment at being forced to give up his plans for the Malouines. It was a voyage of exploration which had long appealed to Bougainville.[4] And from the point of view of the French government, it was a relatively straightforward and inexpensive way of regaining some of France's lost status and claiming a place in Pacific exploration – or, if one takes into account the significant activities of French explorers and navigators along the western coast of South America and in Pacific waters at the beginning of the eighteenth century, reclaiming it. One therefore has to assess the achievements of the circumnavigation in

---

[1] *Supplément au voyage de M. de Bougainville, ou Journal d'un voyage autour du monde fait par MM Banks et Solander en 1768, 1769, 1770, 1771, traduit de l'anglois par M. de Fréville.*

[2] *Relation des voyages entrepris par ordre de Sa Majesté Britannique actuellement régnante, pour faire des découvertes dans l'hémisphère sud et successivement entrepris par le commodore Byron, le capitaine Carteret, le capitaine, Wallis et le capitaine Cook, dans les vaisseaux le* Dauphin, *le* Swallow *et l'*Endeavour, *rédigé d'après les journaux tenus pas les différents commandants et les papiers de M. Banks, par J. Hawkesworth, traduit de l'anglois par J.B.A.,* published by Suard. 5 vols, Paris, 1774. There were other translations at around this time, and Fréville appears to have collaborated with Suard.

[3] Some titles underline this struggle, e.g. Claret de Fleurieu, *Découvertes des Français en 1768 et 1769 dans le Sud-Est de la Nouvelle-Guinée et reconnaissance postérieure* [sic] *des mêmes terres par des navigateurs anglais qui leur ont imposé de nouveaux noms* [sic]…, Paris, 1790.

[4] Not only had Bougainville put forward a proposal for a voyage of exploration back in 1761, but he had envisaged several years earlier that 'if Canada remained French, it would be possible for him to set out to explore the South Sea by crossing the American continent:' Martin-Allanic, *Bougainville*, I, p. 65.

a broad political and geographical context, and accept that, to some extent, it was a modification of an earlier and relatively simple programme.

This said, one needs to appreciate the scientific context of the period of the Enlightenment, the thirst for knowledge, the ongoing discussions of scientific societies in Paris and in the provinces, the influence of men such as De Brosses and Buffon, Bougainville's own personality and early background in Paris and London circles, and Europe's eagerness to know more about those South Seas that, for so long, had been the preserve of the Spanish and the Dutch. The voyage was not an isolated expedition deriving from a failed attempt at colonization in the South Atlantic, but an undertaking conceived in the spirit of the times. It was far from an easy one to carry out, and its achievements were far more impressive than some commentators have been willing to concede.

Once the Falklands transfer was completed, the Bougainville expedition set out for a difficult voyage through the Straits of Magellan and into little-known areas of the Pacific Ocean with generally sketchy and inadequate charts, and with two ships which, although one was new and the other adequately maintained, were not really suited to the task of exploring unknown coasts in reef-strewn waters. It completed a twenty-eight month circumnavigation with no more than seven deaths from sickness or accident. And its commander, however impressive his background, his intellect and his ability, was an army colonel with little naval background or experience. The quiet and self-effacing assistance of Duclos-Guyot no doubt helped a great deal in ensuring a safe navigation under trying circumstances, but the skills and adaptability of Bougainville should not be overlooked.

The winding Straits of Magellan were ably mastered, but hydrographic surveys in this area were not part of Bougainville's programme. Some geographical clarification was achieved, incidentally to the navigation, and for this cartographers were grateful. The still vexed question of the Patagonian giants was, with the work of Wallis and Carteret in the previous year, solved to the satisfaction of all but the most adamant supporters of the giant theory.[1]

Then followed a careful navigation through poorly charted waters, with the discovery of several atolls in the Tuamotus: Vahitahi, Akiaki, Marokau, Hikueru, Reitoru, Haraiki and Anaa. The verification of the tracks of early Spanish explorers, right up to Quiros's Espiritu Santo and the rediscovery of the long-lost or, more correctly, mislaid Solomon Islands, could not be achieved in a single voyage. The problems of determining longitude in the late fifteenth century and throughout the sixteenth meant that scholars would continue to puzzle over the correct identification of the various Spanish landfalls and sightings, but Bougainville made a substantial contribution towards providing some of the answers.[2]

The expedition then reached Tahiti, unaware of Wallis's discovery of June 1767. Had Bougainville been the island's first discoverer, he would probably have been received with the mistrust and hostility the Tahitians directed at Wallis. In any event,

---

[1] On the controversy over the existence or otherwise of a race of giants at the southern tip of South America, including French interest in the question, see Wallis, *Carteret's Voyage*, II, pp. 322-6, and Gallagher, *Byron's Journal of his Circumnavigation*, pp. 185-96.

[2] The works of Colin Jack-Hinton, *The Search for the Islands of Solomon*, and Celsus Kelly, *La Austrialia del Espíritu Santo,* are good examples of the minute analyses required and of the magnitude of the task.

relations would have been far more strained, and the myth of Tahiti, with all the embellishments of Bougainville's classical education and Commerson's Rousseauist enthusiasm, would probably not have emerged. In all probability, had the Tahitians not learnt from Wallis that the strange white men in their great ships possessed superior and lethal weapons, that their intentions however were not to conquer and settle, and that they could be mollified by the relatively simple process of making women and fresh food available, and promptly sent on their way, Bougainville and his companions may well have made about Polynesians the kind of comments they directed at Melanesians.

It is fair to say that New Cythera, together with the presence of Ahutoru in Paris, acquired an importance that was somewhat out of proportion. The publication of Bougainville's elegantly written *Voyage* firmly established his reputation, making up for a relative initial indifference towards the expedition among scientists, but its success helped to misdirect public attention from his other achievements. 'When all is said and done', wrote one historian, 'the voyage of Bougainville was essentially successful in the field of literature.'[1] This comment is harsh and somewhat unfair, as others than Bougainville were responsible for the Cytheran myth, and arguments about the innate goodness of primitive man both preceded and followed the voyage. But the judgement has an element of truth, because wooing the literary public compels the explorer to present his travels in an appealing manner and to omit scientific data. There are few nautical details in the *Voyage*; longitudes and latitudes are often omitted, there are few truly valuable reports on the fauna and flora of the islands visited. Nor did the expedition bring much back for the museums the scientific societies were building up in Paris and the provinces—although much of the responsibility for this lays on the shoulders of Commerson who stayed behind at the Isle de France, had little bent for artistic drawing and had not started on the cataloguing of his collections before death caught up with him; when finally his documents and specimens reached Paris, no one had the time or the patience to do much with them.

But for a balanced assessment of the expedition's achievements one needs to look at the geographical and ethnological contributions.

Sailing on, the French came upon the Samoan archipelago, where Roggeveen had preceded them. It was not until they reached the New Hebrides that they could claim new discoveries: Pentecost, Aoba, Malakula. Quiros may have sighted some of these from a distance, but Bougainville was the first to bring back clear and precise information. His skills as a mathematician, combined with those of the astronomer Véron and the young officer Du Bouchage, enabled him to calculate longitudes with greater accuracy than any of his predecessors. In developing a new technique of calculating longitude by means of lunar distances, they all three made an important contribution to the science of navigation.

The French sped on towards the Australian continent, and were fortunate to come upon early warnings of the presence of the Great Barrier Reef. Their experiences as they neared the coast of what was to become Queensland told the world of navigators how extensive the Reef could be, not only close to the shore but also well out to

---

[1] Devèze, *L'Europe et le monde à la fin du XVIIIe siècle*, p. 232.

sea. The expedition consequently veered north, thereby discovering a number of features of the Louisiade Archipelago, notably discovering Rossel Island, and finally coming upon the famed Islands of Solomon. It would be left to geographers to make the final identification, but Bougainville's contribution is substantial: the discovery of several smaller islands, probably the Treasury Islands, Shortland Island and Vela Lavella, and the large island of Bougainville. He named, but did not actually discover, the islands of Choiseul and Buka.

Conditions on board both ships had by then worsened to the point where further time-consuming exploration was too dangerous. A call for supplies and rest in New Ireland was as unproductive from the point of view of geography as it was in respect of supplies. Carteret had preceded the French, as they discovered, and even had conditions been more favourable there was little incentive for survey work. The ships sailed on north and west, and made as speedily as possible for the Dutch East Indies. Here again, new discoveries or at least first precise sightings were made: the Kaniet Islands and the Ninigo Islands.

Bougainville's influence on other navigators was considerable. The French government, although aware that circumstances had led to Bougainville altering his original plan to sail to China after seeking traces of the southern continent and Espiritu Santo, was impressed by the overall success of the circumnavigation and the positive reaction of European opinion, at a time when French prestige was still suffering from the after-effects of the Seven Years War. England had now sent out James Cook on the first of his great voyages. After Bougainville's successful voyage, France felt confident that she could carve out her own share of Pacific exploration and, hopefully, eventual Pacific colonization.[1] The expedition of Marion Dufresne was a direct result of Bougainville's, since Ahutoru needed to be sent home, but other voyages would soon follow. Surville had already sailed from French India and would play a significant role in finally solving the Solomon Islands enigma. Kerguelen was sent on two major and costly voyages, although the results were meagre indeed. La Pérouse finally would set out on a voyage intended to complete, if not equal, those of Cook. D'Entrecasteaux would go next, to search for the lost ships of La Pérouse, and Baudin would undertake a lengthy voyage of exploration to Australia. In almost every case, Bougainville would be consulted, would advise, and his advice would be received with all the respect due to an experienced senior officer whose achievements were widely recognized.

A number of geographical features bear his name. Bougainville Island and Bougainville Strait in the Solomons are the best known, but there are others: Bougainville Bay (Bahía Bougainville) in the Straits of Magellan, Mt Bougainville on Alofi in the Hoorn Islands, Bougainville Strait in the New Hebrides (Vanuatu), Bougainville Reef off eastern Australia, Bougainville Peak and Bougainville Strait in New Guinea.

Commerson had honoured him by naming a plant after him, the bougainvillea. He had come across it in the neighbourhood of Rio de Janeiro in July 1767 and labelled it in a somewhat latinized form as *Buginvillaea novissima planta*. The name appeared as

---

[1] 'He made a famous voyage [and] he installed the French presence:' Beaglehole, *The Life of Captain James Cook*, p. 124.

the *Buginvillée* in around 1808-10 in Poiret's supplement to the *Encyclopédie méthodique*. By then, Commerson's collections had reached Paris, and the plant was renamed by Antoine-Laurent de Jussieu the *bougainvillea*; it was listed in the *Species Plantarum* in 1799 by the botanist Willdenow. The beauty of its flowers and bracts soon ensured its popularity: by the 1840s it was sufficiently well known in France to be included in French language dictionaries (the *bougainvillier* for the plant and the *bougainvillée* for the flower). It took a little longer for it to be popularized in England, no doubt because it needs a moderately warm climate: the Oxford dictionaries included it from 1866, while in the United States Webster's dictionaries allowed it in from 1881. It is now widely cultivated in milder and warm climates.

Another, far less known tribute was paid to Bougainville by the world of zoology – a medusa or jellyfish bears the name *Bougainvilleia* Lesson.

Bougainville's later career reflected the status he had acquired among the French people. He became a member of the French Senate and Napoleon I made him a count of the Empire. When he died in 1811, France was at the height of her powers. He was interred in the Pantheon where today visitors may be allowed a glimpse of an austere stone tomb bearing simply the name 'BOUGAINVILLE'.

THE JOURNAL

Departure from the River Plate [Rivière de la Plata]

Saturday 14 to Sunday 15[1] [November 1767].
The winds NW to N. The frigate draws 13 feet 2 inches at the stem, 13 feet 6 inches at the stern. We began to tack at 3.30 a.m., sailed under topsails[2] together with the *Étoile* at 5 a.m. Until 6 steered SE5°E, from 6 till 8 E¼NE and ENE, at 8 sighted Montevideo bearing W5°N, the island of Flores[3] NNE distant 1 league. Sounded N and S of the island of Flores 8 fathoms mud and shells.

At 10 sighted Flores bearing W and W¼SW, its centre distance 4 leagues. 9½ fathoms, same ground.

At midday the most E land bore NE¼E, the most W NW¼N, 10 fathoms. From 8 to midday steered E¼NE to E¼SE. At midday we were 12 leagues E and E¼SE of Montevideo from which I am taking my point of departure in latitude 34°58′ South and longitude 58°18′. I am using Mr Verron's observations.[4] I will add at this point that this astronomer checked my octant which he found to be very accurate. At 2 sounded 14 fathoms, from 6 to 8 sounded twice 11 and 13 fathoms, at 9 and 11 p.m. 14 fathoms sandy ooze, at 1 a.m. 22 fathoms sandy ooze, at 3.30 28 fathoms ditto with broken shells, at 6.30 35 fathoms grey sand.

From Saturday midday until Sunday midday the weather has been fine and the winds N to NW fresh breeze, fair [i.e. calm] sea, at midday estimated run SE4°45′E, distance 98′.[5]

Latitude estimated and observed 36°1′.

Longitude estimated: 56°43′.

Sunday 15 to Monday 16.
Winds: NW to S and SW. Light gale in the afternoon, stormy gale during the

---

[1] The nautical day ran from noon to noon.

[2] The *Boudeuse* carried three masts. The fore mast had three sails, the fore sail, fore topsail and fore topgallantsail; the main mast carried the main sail, main topsail and main topgallantsail; the mizzen had two sails, the mizzen sail, which was lateen (triangular) and the main topgallantsail. There was a sail on the bowsprit and various staysails spread from the stays between the masts. The lower sails (fore sail and main sail and mizzen) were known in English as the 'courses'. Studding sails were carried outside the topsails and the fore and main sails.

[3] The island of Flores lies in the wide estuary of the River Plate, some 22 km south-east of Montevideo.

[4] Montevideo is in 34°50′ south and 58°32′ west of Paris. Véron's instruments were up to date and of excellent quality, but did not include a chronometer, no accurate time-keepers being yet available at sea at this date.

[5] See 'Editorial Note' on 'Navigational Practices', for the procedure of estimating the ship's position and 'run'.

night and on Monday. Very heavy seas, under courses and close-reefed topsails, close hauled on the starboard tack.

Estimated run at midday: SE5°30'S, distance 76'30".
Latitude estimated: 37°. Longitude 55°42'.
The *Étoile* sounded at 2 p.m. and found 60 fathoms grey sand.

Monday 16 to Tuesday 17

Winds: SW to W. Stormy gale and rough seas, steered close hauled on the starboard tack, the topsails furled to keep in touch with the *Étoile*.

At midday the corrected run was SE, distance 67'
Latitude estimated: 38°5', observed 37°47'.
Longitude corrected: 54°43'.

Tuesday 17 to Wednesday 18.

Winds: from W to SW, SE, S, SSW. Squally with rain, rough sea, deplorable weather. Nevertheless this is the beginning of summer in this cursed climate which is certainly more affected by winds than our northern hemisphere; steered close hauled on both tacks, keeping the *Étoile* at a short distance with our courses, fore staysail and the main topsail, all reefed.

At 4 p.m. last night my estimate placed me in latitude 38°3' and longitude 54°32'. Mr Verron, starting from his longitude as observed at Montevideo and using in order to calculate the difference in real time from that of our location at 4 p.m. a timekeeper graduated in seconds checked on land as we were leaving that town, having observed the height of the sun, obtains at that same moment a longitude of 54°33', an impressively close concordance between these two methods.

Sounded at midday, no ground.
At midday our run for the 24 hours is SSE3°S, distance 60'.
Estimated latitude: 38°44'. Estimated longitude: 54°17'.

Wednesday 18 to Thursday 19.

Winds: SW, SSW, S. Always the same winds, the sea heavy, unbearable rolling. We had E winds in the river. Close hauled, starboard tack, the *Étoile* close to us.

At midday run gives SE¼S5°15'E, distance 41'30".
Estimated latitude: 39°18', observed 39°16'.
Estimated longitude: 53° 43'.

Thursday 19 to Saturday 21.

Winds: SW, SSW, S, SSE. New moon the 21st at 1h 02 at night. The weather unchanged with heavy seas. Sailed close-hauled, tacking, the *Étoile* beside us and equally troubled, I suspect, by this situation which undermines our moves, and destroys the cattle, the poultry and the most Christian patience. Run corrected for these 48 hours: S, distance 78'.

Estimated latitude: 40°12', observed: 40°24'.
Corrected longitude: 53°43'.

Saturday 21 to Sunday 22.

Winds: NW. Light gale during the afternoon, steady gale during the night, stormy gale in the morning, steered SW and SW¼W to stand in for the land from which we are now far distant.[1]

Estimated latitude: 41°31′, observed 42°3′.

Longitude: 56°19′.

Variation observed NE 18°.

While on land Mr Verron checked my octant's alidade and the division of the edge. He made a great number of observations with my instrument on different days, taking apparent distances of the Eagle and the Lyre Wega[2] along the same vertical, and comparing these observations with the astronomical position of these two stars, he found the height recorded by my octant to be 2′ short. However, he found the measurements of the rim perfectly satisfactory, checking them with a good beam-compass as also the 120° arc equal to the radius.[3]

Sunday 22 to Monday 23.

Winds: NW, NNW, WNW, SW, S. High winds, frightful sea until 10 p.m., squalls, storm, lightning for part of the night. At 4 p.m. the *Étoile*'s fore-topsail yard broke. We remained with the two courses and even the lower section of the main-sail reefed to wait for her. In the morning the winds and the sea moderated. Steered SW until 8 o'clock when we sailed close-hauled on the port tack, the winds being SSW and SW¼S. Sounded at 6 a.m., letting out 120 fathoms without finding ground.

Run at midday SW¼W2°30′S. [distance] 113′.

Estimated latitude: 42°58′, observed: 43°10′.

Longitude: 58°22′.

Monday [23] to Tuesday 24.

Variable winds SW to NW through W. Fine weather, fair sea, close-hauled on both tacks. At midday our run gave me S¼SW2°S, distance 80′.

Estimated latitude: 44°12′ observed: 44°29′.

Our estimated longitude: 58°41′ and 57°46′ basing ourselves on Bellin's chart[4] which shows Montevideo in 58° of longitude whereas Mr Verron, whose calculations I am adopting, has placed it in 58°55′30″. Observed variation: occiduous 18°53′, ortive 18°15′.[5]

---

[1] 'The bad weather had driven the ships some five degrees to the east' (ET).

[2] These terms were in common use at the time to describe the constellations Aquila (Arcturus) and Lyra (Vega).

[3] At this time the scale of an octant or sextant was divided by chords, i.e. the length of the chord subtended by the required angle was calculated (or taken from a table), the distance measured with a beam compass on a scale, and then the arc scribed on the scale round the edge of the instrument. Since the alidade (moving arm) actually moves through half the angle measured, the scribed position of the 120° line would actually be at an angle of 60°, hence the arc should be equal to the radius.

[4] Jacques-Nicolas Bellin (1703-72) was a hydrographer and cartographer employed by the navy. He spent all his working life in the Dépôt des cartes, plans et journaux which was founded in 1721, and never actually went to sea. He published *L'Hydrographie françoise* in 1753 and the five-volume *Petit Atlas maritime* in 1764. Since all his work was based on second-hand material, his maps, although useful pioneering works, contain a number of errors, and Bougainville will be found inveighing against him on a number of occasions.

[5] See Editorial Note on 'Navigational Practices'.

Tuesday [24] to Wednesday 25.

Winds: NW, W, SW, S. Fine weather, fair sea, pleasant fresh gale; steered SW and SW¼W until 8 a.m. when we altered tack and steered close-hauled, the winds S.

Run SW¼W 3°W, distance 138′30″.

Estimated latitude: 45°5′, observed: 45°40′.

Calculated longitude: 61°28′. Observed ortive variation: 19°45′.

For the last 5 days the currents have carried us a considerable distance S, I am not amending the longitude.

At midday yesterday, we spoke to the *Étoile*. In the recent storm they lost 3 of the foremast shrouds, 1 of the main starboard shrouds, 4 stays broke as did the fore-topsail yard and they lost almost all their cattle. Currently the flute[1] is sailing much better than we are doing, who are overloaded and down by the bow.

Wednesday [25] to Thursday 26.

Winds: S, calm, SE, ESE. Fine weather, fair sea, calm almost all night, steered close-hauled port tack then, the winds being favourable, SW.

Run at midday W¼NW 5°30′W, distance 59′.

Estimated latitude: 45°35′, observed 45°34′.

Calculated longitude: 62°37′.

Observed variation occiduous: 19°56′.

At 5 p.m. sounded 200 fathoms no ground. At 3 a.m. ditto.

Thursday [26] to Friday 27.

Winds: E. Fair weather, fair sea, fresh breeze, sailing under top topsails and fore-sail to await the *Étoile* which this time was sailing worse than us. We moved 5 or 6 barrels from the bow to the stern which gave us a little more speed. Sailed SW until 4 a.m. then SW¼S then SSW at 8 o'clock. There is a difference of 4 or 5 degrees between the *Étoile's* compass and ours.

Run at midday gave me WSW 3°45′S, distance 115′.

Estimated latitude: 46°39′, observed: 46°25′.

Estimated longitude: 65°4′, by Bellin's chart: 64°9′.

Observed variation ortive: 19°16′ and 19°24′ by the azimuth.

Sounded at 7 p.m. no ground.

Note on the currents: we have already noticed on other voyages that around 45° the tide runs N just as further N it runs S. I believe that at the same time they bear E, otherwise if the difference had been W in relation to the S differences noticed in recent days we would have found ground.

Friday 27 to Saturday 28. First quarter at 8.02 p.m.

Winds: E to N, NW, W, WSW. Fine weather, fair sea, steered S¼SW and SSW under reduced sails to wait for the *Étoile* over which when we are running free with the wind aft we have as great an advantage as she has when sailing close to the wind.

---

[1] I.e. the *Étoile*. 'Flute' (from the Dutch *fluit*) is a term formerly used to describe a naval vessel serving mainly as a supply ship. The term was more widely used in France (where a ship adapted for this purpose had its armament reduced and was then known as a *vaisseau en flûte*) than in Britain.

Run at midday SW¼S2°30'W, distance 83'.
Estimated latitude: 47°29', observed 47°32'.
Estimated longitude: 66°16', and 65°21' by Bellin.
Observed variation: 19°37'.
Found the bottom at 7 p.m. and at midday 70 fathoms mud and fine grey and black sand.

Saturday 28 to Sunday 29.
Winds: W, SW, S, SSE, E, NE, N. Variable S during the afternoon, which led us to change to the port tack at around 2 p.m., fairly high seas, good breeze, then altered tack at about 10 p.m. until 1 a.m. so as to let out more sail. At 1 changed tack once more, the weather almost calm, then calm until 7 a.m. when we steered S.
Run at midday SW 5°W, distance 40'30".
Estimated latitude: 47°58'55", observed 47°58'.
Calculated longitude: 67°2' and by Bellin 66°7'. Variation corrected: 20°.
Soundings: at 8 p.m., 70 fathoms fine sand with mud; at 10 o'clock, 67 fathoms ditto; at 1 a.m., 72 fathoms ditto.

Sunday 29 to Monday 30.
Winds: NNE, N, E, SE, SSE, S. Fine weather, fair sea, good breeze until midnight; steered S¼SW straining to starboard. At 1 o'clock light rain, light airs, the winds changeable SSE to S; close-hauled port tack.
Run at midday SW 4°S, distance 101'.
Estimated latitude: 49°9', observed 49°14'.
Estimated longitude: 68°39' and by Bellin 67°44'.
Variation observed at 5 p.m. by azimuth: 20°54'.
Soundings at 8 p.m., at midnight and at midday: 65 fathoms fine yellowish sand. According to Bellin's chart I am at midday 15 leagues E of the entry to St Julian Bay [Baye St Julien].[1]

Monday 30 to Tuesday 1.
Winds: S, calm, NW, WNW, W, NW, N, NNE. The finest weather in the world and the fairest sea, as warm as in France, steered close-hauled port tack until 4 p.m. when a light wind arose from the NNW we steered S¼SW then SSW at daybreak.
Run at midday SW¼S2°40'W, distance 62'15".
Estimated latitude: 50°6', observed 50°4'.
Calculated longitude: 69°36', by Bellin 68°41'.
Variation observed occiduous: 20°55', ortive: 21°15'.
The closest land according to Bellin's chart bears from us W 12 leagues.
Soundings: at 3.30 p.m., 60 fathoms grey and yellow sand, some gravel; at midnight, 58 fathoms ditto.

---

[1] Bahía San Julián, Argentina, sheltered by Punta Desengaño and Cabo Curioso, lies in latitude 49°18' south and in longitude 70°03' west of Paris. Bougainville's reckonings were therefore reasonably accurate, allowing for drift and the bad weather encountered since the departure from Montevideo.

December 1767

Tuesday 1 from midday to midnight.

Winds: N. Light gale then stormy gale, fine weather, steered SSW until 8 o'clock then S until midnight, the topsails close reefed. Run at midnight SW¼S3°W, 77′30″.

Estimated latitude: 51°6′.

Longitude: 70°50′, Bellin: 69°55′. Variation amended: 22°15′.

Soundings: at 8 p.m. 60 fathoms grey sand small yellow gravel. At midnight 50 fathoms fine grey sand.

Wednesday 2 to Thursday 3.

Winds: NW, W, SW, WNW. From midnight to four o'clock steered SSE, stormy gale, reducing sails to await the *Étoile*, then S and SSE then close-hauled, the weather foggy with rain until we could no longer see ahead, sounding every 4 hours.

Run for the 24 hours at midday SW¼S2°30′W, distance 163′.

Estimated latitude: 51°59′40″, observed 52°.

Calculated longitude: 71°12′. Bellin: 70°17′.

At midday the Cape of Virgins [Cap des Vierges] bore from me according to Bellin 11 leagues to the SSW.[1]

Soundings: at 4 a.m. 55 fathoms grey and black sand with some small red and black gravel. At 8 o'clock 47 fathoms same ground.

Continuation of Wednesday 2 to Thursday 3.
View of the Cape of Virgins

Winds: WNW, SW, SSE. Fog and almost calm until 2.30 when we saw the land and recognized the Cape of Virgins bearing south 7 leagues away. We were then in latitude 52°3′30″ and longitude 71°12′20″ according to Verron, which taking our bearings into account would place the Cape of Virgins in latitude 52°23′ and longitude 71°25′20″.

From the time we took our bearings we have had calm and light SE breeze, steered SSE until 4 o'clock when we altered course and sailed close-hauled on various tacks according to the wind.

Run at midday SSE5°E, distance 43′.

Estimated latitude: 52°41′, observed 52°44′.

Longitude: 70°39′, by Bellin: 69°57′.

Sighted Tierra del Fuego [Terre de Feu] stretching from SSW to W¼SW.

Soundings: at 1.30 40 fathoms grey oozy sand, at midnight 50 fathoms black sand yellow and black gravel with pebbles.

Thursday 3 to Friday 4.

Winds NW to SW, strong wind, fog. We tacked between the coast of Tierra del Fuego and of the Cape of Virgins, stormy gale. The fore staysail sheet snapped and the sail flapped loose. We hove to at 8 a.m. under the mizzen and foresail.

---

[1] Cabo Vírgenes is the northern point at the entrance to the Strait of Magellan, leading into Bahía Posesión. It forms the south-eastern extremity of Argentina. More correctly, the name is Cape of the Eleven Thousand Virgins, as Magellan sighted it on 21 October which was the feast day of St Ursula, reputed British leader of a large group of pious virgins murdered in Cologne in the third or fourth century.

At 7 p.m. sighted the Cape of Virgins bearing NW¼N, 6 leagues. At 5 a.m. it bore NNW 4 leagues and the starboard point of the Strait of Magellan as you enter bore NW¼W.

Run at midday WNW1°W, [distance] 14ʹ.

Estimated latitude: 52°32ʹ40ʺ. Estimated longitude: 71°1ʹ55ʺ.

Soundings: from 40 to 45 fathoms gravel. The gravel indicates that we are closer to Tierra del Fuego than to the mainland. Near the latter one finds fine sand and at times ooze.

Friday 4 to Saturday 5. Full moon the 5th at 2.12 p.m.

Winds: SW to WNW, storm. High winds, heavy seas, hove to under foresail and mizzen. At 1 o'clock the foresail tore, we reefed it and ran under bare poles. The lead having given us only 20 fathoms, grey sand, red gravel and shells, the fear of the breakers that stretch out to the SSE of Cape of Virgins decided me to remain under bare poles, all the more so because this enabled us to replace the foresail. At 4 we set out on the port tack with the main and mizzen staysails. At 7 o'clock set the courses; at midnight changed tack until 3. At 6 a.m. the wind died down, trimmed the topsails, at 8 we tacked.

Run at a rough estimate for the 24 hours N¼NE2°E, 37ʹ. I reckon that I am 9 to 10 leagues WNW from the Cape of Virgins.

Observed latitude: 51°51ʹ.

Corrected longitude: 70°25ʹ30ʺ.

Amended run N¼NE1°N, distance 51ʹ.

Soundings: 20 fathoms as mentioned above, 50 fathoms grey sand, black stones, 60 fathoms. Ditto as we tacked off shore, 45 fathoms grey sand, 48 and 50 fathoms ditto returning towards the shore.

Note: This sounding of 20 fathoms which caused me to heave to represents no danger. It is the depth of the channel. I was unaware of this at the time and Mr Giraudais told me he had found the same when he was in sight of land, i.e. with a clear view.

Strait of Magellan

Saturday 5 to Sunday 6.

'*Nimborum in patriam, loca freta furentibus austris*'[1]

Winds: WSW, calm, NW, NNE, NW, W¼NW, W. Close-hauled starboard tack until 2 o'clock when we sighted the land bearing S to SW¼W distance 10 leagues. The winds E, NE and N, steered S. At 8 sounded 25 fathoms mud and sand, saw the Cape of Virgins bearing SW¼S 2 to 3 leagues. At 9 sounded 17 fathoms small pebbles, broken shells, the said cape bore from us SW¼W 3 leagues, 14 fathoms small black pebbles, 22 fathoms mud and sand, pieces of shell, the N point of the entrance to the strait bearing N¼NW, the Cape of Virgins NW¼N. We were sounding constantly

---

[1] Bougainville is fond of sprinkling Latin tags in his text, but he does so from memory and at times misquotes or changes the original. He draws his inspiration here and on a number of other occasions from Virgil's *Aeneid* and in particular the first book in which a wild storm drives Aeneas and his men off course. 'In the homeland of the clouds, a region pregnant with raging south-easterlies:' I, 51.

and tacking to avoid the shoals that lie to the SSE of this cape: nevertheless we passed over the end of this shoal: for, between two soundings, one of 25 the other of 17 fathoms, the *Étoile* following in our wake got a sounding of 8 fathoms. Steered until 11 p.m. from SW¼W to SSW, the winds NNE, good breeze. I wanted to sail in under reduced canvas, as the night would only last 4 hours and having 13 leagues to cover from the Cape of Virgins to the first narrows.[1] Mr de La Giraudais hailed me and suggested that we remain athwart until daybreak. Caution led me to follow his advice but I was wrong: prudence in certain cases is a lame virtue; this time it caused us to lose a good deal of valuable time.

At 11 p.m. hove to, the wind on the port side, with the fore and mizzen staysails, at 11.30 set the foresail, it appears that while we were hove to the tide caused us to drift SSE. At 2 a.m. changed tack, at 6 a.m. the wind having changed to W¼NW and W, we adopted various tacks.

Soundings: at midnight 45 fathoms grey sand, at 1 a.m. 43 fathoms ditto with black, at 7 o'clock ¾ of a league from land 20 fathoms mud and sand.

Bearings: at 8 a.m. the Cape of Virgins NE 5 leagues, Cape Entrana W 5°N 6 to 7 leagues, Cape Possession WNW 5°W 2 to 3 leagues, the easternmost of Tierra del Fuego SE 6 to 7 leagues.

At midday the Cape of the Virgins bore NE 5°E 9 to 10 leagues, the Cape of the Holy Spirit [Cap du St-Esprit] ESE 3°S 6 to 7 leagues, Cape Possession NW¼W, 2 leagues, the 4 little hills of Possession Bay which I have named the 4 Aymond Sons [fils Aymond] W¼NW 3°W. I named Father Aymond [Père Aymond] the large hillock that lies further N than the 4 small ones.[2] The ship being in the above position [I] observed 52° 30′ of latitude. This observation together with the bearings gives us the following latitudes: Cape of Virgins 52°21, Cape Possession 52°25′, Cape Holy Spirit 52°44′. Estimated longitude: 71°23′30″ which does not agree with the estimated longitude of the Cape of Virgins and as I believe the former to be the good one I am accordingly amending that point's longitude to 72°7′28″.

Run these last 24 hours: SW 3°30′W, [distance] 19′.

Sunday 6 to Monday 7.

Winds from WNW to W, fresh gale, high wind, rain, hove to under the fore and mizzen staysails. At 3 o'clock Cape Possession (30 fathoms grey sand with yellow specks) bore from me NW¼W 2 leagues, at 6 p.m. sighted the Cape of Virgins bearing NE, Cape Possession NW 2½ to 3 leagues. Set the courses and plied to windward all night changing tack every hour and a half. At 4 a.m. sighted Cape Possession bearing NW¼W 4°W. The tides here are very strong and seem fairly regular.

At midday (28 fathoms black sand) sighted Cape Possession bearing N¼NE 1½ leagues, Cape Orange [Cap d'Orange][3] SW 5 to 6 leagues, Father and the 4 Aymond

---

[1] He has been sailing through Bahía Posesión and is planning to pass through the narrows (Primera Angostura) which lead into Bahiá San Felipe. Posesión is a large bay which provides valuable shelter for ships waiting for favourable winds and tide.

[2] Père Aymond is present-day Monte Aymond; the four 'sons' are four small hills on the west side of Bahía Posesíon which are now known, less poetically, as Orejas de Burro (the donkey's ears).

[3] Cabo Orange is a sandy rise which forms with Punta Anegada the north-eastern extremity of Tierra del Fuego.

Sons W¼NW 3 to 4 leagues. The Savages[1] of Tierra del Fuego and those of the main-land have lit great fires at various locations. In 1580 Sarmiento is reported to have built a fort on behalf of Spain on this Point of Possession, which was devoutly called Name of Jesus.[2] From the Cape of Virgins to this Point of Possession I estimate 9 leagues.

Variation NNE observed occiduous: 21°45', ortive 21° 40'.

From Monday 7 to Tuesday 8.

At 1.15 p.m. (13 fathoms grey sand and pebbles, Cape Possession bearing NE¼N, the four Aymond Sons NW¼W 5°W) the wind NNW, good breeze, all sails set, steered W¼SW 4°W from midday to 1.30 (20 fathoms rocks and small pebbles, the 4 Aymond Sons bearing NW¼W 4°N, Cape Orange SW¼S and SSW) and from 1.30 to 5.30 WSW, the tide then running E. At 6 calm, the wind then altered SW, we tacked to go and anchor in Possession Bay and at that moment sighted the point of the narrows on the port side as you enter bearing S¼SE and the starboard one W¼SW.

At 7 p.m. we dropped anchor in Possession Bay in 20 fathoms sandy ooze and gravel. We had taken soundings for about an hour, finding only a rocky bottom, the *Étoile* even passed within moments from 20 to 5 fathoms. Bearings from the anchor-age: Cape Orange S, Cape Entrana SW, the highest mountain in Possession Bay NW 3°W, the easternmost point of Possession Bay ENE, about 2 leagues offshore.

The Savages lit a fire in front of the ships at the back of the bay. At 9.30 the tide turned, running W. At 10 the wind freshened coming from WSW. At 3 the winds NW. At 4 a.m. we began to raise the anchor and at 5 we were under sail. From then until 10 a.m. steered SW and SSW. At that time the winds changed to S and SW. But as a very strong tide was running SW into the narrows which was our direction we proceeded backing and filling, under close reefed topsails, and followed the tide on various tacks.[3]

At midday sighted the headland on the port side of exit from the narrows bear-ing ENE, the starboard one N, Cape Gregory [Cap Grégoire] WSW.[4] In the middle of the narrows sounded, letting out 50 fathoms of line without finding ground.

At 8 a.m. the Patagonians raised a white flag near the narrows, no doubt the same one that Mr de La Giraudais had given them in June 1766.[5] We answered by

---

[1] See introductory 'Editorial Note' for interpretation of this term.

[2] Pedro Sarmiento de Gamboa (1532–92) left for South America after a period in the army. Author of *Historia de los Incas* (1572), he advocated the exploration of the Pacific by the Spanish and was instru-mental in organizing the expedition of Alvaro de Mendaña with whom he sailed to the Solomons. In 1579 he sailed in the *Nuestra Señora de Esperanza* for a survey of the south Chilean coast and continued through the Strait of Magellan, going on to Spain. He set off again in 1581 with orders to take steps to curb English incursion; he reached the Strait of Magellan in 1584 with plans to erect defensive forts but he was captured in 1586 and taken to England. Bougainville's comment on the name of Sarmiento's outpost contains a touch of irony, the 'devout' name of Nombre de Jesus being rather inappropriate for a military structure.

[3] A vessel passing through a narrow channel, carried by the tide against the wind, would fill her sails to make headway and back them to make sternway to keep in mid-channel; this would appear to be the action Bougainville is indicating here.

[4] Cabo Gregorio, on the western side of the bay of the same name, is a small sandy point which sig-nals the entrance to the second narrows (Segunda Angostura).

[5] Chesnard de la Giraudais had sailed to the strait in the *Étoile*, leaving the Falkland Islands on 24 April 1766 and returning on 26 June. An account, 'Extraits du journal du sieur de la Giraudais…' was included in Pernetty, *Histoire d'un voyage aux isles Malouines*, II, pp. 273–83.

raising the ships' ensign. The Savages of Tierra del Fuego also came, numbering approximately 20, on the port side as you enter the narrows. They were on foot and dressed in skins. They followed us and gestured as though calling us.

Advice for navigators. 1°. When one wishes to enter the first narrows one should pass 1 or 1½ leagues off Cape Possession, then sail SW¼W, taking care not to sail too far S on account of the breakers which extend for more than 3 leagues NNE and SSW off Cape Orange. Sometimes one can see the waves breaking over them, but the sea is not always that kind towards sailors. Once N and S of Cape Orange one leaves those breakers behind. And do not come too close to the SW shore of Possession Bay facing Father Aymond and his 4 Sons on account of a reef which, in this area, stretches along the coast 2 leagues off and is sometimes awash. When the 4 hillocks known as the 4 Aymond Sons merge into 2 forming a portal you are abreast the said reef.

2°. When you find yourself at the entrance to these narrows, with the tide running into it, fairly straight and running at 7 or 8 knots, you should enter in spite of opposing winds even if these are violent. Sailors will know what must be done. In general, the tides make this strait quite navigable and make up for the ill will of the winds that seem united in their efforts to guard the entrance.

Distances: from Cape of Virgins to the entrance of the first narrows my estimate is 14 leagues. The N coast is safe, 20 fathoms 2 miles from land.

Tuesday 8 to Wednesday 9.

At midday, being inside the pass, we let out the topsails, set the courses and steered WSW. At a ¼ past midday 34 fathoms rocky bottom, at 12.45 15 fathoms rocky bottom, a few minutes later 14 fathoms same ground.

Sighted Cape Gregory bearing SW¼W. At 2.45, with a slack tide and almost calm, we anchored in 18 fathoms mud about 1½ leagues from land. Bearings from the anchorage in Boucault Bay: the knob of Cape Gregory W¼SW 5°S, the starboard point of the second narrows SW, the NW point of St George Island [Isle de St-Georges] or the one on the port side SW¼S.

Sounded at slack tide and found 21 fathoms; the tide since the time we anchored ran E.

Distances: the first narrows bears NE and SW and is probably some 2 to 3 leagues long. Its width varies from one and a half to one league.

Meeting with the Patagonians

As soon as we had anchored, I lowered a boat and signalled the *Étoile* to do the same. We landed opposite the ships and noticed as we did that the tide was low and beginning to turn. Hardly had we walked a few steps when we saw some savages, 6 in number, racing towards us on horseback. They dismounted at 60 paces and gaily and confidently marched towards us, shouting *Chaoua, chaoua*. We made a great fuss of them; in the two hours we spent ashore their number grew to 30, I had bread and biscuit given to them. This group is the same as the one seen in 1766 by the *Aigle* and the *Étoile*.[1] They were extremely welcoming in their gestures, clutching us in their arms

---

[1] There are a number of references made to the March–June 1766 meetings of the French with the local inhabitants in such works as Martin-Allanic, *Bougainville*, I, pp. 211–14, and in the journals of those

and expressing the greatest joy at seeing us. They were most eager for us to spend the night with them and to wait for the others who would soon be coming. They all asked for tobacco and the colour red seemed to please them greatly. When we returned to our boats they accompanied us and without any ado took anything they found; it is true to say that they gave back what they had taken without any fuss. As we rowed off, great shouts of *chaoua* arose from both sides.

So such are those Patagonians whom some travellers described to us as giants[1] and whom in 1765 the English on Mr Byron's ships reported having seen as giants.[2] I am sorry for Dr Mati[3] who was displeased when I declared upon my return from my first voyage to the strait that I had not encountered any giants. However, even though this may cause him once again to feel displeased with me, [I can only say that] these men are of a good height, but the tallest one I met attained scarcely a height of 5 ft 9 in. Several were of my height and thanks be to God I am only 5 ft 6 in. Anyhow the English should have accepted the deal which, so they say, these giants suggested, namely to exchange their tall women for small men, a most advantageous deal which would have brought to Europe living proof of the existence of a race of giants, which is such a contentious subject. What does distinguish these people and what one can really call gigantic is their frame, the size of their heads and the thickness of their limbs. They seem to be strong and well fed, their flesh is firm, their sinews are taut. The features of these Americans are not harsh, and some are pleasing to look at. Their

---

who took part, including Bougainville and Alexandre Guyot, the captain of the *Aigle*. Thus on 18 March Bougainville wrote: 'I went back to see the Savages … their number had increased. I gave them blankets, shirts, hoods, belts, coppers and pans, axes, handles, knives, mirrors and some cinnabar.' 'Journal', BN, NAF 9407, f. 44.

[1] The size of the Patagonian people was much argued over during the eighteenth century. While the Indians of Southern America are not short (the Tehuelches, which are the main group encountered by the French, average 1m 78, slightly under 6 ft, and a number of them would have been taller than the average French sailor), reports that the Strait of Magellan was inhabited by giants resulted from a series of misapprehensions and exaggerations. The name *Patagones* actually derived from a Spanish term which referred to the natives' large feet (*pata*, foot). It should also be borne in mind that a man standing on a rock or an islet, seen from a distance, can appear inordinately tall. Thus Abel Tasman, while sailing off the Three Kings Islands north of New Zealand, reported seeing Maori warriors 'in various places, on the highest mountains… people of tall stature'. Sharp, *The Voyages of Abel Janszoon Tasman*, p. 144.

[2] John Byron who sailed through the Strait with the *Dolphin* and the *Tamar* played a significant part in reviving the legend of Patagonia's giants. The British government found the attention paid to this story quite useful in deflecting Spanish suspicions about British plans in South America and the Pacific. See on this voyage Gallagher, *Byron's Journal of his Circumnavigation 1764–1766*, and especially the appendix, pp. 185–96. Philip Carteret in January 1767 reported on the Patagonians, stating that he had found no giants and that Samuel Wallis who had measured several of them had found only a few who attained six foot. This information, however, was not published in London until 1771, by which time tales about Patagonian giants had gained wide currency in England. A useful summary of the controversy will be found in Wallis, *Carteret's Voyage*, II, pp. 322–6.

[3] Matthew Maty (1718–76) was Secretary to the Royal Society. He had been fascinated by Byron's report on the Patagonians and had promptly passed it on to his French scientific colleagues. Maty had a fairly uncritical approach to the reports of navigators, including those of Byron's expedition. He pressed for the publication of Jacob von Stählin's account of Russian discoveries in northern waters and translated from the original German his *Account of the New Northern Archipelago, lately Discovered by the Russians in the Seas of Kamtschatka and Anadir* which was published in London in late 1774. Cook was to find this account of doubtful value during his third voyage. Maty had been the subject of some criticism in France. Bougainville did not include any reference to Maty in the printed account of his voyage.

face is round and flat, their eyes bright, their teeth extremely white. They wear their black hair long, tied up on the top of the head, I saw some who wore a pair of mustachios that were more than lengthy. Like all Savages they paint their bodies. They wear the skins of various animals. A leather codpiece covers their natural parts, a great cloak of guanaco and small lizard skins tied around them with a belt reaches down to their heels. They have a kind of leather riding boots opened at the back. A few wore copper rings around their legs. None of those who came up had any other weapons than those bolas we have already seen in France. Their horses were harnessed and saddled in the way they are in the River Plate, the poor animals as thin and puny as their masters were big and fat. Several had haunches of vicuña across their saddles. They did not offer us any, they merely bartered vicuña skins for handkerchiefs and hats. These people lead a life similar to that of the Tartars. Roaming across these enormous South American plains, on horseback, men, women and children, tracking game, using skins as clothing and for shelter, they bear an additional resemblance to the Tartars in that they go off robbing travellers who cross over from Buenos Aires to Chile.

One Patagonian had a saddle with golden studs, some gilt copper stirrups, a braided leather bridle, an entire set of Spanish harness. Several used a few words of Spanish, such as *mañana, muchacho, bueno chico, capitán*. Moreover I do not think they should be called brigands because they would rob the Spaniards, whom all the Americans call brigands who have invaded their homeland. The Patagonians have a bronzed complexion like all the natives of both Americas. They piss in a crouched position, would this be the most natural way of passing water? If so, Jean-Jacques Rousseau who is a very poor pisser in our style, should have adopted that way. He is so prompt to refer us back to Savage Man.[1]

The terrain of our landing place is absolutely similar to that of the Malouines. Mr de Commerçon did not fail to gather a good collection of plants. The Savages kept fires going all night and there was almost no wind. At 3 o'clock slack tide.

At 4.30 a.m. with the winds NNW we weighed anchor, all sails set against the tide, steered SW¼W and covered half a league. At 6.30 the winds turned SW stormy gale; I dropped anchor in 19 fathoms mud and broken shells.

Bearings at the second anchorage in Boucault Bay:

Cape Gregory W¼SW2°S, the starboard point of the second narrows SW4°S, the port side point SW¼S2°S.

The Savages appeared on the shore, numbering over 200. They are waiting for us.

Wednesday 9 to Thursday 10.

SW to SSW winds, stormy gale. At dawn a strong W wind. The Savages lit great fires and seem to have erected a few huts. I believe the camp has moved opposite us. The high wind and our distance from the coast will prevent me from taking presents to the Patagonians as I hoped to do.

At 4 p.m. the tide slack, at 10 p.m. it began to turn W. Sounded 18 fathoms; at 4

---

[1] Not surprisingly, this passage was omitted from Bougainville *Voyage*, the published version of his circumnavigation, which was intended for a large reading public. The note of sarcasm evident in Bougainville's comments reflects his scepticism about Rousseau's theory of the Noble Savage and reports of Rousseau's poor health at the time.

a.m. the tide turned E; sounded 19 fathoms; at 10.30 slack tide. Mr Verron made observations of the distance between the moon and Regulus and concluded that the longitude of our anchorage was 73 °26′30″, that of the eastern entrance of the second narrows 73°34′30″ and its latitude 52°40′.

Thursday 10 to Friday 11.

SW to WSW gale. At midnight NE breeze, I gave instructions to raise the anchor and began heaving in. We made fruitless efforts to raise our anchor, even after setting up our best ropes and tackle. A 2.30 a.m. the cable snapped between the bits[1] and the hawse-hole[2] and we lost our anchor. It was our 3rd, weighing [blank].

At 4.30 the tide turned E. At 11 p.m. it began to run W, calm during the night, light airs from the N.

We sailed at 3 a.m. all sails set; until 4 steered SW¼S3°S; from 4 until 8 WSW to SW¼S. At 5 a.m. sounded 35 fathoms gravel and broken shells. At 5.30 a.m. the tide turned E. The cape of St George Island [Isle St-Georges][3] which forms the port head on entering the second narrows bore from us SW¼S ¾ of a league, NW wind, fresh breeze. At 8 a.m. bearings, the middle of St Elizabeth Island [Isle Ste-Elizabeth][4] SW5°S, the point of St George Island at the port entrance to the narrows SW¼S2°W, the starboard head as one leaves SW¼S3°W.

From 8 to midday light NW breeze, steered SW¼W to clear the second narrows, made little progress, lost ground even on account of the tide running against us although it is far from being as strong as in the first narrows.

Bearings at midday St George's Point on the port side as one enters the narrows SSW3°W, the starboard point as one leaves the narrows SW¼W5°W.

Distances: From the exit of the first narrows to the entrance of the second there are possibly 6 leagues. The second narrows seems to be 3 to 4. It lies W¼SW and E¼NE adjusted, its width is approximately 1½ leagues.

Advice for navigators: as you pass through the second narrows you should borrow[5] along the Patagonian coast because when you leave the tides run S and although the headland of St George Island as you emerge from the pass is high and sheer, a low point appears below it which stretches towards the W.

St Elizabeth Island lies NNE and SSW in line with the W head of the narrows on the Patagonian side. St Barthélémy Island [Isle St Barthélémy] and Lions [Island] also lie in line with the W point of the narrows by St George Island, NNE & SSW. The islands of St Barthélémy and Lions are linked by a reef. Two other reefs extend for some 1 or 2 leagues, the one from the SSW of Lions Island [Isle aux Lions], the other NNE of St Barthélémy, so that these 3 reefs and 2 islands form a chain between

---

[1] A frame fixed vertically on the fore part of the ship to which cables are attached when riding at anchor.

[2] A hole cut through the bows adjacent to the stem through which the anchor-cable passes.

[3] The cape is Punto San Isidro. 'Isles St-Georges' is not in fact an island, but a peninsula standing NW from Tierra del Fuego and almost separated from it by Bahía Gente Grande.

[4] Now Isla Isabel, a substantial island some seven miles south-west of the second narrows. A little further on is Isla Magdalena which may well be Bougainville's Barthélémy and Lions which he reports as being linked together. A number of the names used by him have vanished from the charts and precise identification in these meandering canals is difficult in modern times.

[5] I.e. hug a shoal or coast in order to avoid an adverse tide.

which, when it bears ESE and St Elizabeth Island bears WNW, is found the channel leading into the strait, a channel running NNE and SSW which I believe to be the only one. The land of St George beyond the point forming the exit from the second narrows runs ESE and WNW.

St Elizabeth Island extends NNE and SSW, is 3 to 3½ leagues long, one to 2 leagues wide, [and] ends in a point at the SW quite close to the mainland. It is high and steep on all sides except at the SW and SE points where the land slopes down. Sounded from the ship to the island from ½ a league 7 fathoms to less than ¼ of a league 10 fathoms ooze and gravel.

Friday 11 to Saturday 12.

At midday the tide began to turn in our favour, the winds were variable and very light, steered until 3 o'clock SW¼W. At 3 stormy gale from SSE to SSW with squalls and rain; we followed both tacks to reach the anchorage N of St Elizabeth Island on 7 fathoms grey sand, gravel and broken shells, good holding.

Bearings at the anchorage near St Elizabeth Island: the E point of St Elizabeth Island SE¼S 1 league; a low point below the previous one, which stretches out further E by the W head of Lions Island SE¼S5°E; the W point of St Elizabeth Island SW; the E point of Lions Island SE5°E; the NE point of St Barthélémy Island E¼SE; the SW point of St Barthélémy Island ESE 2 leagues; the point of St George Island as you leave the narrows NE 3 leagues; the centre of the entrance of Port Crayfish [Port des Ecrevisses] on the Patagonian coast N3°W 3 leagues; the N point of Port Vaughan on the same coast WNW5°W 2½ leagues; the S point of the same port W 2 leagues. To the SW a wide gulf and some lowlands spread out to our view. The *Étoile* dropped anchor ¼ of a league SE of us in 17 fathoms.

Note: The two above-mentioned harbours, so named by Wood and situated on the N coast, one 2 leagues and the other 3½ leagues from the second narrows, are only suitable for small vessels.[1]

The winds ranged from SW to WNW, stormy gale, squalls, rain, hail.

Saturday 12 to Sunday 13.

Last quarter at 8.56 a.m. Stormy W and WSW gale, nasty weather; Saturday at midday I had the boat lowered and went with several of our gentlemen to explore St Elizabeth Island. It is arid, without timber, covered with fairly good grass. Its terrain is similar to that of the Malouines, with the same plants and even the gumtree.[2] Above fly some extraordinarily timid bustards. The Savages who hunt them sometimes land on this island. We found traces of their fires.

Sunday 13 to Monday 14.

WNW winds, fresh breeze. At 1 p.m. I set sail together with the *Étoile* under topsails and foresail and entered the channel between St Elizabeth Island and the other

---

[1] The reference is to John Wood who sailed through the Strait in 1669 as part of English attempts led by Sir John Narborough to carry out raids on Spanish settlements on the Pacific coast. *Capt. Wood's Voyage thro' the Streights of Magellan* was published by William Hack in 1699 as Part III of his *A Collection of Original Voyages*, and reprinted in William Dampier, *A Collection of Voyages* (1729).

[2] Bougainville has in mind the resin-producing shrub (the term 'gumtree' is somewhat misleading) *Hydrocotyle gummifera,* found in marshy areas in the Falkland Islands.

two. Soon after, the tide turned in our favour. It is strong in this pass. We ranged along St Elizabeth Island and steered SSW, Lions Island reef stretching well towards the SW, then steered S¼SW once we had the point of St Elizabeth bearing N. There is a depth of 30 fathoms in the channel and one can anchor in 20 fathoms and less along St Elizabeth Island. We came up towards the coast of Patagonia in the direction of Black Cape [Cap Noir].[1] At this point the mainland trends SSE and the coast becomes forested. We sailed along it at a distance of approximately one league. Very strong squally wind until 6 in the evening.

From 1 o'clock until 4.30 steered SW¼S to S¼SE, squally weather requiring care with our halyards and sheets.[2] From 4.30 to 5.30 steered S. At 6 p.m. sighted St Anne Head bearing SSE5°S, the southernmost of Tierra del Fuego SE¼S, Black Cape NW¼N and NNW, a headland of Tierra del Fuego which we believe to be Cape Monmouth [Cap Montmouth] E3°S.

At 8 p.m. sighted St Anne Head [Pointe Ste-Anne] bearing S¼SE, Freshwater Point W5°S, Cape Monmouth E¼NE, the most southerly of Tierra del Fuego SSE. The winds varied until midnight from WNW to WSW; we sailed with the topsails close reefed and the foresail, steering SSE to round St Anne Head. From 10 o'clock the winds were very weak, calm at half past midnight, rain then right away stormy SSW and SW winds. We tacked until 8.30 with frightful weather and repeated squalls. But as our tacks were short and we had to run before the wind with a wild current carrying us strongly towards a wide bay on Tierra del Fuego of which Cape Monmouth was one of the E headlands, we lost almost 3 leagues. I am determined to make for the anchorage provided by the bay some 2 leagues S of Freshwater. I named this bay Duclos Bay [Baye Duclos], after Mr Duclos-Guyot, fireship captain, my second-in-command who has been sailing with me for 9 years; having already rounded Cape Horn 4 times he well deserves to leave his name in this region.[3]

Having approached within 3 or 4 miles of the shore we sounded several times at 50 fathoms without finding ground, within two to one miles [found] 25, 20, 18, 12, 10 fathoms, gravel bottom then sand and broken shells, then sand and ooze. At 10.30 we dropped anchor in 8½ fathoms, sandy ooze, with the following sightings:

Bearings at the anchorage in Duclos Bay: St Anne Point SE¼S, Cape Monmouth ENE5°E, the N head of the said bay NNW5°W, another large headland being the most N land N¼NW, the furthest land in this channel which belongs to Tierra del Fuego SE.

---

[1] Cabo Negro is a high, dark cliff at the north-eastern extremity of Brunswick Peninsula.

[2] Haliards are the ropes used to hoist or lower the sails; sheets are ropes secured to the lower corners of the sails (the clews) and are used to trim them to the wind.

[3] The expedition has now sailed down the wider part of the Strait, past the site of present-day Puntas Arenas, with the wide opening of Bahía Inútil appearing on Tierra del Fuego past Cabo Monmouth and the islands separating the true strait from canals south of Tierra del Fuego. Freshwater (today's Aqua Fresca and so named on account of a small river which tumbles down from the heights of Brunswick Peninsula) opens out on the Patagonia shore before Punta Santa Ana (Pointe Ste-Anne) is reached. A little further on is Gamboa de Sarmiento's ill-fated settlement of Rey Don Felipe, founded in March 1584 and known in English as Port Famine (present-day Puerto del Hambre). Sadly, Duclos' name has vanished from the charts; it was probably present-day Bahía Carreras.

We are 1 good league from land. As soon as we dropped anchor I sounded up to the land, 7, 6, 5 and 4 fathoms as far as a musket shot from the shore.

Distances: from the exit of the second narrows to the NE point of St Elizabeth 4 leagues, St Elizabeth Island about 3½. The charts do not give good locations for the islands of St Elizabeth, Barthélémy and Lions. When one has emerged from the second narrows, one should hug the mainland as much as possible because the current runs strongly towards St Barthélémy reef. When the NE head of St Elizabeth bears S one then steers to enter the channel while still hugging the island.

Comments on tides: I am still inclined to the view that in all this part of the strait the current comes from the SW. We noticed this on our first voyage and we daily repeat the same observation, which is that when the current is favourable to our entering, the tide is high along the coast but low when the current becomes unfavourable. The strong current we encountered on reaching the opening of which Cape Monmouth is one of the points leads me to believe that this opening is a channel leading to the neighbourhood of Cape Horn and that possibly the tide which makes itself felt as far as the eastern entrance to the strait runs from the S by way of this channel. Generally, if several channels cut through Tierra del Fuego they must cause considerable variations in the currents one finds in the strait.

Distances: from the SW point of St Elizabeth Island to Black Cape possibly 1½ leagues; from Black Cape to Freshwater Point 6 leagues; from the S head of Freshwater to St Anne Point 6 leagues, sailing SE¼S. From St Anne Point to Round Cape [Cap Rond] 5 leagues, sailing S¼SE; from Round Cape to Observatory Island [Islet de l'Observatoire] 4 leagues, WSW and SW¼W; from there to Cape Forward 5⅔ leagues distance. The strait before the first narrows may be some 7 leagues wide, [and] as much from the first as to the second. Then its width varies from 7, 6 to 5 leagues until it narrows near Round Cape where it is no more than 3. It then widens until Cape Forward. Facing Cape Forward Tierra del Fuego forms an arc with a radius of some 4 leagues, which merging in the distance with the mainland gives no inkling of any exit. Several [exits] appear on Tierra del Fuego, bays and what looks like harbours are visible on all sides. These mountains, snow-covered all the year round, which make up this land, are all jagged and broken up by inlets.[1]

Cape Forward presents a surface with 2 headlands approximately ¾ of a league apart, the E one being higher than the W. There is almost no bottom, however between these two points one could anchor in a sort of small bay in 15 fathoms sandy ground. The entire cape consists of bare rock, but nevertheless covered with trees almost to its summit. On the most S extremity of this cap where ends or begins an immense continent, we raised the French flag and shouted 3 times Long Live the King.

Distance: Sighted from above Cape Forward Cape Holland [Cap Holland] bearing W4°N distant 5 leagues.

---

[1] This somewhat confusing passage is a description of the channel between the South American continent, of which Cape Forward [Cabo Forward] is the southernmost point, and the island that lies off Tierra del Fuego, Dawson Island. The mountains referred to are the Darwin Mountains to the south. Round Cape is now known as Cabo San Isidro.

Monday 14 to Tuesday 15.

In the afternoon we went ashore, the weather was mild, a kind of grassland stretches along the landing place which is very attractive and sandy. Then woods rise up in the shape of an amphitheatre. There are two streams in this bay but it is very difficult to get water from them, the water close to the shore being brackish. One can find here some plants from the Malouines and others that are peculiar to the strait. We found on the shore 8 native huts made of branches very ably interlaced in the shape of an oven. Each hut could hold 8 people. Winds variable SSW to WSW, fine weather on land.

Mr Verron determined the latitude of the NNW point of Freshwater shown above as being 53°17′, the ship being in 53°20′.

Tuesday 15 to Wednesday 16.

Fine weather in the afternoon, calm in the evening and throughout the night; since the slight breeze was coming from the E, I anchored apeak[1] at 9 p.m.

At 4 a.m. the winds being N, the sky overcast with mist, we hoisted in our boats and sailed with all sails high set.

From 5 to 8.30 steered SE¼S and from 8.30 to 9 SSE, the tide being favourable and the ship making 4, 5 and 5½ knots an hour.

At 8.45 we were abreast a reef which is not more than a mile from shore.

Sighted St Anne Point bearing S5°E distant 1½ leagues, the middle of Round Cape S¼SE3°S. The reef runs from the WSW up to S¼SW. Sounded at that same time being ½ a league from land and let out 30 fathoms without finding ground.

At 9 being E and W of St Anne Point ½ a league steered S¼SE until 11.

Bearings at midday Round Cape N¼NE5°E, the middle of French Bay [Baye Françoise][2] W¼NW, a cape which is part of Cape Forward SW3°S, the nearest of Tierra del Fuego E 3 leagues.

Wednesday 16 to Thursday 17.

Since midday the winds have been constantly changeable first from the NE to the NW light breeze. Steered SW and SW¼W until 4 making little progress. Bearings at 4 p.m. the most westerly land of Tierra del Fuego to the W¼SW, Cape Forward W3°S, the point of French Bay at NE¼N and NNE.

Calm from 4 until 6.30. During this time I went in the jolly-boat to take soundings near Cape Forward and below the cape. The depth is enormous. Herewith is the description of this cape.[3]

At 6.30 the breeze came from the SW. I tacked and dropped anchor in French Bay in 10 fathoms, muddy bottom.

Bearings in French Bay anchorage: the N point of the bay NE¼E5°N, the islet in the middle of the bay NE; the most N point of an island that lies E of Bougainville

---

[1] I.e. on a short cable, so that the ship is vertically above the anchor.
[2] Present-day Bahía San Nicolás.
[3] 'The description referred to here, which must have been written out on a separate sheet that now cannot be found, does appear in Bougainville, *Voyage*, p. 140' (ET).

Bay [Baye Bougainville][1] named Nassau Island [Isle Nassau] NE¼E; the S point of French Bay S5°W.

During the night and throughout the morning the winds were WSW and SW stormy gale, squalls and rain.

Notice: In French Bay it is necessary to moor astraddle,[2] the wind gusts between the mountains causing the ship to turn a hundred times on its anchor, which can cause her to drag,[3] anyhow the bottom provides a good holding ground. Moor with a kedge anchor almost at the end of the cable as it will be necessary to adjust it frequently.

As one can obtain water from De Gennes River [Rivière de Gennes][4] only with the greatest difficulty, on account of a bar at the river mouth which allows boats to pass through only at high tide, and even then they have to go up for quite a distance before finding drinking water, the sea being very unfavourable for a landing and because of the possibility of ships being adversely affected by the SSE and SE winds, I decided to leave the French Bay anchorage in favour of my bay, which is as good as any port,[5] since I needed to obtain water and wood, to clean the ship and change all the stowage. During the morning I went with Messrs Duclos and La Giraudais to look for a watercourse and take sounding of the route between French Bay and mine.

At midday, Mr Verron observed by means of a quadrant on the islet in the middle of French Bay a latitude of 53°51′15″ south.

Thursday 17 to Friday 18.

In the afternoon SW winds, light breeze and squally, we sailed from French Bay under fore and mizzen topsails, we passed between the mainland and Nassau Island and dropped anchor in Bougainville Bay where we are riding land astern with one

---

[1] This name dates back to the voyage of the *Aigle* which sailed here from the Falkland Islands in February 1765 under the command of Bougainville and of Alexandre Duclos-Guyot. The French anchored in a bay 'to which the crew has given my name' wrote Bougainville in his log (BN, NAF 9407), and remained there from 21 February to 15 March, loading timber, after which they returned to the Falklands. The name Bougainville has remained in use.

[2] I.e. to use two anchors.

[3] I.e. drag the anchor along the seabed.

[4] The name Río de Gennes remains in use. It was named after Jean-Baptiste de Gennes (d.1705), a French naval officer who sailed to the South Seas from Rio de Janeiro on 5 January 1696, on the first French attempt to sail into the eastern Pacific. He entered the Strait of Magellan on 12 February, but lacking reliable charts, his crew affected by scurvy and reduced to eating rats and wormy biscuit, he was forced to admit defeat after six weeks of struggle and turn back in the neighbourhood of Cabo Gallant. His failure however did not discourage the French and plans for further expeditions were made almost immediately. See Dunmore, *Visions & Realities*, pp. 19–25.

[5] The term 'port', from Latin *portus*, is used in French for sheltered bays and inlets suitable as anchorages and places of call for vessels. It does not imply, as modern English usage tends to, that it contains any docking or other facilities or a town. Other Latin-derived languages similarly use *puerto* or *porto*. Thus Bougainville's 'Port Beaubassin' is now known as Puerto Beaubassin. English navigators often used 'port' for bays that had the potential of developing into a true port: James Cook named 'Port Jackson' the vast and sheltered bay now part of the great Sydney Harbour when he first sighted it, but he called the nearby and less promising harbour Botany Bay; George Vancouver optimistically named 'port' a number of often bleak inlets on the northwest coast of Canada: Port Althorp, Port Fidalgo, Port Houghton. Philip Carteret more modestly called 'English Cove' what Bougainville would shortly after name 'Port Praslin', but he took possession of it in a more grandiose style: 'taking possession of it, Lands, Islands, Bays, Ports and harbours here abouts in the Name of his Majy George ye 3d' (Wallis, *Carteret's Voyage*, I, p. 183).

anchor ahead in 23 fathoms, muddy sand, a kedge anchor almost on the shore, two cables tied to trees on the port side and two on the starboard side tied to the *Étoile* which is similarly moored.

Bearings at the anchorage in Bougainville Bay: the N point ENE5°E, the S one SE4°S, Observatory Islet E2°S, the islet off Nassau Island SE¼E3°S.

We spent the day securing the ships.

Friday morning we started collecting wood for the two ships, which will be quite easily done as it is only a matter of splitting pieces of wood I got felled during my first voyage. As for water, we will obtain our daily needs from this very bay, for the *Boudeuse* in a small river in a bay 2 miles away which I have named Bournand Bay [Baye de Bournand] after La Motte de Bournand, first lieutenant on the *Boudeuse*, and for the *Étoile* in a stream that flows into the small bay to the S of this one. We shall also have to restow everything in the frigate in order to recover our speed, give her a boot-topping[1] and bring up 18 guns.

I was unable to find the hut I had built when we left this bay, nor the various items suitable for the natives which I had decorated by tying a white ensign. They have taken everything away, including the ensign and the clothes, and even the bark covering the hut.[2]

Friday 18 to Saturday 19.

We are continuing with our work. Mr Verron has set up his instruments on this islet which, because of this, I am naming Observatory Islet. Mr de Commerçon is adding to his botanical treasures and daily finds new plants. Mr de Romainville, the engineer,[3] for his part works on correcting errors in previous charts of the Strait of Magellan. On a simple through route, one can only aspire to correcting a few imperfections.

The winds N outside although the clouds have always travelled to W and SW.

Saturday 19 to Sunday 20.

The winds N like yesterday, the clouds from the SW, squally weather with intervals of rain. Continuation of the same tasks. One of my blacks[4] got lost in the woods, several parties went out looking for him and did not find him until 10 at night almost a league inside the forest. The poor man did not know where he was. I had had the

---

[1] I.e. heel the ship over as far as it is practicable and safe to do, first one way and then the other, in order to scrape the weeds, barnacles etc. off the side from the waterline as far down as possible and then to cover the side with tallow and sulphur to reduce the drag of the ship moving through the water. It might also be done by allowing the vessel to sit on the bottom near high water and clean the side as the tide fell.

[2] On 15 March 1765, Bougainville had gone ashore to leave various presents for the natives he had met but who had now gone away. '…he had a small tent erected and covered with tree bark to protect it from the rain and the bad weather. In it he had placed various items of clothing, some kitchen implements, some carpenters' tools, some vermilion of which the savages are very fond, so that they would find it all on their return. A white flag had also been placed next to the said tent.' Michel de Sirandré, lieutenant on the *Aigle*, 'Journal', BN, Dépôt des Cartes, MS 1560 A.

[3] French *ingénieur*: actually *ingénieur-cartographe*, employed to survey coasts and draw charts.

[4] 'The man in question was one of Bougainville's servants. This incident is not mentioned in the *Voyage*' (ET).

drums beaten, the bells rung and a gun fired. This latter sound reverberating on all sides echoing in the nearby mountains gave him no real information about the bay.

Sunday 20 to Monday 21. New moon at 4.26 a.m.

Variable SW to W winds, light breeze. We heeled the *Boudeuse* and gave her a boot-topping.

Monday 21 to Tuesday 22.

I went with Messrs de Bournand and du Bouchage to reconnoitre possible anchorages between Cape Forward and Cape Holland and assess the lie of the coast. The weather was serene and calm when we left and we arrived at a small river which I estimate to be 3½ leagues from Bougainville Bay and 2 good leagues from Cape Forward, that is from its southernmost point. The powerful SW wind forced us to spend the night there, in considerable discomfort, violent wind gusts and rain.

Bearings: being ¼ of a league to the SE of the islet in French Bay, sighted the low headland W of the said bay bearing SW distant ⅓ league; a round headland from W to SW5°S 1 and a half leagues; a low point sandy like the low land running from there to Cape Forward SW¼S2°W 3 leagues.

Tuesday 22 to Wednesday 23.

We returned to the ships on the Wednesday morning, the bad weather continuing; stormy from SW to WNW with almost constant rain.

As the *Étoile* had begun to make as much water from the time of her departure from Montevideo as she did before her partial careening at La Encenada, I took the opportunity provided by our stay to order a new survey. We put her on the bottom so as to raise the bow 12 feet.[1] We believe that we have at last found the leak by the stem; at least since these repairs were carried out she has stayed watertight.

Our restowing is complete, our timber is cut and our watering almost done; we shall be ready in a couple of days, may God grant us fair weather.

Observation of the latitude: according to Mr Verron's observations made on Observatory Islet on the 20th, 21st and 23rd its southern latitude is 53°50′25″. Until now the sky has not allowed him to attempt any observation of the longitude.

Note: On the 22nd of this month, Mr Verron, being on Observatory Islet at 17h23′ after the new moon observed that the tide was high at 1.25 a.m., which places it at the entrance to my bay at 0h59′. The greatest rise and fall is 10 feet.

Wednesday 23 to Thursday 24.

Continuation of the bad weather from W to WNW, high wind, squalls, rain, snow and hail. The mountains have turned white again over the last 2 days. In fact the conjunction of the new moon and the solstice can be the cause of this bad weather. These circumstances arise every 19 years.

Thursday 24 to Friday 25.

Nasty weather from W to SW, high wind, snow and rain, I left this day free for the crew to go strolling ashore and wash.

---

[1] I.e. the ship was beached (probably after adjusting the trim to raise the bow, and probably shortly after high water to ensure than she would float again on the following tide) and when the tide went out the bow was 12 feet clear of the water.

Friday 25 to Saturday 26.

Winds WNW fresh breeze, cloudy sky with rain.

Saturday 26.

Variable winds, weather foggy with rain.

Sunday 27.

I went with the longboat armed with swivel guns and blunderbusses and a boat from the *Étoile* similarly armed to investigate several openings on Tierra del Fuego which could enable us to round Cape Forward.

I left at 4 a.m. and made for the W point of French Bay to effect the crossing. As the boat was making much more rapid progress than the longboat I changed to it when I was roughly in mid-channel. We went ashore by a small watercourse in a cove on Tierra del Fuego, bad even for boats and from where we sighted the W point of French Bay bearing NW¼W 5°W some 5 leagues away. The longboat arrived later. In bad weather, boats could go into the small watercourse at high tide and shelter there.

In the afternoon we started again, rowing along Tierra del Fuego, the wind being a light W gale and the sea extremely choppy. We crossed a wide gulf that seemed to me to be a strait leading to the open sea, of which the E point was some 2 leagues from the small watercourse. The opening of this gulf could be 2 leagues, an island cuts it almost in the middle[1]; abreast of it we sighted the W point of French Bay bearing NW, 4 or 5 leagues distant. The large number of whales we saw in this area confirms that it is a strait.

Being about to land on the other side of the strait, we saw several fires appear and go out; then they remained lit and we saw clearly some natives on the low headland of a bay I intended to visit. I went to speak with the savages whom I recognized as belonging to the same group as those I saw on my first voyage; great displays of friendship between us. The longboat came to join me. These Savages had 6 bark boats in a neighbouring cove. As the sun was about to set, I left the natives to visit the bay and a fine bay that we saw provided an excellent harbour where we spent the night on the bank of an attractive river.

I named the bay and the port Beaubassin [Baye et Port de Beaubassin].[2]

Bearings taken in Beaubassin Bay: being N and S of the narrows and very close to the islet bearing NNW, the W point of the bay called Savages Point [Pointe des Sauvages] NW¼N, the large cape E of the bay NNE. Soundings of the port and bay of Beaubassin:

Soundings of the port: abreast a small hill near the river starboard as one enters: 4 fathoms mud; in front of a small inlet on the same side: 10 fathoms ooze; at the point of the narrows on the port side as you enter very close to shore: 4 fathoms; in the middle of the narrows: 5 fathoms rocky bottom; by the starboard point as you enter: 6 fathoms rocky bottom.

Soundings in the bay: before entering the narrows: 12 fathoms grey sand; between the islet and the narrows at an equal distance: 18 fathoms broken shells; very

---

[1] Dawson Island.

[2] 'Fine Basin'; the name has remained in use as Puerto Beaubassin. It lies south of Greenhough Peninsula (Islota Periagua) on Isla Capitán Aracena.

close to the islet: 37 fathoms small shells; slightly east of the islet: 18 fathoms mud; to the N of the islet and of Savages Point: 24 fathoms gravel.

Note: There are reefs between the islet and Savages Point and there is no passage.

N of the islet close to land: 8 fathoms fine sand and gravel; NE of the islet: 10 fathoms large gravel and shells (when sounding here, the entrance of the haven bears SSE); further from the islet with the entrance bearing S¼SE: 40 fathoms gravel

Estimated a crossing of 5 leagues from Port Beaubassin to the W point of French Bay, course NW¼N.

Monday 28. First quarter at 3.30 a.m.

I sent back the longboat which reached the ship at 5 p.m. and continued my investigations in the boat working along the shore. We saw a native canoe leaving a small island which is approximately 1 league offshore, which I named Canoe Island [Isle à la Pirogue], and making for the mainland. This canoe seemed to be going very swiftly. We went around a fairly large and very high island, which I named Turned Island [Isle Tournée], around which one could anchor.

Bearings: being W of the W point of Turned Island distant ½ league; a point of the mainland which I believe to be Nassau Island N, Cape Forward NW¼N, the most N point of Tierra del Fuego NNE.

Sounded E of the said island near the mainland: 18 fathoms gravel. We dined on Turned Island where there were Savages fishing for mussels.

In the afternoon we continued working along the coast and shortly before sunset we reached a bay which offers an excellent anchorage, which I named Cormandière Bay [Baye de la Cormandière][1] on account of a rock which lies uncovered on the seaward side, and we slept there.

Bearings while at the entrance to Cormorandière Bay:

Cormorandière ESE 1 league, the outer head of Turned Island E, Canoe Island with the point of the bay W¼SW and E¼NE.

Sounded 15 fathoms at the entrance to the bay and 9 fathoms at the anchorage.

Tuesday 29.

At daybreak we left Cormorandière Bay with the tide running strongly in our favour and covered almost 5 leagues; after which we found a fine bay and a superb port which I named Cascade Bay and Port [Baye et Port de la Cascade] on account of a waterfall at the back of the port. We passed between two islands a short distance from the shore, which I named the Two Sisters [Deux Soeurs] bearing NNE and SSW with the middle of Cape Forward about 3 leagues. Cascade Bay lies NE and SW with the said cape and a very remarkable sugarloaf on Tierra del Fuego almost abreast the Two Sisters bearing NNE and SSW of the most S point of the said cape.[2]

Note: An island lies almost in the middle of the bay a little outside.

---

[1] This is a misspelling for Cormorandière (Bougainville corrects this later) and presumably refers to the cormorant-like shape of the rock. It is probably Bahia Mazaredo, the rock being Isla Harrison.

[2] Bougainville erroneously believes himself to be off Tierra del Fuego, as the entry of the 30th indicates, whereas he was in fact exploring the coast of the Isla Capitán Aracena. Bahía Cascada was named on account of the waterfall which, according to other journals, Bougainville intended to name Montmorency Falls after the Montmorency River, north of Quebec, and honouring at the same time one of

Bearings while at the foot of the cascade: the starboard point of the port as you enter by a small islet bears N¼NW4°W; the outside head of the bay N¼NW2°W; the W point of the island in the pass N¼NW; the E point of the said island N7°W; the portside point of the port as you enter N4°W; the starboard point of the small haven N2°E, its portside point N6°E; the portside point of the cove NNE, its starboardside point NE¼E.

Bearings while at the portside of the cove: the small islet W2°N, the starboard point of the port NW¼W3°N.

Bearings on the islet: the W point of the island at the pass N5°W, its E point N2°E by the portside point of the haven.

Bearings on the island at the pass at its E point: the portside point of the haven N3°E, the portside point of the small port E¼SE3°S; the starboard point of the haven ESE.

I climbed the cascade mountain; it is formed by a stream that meanders in the gully of several quite high mountains. The ground is a combination of groves and small plains of spongy moss. Anyhow all this part of Tierra del Fuego from St Barthelémy to Lions and St Elizabeth seems to be nothing but a shapeless group of large, unequal, quite high islands topped with perpetual snow.

I have no doubt that there are a large number of openings to the open sea between them. The trees and plants are the same on Tierra del Fuego as on the coast of the Patagonians and excepting only the trees the land here is similar to that of the Malouines and grows the same plants.

I spent the night in Port Cascade where the water is excellent; bad weather and rain in the afternoon and throughout the night.

Wednesday 30.

SW to W gale with rain. I left Port Cascade at 5.15 and crossed the 2 leagues to the W of Cape Forward, the sea being very rough, then sailed along the coast with the wind aft, very strong gusts. I reached the ships at 10 a.m. I then began to unmoor.

I had a small chart made of the country we have visited which is all the more interesting in that no anchorage was known in that part of Tierra del Fuego. I now advise those who are passing through the Strait of Magellan to endeavour to reach one of the 3 above-mentioned anchorages instead of anchoring on the mainland. They will thus find it easier to round Cape Forward, the S and even the SW winds then becoming favourable with the tide running in their favour.

Our draught corresponds to that required by the builder.

Thursday 31.

NE winds, light breeze with rain. Sailed from Bougainville Bay at 6 with our boats ahead warping us out, then the breeze freshened; steered SE¼E to SSW until 8. At 8 sighted the headland of Nassau Island bearing NNE, the W point of French Bay N¼NE, Cape Forward WSW. At 9.30 sighted Cape Forward bearing NW5°W, the

his protectors, the duc de Montmorency-Luxembourg. The sugarloaf is quite distinctive, still called Pan de Azucar, rising almost 300 m. at the entrance of Bahia Cascada. The Two Sisters have been renamed and undergone a sex change: they are now Dos Hermanos.

Sugarloaf SW5°S, the middle of the Two Sisters SW¼S2°S. Steered SW5°S from 8 till 9, WSW from 9 to 10, W5°S from 10 to 11, W5°N from 11 to 12. At midday sighted the Sugarloaf [Pain de Sucre] bearing ESE2°S, Cape Forward E¼NE, Cape Holland WNW4°W. From midday until 6 steered W¼NW, from 6 to 9 W5°N.

At 9.30 I anchored in Port Gallant [Port Galant] in 16 fathoms coarse gravel, sand and coral, with the following marks: the E point of the bay E¼SE, the E head of the peninsula that closes the port N¼NE5°E, the middle of the island in front of the peninsula N¼NE, the low land of Cape Gallant SW3°W, the E point of Charles Island [Isle Charles] S¼SW. Bad weather with a great deal of rain.

1768
January

Friday 1st.

Strait of Magellan. '*Et nos jam tertia portat omnibus errantes terris et fluctibus aestas.*'[1]

Note: It is easy to miss Port Gallant when one does not sail along very close to it. Two fairly large capes beyond the W point of the Port Gallant roadstead and quite close cause one to overlook the latter. In order not to pass Port Gallant one needs to know that it lies E of the first large cape which can be seen beyond Cape Holland distant 7 to 8 leagues and that Cape Gallant almost touches the W point of the roadstead. Moreover the islet lying E of Charles Island bears SSE and NNW of the W point. Port Gallant Bay is the one named Fortescue Bay [Baye de Fortescue].[2]

WSW, W and WNW winds, squally weather, cold and cloudy with rain. The map of the bay and Port Gallant reproduced in Mr de Gennes's book is most accurate.[3] The details given by his book on the part of the strait that he saw are thin and of little value to a sailor. The best journal on the strait is Sir Narborough's.[4] It is still instructive, in spite of the Abbé Prévost's attempts to edit it down and disfigure it. Frankly, the way in which fine style writers render sailors' journals is pitiful. They would blush at the stupidities and absurdities they make them say, if they had the slightest knowledge of naval terminology. These authors take great care to cut back every detail that has to do with navigation and that could help to guide navigators; they want to make a book that appeals to the silly women of both sexes and end up

---

[1] The quote is from Virgil, *Aeneid* I, 755–6, but amended (the original is '*Nam te iam septima portat omnibus errantes…*') to read: 'And now a third summer has begun, to carry us on our wanderings through all lands and seas'.

[2] Bougainville has now rounded Brunswick Peninsula and entered the long and fairly narrow strait that will lead him to Cabo Pilar and the Pacific Ocean some 150 miles away. A number of inlets and small islands off the continent and Isla Desolación present dangers to navigators which Bougainville will take care to note as he sails through, battling appalling weather in what is after all the beginning of summer in the southern hemisphere.

[3] Froger, *Relation d'un voyage fait en 1695 … par une escadre de vaisseaux du Roy, commandée par M. de Gennes* (1698).

[4] Sir John Narborough sailed from Deptford in 1669 in the hope of sailing into the Pacific. He was unsuccessful in this endeavour but entered the Straits of Magellan in October 1670 and reached Port Galant before being forced to turn back. A narrative of this expedition, *An Account of Several Late Voyages and Discoveries to the South and North*, appeared in 1694. One of the expedition was John Fortescue, after whom Bahía Fortescue is presumably named.

writing a book that every reader finds boring and no one finds of any use.[1] Jean Jacques Rousseau states positively (*Traité de l'égalité des conditions*) that one can ask sailors whether they are men or beasts; he judges them no doubt according to the way in which his colleagues misrepresent these sailors, who might in their turn ask Jean Jacques how he would go about going around the world in 10 years on 60,000 *livres*, as a couple of friends, one being rich and stupid, the other poor and a pretty wit. The list of what has been spent on this interesting voyage should then be posted up in every port to serve as a model of invaluable economy. However, may God save me from going as a third party with these two friends.[2]

Abbé Prévost in the 1st volume of *Histoire des Voyages*, p.260, writes about William Towtson arriving at the Isle of Wight with only an old bonnet hanging from his mainmast to serve as a sail.[3] Bad weather had carried all his sails away and the Englishman had used an old *bonnette* as a topsail. Those poor sailors, the Abbé must have thought, these good people do not know their own language, and as a good academician he turned the old *bonnette* into an old bonnet.[4]

I sent an officer in our longboat to examine the coast of the mainland as far as Elizabeth Bay and the *Étoile*'s boat to survey the Royal Islands [Isles Royales], in order to see whether there was an anchorage between them. The longboat came back in the evening and Chevalier du Bouchage brought back the following bearings taken between Cape Gallant and the anchorage in Elizabeth Bay.

Bearings taken in the longboat: the W point of Fortescue Bay to Cape Gallant SW¼W5°W distant ⅙ of a league, Cape Gallant to the next headland W, E and W 1¼ league, headland 3 to 4 W¼NW and E¼SE 1 league, headland 4 to 5 WNW5°N 1½ league, headland 5 (which is Passage Point [Pointe du Passage]) to the E head of Elizabeth Bay [Baye Elizabeth] WNW5°N ½ a league, E head of Elizabeth Bay with the

---

[1] This form of attack against popular writers and compilers of voyages will recur at intervals in Bougainville's journal, but he will tone them down when writing his own *Voyage* which, somewhat ironically, was a great literary success – and intended to be. The literary cleric Antoine-François Prévost d'Exiles (1697-1763) is better known as the author of the novel which inspired Massenet to compose his opera *Manon* and Puccini his *Manon Lescaut*, but he gained wide popularity in his day for his 20-volume *Histoire générale des voyages, ou nouvelle collection de toutes les relations de voyage par mer et par terre publiées jusqu'à présent dans les différentes langues* (Paris, 1746-69).

[2] Again, Bougainville will omit most of these critical comments from his published *Voyage* and leave out the mention of Rousseau. His attack was directed at the latter's *Discours sur les origines et les fondements de l'inégalité* of 1755, in which he wrote: 'There are really only four types of men who travel long distances, sailors, merchants, soldiers and missionaries. One can hardly expect the first three of these groups to provide good observers' and 'I find it hard to believe that in a century that values highly advanced knowledge, there are not two men who are of one mind, both being rich, the one in money and the other in intelligence, both aspiring to glory and immortality, of whom one would sacrifice twenty thousand *écus* from his fortune and the other ten years of his life to travel around the world in order to study … men and customs.'

[3] The reference is to William Towrson, who, returning from his third voyage to Guinea in 1577, limped back to England with his crew in a state of 'extreme weakness'; after encountering a severe storm he put together some makeshift sails and 'we put an old bonnet to our foreyard, which by the good blessing and providence of God, brought us to the Isle of Wight': Hakluyt, *Principall Navigations*, I, 25, p. 130.

[4] This passage makes no sense until it is realized that 'bonnet' is used in English for the strip of canvas laced to the foot of a sail to increase its size. In French, in contrast, the term *bonnette* used for a studding-sail is distinct from that for a cap (*bonnet*).

W headland WNW 5°W ¾ of a league, the first point W of Cape Gallant with Rupert Island [Isle Rupert] W¼NW 5°W and E¼SE 5°E distant 2 to 2½ leagues.

There is no anchorage off the mainland shore from Port Gallant to Elizabeth Bay.

Bearing between Cape Gallant and Cape Holland E2°S and W2°N and Cape Coventry [Cap Coventry] E and W.

Bearings for the anchorage one needs to select in Elizabeth Bay being in 9 fathoms sand gravel and coral: the E head of the bay SSE 5°E, the W head W¼NW, the E point of Louis le Grand Island [Isle Louis le Grand] SSW 5°S, the reef NW¼N.

The boat could not find any anchorage between the islands, the sea and the tide being against it. A very strong tide was seen to be running in the channel. Two native canoes coming out from the islands came up to the boat. The officer in charge reckons that they belonged to a different group from the Pêcherais.[1] They seemed stronger and a little less wretched. Everything went off amicably.

Saturday 2.

The winds WSW to WNW in gusts with rain preventing us from leaving today, we moored SW and NE with a kedge anchor.

We found in this port fresh traces of the arrival of ships and a stay ashore, recently cut and sawn timber, a few lengths of rope, a short length of wood of the type used in dockyards to affix to ropes and sail-cloth with an inscription one could still make out to be Chatham March 1766, and on the bark of a tree we found carved D T 1767. I am keeping these two items.[2] There is little doubt that the English are setting up a post in the South Sea, possibly on the very coast of Chile in that port where the pink[3] *Anna* put in during Mr Anson's voyage.[4] I have no doubt that the two vessels which put in here were the two frigates which a Spanish ship on her way to Buenos Aires met in the tropics transporting families which these frigates said were intended for Cod Island [Isle de la Morue].[5]

Observation of latitude. Mr Verron, on the point of Port Gallant peninsula, observed at midday today by means of a quadrant 53°40'41" and variation by azimuth of 22°35' NE.

He has not yet been able to carry out any lunar or planetary observations, the nights are as bad as the days. Nevertheless all the shrubs and plants are in flower, the eggs have hatched and the greenery is fairly bright. Such are the effects of spring, but where is the zephyr?

I sent a boat to find out how the tides run in the channel; the officer reported

---

[1] This name was bestowed on the inhabitants of southern Patagonia.

[2] Carteret's *Swallow* was repaired and fitted out at Chatham Dockyard in June 1766. On 19 January 1767 'we anchored a little to the Eastward of Cape Holland', Wallis, *Carteret's Voyage*, I, p. 113.

[3] The name given to a ship with a very narrow stern.

[4] George Anson led several ships into the Pacific Ocean in 1741 intending to raid Spanish settlements and shipping. This and other English challenges to Spanish supremacy in the ocean led to much speculation in France about English intentions, including settlements in the Juan Fernandez Islands, which were uninhabited at the time and on islands of the Chilean coast such as Chiloe.

[5] On 17 November 1766, sailing to the Falkland Islands, Wallis's expedition 'Saw two Strange Sail in the NW Quarter steering about NE. I suppose they are Spaniards from the river of Plate bound home.' Carrington, *The Discovery of Tahiti*, pp. 13-14.

that the rising tide, which runs E, is very strong and that the ebbing tide is hardly felt between the islands and the mainland, while being very strong between the islands and Tierra del Fuego.

We have again found a number of English inscriptions on the trees, all with the date 1767.

Notes. Distances: the latitude observed today together with the bearings taken at Cape Holland on 16 December at the head of Cape Forward and the bearings of the said Cape Holland taken today from here give us a distance of 12 leagues from Cape Forward to the place where the observations were made.

We continue to observe the rising tide running E and the ebb flowing W. It is quite extraordinary to find narratives stating the opposite. I am led to believe that this is not an error in the journals but one made by those who transcribe them.[1]

Sunday 3.

Rain throughout the night and the day, squalls from the W, deplorable weather. I sent a boat to look for an anchoring place on Tierra del Fuego. One was found about 4 leagues and SSW of that bay at the entrance to the appearance of a channel which we believe to be Ste-Barbe Channel [Canal de la Ste-Barbe].[2]

There was an excellent opportunity of determining this bay's longitude by means of an eclipse of the moon which began at 10.30 p.m. The weather conspires against any observation.

Monday 4. Full moon at 4h 27′ 25″ a.m.

Frightful weather during the night and day, rain, snow, wind WSW to WNW gale. I do believe this is the worst climate in the world. It is worthy of its reputation. Mr Duclos and several of our gentlemen climbed the mountain at the back of the port and from its summit they thought they could discern Ste-Barbe Channel bearing WSW. Another may be seen further E with a scattering of islands, which could be the one named Jelouzel.

Bearings taken on the mountain: the starboard entrance of the supposed Ste-Barbe Channel WSW, the port entrance SW¼W, distant 5 to 6 leagues.

Tuesday 5.

Continuation of the worst weather in the world. Strong wind, rain and snow. During the day the winds varied from W to SSW and even S. It is as cold as it would be in Paris in the middle of winter. '*Nix, glando, glacies, spiritus procellarum*'.[3]

Wednesday 6.

It snowed all night. Very cold, WNW to WSW winds. It calmed during the

---

[1] The tidal streams in the Straits of Magellan flow inwards on the rising tide and outwards on the ebb from both ends.

[2] It should be borne in mind that what Bougainville takes to be a continuation of Tierra del Fuego is in fact Isla Santa Ines. At this point, the strait is split into several channels by the presence of islands of various sizes. The French are now exploring Canal Bárbara opposite Bahía Fortescue. It had obtained its original name from the ship *Sainte-Barbe*, Captain Marcand, one of the numerous merchant vessels that sailed into the strait at the beginning of the century.

[3] Taken from Psalms 148:8: 'Fire and hail; snow and vapour; stormy wind, fulfilling his word.'

morning and the day was fairly fine. A 6 p.m. with a breeze from the S and SSE we unmoored and endeavoured to sail. The winds veered back to SW, W and WNW.

We had a visit on board from several native canoes. They belong to the Pêcherais group, small, ugly, knock-kneed and the most wretched of men. However they seem to be nice enough, but they are so weak that one is tempted not to appreciate it. They go naked, wearing nothing more than a scruffy skin that can hardly cover them. This practice of going about naked does not render them immune to the cold from which they seem to suffer a great deal … They always have a fire in their canoes which are made of bark roughly held together by reeds with moss in the seams. They live on shellfish which they eat almost raw although scalding. However they do at times catch game as they have snares made of whale fibrils. I did not see them with any weapons. What would they do with them? There is no hunting in their part of the Strait and they are not at war with anyone: nor do they seem to have any chiefs. They expressed no surprise at the various objects they saw in the ships. That is because in order to be surprised by some craft item one has to have some elementary notion of them. They certainly admire neither the laws of nature nor its phenomena. To be frank, when one sees these Savages, however much one would like to philosophize, one could not express any preference for man in that state of nature over civilized man unless one is looking at them in the way Tacitus presents the Sarmatae: '*securi adversus homines, securi adversus deos, rem difficilem assecuti sunt ut illis ne voto quidem opus sit.*'[1] Being in a condition that challenges the greed of other men and the anger of the gods, they have reached the most difficult point of all: being beyond even the need to express any wishes. I noticed that all these Pêcherais' teeth were bad, whereas the Chaouas have very good ones. The former live on grilled and scalding shellfish, the latter on almost raw flesh. Is it in this difference in their diet that Bourdet will find an explanation for that in their teeth?[2]

Thursday 7.

Dreadful weather night and day. Strong gale, very cold, rain and snow and always the infernal WNW to WSW wind. Réaumur thermometer 2½° above freezing point.[3]

Friday 8.

Strong gale at night, we had to moor a 2nd anchor. Snow, rain, cold like in winter. Endless squalls. The imagination cannot grasp this execrable climate.

Variation observed by azimuth: 22°35′30″ and the inclination towards the austral pole 11°11′.

---

[1] The Sarmatae originated in eastern Scythia, but gradually moved into south-eastern Europe, where the Romans encountered them. 'With no cause for worry from men or gods, they have achieved a thing extremely difficult to attain: they do not even have to resort to prayer': Tacitus, *Germania*, 46.5.

[2] 'Bourdet was a celebrated Parisian dentist. He was dental surgeon to the king, the queen, the king's eldest brother and the Comte d'Artois. Metra mentions him in his *Correspondance secrète*, vol. IV, p. 119' (ET).

[3] Equivalent to approximately 3° centigrade or 37° Fahrenheit.

Saturday 9.

Thermometer 4°. Frightful night, in the morning all the land was snow-covered, it continued to snow during the day. The winds SW to W. It is unbelievable that this is summertime in this country.

On board the *Étoile* they gave some pieces of glass and looking-glass to the natives. Their custom is to hide lumps of talcum at the back of their mouths and in their nostrils. They no doubt wanted to do the same with the glass.[1] A child of 10 or 12 years of age, with an interesting face, became the victim. He swallowed some glass. His tears, the blood clots he was spitting out frightened the natives. Two jugglers wearing a ceremonial hat took him over, jugglers are the Savages' doctors and you will notice that doctors are about the same everywhere. They pressed down on his stomach enough to hurt him with a number of incantations and at least this part of their medicine is neutral. They tore off him a waistcoat we had given him, no doubt considering this as the instrument of the fate that had befallen him. I went ashore with our medical officer to take him some milk and an emollient infusion. The mother and the father's pain, their tears, the keen interest of the entire group made quite evident by their behaviour, the child's patience, his docility in accepting the remedies, created a most moving spectacle. We had to taste the milk and the infusion in their presence before they agreed to make use to it, and even then they did so only reluctantly. The jugglers and the father were sucking the child's bleeding mouth and this practice which we call tending the secret sickness is common among all the Savages I have seen. In addition they were clearly calling upon the divinity, respectfully pointing to the heavens, and speaking to it in a supplicatory tone. In the evening, when I left, the child was in less pain. However his almost constant retching made me fear that the glass had passed into his stomach. Our chaplain baptized him. '*Multi vocati, pauci vero electi*.'[2]

Sunday 10.

Thermometer 4° in the morning, 8° at midday.

The night was less bad, but a stormy gale nevertheless from WNW. I am very much afraid that our child is dead. Two hours after midnight we heard from the ship some repeated howls and the Savages went away at daybreak. They flee from a place death has soiled, they flee from nefarious strangers whom they believe to have come only to destroy them. The heavy seas prevented them from rounding the W point of the bay. When a calmer break occurred they raised their sails, their skins serve this purpose, and a violent squall dispersed their feeble embarkations. How they hurried to get away from us! They abandoned one of their canoes that needed repairing. '*Satis est gentem effugisse nefandam*.'[3] And yet God is my witness to our humane intentions towards these individuals who are part of our species. But I forgive their feelings in these circumstances. What a loss in such a small society is an adolescent who had survived all the hazards of childhood!

[1] In his published narrative, Bougainville speculates that this practice may be talismanic or connected with some belief in the medicinal value of such items: *Voyage*, p. 116.
[2] 'Many are called, but few are chosen:' Matthew, XXII, 14.
[3] Virgil, *Aeneid*, III, 653. 'It is enough to have escaped from a race of evil men.'

Rain, squalls, SW and WNW winds.

Monday 11.

Frightful night, deplorable day, rain, squalls, violent WNW wind. What a sequence of bad weather. Oh ye banks of the Seine, glow of a fine dawn, gentle scent of flowers, charm of greenery, enamel of our grasslands, sparkling network of life-giving dew, bird songs, oh spectacle of smiling Nature, when will you come to refresh our senses saddened by the awful aspect of this land against which its Maker seems to be angry. One cannot live in this horrible climate which is equally shunned by quadrupeds, birds, and fish and where only a handful of Savages live, whose wretchedness has been increased by their dealings with us.

Tuesday 12. Last quarter at 4.14 a.m.

Same weather as yesterday. It rains almost without cease. Another plague is undermining us. Rats are eating almost as much as a 5th of our crews do. The *Étoile* is infected by the same plague.

Wednesday 13.

Bad night, mild day, almost calm, the breeze itself variable up to the NE. In the afternoon we held some hope of being able to sail. Maybe it will be at daybreak tomorrow.

Note: Nearly all these mountains offer the likelihood of marble quarries.

Thursday 14.

Calm during the night: at 2.30 we unmoored and anchored apeak, but we had to moor again at 6 a.m., the wind having returned to W and WNW, fairly fresh breeze. My God, how much patience this navigation requires! Wind and rain squalls.

Friday 15.

The weather was fairly mild, the sun shone almost all day, but the W winds did not let up. We collect wood and water daily for our consumption. Mr de Gennes lost patience on the 14th day of his detention in this roadstead:[1] this is the 16th day we have been held here and imperious necessity holds us under her thumb. Rainy squalls and gusts during the afternoon.

Saturday 16.

Fine weather during the night, almost no wind. At 5.30 a.m., with the breeze coming from the N, I ordered the ships unmoored and we set sail. We steered W¼NW to WNW, the tide ran in our favour until midday, but the winds altered to the NW, W and WSW. We tacked throughout the day. The *Étoile* gained 2 leagues on us with the wind in 3 hours so that at midday she could have put into Dauphine Bay [Baye Dauphine]. We were never able to round Rupert Island. From midday I continued to tack to await the ebbing tide with which I hoped to reach the anchorage in Dauphine Bay; at around 5 o'clock while tacking towards Passage Point the frigate failed twice to go about, no ground, quite close to land towards which the currents

---

[1] De Gennes reached Port Gallant with five ships on 20 March 1696; after a fortnight, his supplies running out and the winds remaining unfavourable, he turned back.

were carrying us with force; as we were bearing away the tide turning made us return to windward and we were still drifting towards the shore. I anchored in 8 fathoms without letting out any cable. The anchor was dragging as the bottom was rocky; then our boats and those of the *Étoile* having gone in front of the frigate to tow her away and the breeze coming from the land, we let out the jibs and topsails, the mizzen and the mizzen top and with the frigate catching the breeze we let go of the cable; I had planned to let all the cable run out, on which a buoy rope had been fixed together with a buoy but when half of it was already out it got jammed in the 'tween-decks, and made the frigate veer into the wind, causing her to run the greatest risks. I ordered the cable cut at once, spilled the sails, and the ship bore away. I left the boats to raise the anchor and returned to Fortescue Bay, anchoring in 20 fathoms green ooze; moored NE and SW. At 10 p.m. our boats returned with the anchor and cable.

I cannot but renew my complaints about the kind of vessel supplied for such a voyage. It could not be less suitable. On her own, the *Étoile* would have cleared the strait long ago; she has over us the advantage that ordinary frigates usually have over other ships. However our draught is precisely what the builder requires. The *Étoile*, a better sailer than we are, offers a quarter more to the wind and drifts a quarter less.[1] This unfortunate frigate cannot change tack without losing a great deal of ground. In truth, we could not have a less effective means of transport; furthermore, requiring a large crew she cannot carry more than three months' food supplies. I do not know what is the point of such vessels, enormous in length, monstrously tumbling home, without depth in the hold and without guns: I would not want to travel or to fight in one. God save us from greater misadventures in this *Boudeuse*.[2]

Bearings at 8 o'clock: the W point of Fortescue Bay bearing E¼NE¼5°N; the W point of Monmouth Island S¼SW; the E point of Charles Island SE¼E; the middle of James Island [Isle Jacques] SW; the middle of the so-called Strait of Ste-Barbe [Détroit de Ste-Barbe] WSW5°S; the E point of Elizabeth Bay NW¼W4°W; the W point of Rupert Island W¼NW2°W.

Between Rupert Island and Passage Point there is a channel not even one league wide and the currents are very strong.

At midday: Cape Gallant bore E 3 leagues; the E point of Elizabeth Bay NW¼N, 1 league; the W point of Rupert Island W¼SW5°S, ⅓ of a league; the E point of Charles Island SE¼E5°E, 2½ leagues. Then sighted a sugarloaf at the entry to the so-called Strait of Ste-Barbe by the E point of Louis le Grand Island bearing W¼SW and E¼NE. The day was fine and the sun warm.

Note: to the NE of the SE point of Louis XIV Island [Isle de Louis XIV], 2 cables from land the bottom is 15 fathoms small gravel; at 3 cables 25 fathoms same ground,

---

[1] I.e. sails a quarter, i.e. a compass point closer to the wind and makes a quarter less leeway. (A point = 11¼°.)

[2] Not surprisingly, Bougainville deleted this passage when he wrote his *Voyage*: overall, his journal contains far more complaints about the slowness of the *Étoile* than about the shortcomings of his own ship. The *Étoile* was a stolid flute intended to carry stores and needing fewer men, the *Boudeuse* was a medium-size naval frigate meant to be more manoeuvrable in difficult conditions and especially in wartime. She was not ideal for the planned work of exploration among unknown islands, as the French officials well knew, but it was hoped that the two ships would complement each other.

and at 4 cables 45 fathoms ditto. This spot can serve an as anchoring place to wait for the tide to turn or to spend a night when the weather is fair, the bottom being too steep to be certain of good holding.

Warning: towards the E point of Elizabeth Bay there is a shoal that can be seen at low tide. There is a pass between the land and this shoal which is covered with seaweed. In general all the shoals in the Strait are covered with these long plants called kelp.

Sunday 17.

The night was fairly good, clear with little wind. The day was stormy with WSW to W winds. During the morning whirlwinds kept on appearing which whipped up the seas to a great height. In the afternoon we dragged our anchor, lowered a large one, and struck the yards and topmasts.[1] Rain and cold.

Note: during our stay in the Strait the Réaumur thermometer stayed usually between 4 and 5 degrees above freezing point. The lowest reading it gave was 3° and the highest 12°, but this latter reading on only two occasions.[2]

Monday 18.

Same WSW to WNW winds with intervals of rain and sunshine. Since the wind had moderated we hoisted the topmasts and sent up the yards again. In the afternoon I sent the boat from the *Étoile* to survey the Ste-Barbe Channel which the extracts from Mr Marcant's journal published by Messrs Frezier and the President de Brosses[3] situate SW and SW¼S of Elizabeth Bay.

Tuesday 19. New moon at 6h 20′45″ a.m.

High wind from the SW to WNW with rain and the usual gusts. I was affected by a severe sore throat for which I was bled three times. Several of our men are also affected. If one did not die of sickness in this cursed climate one would die of impatience and boredom. 'Patience, *forsan et haec olim meminisse juvabit*.'[4]

Wednesday 20.

SW to NW winds with rain and squalls. The boat from the *Étoile* came back at 11 a.m. Mr Landais told me that after following the route and the instructions I had given him he found no opening but only a narrow channel ending in ice floes and

---

[1] I.e. the yards and topmasts were lowered to reduce the top weight, a precaution frequently taken by ships at anchor, and sometimes at sea, experiencing or anticipating strong winds.

[2] In centigrade terms this gives a minimum of just under 4° and a maximum of 15°, or a range of just under 39 to 59° Fahrenheit.

[3] Marcand was the captain of the tartan *Ste-Barbe*, one of a number of French merchant vessels sailing to the South American coast in the early eighteenth century. He discovered the channel to which he gave his ship's name on 25 May 1713. Amédée-François Frézier (1682–1773) was an army officer specializing in defence works who was sent to report on the state of Spanish fortifications and outposts in Chile and Peru; he sailed from France at the end of 1711 and returned five years later. His *Relation du voyage de la Mer du sud aux côtes du Chily et du Pérou* was an instant success and promptly translated into English, German and Dutch. His account of Marcand's discovery appears on pp. 37 and 263 of his *Relation*. On French voyages to South America from 1695 to 1725, see Dunmore, *French Explorers*, I, pp. 10–31, and *Visions & Realities*, pp. 29–35. Isle Louis le Grand, so named by the French after Louis XIV, is present-day Isla Carlos III.

[4] Virgil, *Aeneid* I, 203: 'The day may dawn when this plight shall be sweet to remember.' The word 'patience' is Bougainville's addition.

land. A channel all the more dangerous to take in that there is no worthwhile anchorage and a bank covered in mussels almost bars it in the centre. He rounded Louis le Grand Island by the S and saw no sign of any real channel other than the usual one. Furthermore, by sailing SW and SW¼S on leaving Elizabeth Bay as the above-mentioned authors stated Marcant did, one would cut through Louis le Grand Island. I therefore believe that one must seek Ste-Barbe Channel further W or else that it lies in front of the very bay in which we now are.

Thursday 21.

WSW to NW winds, fairly good weather; since we had again dragged and come much closer to the *Étoile*, we raised our anchors and anchored at a good distance away. In the afternoon the weather spoiled with rain.

Note: I am firmly of the opinion that Ste-Barbe Channel faces Fortescue Bay. Re-reading the passage in Frézier and combining it with the map of the strait that he provides, one can see that Frézier places Elizabeth Bay from where Marcant sailed to enter his channel at 10 or 12 leagues from Cape Forward, and Frézier cannot have failed to question Marcant about the distance which he estimated Elizabeth Bay was from Cape Forward, one of the most noticeable features of the strait. My conclusion is that Marcant mistook Green Bay [Baye Verte] or Cordes Bay [Baye de Cordes] for Elizabeth Bay, the former being 12 leagues from Cape Forward, that sailing from that bay and steering SW and SW¼S he sailed round the eastern point of the Royal Islands which he thought were one and the same, mistaking it for Louis le Grand Island, and entered the channel which several of our gentlemen saw from the mountains at the back of this bay which seemed to them to be an opening with islands scattered within it.

Friday 22.

At 2 p.m., the NNW winds shifted to SSW with such fury that one could not look into the wind. Hail came with this storm which was the wildest I have ever experienced in my life. We had to drop a 2nd anchor and keep our large anchor in a state of readiness. We dragged, as did the *Étoile*. Our mizzen sail tore loose and we had to lower the yard in order to save what was left of it. We raised our 2nd anchor as soon as the weather returned to normal. Throughout the day, stormy gale from WSW to WNW with squalls and rain. We struck our yards and topmasts.

Saturday 23.

In the night SSW gale, during the day stormy SW, WSW and W gale. We took in 10 or 11 quintals of ballast. Occasional rain.

Sunday 24.

Rain and W squalls during the night and part of the morning; fine weather in the afternoon and light S to SE to breeze. This appearance of good weather prevents me from sending a boat to examine the channel that I believe to be Ste-Barbe.

Draught on leaving Port Gallant: 13 feet 11 inches aft, 13 feet 4 inches at the bow.

Monday 25.

At 1.30 a.m. we unmoored and anchored apeak. At 3 we set sail, the winds N almost calm; we left Fortescue Bay, towed by our boats. At 6 o'clock the wind freshened.

Bearings at 7.30: the E point of Rupert Island in line with the E point of Louis le Grand Island SW¼W1°S, the E point of Charles Island SE¼E, Cape Gallant E5°S.

At 8 o'clock: the reef that lies to the E of Elizabeth Bay and the E point of that bay bore N; the W point of Rupert Island S¼SW.

From being abreast of the W point of Fortescue Bay we steered W¼NW until 8 o'clock; from 8 till 9 steered WNW, from 9 till 9.45 NW¼W. We then steered W¼SW to round the elbow forming the S point of St-Jerosme Channel [Canal St-Jerosme],[1] then W and W¼NW until 10.45, then W5°S until midday when we sighted: the E point of Isle Louis le Grand bearing E¼NE distant 2 leagues; the E point of St-Jerosme Strait NE2°E 3 leagues; Cape Quad or Gate bearing W¼SW 2 leagues.[2]

From midday we steered W¼SW until abreast Cape Quad; at that time bearing between the said cape and Split Cape [Cap Fendu] was E5°N and W5°S; Cape Quad and the W point of Louis le Grand Island E¼NE2°E. Cape Quad with the middle of Fontainebleau Bay [Baye de Fontainebleau][3] the entry to which is barred by an island lying across it, E¼NE and W¼SW distant 3 leagues. In general, from Cape Quad to the exit, the two shores display nothing but bare rocks piled over each other, and consequently Narborough called this area Southern Desolation [Désolacion du Sud].[4]

Our track from abreast Cape Quad until 4 o'clock was W5°N, from 4 until 6 WNW, from 6 until 11.30 WNW2°W, in all about 12 leagues.

Note: Without fear of contradiction, the incoming tide flows E and N and the ebb W and S throughout the strait.

Bearings at 4 p.m.: Cape Quad E5°S, distant 5 leagues; the island that bars off Fontainebleau Bay SE¼E 2 leagues; Cape Monday [Cap Monday] W¼NW 8 to 9 leagues; Cape Boudeuse [Cap Boudeuse] W¼NW5°N 6 to 7 leagues; Cape Etoile [Cap Etoile] WNW about 3 leagues; Split Cape NW¼W 1 league. E of this cape lies a bay that seems to be good.

At 6 o'clock: Cape Quad bore E¼SE3°S 6½ leagues; the W point of Split Cape E¼SE3°E 2 leagues; Cape Monday W¼NW3°N 8 leagues; Cape Boudeuse W¼NW5°N 5½ leagues; Cape Etoile WNW 1 league.

At 10.30 we clewed up the foresail.[5] At 11.30 with the winds blowing from the E, stormy gale, rain, very black sky, we hove to with the mizzen topgallant, mizzen, fore topsail and fore staysail, the wind to starboard, lying to NNE to NNW, which gained us during the night about 2½ leagues to NNW.

Table of distances, courses, latitudes and longitudes of the main features of the strait corrected for the variation.

---

[1] Saint Jerome Channel (Canal Gerônimo) or False Strait opens out to the north of Carlos III Island (or Louis-le-Grand Island).

[2] Bougainville is choosing the northern channel, sailing along Penínsul Brunswick, or Crooked Reach. This is leading him past Canal Jerónimo which, if he had unwisely ventured into it, would have led him into Seno Otway and beyond. Instead, he continued past the channel entrance towards Cape Quad (Quod) and, rounding Punto Jerónimo, entered Paso Largo, the final and fairly straightforward route to the Pacific Ocean. His 'Split Cape' mentioned a little later is the prominent Cape Notch.

[3] Apparently Bahía Swallow.

[4] Cf. Narborough, *Account of Several Late Voyages*, p. 78.

[5] I.e. hauled it up to the yard using the brails, in order to furl it.

Latitude of the Cape of Virgins: 52°21', its longitude: 71°25'20".

Longitude of the Cape of the Holy Spirit: 71°48', latitude: 52°44'.

Latitude of Cape Possession: 52°25', longitude of the same: 72°7'28".

Latitude of the eastern entrance to the second narrows: 52°40', its longitude, observed: 73°34'30", estimated: 73°19'20".

Length of the second narrows: 9' W¼SW. From there to the anchorage N of Elizabeth Island: 10' WSW. From there the length of the island: 10' SW. From there to the W point of Duclos Bay: 22' S. From there to St Anne Point 16': S¼SE. From there to Round Cape: 15' S¼SW.

Estimated latitude of Round Cape: 53°45'10".

From Round Cape to Observatory Islet: 12' WSW 2°45'S.

Observed latitude for Observatory Islet: 53°50'25".

From this islet to Cape Forward: 17' WSW 5°S, corrected: 21'30" SW.

Estimated latitude of Cape Forward: 53°58'35", corrected: 54°5'45".

Its estimated longitude: 75°00'10", corrected: 75°1'47".

From Cape Forward to Cape Holland: 15' WNW 4°N.

From Cape Holland to Cape Gallant: 24' WNW 2°N.

Latitude observed at the point of Port Gallant Peninsula: 53°40'41", estimated longitude: 76°1'23", corrected: 76°3'.

From Cape Gallant to N and S of Batchelor River: 15' WNW.

From there to Cape Quad: 16' W¼NW 5°W.

From Cape Quad to Cape Monday: 43' NW¼W.

From Cape Monday to N and S of Cape Pillars [Cap Pillars]: 42' NW¼W, corrected 48'.

Estimated latitude of Cape Pillars: 52°50'50", corrected: 52°54'10", its estimated longitude: 78°55'47", corrected: 79°3'47".

Estimated latitude of Cape Victories [Cap des Victoires]: 52°24'25", corrected: 52°27'45", its estimated longitude: 79°6'7", corrected: 79°14'7".

Note: Sir Narborough had estimated the latitude of Cape Pillars at 52°58'. Entering the Strait from the W, Cape Pillars being, he said, S of the ship distant 1½ leagues, he observed a latitude of 52°51'. From this I deduce the accuracy of my above-mentioned estimation.[1]

Note on the Strait of St Jerosme. When one is abreast the Batchelor River [Rivière de Batchelor], one can no longer see this St Jerosme Strait which runs towards the WNW. However, upon later examining the trending of the coast one can see that at the back it trends NNE. To avoid any error, one needs to keep along Louis le Grand Island, then steer W¼SW, and shortly afterwards one has a good view of Cape Quad.

Tuesday 26. First quarter at 11.45 a.m.

At 3.30 a.m., after signalling to the *Étoile*, I set the sails and my run until 8

---

[1] Narborough gives the latitude of Cape Pillar as 53° 5': *Account of Several Late Voyages*, p.78. However, in the passage quoted (ibid., pp.115-16) Bougainville appears to have mistaken 'about a mile and a half' for 1½ leagues, i.e. 4½ miles. He did not have access to the English edition (although he could read English), and had to use an unreliable translation: see his entry for 1 January.

o'clock was W¼NW3°W and from 8 till midday W¼NW3°N distance approximately 8 to 9 leagues.

Bearings at midday: Cape Pillars W5°S; an islet shaped like a haystack NNW5°W.

The winds then veered to WSW fresh breeze, we sailed NW until 2.30.

Bearings at 2 o'clock: Cape Victories NW 8 leagues; Cape Pillars S3°W 2 leagues; these bearings enabled me to determine the above latitudes and longitudes of Cape Pillars and Cape Victories. From 2.30 until 8 o'clock, we followed various tacks, the winds variable W to WNW then N fresh breeze, all sails set to round the land. Bearings at 6 o'clock: Cape Victories NW, the most W island of the Evangelists [Evangélistes] WNW4°W, Cape Pillars S¼SE3°S.

At 7 o'clock we had rounded Cape Pillars.[1]

Sighted at 8 p.m.: Cape Pillars bearing ENE5°E 3 leagues, Cape Desired [Cap Désiré][2] SE¼E4°E 6 leagues, the large island of the Evangelists NW¼N 4 leagues, Cape Victories NNW4°W 7 or 8 leagues. I am taking my real point of departure from these bearings in accordance with the above estimates in latitude 52°50′ and longitude 79°9′ from Paris.

Note: 'Tandem ad superas evasimus auras.'[3] I advise those who may wish to cross the Strait of Magellan to do so in October and November. That will be, I believe, the period of the quickest crossings. The journal of 3 years [of voyages] to the Malouines and our experience shows that December and January in these waters are the windiest months of the year. Magellan entered on 21 October and came out on 28 November 1520.

From Tuesday 26 at 8 p.m. to Wednesday 27.

I am taking my point of departure from the above bearings on Bellin's map and following that map in latitude 52°58′ and longitude 77°44′ Paris meridian. My true point of departure being as above in latitude 52°50′S, longitude 79°9′ west of Paris.

Stormy gale all night, the topsails reefed, steered WNW, W¼NW and W, the winds N to NNW, fresh gale.

At midday my estimated run since my point of departure is W¼NW5°30′N, distance 60′30″.

Estimated latitude per Bellin: 52°41′30″, est. longitude: 79°19′. Estimated latitude per Véron: 52°32′30″. Estimated longitude per Véron: 80°45′37″.

Note: We are still using a log of 47 feet 6 inches.[4]

---

[1] The expedition had now passed the final extremity of Isla Desolacíon, which is Cabo Pilar, and finally entered the Pacific Ocean. The struggle through the Straits of Magellan had taken 52 days, which as Bougainville notes later was considerably longer than Magellan's own crossing which had taken him 27 days. By comparison, Byron had taken 51 days and Wallis almost four months.

[2] Cabo Deseado, the western extremity of Isla Desolacíon, some 2 km from Cabo Pilar.

[3] From Virgil, *Aeneid*, VI, 126: '… and escaping to the airs above'. The passage from Virgil refers to the Sibyl's instructions to Aeneas to guide him into the underworld to consult his father; she tells him the journey down is no trouble and the door stands open; it is getting back and out into the open air that presents the difficulty. Bougainville adapts the original, adding *tandem* (at length) and replacing the infinitive *evadere* with a past tense.

[4] See 'Editorial Note' on 'Navigational Practices'.

Wednesday 27 to Thursday 28.

Variation corrected 18° NE.

Almost no wind, the breeze from the SW to WSW, we steered NW to NNW. During the morning I had the boat lowered to collect Mr de Romainville, the engineer, who is on board the *Étoile*, to discuss with him the map of the Strait of Magellan. The *Étoile* rigged up her topgallant-sails. Since we put ourselves on the bottom and set up our main mast to rake aft,[1] we have regained our superiority of speed. At midday estimated run is NW 1°N, corrected NNW 5°30′W, distance 24′30″.

Estimated latitude Bellin: 52°22′ observed: 52°14′. Corrected longitude Bellin: 79°44′.

Estimated latitude Veron: 52°13′, observed: 52°14′. Estimated longitude Veron: 81°10′37″.

Note: Dom Georges Juan and Dom Antoine de Ulloa went to the trouble of drawing up a chart of the western shore of southern America on which they show the coast trending SW and NE from Cape Corse [Cap Corse][2] to Chiloé, and this following rumours and guesswork. This correction is little consolation for those who, having emerged from the strait, try to sail north with winds that usually vary from the NW to the SW by W. Fortunately, this correction deserves another. Sir Narborough, who in 1669, emerging from the Strait of Magellan, sailed along the coast of Chile, ferreting in and out of coves and inlets as far as the Baldivia River [Rivière de Baldivia],[3] into which he sailed, states that the course from Cape Desired to Baldivia is N6°46′E. That is much more definite than the assertion of the Spanish gentlemen. Had they been right, we would have come upon the land on the course we have been following.[4]

Thursday 28 to Friday 29.

Variation corrected: 17°NE.

The winds were variable in the afternoon NW to WSW, we sailed close hauled on the port tack until 7 p.m. when the breeze having shifted WNW and then NW we altered tack. At 4 a.m., the wind having altered from NNW to SW, we tacked wind astern and steered NW and NW¼W; cloudy sky with squalls and rain at intervals, fairly heavy seas. From time to time we had to reduce sail in order to wait for the *Étoile*.

At midday our run corrected gives me NW, distance 52′20″.

---

[1] This may refer to the work done in Bougainville Bay, 18-20 December, when they restowed their stores and cleaned the ship's side. The shrouds of the main mast could have been set up again and given the mast a slight tilt aft, which together with the re-stowage of stores could have brought considerable weight aft and improved the trim.

[2] Possibly Cabo Primeiro, at the southern extremity of Península Corso on the north side of the entrance to Golfo Trinidad.

[3] Río de Valdivia.

[4] Jorge Juan y Santacilla (1713-73) sailed in 1740 with Antonio de Ulloa (1716-95) for the Pacific coast of South America. The expedition, which was to last ten years, had scientific and geographical aims, but was also intended to provide information for the Spanish court on the administration and potential of its distant outposts. They wrote together *Noticias secretas de América*, which contains a number of critical comments about local officials and the colonial system itself; the full text was not published until 1846. Ulloa's *Relación histórica del viage a la América meridional* appeared in 1748. Their collaboration continued in later years and they built up a solid international reputation, both becoming members of the London Royal Society.

Estimated latitude: 51°33′, observed: 51°37′.
Longitude Bellin corrected: 80°44′, Veron corrected: 82°9′.

Friday 29 to Saturday 30.
Variation corrected: 17°NE.
The winds SW, good breeze with squalls, rough sea. We steered NW and NW¼W under the 4 main sails, the topsails reefed. At 10 a.m. a young sailor fell overboard. We threw him a [hen] coop and a rope sling and I altered course; all our efforts were in vain, the wind and the sea were too fierce.[1]
Estimated run NNW4°N, distance 103′.
Estimated latitude: 49°59′18″.
Estimated longitude Bellin: 81°35′, estimated Veron: 83°1′37″.

Saturday 30 to Sunday 31.
Variation corrected: 15°NE.
The winds SW to SSW, stormy gale with squalls and rain, very high seas; steered NW¼W then at 10 a.m. NW; under the two courses all the topsails reefed, almost bare poles, clewed up frequently.
At midday estimated run NW¼N1°W, distance 160′.
Estimated latitude: 47°46′33″. Longitude Bellin: 83°52′, Veron: 85°18′37″.

February

Sunday 31 January to Monday 1 February.
Variation corrected: 12°.
SSW to SSE winds, fresh breeze, still a few squalls but the sea moderating; in the morning, let out a reef in the topsails, steered NW and NW¼W.
At midday my run corrected gives me NW¼N3°45′N, distance 200′.
Estimated latitude: 45°13′48″, observed: 44°53′30″.
Longitude Bellin corrected: 86°16′, Veron corrected: 87°41′.

Monday 1 to Tuesday 2.
Full moon the 2nd at 8.45 p.m.
S to SSE winds, fresh breeze, fine weather, fair sea, all sails high, steered NW and NW¼W.
At midday the run corrected gives me NW¼N2°N, distance 171′.
Estimated latitude: 42°36′, observed: 42°27′.
Longitude Bellin corrected: 88°11′, Veron longitude: 89°46′.
Note: for the last fortnight there have been numerous cases of sore throats, as we put them down to the cloud-covered seas of the Strait, we are placing daily two red-hot canon balls in the drinking-tank and a pint of vinegar.

---

[1] He was Jean Chenet, from St Servan. According to Walsh's journal, Section II, the man was washed overboard by a sudden wave that swept over part of the ship. Various other objects were thrown after him, planks, etc., but to no avail. Fesche expressed the view that the man's winter clothing and especially his boots weighed him down and that he had little chance of survival in the heavy seas, 'Journal', f. 112, 29–30 January.

Tuesday 2 to Wednesday 3.

SSE to S winds, fine weather, fair sea, light breeze, all sails set, steered NW¼N.

Run at midday NNW1°30′, distance 99′.

Estimated latitude: 40°58′, observed 40°55′.

Longitude Bellin corrected: 89°9′, Veron corrected: 90°34′.

Variation observed by azimuth: 12°30′ NE.

Wednesday 3 to Thursday 4.

Variable N to WNW winds, light breeze, fair weather, fair sea; sailed close hauled port tack, all sails set.

Run at midday: N¼NE2°50′E, distance 116′.

Estimated latitude: 39°, observed: 39°2′.

Calculated longitude Bellin: 88°33′, longitude Veron: 89°58′.

Corrected variation: 12°NE.

Thursday 4 to Friday 5.

Fair weather, fair sea, little breeze variable from SW¼W to SSW, S, SSE. In the afternoon we furled almost all the sails to wait for the *Étoile* which did not join us until 9 p.m. In the morning we rigged our topgallant-sails and finally got rid of those lugubrious winter bare poles. Steered NW¼N until 8 a.m. when we sailed NW.

Run corrected at midday NNW1°W, distance 89′.

Estimated latitude: 37°52′, observed: 37°40′.

Longitude Bellin corrected: 89°17′30″, Veron corrected: 90°42′30″.

Corrected variation: NE11°15′.

Friday 5 to Saturday 6.

Fair weather, sea of the Tropics, light breeze SSW to SSE, steered NW all sails set. The night was the finest in the world and Mr Verron took advantage of it to make his observations.

Run at midday gives me NW¼N, distance 87′.

Estimated latitude: 36°29′, observed: 36°27′.

Veron calculated longitude: 91°45′.

Observed variation occiduous: 10°21′NE.

Astronomical observations: apparent distances of the north closest to the moon by Virgo's Spica, measured during the night of the 5th to 6th by means of the reflecting octant, run on to midday on the 6th, give for the Paris longitude: 91°14′.

Saturday 6 to Sunday 7.

Variable weather squally with a storm. The winds from the SW to the NW through W, then SW and SSE, good breeze, fairly stormy sea; steered close hauled port tack then NW.

At midday my run gives me N2°25′W, distance 124′.

Estimated latitude: 34°27′, observed: 34°23′.

Calculated longitude: 91°51′.

Corrected variation: 9°50′.

Sunday 7 to Monday 8.

Fine weather, good SSW to SE breeze, we steered NW, all sails set until 8 a.m.

when we sailed W¼NW. I spoke to Mr de La Giraudais and instructed him to sail away to port each morning to a distance appropriate for the appearance of the weather and to come back close to us each evening. In this way we shall cover fairly close to one degree of latitude.

My run at midday gives me NW 5°40′N, distance 171′. Estimated variation: 9°. Estimated latitude: 32°16′, observed: 32°12′. Corrected longitude: 93°59′.

The upper works of the frigate had been overhauled from end to end in the strait. 4 or 5 days of heavy seas since we emerged from it have reopened all the seams and it is all to be done again.

Monday 8 to Tuesday 9.
Fair weather, fine sea, fresh breeze from the S, SSE, SE, it is the real trade wind. Steered NW and NW¼N.

My run at midday give me NW 4°N, distance 113′.

Estimated latitude: 30°51′, observed: 30°47′. Calculated longitude: 95°26′.

Observed variation at 8 a.m. by azimuth: 6°58′NE, observed 8° at 5 p.m., average 7°.

Tuesday 9 to Wednesday 10.
Fair weather, good breeze, all sails high, steered N¼NW and NNW.

At midday my run gives me N¼NW 2°W, distance 150′.

Estimated latitude: 28°25′, observed doubtful: 28°21′. Calculated longitude: 96°6′.

Observed variation ortive: 6°47′.

Wednesday 10 to Thursday 11. Last quarter on the 11th at 0h 37′ a.m.
Fair weather, fine sea, light breeze, the winds unchanged. Steered W and W¼NW. At midday my run gives me W¼NW 3°W, distance 118′.

Estimated latitude: 28°4′, observed: 28°3′. Calculated longitude: 98°19′.

Variation observed at 7.30 a.m. by the azimuth: 6° NE.

Astronomical observation: the observations on the mornings of the 10th and 11th of the distances from the visible rim of the moon to Antares run on to midday on the 11th give a longitude of 98°56′45″.

Thursday 11 to Friday 12.
Fair weather, light breeze, steered W, W¼NW and WNW.

At midday my run gives me W¼NW, distance 94′.

Estimated latitude: 27°48′, observed 27°45′. Calculated longitude: 100°3′.

Observed variation: 4°30′.

Friday 12 to Saturday 13.
Fair weather, light breeze, fair sea, E to NE winds, steered WNW.

At midday my run gives me WNW 3°15′W, distance 97′.

Estimated latitude: 27°24′, observed: 27°29′. Longitude: 101°40′.

Observed variation ortive: 3°40′.

Saturday 13 to Sunday 14.
Same weather, steered WNW and W¼NW all sails set.

At midday my run gives me WNW4°W, distance 80′.

Estimated latitude: 27°1′, observed: 27°. Calculated longitude: 103°6′.

I passed over David's Land [La Terre de David] without seeing it.[1]

Estimated variation: 3°.

The sore throats have gone. We have no men on the sick list and have only 4 sailors with signs of scurvy. May God keep us in that situation.

Sunday 14 to Monday 15.

Slight E to ENE breeze, all sails set.

At midday the course is W4°S, distance 59′.

Observed latitude: 27°4′. Longitude: 104°12′.

This morning we caught about twenty big-ears.[2] These are the first ones.

Monday 15 to Tuesday 16.

Little wind from NE to SSE, steered W¼NW.

My run at midday give me W¼NW15′N, 70′. Heavy swell from the SW.

Estimated latitude: 26°51′, observed: 26°50′. Calculated longitude: 105°28′.
Observed variation occiduous: 3°48′, ortive: 3°NE.

Tuesday 16 to Wednesday 17.

Same weather, steered WNW. My run WNW2°30′N, distance 75′, corrected WNW4°W, distance 72′, heavy swell.

Estimated latitude: 26°18′, corrected: 26°27′. Calculated longitude: 106°44′.

Observed variation ortive: 3°45′, by azimuth: 4°16′.

Wednesday 17 to Thursday 18. New moon on the 18th at 6 a.m.

Same weather, same sea, steered WNW.

At midday my run gave me WNW3°43′N, distance 70′.

Estimated latitude: 25°56′, observed: 25°58′.

Calculated longitude: 107°54′.

Observed variation ortive: 3° and by azimuth: 3°43′.

Thursday 18 to Saturday 20.

Very slight wind E to SW through S then ESE, we steered WNW and NW¼W.

My corrected run for the 48 hours gives me NW¼W3°N, distance 119′.

Estimated latitude: 24°52′, observed: 24°47′. Calculated longitude: 109°40′.

Observed variation occiduous on the 18th: 3°39′NE, occiduous on the 19th: 3°7′.

Since the 14th we have been continuing to catch enough fish to supply at least a third of the crew. This morning we rationed water to one *pot* per man per day of which half is for the boiler.

Since we came out of the strait we have caught more than 200 rats. I am giving 3 *sols* for each one caught.

Saturday 20 to Sunday 21.

Same weather, too fine. There has been little wind for [?] days, steered NW¼W.

[1] This is Davis Land, which was believed by some to lie in latitude 27°30′; the longitude was much more uncertain. On Davis Land, see Introduction, pp. liii–lv.

[2] The reference is to the prominent fins of fish belonging to the *Scombridae* family, the bonito or the larger tunny fish, especially the *Thunnus alalunga*.

At midday my run is NW¼W 3°4'N, distance 88'.

Estimated latitude: 23°59', observed: 23°54'. Calculated longitude: 110°57'.

Observed variation occiduous 3°39'NE, ortive 2°40', by azimuth in the morning 3°4'.

Sunday 21 to Tuesday 23.

Sad little weather, almost calm, great heat, steered NW¼W. Run these 48 hours is NW¼W 1°30'N, distance 59'.

Estimated latitude 23°17', observed 23°20'. Calculated longitude: 111°50'.

Variation observed by azimuth: 3°4'NE, occiduous 3°, ortive 2°53'.

Observations of longitude: Monday 22 at midday I was according to my calculations in longitude 111°30' and Mr Verron following 4 lunar distances,[1] compared with the tables of the *Connaissance des temps*,[2] finds at midday that the ship's longitude is 110° 47' 30".

Tuesday 23 to Wednesday 24.

Almost calm, the slight wind there is comes from the NW to W, we sailed close-hauled on both tacks.

At midday the run corrected is N, distance 22'.

Estimated latitude: 23°2', observed: 22°58'. Longitude: 111°50'.

Wednesday 24 to Thursday 25. First quarter on the 24th at 9 p.m.

Same weather as yesterday, the winds variable from SW to NW and calm. We sailed close-hauled on both tacks.

At midday the run gives me NW¼W 3°N, 25'.

Estimated latitude: 22°43'. Calculated longitude: 112°11'.

Variation observed by azimuth: 2°36'NE.

Thursday 25 to Friday 26.

Same weather, variable winds NW to WSW through W. Steered close-hauled on both tacks.

Compass point[3] corrected: SW¼W 1°45 W, distance 70'.

Estimated latitude: 23°19', observed: 23°23'. Calculated longitude: 113°21'.

Observed variation ortive: 2°39'NE, 2°20' the azimuth.

The bonitos abandoned us yesterday.

Friday 26 to Saturday 27.

Squally NW to WSW winds with rain. Steered close-hauled on both tacks.

Compass point: WSW 5°S, distance 86'.

Estimated latitude: 24°3'. Calculated longitude: 114°44'.

Observation of the longitude: by the occultation of the star E of Gemini to below Alpha Geminorum behind the dark side of the moon on the 26th at 9h45'41" in

---

[1] The calculation of the distance from the moon to the sun, as in this case, or to another star, was used to determine a ship's longitude in the absence of known landmarks; frequently, several calculations would be made to ensure greater accuracy.

[2] This publication, which still exists today, dates back to 1679 and was published under the supervision of the *Bureau des longitudes*.

[3] French *air de vent*, i.e, the compass point in which the wind sits; a variant for the 'run'.

the evening, Mr Verron calculated the longitude of the ship at that time at 112°45′15″, which gives 113°19′ at midday and consequently a difference of 1°25′. The calms and the constant changes of tack must be the cause of it.

Saturday 27 to Sunday 28.

Same NW to W winds with rain and brief calms, steered close-hauled on the 2 tacks.

Compass point: WSW 3°S, distance 25′.
Estimated latitude: 24°13′, observed: 24°14′. Calculated longitude: 115°9′.
Variation observed by azimuth: 2°30′.

Sunday 28 to Monday 29.

And still unfavourable weather, head wind and calms.
Compass point: NW 2°N 18′20″.
Estimated latitude: 23°56′, observed: 23°58′. Calculated longitude: 115°22′.

Monday 29 to Tuesday 1.

Variable WNW, SW, NNE, SE and calms. We steered NW when we could.
Compass point covered NW 1°W, 47′30″.
Estimated latitude: 23°27′, observed: 23°25′. Calculated longitude: 115°59′.
For three weeks I have been carrying out gun practice morning and evening.
Observed variation ortive: 2°30′ NE.

March

Tuesday 1 to Wednesday 2.

Fine weather, light NE to SE breeze, steered NW.
Compass point covered: NW 2°N, distance 48′.
Estimated latitude: 22°47′, observed: 22°51′. Calculated longitude 116°34′.
Observed variation azimuth: 2°40′.

Wednesday 2 to Thursday 3.

Light breeze and calms. Steered NW.
Compass point adjusted: NW, distance 68′.
Estimated latitude: 21°59′, observed: 22°3′. Calculated longitude: 117°26′.
Observed variation azimuth: 2°10′ NE.
Longitude observation: following distances from Regulus to W of the moon and from Spica to the E, measured by the octant on Wednesday 2 9h38′ in the evening true time,[1] Mr Verron deduced the true longitude of the centre of the moon to […] 5 signs[2] 9°2′31″. This longitude gave him at Paris, following the calculations of the *Connoissance des temps*,[3] 17h23′52 which is to say the morning of 3 March at 5h 23′ 52″.

---

[1] I.e. the ship's time.

[2] This probably means the mean of five observations, but it may be his assessment of the accuracy of the observation.

[3] Bougainville normally uses the more modern 'Connaissance', but on this occasion he reverts to the older spelling. The change from *oi* to *ai* in many French words, especially verbs, was gradual but became more generalized in the 18th century under the influence of such writers as Voltaire; the *Académie française* did not make the *ai* compulsory until the early 19th century.

But on the ship it was only 2 March at 9h 38' at night, which gives a difference west in time of 7h45'42" or 116°25'30" W of the Paris meridian for the Wednesday 2 at 9h38' in the evening and for the Thursday at midday of 117°4'30". Difference from my position: 21'30"E.

Thursday 3 to Friday 4. Full moon on the 3rd at 2.24 p.m.

Fair weather, NE to ENE winds, fresh breeze with squalls and rain. We steered NW for 5 hours then NW¼W, having as usual to reduce sail in order to keep in touch with the *Étoile*.

Compass point corrected: NW¼W 1°30'W, distance 116'.

Estimated latitude: 20°53', observed 21°1'. Calculated longitude: 119°10'.

Last evening using charcoal we lit a fire with under the cucurbit[1] to desalinate sea water. When this operation began, the force of the boiling and the lack of a tie caused the cap to snap off, we will start again this evening.

Friday 4 to Saturday 5.

Fair weather, fair sea, fresh breeze, steered 5 hours NW¼W then WNW.

Estimated compass point covered: WNW4°30'N, distance 139'40", corrected: WNW3°N, distance 128'.

Estimated latitude: 19°56', observed 20°6'. Corrected longitude: 121°13'.

We incline to the view that the currents run SE, the differences S observed over several recent days and those of my estimated longitude with Mr Verron's observations support this conjecture. Observed variation by azimuth: 2°20'NE.

The cucurbit under which we had a fire going from 4 p.m. yesterday to 5 this morning produced one barrel and a bucket of water. I am keeping this water which can only improve.

Here is the certificate established by the frigate's surgeon on this matter:

The fourth March 1768.

The fire was lit under the cucurbit at 4 p.m., at 11 in the evening it was filled and distillation continued until 5 a.m. This produced 135 *pots* of water. To obtain this we used a quarter barrel of poor quality charcoal and a quarter of a cord[2] of very porous wood from the Strait.

On the 5th. Fire lit at 5 o'clock, filled at midnight and distillation continued until 8 o'clock. This produced one barrel of water, same consumption of combustible material.

On the 6th. Fire lit at 4.30, filled at midnight, same production and same consumption.

---

[1] The distillation of seawater was obtained by means of a machine known as a 'cucurbit', of somewhat clumsy appearance, being gourd-shaped, whence its name (from *cucurbita*, a gourd). Bougainville had hesitated at first to take on this substantial piece of equipment, but it had been designed by Pierre-Isaac Poissonnier, the Inspector-General of the Navy's hospital service, a highly-regarded medical man and scientist who was also an enthusiastic supporter of Bougainville's plans. As it turned out, the cucurbit soon proved its worth. Having agreed to take on the curcurbit, Bougainville decided to make space on the *Boudeuse* by removing excess adornments, statues, ornamental corbels and gilded carvings along the stern, an impressive but 'useless burden'.

[2] A quantity of wood usually, but not always, equivalent to 128 cubic feet. It was not always a reliable measurement and could vary from one region to another. It is now largely obsolete.

On the 7th. Fire lit at 5 o'clock, filled at midnight, producing 130 *pots* until 8 a.m., same consumption. Distillation stopped until the 12 for lack of coal.

On the 12th. Fire lit at 5 o'clock, distilled until midnight when we stopped because of insufficient sealing. Production was 70 *pots* of water. We took a sack of charcoal, which lasted two days and some wood.

On the 13th. Fire lit at 5 o'clock, filled up at midnight, continued distillation until 8 a.m. the following day, produced 130 *pots*, coal and approximately one third of a cord of wood.

On the 14th. Fire lit at 5 o'clock, filled up and finished distilling at the same hours. We took a sack of coal to light the fire, same production and same consumption.

On the 15th. Ditto. On the 16th. Ditto.

On the 17th. Ditto except that distillation produced 135 *pots*.

On the 18th. Ditto. 130 *pots* of water.

On the 19th. Ditto. Produced 135 *pots*. We gave some wood from Montevideo, which burns and heats better than the Strait wood.

On the 20th. Ditto. Produced 130 *pots*. Took a sack of coal only to light the fire. On the 21st. Ditto.

On the 22nd. Stopped distilling until the 24. Same times, same production.

On the 25th. Ditto, produced 130 *pots*.

On the 26th. Ditto, produced 135 *pots*.

On the 27th. Ditto, produced 130 *pots*.

On the 28th. Ditto, produced 130 *pots*.

On the 29th. Ditto, produced 140 *pots*.

On the 30th. Ditto, produced 145 *pots*.

On the 31st. Ditto, produced 135 *pots*.

On the 1st April. Ditto, produced 140 *pots*.

Mr de Bougainville who had ordered the distillation instructed us to dismantle the cucurbit when we sighted land which we did on that day. One should add to all the distillations at least 5 or 6 *pots* of water that were taken during the night.

Signed Laporte surgeon Master of Arts.

Saturday 5 to Sunday 6.

Fine and good weather, good breeze, steered all sails high WNW. Estimated compass point: WNW 2°N, distance 153′, corrected 129′.

Estimated latitude: 19°3′, observed: 19°13′. Corrected longitude: 123°19′.

Sunday 6 to Monday 7.

Same weather as yesterday, steered W and W¼NW.

Corrected compass point: W6°N, 143′.

Estimated latitude: 18°54′, observed 18°58′. Calculated longitude: 125°49

Variation observed by the azimuth on the Monday morning: 2°58′NE.

Monday 7 to Tuesday 8.

Same weather, E winds, good breeze, steered W.

Compass point: W2°N, distance 140′.

Estimated latitude: 18°51′30″, observed: 18°54′. Calculated longitude: 127°55′.

Observed variation occiduous: 2°30′ NE, by azimuth in the morning: 3°6′.

Tuesday 8 to Wednesday 9.
    Same weather, good E breeze, steered W.
    Corrected compass point: W, 111'.
    Estimated latitude: 18°49', observed: 18°54'. Calculated longitude: 129°52'.
    Observed variation occiduous: 2°34', ortive: 2°32' azimuth 3°.

Wednesday 9 to Thursday 10.
    Same E wind, a slight breeze, steered W.
    Corrected compass point: W1°N, distance 175'.
    Estimated latitude: 18°51', observed: 18°53'30". Calculated longitude: 131°12'.
    Corrected variation occiduous: 2°48', ortive by azimuth: 3°.
    Longitude observation: by means of three distance from Antares to the closest edge of the moon taken by means of the megameter,[1] at midday on the 10th Mr Verron found that the longitude of the ship was 128°6'. Difference from mine 3°6' E.

Thursday 10 to Friday 11.
    Almost calm, light airs from the E, courses W.
    Compass point: W3°N, distance 37'.
    Estimated latitude: 18°51', longitude: 131°51'.
    Observed variation occiduous: 3° NE.
    I went on board the *Étoile*, where I had the satisfaction of finding everyone in the best possible condition. There is not a single case of sickness. I drew from the *Étoile* 12 *couaros*[2] of flour, a quarter of peas and some coal to continue the work of the cucurbit interrupted for the last three days.

Friday 11 to Saturday 12. Last quarter on the 11th at 6h 18' in the evening
    Almost calm, steered W.
    Compass point estimated: W3°N, corrected W¼SW3°W, 50½'.
    Estimated latitude: 18°48', observed: 18°40'. Calculated longitude: 132°44'.
    Observed variation ortive: 2°54'.
    It seems that the currents are beginning to turn.

Saturday 12 to Sunday 13.
    Almost calm, steered W.
    Compass point: W3°45'N, distance 44'.
    Estimated latitude: 18°41'25", observed: 18°43'. Calculated longitude: 133°30'.
    Observed variation occiduous: 3°2', ortive azimuth: 2°57'NE.
    For the last two days, we have been using water from the cucurbit for the soup and the vegetables.

Sunday 13 to Monday 14.
    Calm. Estimated compass point: W3°N, 17', corrected: W¼SW5°S, 16'.
    Estimated latitude: 18°42', observed: 18°48'. Calculated longitude: 133°46'.

---

[1] This was a type of heliometer developed by the naval officer and astronomer Charles-François de Charnière in association with Véron himself

[2] On 13 November Bougainville had bought 60 *cuertos* or leather sacks of flour from the Spanish in Buenos Aires. A *cuerto* held approximately a quintal.

Observed variation occiduous: 3°14′, azimuth ortive: 3°15′.

We are cooking the meat and making bread with sea water, it saves 2 barrels of water a day.

Monday 14 to Tuesday 15.

Calm. Estimated compass point: W 5°N, corrected: W 1°30′S, 25′.

Estimated latitude: 18°46′5″, observed: 18°48′30″. Calculated longitude: 134°12′.

Observed variation occiduous: 2°53′, ortive: 3°.

I again sent to the *Étoile* for 12 *couaros* of flour.

Tuesday 15 to Wednesday 16.

Calm. Compass point: W 4°N, distance 34′.

Estimated and observed latitude: 18°46′. Calculated longitude: 134°47′.

Observed variation occiduous: 3°18′, ortive azimuth: 4°12′ NE.

Wednesday 16 to Thursday 17.

Calm. Compass point W 3°42′, distance: 31′.

Estimated latitude: 18°44′. Calculated longitude: 135°17′.

Observed variation ortive: 3°42′.

Thursday 17 to Friday 18.

Calm with storm and rain. Compass point: W 4°N, 44′30″.

Estimated latitude: 18°41′. Calculated longitude: 136°2′.

Observed variation occiduous: 3°40′.

Friday 18 to Saturday 19. New moon on the 18th at 4 p.m.

Calm with a few squalls, steered W always losing a little ground in order to wait for the *Étoile*.

Compass point: W 3°42′N, distance 51′.

Estimated latitude: 18°37′40″. Calculated longitude: 136°54′.

Observed variation occiduous: 3°33′, ortive: 3°42′.

Saturday 19 to Sunday 20.

Light breeze with a few squalls, steered W.

Compass point estimated: W 3°50′N, distance 70′, corrected: W¼SW 4°S, 70′.

Estimated latitude 18°34′20″, observed: 18°53′. Calculated longitude: 138°.

Observed variation ortive: 3°50′.

Sunday 20 to Monday 21.

Light breeze, steered W and at night W¼NW to compensate for the effects of the currents.

Estimated compass point: W¼NW 2°15′W, distance 95′, corrected W 3°30′N, 94′.

We caught a tunny fish weighing 70 pounds absolutely similar to those of the Mediterranean.

Monday 21 to Tuesday 22. Sight of land. Quiros's archipelago [Archipel de Quiros].

No. 1. Estimated position of the middle of the first 4 sighted: latitude 18°50′, longitude 140°26′37″.

Fresh E breeze. We steered W¼NW until midnight when we set our course W.

At nightfall we reduced to the fore topsail to await the *Étoile* who joined us at 8.30. At 6 a.m. the *Étoile* signalled the presence of 4 islands approximately to the SE and at the same moment we saw one to the W distant approximately 4 and a half leagues. We sighted the islands signalled by the *Étoile* bearing SE¼S, SE and SE5°E, distant 7 to 8 leagues.[1]

We then sailed W¼NW to approach in a circuitous way the one bearing from us W. At 9 a.m. we were N and S of it, distant approximately ¾ of a league. I gave the *Étoile* a signal to take soundings and at almost one and a half leagues from land she let out 200 fathoms of cable without finding bottom. This small island, almost round and completely wooded, is surrounded by sand. It may be a league in diameter.[2] I considered it to be uninhabited and seeing some coconut trees I would have liked to send the boats ashore to get wood, some fruit and possibly some water. A great number of birds flew around the island and on its shore and seemed to promise a coast with a plentiful supply of fish, but the sea broke everywhere and no beach offered anywhere to land. I had consequently ordered the ship to continue on her way, when someone called out that two or three men could be seen running to the edge of the sea. We had all previously noticed a kind of fairly high pillar, isolated on the SE point, and I thought that possibly some Europeans had been shipwrecked on this desert. I hove to at once, determined to make every effort to rescue them. The men had gone back to the woods and shortly after they came out again, approximately 15 in number, carrying very long sticks that they came brandishing along the seashore and after this parade they retired to the trees where, from the *Étoile*, they sighted a few huts. These men seem to be fairly tall. Who will tell me how they have been transported to this place and what links them to other human beings? Have Deucalion and Pirrha's [sic] stones been flung as far as this isolated lump of earth?[3] Whatever the answer may be, these islands are those seen by Quiros in 18°40′ of latitude and which Bellin places one degree further S.[4]

At 6 a.m. we were, according to my reading, in latitude 18°35′ and longitude

---

[1] Bougainville had reached the eastern edge of the Tuamotu archipelago, although in the night he failed to see the two easternmost atolls of Pukarua and Réao. The *Étoile* sighted Vahitahi and soon after the *Boudeuse* saw Akiaki. Both these are discoveries that can be credited to Bougainville: Sharp, *Discovery*, pp. 155, 119. Akiaki he called Lancers Island (Isles des Lanciers), on account of the long spears that he saw the islanders carrying; the four islands that make up Vahitahi he called 'Les Quatre Facardins' after a popular story of the time written by Anthony Hamilton (1646?–1720), the grandson of the Earl of Abercorn and a Jacobite who emigrated to France during the first exile of the Stuarts and finally settled there in 1688. His writings include the *Mémoires du Comte de Gramont*, possibly written in collaboration with his brother-in-law Philibert de Gramont; the short story in question however was published posthumously in 1730.

[2] This is Akiaki, which Cook was to name Thrum Cap. Vahitahi, which he described as 'an Island of about 2 Leagues in circuit and of an Oval form with a Lagoon in the Middle', he called Lagoon Island. Beaglehole, *Journals of Captain James Cook*, I, pp. 69–70.

[3] In Greek mythology, Deucalion, son of Prometheus and husband of Pyrrha, was warned that Zeus intended to destroy the human race by a flood. Having constructed a boat which landed on Mount Parnassus he cast behind him stones from the hillside; those he threw became men, those thrown by Pyrrha became women.

[4] Bougainville is referring to Quiros's 1606 voyage, but the Spaniard in fact sailed further south and Bougainville would cut his track later, near Hao atoll. Quiros had sailed towards Ducie and Henderson islands and was proceeding on a north-westerly course hoping to find land south of the Marquesas. See Kelly, *La Austrialia del Espiritu Santo*, I, pp. 40–42.

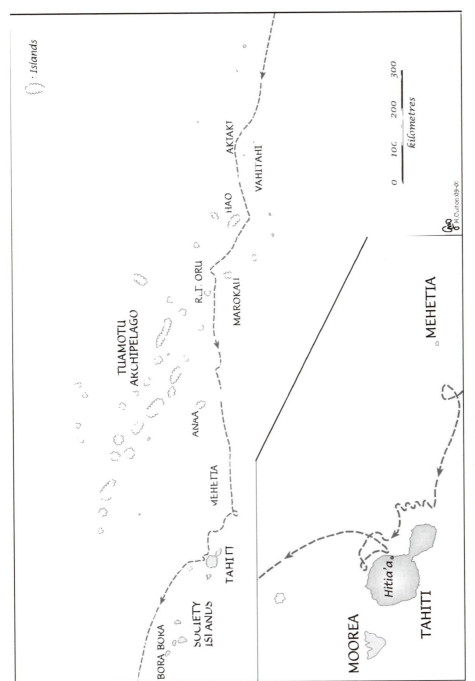

2. Tuamotu Archipelago to Tahiti, 22 March–18 April 1768.

140°41', which places the middle of the 4 islands in latitude 18°50' and longitude 140°25', and the 5th in latitude 18°33'30" and longitude 140°56' (a difference with Bellin of 50' of longitude).[1]

At 9.30 we let out all the sails and steered W. Run for the 24 hours at midday W¼NW2°15'W, distance 90'. Estimated and observed latitude: 18°33'. Calculated longitude: 141°6'.

At midday the island bore from us E5°N, which places it in latitude 18°32 to 33' and longitude 140°57'.

Tuesday 22 to Wednesday 23. Land sighted.

Estimated position of the SE point of this island: latitude 18°17'30" and after correcting by the elevation observed on the 25th: latitude 18°22', longitude 142°9'. Wednesday afternoon, light E breeze, steered W. At nightfall I had the studding sails reefed and we stayed under the main sails. During the night the weather turned to storm, high wind, rain, lightning, thunder: we brought to under the fore staysail and the mizzen until dawn when we saw the land bearing NE¼N distant approximately 4 leagues.

I was then according to my reckoning in latitude 18°27' and longitude 142°16' which would situate the SE point of this island in latitude 18°17'30" and longitude 142°8'.

As the weather was hazy and stormy, I made for the open sea, the winds S and heading SSE until 10 o'clock when the weather having cleared I made for the land.

Bearings at midday the E point NE¼E distant 3 leagues, the W point that seemed to advance the furthest NW 4 to 5 leagues.

Run these 24 hours gives me W6°N, distance 62'.

Estimated latitude: 18°27'. Estimated longitude: 142°11'.

We have caught sufficient bonitos to feed the entire crew. Sounded at one league from the coast, let out 100 fathoms without finding ground.

Wednesday 23 to Thursday 24.

Note: I am convinced that what we took yesterday for a high pillar on the point of the first island was the sail of a canoe hidden by this point.

Note: This land is like a harp the middle part of which would be in the water.[2]

We sailed along the coast throughout the day. The island is low, drowned in its middle, so that it presents only two very narrow stretches of land with small sandy wooded hillocks at intervals with the sea between these stretches. We clearly saw several canoes with sails between the two stretches and Savages on foot in a few places. The waves break all along the beach so as to leave no landing places. The two stretches

---

[1] Vahitahi, or the Quatre Facardins, is situated in latitude 18°43' and longitude 141°13' west of Paris, and Akiaki, or Isles des Lanciers, in latitude 18°30' and longitude 141°34' west.

[2] Accordingly, Bougainville will name it Harp Island (Isle de la Harpe). Cook would call it Bow Island. It is present-day Hao. Its first discoverer was Quiros who came upon it on 10 February 1606 and named it La Conversión de San Pablo: Kelly, *La Austrialia del Espiritu Santo,* I, pp. 40, 42, 55–6, 159n. Its latitude is 18° 10' to 18°30' south, which corresponds to Bougainville's 18° 17'; its longitude is 140°41' to 141°5' west of Greenwich or 143°14' west of Paris, so that at this point Bougainville's estimate of 142°11' was approximately one degree out.

in their larger extent are certainly not 50 paces wide and one can see nothing but stones and sand. The wooded patches are low and only a few coconut trees raise their tops above the other trees; on one part of the stretch of land, slightly wider in the SE of the island, a bunch of coconut trees gave us the impression of having been cultivated by the hand of man. And anyhow, is this quite extraordinary place growing or decaying? How is it peopled? Its inhabitants seem to be tall and well proportioned. A fine field for conjectures.

At 6 o'clock we sounded without finding bottom quite close to the coast; to sail along it we set course NW and NW¼N.

Bearings at sunset: the W point NNW 3 leagues, the most E SE¼E5°E 6 to 7 leagues, the closest to us N5°E about 1 league.

During the night, we followed a SSE course, the winds being E, with the topsails double-reefed and the mizzentopgallant. At 3.30 we tacked towards the land. At day-break sighted the most W land bearing N¼NE distant 4 to 5 leagues and at 7 o'clock bearing NNE about 3 leagues. Having then reached the end of this wretched sterile and inaccessible tongue of land, I reverted to my original course and steered W¼NW, all sails high.

What is this island? Did Quiros see it? It is very much like a great island 30 leagues round drowned in its centre that he describes, but the latitudes do not tally.[1]

At midnight I had 4 rockets fired to which the *Étoile* replied by letting off a similar number. This spectacle will have greatly astonished the islanders.

Sighted at 9.30 the NW end of the island bearing ENE5°E distant 6 leagues. At midday the run for the 24 hours gives me NW¼W distance 44′.

Estimated latitude: 18°2′. Longitude: 142°39′.

Thursday 24 to Friday 25. First quarter on the 25th at 8 a.m.

Sight of land No. 4. Observed variation occiduous: 5°NE.

Weather overcast. Good S to E breeze, steered W¼NW for 1 hr then W until 5 o'clock when, the weather having cleared, we saw land from the mast-head bearing S¼SW as far as NW¼W. We let our more sail to reconnoitre it better. At sunset we could see it from below, a low and wooded land, with intervals, like yesterday's. As the weather was stormy we reefed the topsails and sailed close hauled on the starboard tack, steering N¼NE. At 10 o'clock, let out one reef and ran SSE, then the weather having improved we let out all canvass. At 2 a.m. resumed our starboard tack and steered N¼NW then NW, the winds variable and almost calm until 9 a.m. when the wind settling in ENE, I steered NW and NW¼N to round the island the NW point of which bore from us W¼SW5°S distant about 3 leagues. At dawn we had sighted this said point bearing W5°N distant about 4 leagues.

When we sighted this island at sunset last night, my estimated position was 17°55′ of latitude and 143°20′ of longitude, which would situate the SE head of this

[1] The Spaniards actually landed on the island and were made welcome by the islanders who 'kissed them and embraced them with signs of great joy'. A description of the island, 'flooded within' and 'coconut palms soaring to the skies and intertwined with them some very fresh green trees which formed a truly peaceful bower', is given by Martin de Munilla, in Kelly, *La Austrialia del Espiritu Santo*, I, pp. 160–61. See also Markham, *The Voyages of Pedro Fernandez de Quiros*.

island in latitude 18°12′ and longitude 143°22′ and its NW point in latitude 17°47′ and longitude 143°31′.

Corrected position of the 3rd island, NW point: latitude 18°3′, longitude 143°28′. SE point: latitude 18°28′, longitude 143°25′. This land is drowned almost everywhere and very dangerous to come upon at night. Navigation in this archipelago requires the greatest precautions. From 10 o'clock to midnight I steered WNW. At midday our estimated run is NW4°30′N, distance 46′. Estimated latitude: 17°32′, observed: 17°48′. Longitude: 143°28′.

Note: There is today a fair difference between the octants. Mr Verron has 17°52′, I have 17°50′. Several have 17°44′ and 45′. I am taking the mean which is 48′.

I am not correcting my run to W, not knowing how the currents are running.

At midday sighted the NW point bearing S3°E [distant] 5 leagues, which places it in latitude 18°3′ and longitude 143°28′. This estimate agrees with the position of the said point as determined by our sightings of 6 p.m. and my position at that time as the estimated position we gave above of latitude 17°47′ will become 18°3′ after adjusting the 16′ of difference S given by the midday observation.

Furthermore this land seems to be cut into two islands and the strip of land that joins them may itself be drowned. Bad country, dangerous archipelago.[1]

Friday 25 to Saturday 26.

Observed variation occiduous: 4° 10′.

Fair weather, good ENE breeze, steered W. At one o'clock we saw from the mast-head another island on the port side bearing from us NW¼W which seemed to stretch out towards the NW. Its estimated position is latitude 17°32′ longitude 143°47′. I made for the WNW to come closer to it. An hour later we saw it from below and altered back to W. When one sees this archipelago's islands from the mast-head, one is no further from them than 6 leagues. This island looked like a tongue of land covered with trees, one could see the sea on the other side.[2]

At 2.15 sighted the starboard island bearing S¼SW distant 6 to 7 leagues.

At 4.30 sighted the SE end of the port island bearing N and S, distant 3 leagues.

At 6 o'clock sighted the most NW part of this same island bearing NW to NW5°N distance of approximately 6 leagues.

According to my reckoning at 4.30 and the bearings, the SE end of this port island would lie in approximately 17°32′ of latitude and 143°46′ of longitude.

At 6 o'clock I had the studding sails, topsails and smaller sails furled and taking advantage of the moonlight we sailed W until 1 a.m., having even brailed up the

---

[1] The two islands are the twin atolls of Marokau and Ravahéré, latitude 18°5′, longitude 144°33′. The descriptive expression 'dangerous archipelago' will eventually become a name applied to the entire Tuamotu group, the Dangerous Archipelago. However, in his first use of the term, Bougainville does not capitalize it. Several navigators use names that stress the dangerous nature of these scattered low atolls: Byron named Pukapuka in the northern Cooks 'The Islands of Danger', Roggeween called Takapoto 'Schadelijk' (Disastrous) and Arutua 'Meerder Zorg' (More Trouble). Vivez, the *Étoile*'s surgeon gave the name of 'Stormy Archipelago' to the whole group, *Voyage*, Versailles MS 126, entry of 24 March.

[2] ET points out that if the island was seen to the north-west Bougainville made a slip in stating that it was seen on the port side. This is probably Hikueru, a relatively small island in latitude 17°35′ and longitude 144°55′ west of Paris.

mainsail and the foresail between 10 and 11 in the evening because the wind had freshened.

At 1 o'clock we hove to, with the breeze to starboard, facing N¼NW to NW¼N.

At 5 o'clock I set out with the topsails and at 5.30 with all the sails and steered W. We then saw from the decks a new island to the S, distant about 4 to 4½ leagues, trending roughly E to W, very low in the W part. Sight of land No. 6: estimated position: latitude 17°52'40", longitude 144°15'.

At 7.30, sighted to the SSE distant about 3 leagues, some breakers that appear to be at the SW tip of the said land and be of a considerable extent.

At 8 o'clock, sighted the NE point of this island bearing SE5°S, distance 5 to 6 leagues.

At 9.30, sighted from the mast-head this land bearing ESE5°S, distant 6 leagues.

At the same time we saw from the mast-head to windward starboard side another island to the NW¼N distant 6 to 7 leagues. Sight of land No. 7: estimated position of its middle point: latitude: 17°24'10", longitude, 144°42'35".

At midday my run for the 24 hours gives me W¼NW3°45'W, distance 67', corrected W5°N, ditto.

Estimated latitude: 17°39', observed: 17°42'. Calculated longitude: 144°38'.

Sighted at midday the starboard land which seems to form two islets the one bearing N¼NW, the other N¼NW5°W 6 leagues and the port island ESE, 8 leagues.

The above bearings when compared and rectified by the midday observations give: 1° for the port island: 17°52'40" of latitude and 144°14' of longitude, 2° for the starboard island for the first islet: 17°24' of latitude, 144°40' of longitude, for the second islet: 17°24'20" of latitude, 144°41'30" of longitude.[1]

Saturday 26 to Sunday 27.

Light ENE to NNW breeze, steered W and W¼SW until 5.30 when the *Étoile* signalled land to the SW. The wind having turned stormy and veered ENE we steered WNW. We were unable to see this land from the mast-head.[2] The weather was squally. At 2 o'clock the moon having set, we hove to with the wind to starboard, blowing ESE until 5 when I let the sails draw and set a WSW course to get out of this labyrinth of islands seen by Quiros and Roggewin,[3] where one finds neither water nor refreshments, as it is impracticable to land on most of them.[4]

At midday, the run is W2°20'S, distance 46'.

Estimated and observed latitude: 17°44'. Longitude: 145°26'.

Sunday 27 to Monday 28.

Observed variation occiduous: 5°2', ortive azimuth: 5°42'NE.

[1] Bougainville has now reached the islands of Reitoru and Haraiki.

[2] This was Anaa in 17°25'S and 145°30'W.

[3] Quiros and Roggeveen both sailed through or along the Tuamotu Archipelago, but their navigational details are insufficient to identify some of the islands they saw. Bougainville's comment should be read as referring to the 'labyrinth of islands' as a whole, and not to the islands he has described and of which he was, with the exception of Hao, the first discoverer.

[4] On this day, 26 March, Bougainville drew up an Act of Possession for the ten islands which he then formally named Archipel Dangereux. This document is held in the BN, NAF 9407, f. 139.

Fair weather, good E breeze. I steered WSW to come out of this dangerous labyrinth of islands found by Quiros and Roggewin. We hove to at the setting of the moon until dawn that is to say from 4 to 5 a.m.

Run at midday WSW 4°W, distance 98'.

Estimated latitude: 18°13'20", observed: 18°15'. Longitude: 147°4'.

Observation: at 7.40 p.m. Mr Verron observed 146°33'15". According to my reckoning I would be at that time in longitude 145°54' which places me 39'15" E of the observation.

Monday 28 to Tuesday 29.

Fair weather, good breeze in the afternoon. Steered WSW all sails set until 6 when we stayed under the two topsails on account of the night and to wait for the *Étoile*. At 5 o'clock let out all sails, squally weather, rain and variable winds from NW to N.

Estimated compass point W¼SW 1°45'S, distance 107'.

Estimated latitude: 18°39'30". Calculated longitude: 148°54'.

We are continuing to use the cucurbit. It makes over a barrel of water every night.

Tuesday 29 to Wednesday 30.

Calculated variation: 5° NE.

Fairly good weather in the afternoon; frightful during the night and the following day, storm, squalls, deluge of rain, headwind SW to NW and W. At midnight we hove to, starboard wind until 4.30 a.m., then sailed close hauled, starboard tack under the 4 main sails. This South Sea is not favourable towards us, almost always calm or rain and head wind; look out for scurvy, several of our people are threatened by it.

Estimated compass point WSW 40"W, corrected SW¼W 1°45'W, distance 65'45".

Estimated latitude: 19°4', observed doubtful: 19°14'. Amended longitude: 149°52'.

Wednesday 30 to Thursday 31.

Corrected variation: 5°.

The most horrid weather in the world, day and night. Squalls and deluge of rain, the winds continually in WSW to NW. Beat about under the four main sails, all the topsails reefed, broached to at night. This morning a coconut, still fresh, floated alongside, an indication that the land is not far away.

My run for these 24 hours gives me S¼SE 2°15', distance 12'.

Estimated latitude: 19°26'. Longitude: 149°50'.

April

Thursday 31 to Friday 1 April.

Continuation of the worst weather in the world. This is where the sky is no more than three ells away: squall upon squall and almost constant rain. The wretched sailor cannot get dry and since damp is the most active cause of scurvy[1] we shall soon be infected by it. The commissary's clerk and one of my blacks are very sick with it.

We kept the wind on our starboard side until 6 o'clock when we hove to on the port tack with the foresail and the mizzen, the wind still W to NW.

At dawn let out sail and continued on a N track. At midday my estimated run was NE3°N which corrected excluding our observation of the 30th gives NE¼N45'N distance 35'.

Estimated latitude: 18°57', observed: 18°54'. Longitude: 149°36'.

Friday 1 to Saturday 2. Full moon on the 2nd at 8.05 a.m.

2nd group. Bourbon Archipelago [Archipel de Bourbon]. Land sighted No. 1 in latitude 17°53' and longitude 149°27' W; the middle of No. 2 in latitude 17°25' and longitude 150°21'W.

In the afternoon the sun came out enabling us to clear the decks, dry things out, clean and disinfect. At 5 the rain and the squalls were back and continued all night, the winds variable from S and SW to W and NW. We tacked variously and spent part of the night broached to.

At 10.30 a.m. we saw two extremely high lands, namely (7th) one islet in the shape of an isolated peak, which I am naming the Boudoir [Le Boudoir],[2] bearing NNE5°E distant about 5 leagues, and a larger stretch of land (8th) equally high bearing W¼NW and WNW, distant approximately 14 to 15 leagues.[3] At midday the estimated run is N¼NE1°50'N, having covered 34', corrected N sailed 52'. Estimated latitude: 18°12' observed: 18°2'. Longitude: 149°36'.

Bearing of the large land to windward, the highest hill[4] W¼NW distant 12 to 13 leagues and the Boudoir bearing NE¼E 5 leagues. Which places the mount of the

---

[1] Scurvy will increasingly prey on Bougainville's mind and contribute to his later decision to hasten through the Samoas and the New Hebrides. Some warning signs had appeared as far back as February and were mentioned again in March. The stay in Tahiti, brief though it was, would help to keep the disease at bay, not only because the weather improved but on account of the fresh food the French obtained there. The real causes of scurvy were still unknown, although James Lind's *A Treatise of the Scurvy* had appeared in a French translation in 1756. See on this, Carré, 'L'Expédition de Bougainville et l'hygiène navale de son temps', pp. 73-5.

[2] This is the small island of Mehetia, which rises to 1,427 ft (435 m). It had been discovered by Wallis the previous year and named by him Osnabrug Island. The term *boudoir* today refers to any small private room; it derives from the verb *bouder*, to sulk, and in earlier times was frequently quite separate from the other living quarters; Bougainville chose this name in view of Mehetia's relative isolation from the other islands of the group. It will be noted that the name of his ship, *Boudeuse*, translates as 'the surly one'.

[3] This is the island of Tahiti, some 60 miles east of Mehetia.

[4] One can only hazard a guess here. Sailing from the east, Bougainville would have been approaching Taiarapu Peninsula, the highest point of which is Mt Roniu, 4,370 ft (1,330 m), but he might have been able to see Mt Orohena, 7,342 ft (2,237 m) further west on the main part of the island. Clouds often obscure the high peaks of Tahiti, so that any identification needs to take weather conditions into account: on the previous day Bougainville had complained of poor visibility and there was intermittent rain during the weekend.

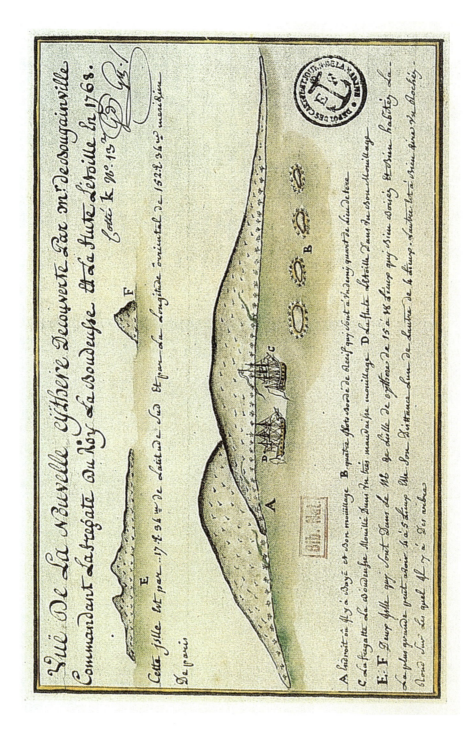

3. New Cythera, coloured drawing believed to be by Charles de Romainville.
Reprinted by permission of the Bibliothèque Nationale, Paris, Service des Cartes et Plans.

windward land in latitude 17°51′30″ and longitude 150°13′ and the Boudoir in latitude 17°53′30″ and longitude 149°26′.[1]

Saturday 2 to Sunday 3. Easter.

Observed variation by azimuth: 5°46′.

Almost calm throughout the afternoon, in the night we steered W, W¼NW even N to reach to windward this high land that seems to promise a place of call suitable for our most urgent needs. At 2 a.m. we stood out to sea for a short tack of two hours after which the calm returned, then a head wind and rain; we sailed close hauled starboard tack, variable NW to N and N¼NE winds.

Run at midday WNW, distance 24′.

Estimated latitude: 17°53′, observed doubtful: 17°53′. Longitude: 150°.

Bearings: at 6 p.m. the Boudoir bore from us ENE4°N, 5 to 6 leagues, the S point of the large land W2°N, distant about 12 leagues. At 8 a.m. the said S point of the large land W¼NW, the peak E¼NE.

At midday the S point of the large land bore W2°N 6 leagues, one further N bore WNW3°W, another point, more advanced, NW¼W5°W.

Sunday 3 to Monday 4.

Observed variation ortive: 4°40′.

Note: There is always a great difference between observation by amplitude and by azimuth. This latter has always given us the greatest variation.

We had variable WNW to N winds. We tacked in accordance with its changes so as to catch the land breeze; weather squally with rain. This seems to be a beginning of a W monsoon that would exist along the land in these parts whereas out at sea there is only the E trade wind. We kept the land in view all night. At around eight in the evening, we saw from the tops a dozen fires lit along the shore. At daybreak we thought we could see a channel that cut through this high land and made it into two islands, but on coming closer we recognized that it was only a wide bay with these low lands at the back.[2]

Run to midday NW¼N5°15′W, distance 26′.

Estimated latitude: 17°37′, observed doubtful from 17°44′ to 17°35′. Longitude: 150°16′.

Bearings: at midday the N point W¼NW, the middle of the bay WSW, the S point SSW5°W about 8 to 9 leagues from the back of the bay.

Monday 4 to Tuesday 5.

Winds N to W, weather fair.

---

[1] 'After sailing from the Strait with an estimated longitude that was too westerly, the *Boudeuse* was affected during her crossing of the South Pacific by the Humboldt Current. During the voyage Véron carried out some ten lunar readings that set fairly good parameters for the dead reckoning. Latitudes are very exact, as is shown by the landfall on the Boudoir, the true position of which is latitude 17°54′S and longitude 150°22′ W of Paris. The dead reckoning placed the *Boudeuse* 56′ too far E. Véron's latest observation, on 27 March, gives a position that was a mere 17′ too far E' (ET).

[2] Bougainville is now in sight of the isthmus of Taravao which joins Tahiti-nui to Taiarapu Peninsula.

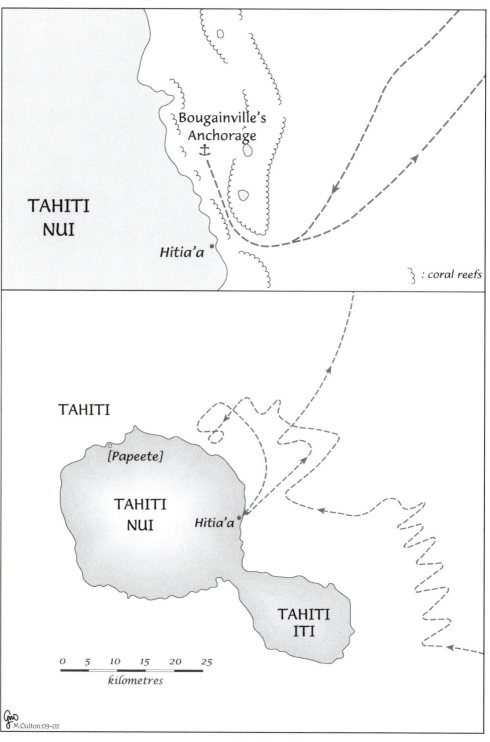

3. Tahiti, 4 April–15 April 1768.

Bearings: at 4 p.m.: the said N point NW5°W, the S point SE about 3 leagues from land.

At 5.45 p.m.: the said N point NW¼N4°N, a low wooded islet by this point NNW4°W, the S point SE¼E4°E 1½ leagues from the N land.

At midnight having sailed N from 6 o'clock: the said N point W5°S, the middle of the bay S¼SW5°W, the S point S¼SE. At 6 a.m.: the said S point SE¼S4°E, another large land to the W bore W5°S.

At 8 o'clock: the most S point SE¼S2°E, the point of the large W land W3°S.

During the afternoon, we stood in for the land. The whole coast rises in an amphitheatre with deep gullies and high mountains. Part of the land seems to be cultivated, the rest is wooded. Along the sea, at the foot of the high country, runs a band of low land, covered with trees and habitations and as a whole this island presents a charming aspect. Over a hundred canoes, of various sizes, but all with outriggers, came around the ships.[1] Several came aboard with demonstrations of friendship, all carrying tree branches, symbols of peace. One of the large canoes, which came up to us with sail and oars, contained a dozen men one of whom was notable for his enormous bristling and spiky hair, presented us with a small pig and several fruit, namely coconuts and bananas. We exchanged these against some small items. We shall describe later these people and what we will have been able to discover of their customs. At 6 o'clock we changed course and tacked variously during the night, during which we gained almost 5 leagues. As soon as night had fallen we saw a cordon of fires laid quite close to each other along the entire coast, fires that seemed to us to be somewhat like oil burning in some kind of vase. We fired off a few rockets in response to this illumination. By and large, this country appears to be very highly populated.

In the morning, same W to N winds, sailed various tacks. The canoes came alongside and brought us some fruit and a few very small red fish. We have to windward a fine waterfall coming down from the mountaintop that makes your mouth water. There is a village at the foot of it, in the shadow of an orchard of trees. The sea does not appear to break along the coast in this part, we shall try to anchor there. At 11 I signalled the *Étoile* to lower her boat.

Run to midday NW¼W4°W, distance 36'.

Estimated latitude: 17°30', observed: 17°30'. Corrected longitude: 150°48'.

Variation observed in the morning by azimuth: 6° NE.

Winds W to N. Fair weather.

Bearings: at 11.30 the most S point bore SE5°S distant 12 leagues, the W point of the bay SSE, the head of the large W land W, distant 12 to 14 leagues, the lowest point close to us W¼SW, the waterfall SW¼W.

Tuesday 5 to Wednesday 6.

I sent the *Étoile*'s boat and my small boat to take soundings leeward and windward of the waterfall. We also took soundings from the ship. The boat found at about ¼ of a league from land a reef with only 4 fathoms of water over it, sandy bottom,

---

[1] The *Voyage* and the journals of those who accompanied Bougainville describe a number of Polynesian canoes, with outriggers and/or sails, single, double-hulled, etc.

between the reef and the coast good sandy bottom with 12, 14 and 20 fathoms; leeward of the reef my boat found a rocky shelf over which the depth of water varies between 9 and 4 fathoms. With the ship, we found 25 and 20 fathoms, rocky ground, then drifting SE we found no ground, then 30 fathoms coral and gravel bottom, then 40, then within a moment no ground after letting out 80 fathoms. I am determined to find a better anchorage by trying to reach a point between the low headland sighted at midday bearing W¼SW and the large land sighted to the W. There would seem to be a large bight that must make a fine bay if the large land is not an island separated from it.

Bearings: At 6 p.m.: the most N point bore W 5°S, the low headland W¼SW; the first point of the fine bay S¼SE, the most S point SE¼S.

A great deal of bartering with the Savages who do not seem to be surprised to see us,[1] and are skilful traders but display good faith. A young and fine-looking young girl came in one of the canoes, almost naked, who showed her vulva in exchange for small nails.[2] We also obtained a few shells, an unknown fruit similar to a pear, a cock, a pigeon, a parakeet and two flutes one with 4 the other with 3 holes.[3] One Indian who seems to be a chief has been on board the *Étoile* since last night and does not want to leave.[4] At nightfall we saw along the coast the same fires as on previous nights. It was fine during the night and almost calm, we followed various tacks both off and on.

In the morning, the wind having freshened, we were able to beat up the coast; we saw that the large land to the W was a detached island, that its lowest N point was drowned and went out a good distance to seaward with a reef and breakers, which decided me to go about and seek an anchorage in the fine bay.

Bearings: (upon arriving to return to the bay) the most N point of the most N

[1] Bougainville is naturally unaware of Samuel Wallis's visit of June 1767. The Englishman's *Dolphin* reached Taiarapu on 19 June and sailed slowly up the west coast in search of a suitable place to anchor. The islanders showed considerable hostility, attacking in large numbers and suffering many casualties when Wallis was forced to fire. Once the superiority of European arms was established, relations became peaceful and indeed quite friendly. Wallis finally sailed on 27 July.

[2] This dry and prosaic comment is far from the famous lyrical passage one will find in his published narrative: 'In spite of all our precautions, one young woman came aboard onto the poop, and stood by one of the hatches above the capstan. [...] The young girl negligently allowed her loincloth to fall to the ground, and appeared to all eyes such as Venus showed herself to the Phrygian shepherd. She had the Goddess's celestial form.' Bougainville, *Voyage*, p. 190. The Prince of Nassau, who had an eye for such matters, does not mention this scene, but refers to an islander who came aboard with his wife who went about 'showing most willingly all the perfections of a fine body' (see below, p. 282). Fesche mentions a woman who came on board 'accompanied by an old man and several of her compatriots' and dropped 'the veil that hid the charms' (below, p. 255). Both state that no one took advantage of her nudity and that she left 'offended'. However, Vivez (below, p. 236) reports that when the *Boudeuse* was about to leave, Bougainville 'opened two portholes opposite the capstan on the starboard side ... in front of the port-holes there were three canoe-loads of women ... [he] signalled to them to display themselves ... the anchor that had been so stubborn was raised immediately'. It is evident that Bougainville constructed the scene out of several incidents when writing his *Voyage* for the wider public, and in so doing helped to create the legend of Tahiti-Cythera.

[3] The flute is the Tahitian *vivo* mentioned by a number of early visitors and reproduced by Sydney Parkinson, in his *A Journal of a Voyage to the South Seas in His Majesty's Ship the Endeavour*, Plates IX and XIII.

[4] Other journals and subsequent comments enable us to identify this individual with Ahutoru, whom Bougainville will subsequently agree to take to France.

island bore WSW 5°W, the S point of the same island WSW 2°S, the low point of the large island WSW 5°S, the N head of the fine bay SSW, the point of the most S island SSE 5°E.

At midday: the N point of the bay [bore] S 3°E, the most S point in view SE 2°E.

Continuation of Wednesday 6.

At 12.30 I sent my small boat and the *Étoile*'s boat to take soundings for an anchorage that looked good to us between a reef and the shore facing a fine river.[1] During this time we backed and filled.

At 1.30 the *Étoile*'s boat gave signal of a good anchorage facing a village and a small river inside the reefs and following a channel that separates them. I ordered a boat to lead the way and we steered a course basing ourselves on the *Étoile*'s boat. At 2 o'clock we dropped anchor in 30 fathoms, bottom of grey sand, shells and gravel, and the *Étoile* anchored a quarter of an hour later. We let out a kedge anchor secured to two cablets[2] and hauled on them to moor further inside the reefs. Moored SE and NW and reefed the lower yards and topmasts.

Bearings of the anchorage: its position: latitude 17°33′ south and longitude 150°36′17″ west of Paris.

Note: The notebook in which Mr Verron had written down the observations made on land for 4 days and 4 nights having been stolen by the Indians,[3] he only had left a few observations made on the final day of the stay ashore according to which, by averaging, he determines the longitude at 150° 23′ 7″ W of Paris. He determined the latitude by using the quadrant at 17°35′3″.[4]

The most S point in view bore SE 5°E, the head of the port reef as one enters SE 5°S; the point of the starboard reef as one enters NNE 5°E; the nearest islet N¼NE 1°N, the middle one N¼NW 3°N; the most NW N¼NW 5°W; two most N point together NNW 3°W; the river mouth SW¼W 4°W; the *Étoile* NW from us 3 cables from land.

After we had secured, I went to examine the watercourse. A crowd of Indians welcomed us on the shore with the most emphatic demonstrations of happiness. Not one carried any arms, not even sticks. The chief of this settlement led us to his home where we all sat down on the ground, they brought fruit, water and dried fish and we had a golden age meal with people who are still living in that happy time. Mr de Susannet having lost one of his pocket pistols, we indicated this to the chief who

---

[1] This is Hitiaa on the east coast of Tahiti-nui, 29 km from Papeete as the crow flies. In 1769 James Cook will anchor at Matavai Bay by Point Venus, some distance north of Hitiaa, on the north coast. The *Instructions nautiques*, 939, Iles de l'Océan Pacifique (Partie centrale et Partie est), 1911, describe the 'Port de Hitiaa' as follows: 'Bougainville anchorage. This anchorage also known as Hitiaa Port is sheltered by a part of the reef that rises slightly above the water at each end of which are the islets of Oputotara and Variararu, low and wooded. One enters the anchorage on the south side by Boudeuse Pass. [....] this anchorage is poorly sheltered and the bottom is littered with coral outcrops.'

[2] A cablet, also called a hawser, is a large rope with a circumference of between 5–10 inches, used for warping and mooring, being easier to handle than the standard heavy cable.

[3] That is, by the islanders. The use of the term Indian to describe a native people encountered west of Europe was fairly widespread and goes back to the days of Christopher Columbus.

[4] The correct position is 17°35′S and 151°38′. Véron's latitude is therefore accurate, but his longitude reflects the difficulties caused by the loss of his notebook and is one degree out.

addressed the people vehemently and the pistol was brought back to the ship the next morning. The cacique[1] took us back to the shore; as we were about to reach it, an Indian lying down under a tree offered us the grassy area on which he was seated, leant towards us and with a gentle expression, accompanied by a three-holed nose flute played by another Indian, he slowly sang us a song that no doubt was Anacreontic; a charming scene worthy of Boucher's brush.[2]

These islanders have various designs imprinted on their skin and the paint used is almost blue.[3] I do not know how they print these ineradicable marks; I think that it is by pricking the skin and pouring over it the sap of certain herbs as is done by the Americans. When Caesar carried out his first incursion into England he found this paint used by the English: '*Omnes vero Britanni se vitro inficiunt quod ceruleum efficit colorem.*'[4]

4 Indians came for supper and slept on board. We played them the flute, cello and violin and put on a fireworks display consisting of a few rockets. Their surprise was mixed with fear.

I shall point out, to the shame of certain Europeans of my acquaintance, that the entire stretch of the reefs surrounding this anchorage is marked out and that at night fires are lit on the islets that form part of these reefs and shown in the above bearings.[5]

Thursday 7

We worked on clearing our water casks and in the afternoon I went to set up a camp on land. The Indians seemed at first pleased to see us, then the cacique came, he held a kind of meeting with the leading men. The outcome was to tell me that we had to go and sleep on board and return in the morning. I explained that we were coming to sleep ashore in order to obtain water and wood, and that for this I needed 18 days after which we would leave. I gave him a number of stones equal to the number of days I expected to stay. Deliberations of the council, they wanted to remove 9 stones, I did not agree. In the end everything was settled, the Indians however still displaying a great deal of mistrust. I had supper in my tent with the cacique and part of his family, each kitchen having supplied its dishes; we did not eat much, His Majesty and

---

[1] Occasionally, Bougainville uses the term 'cacique' or Indian chief, which is not inappropriate in the context of his referring to the islanders as Indians.

[2] Anacreon was a Greek lyric poet of the 6th century BC Anacreontic poems, short lyrical pieces were popular in the 17th and 18th centuries. François Boucher (1703–70) was a French painter of pastoral and mythological scenes in a light rococo style.

[3] Tattooing is widespread in Polynesia. Tahitians might have their entire body tattooed with the exception of the head. The skin is lightly punctured by means of a fine-toothed bone or wooden comb and the colouring matter is then rubbed over the area; the process is slow and can be quite painful. Tattoos usually reflected the individual's rank or status, and of themselves were a badge of courage and endurance.

[4] 'But all the Britons dye themselves with woad because it produces a blue colour:' Caesar, *Commentarii de bello gallico*, V, 14. The ancient Britons painted their bodies with blue woad, the juice of the *Isatis tinctoria*.

[5] The somewhat cryptic allusion contained in these remarks is a dig at the Spanish authorities at the River Plate which was not marked out and lacked any indications that could guide navigators and fishermen, so that shipwrecks were frequent. It was and remains customary for Polynesians to mark out reefs and other dangers, usually by means of long poles inserted into cracks in the coral. That the fires Bougainville saw at night were intended as guides to navigation is more doubtful: they are more likely to have been lights in temporary homes or torches carried by Tahitians fishing at night.

the princes of the blood having an appetite that we could not emulate. After supper, I had 12 rockets fired on land in front of the guests. Their terror was indescribable. During the night I spent in the camp with the soldiers and armed boat in front of the camp, the Indians did not sleep a great deal. There was much coming and going among them, they had some kind of bodyguards situated at intervals. Fear makes them vigilant. Towards the middle of the night, they brought us two pigs. There was a loud altercation between the king, his brothers and people who appear to be their servants, over some opera glasses that had been stolen from my pocket during the supper. The king was accusing his subjects and threatening to kill them; the poor devils were, I think, less responsible for this theft than he was. It all quietened down and the end of the night was more peaceful. During the following day, several Frenchmen had cause to praise the country's customs. As they went into houses, they were presented with young girls, greenery was placed on the ground and with a large number of Indians, men and women, making a circle around them, hospitality was celebrated, while one of the assistants was singing a hymn to happiness accompanied by the sounds of a flute. '*O Venus, hospitibus nam te dare jura loquuntur, hunc loetum Tyri-isque diem Trojaque profectis esse velis nostrosque hujus meminisse minores.*'[1] Dido then offers the delights of her home in the style of this island. Married women are faithful to their husbands, they would pay with their lives any unfaithfulness, but we are offered all the young girls. Our white skin delights them, they express their admiration in this regard in the most expressive manner. Furthermore the race is superb, with men 5 feet 10 inches tall, many reaching six foot, a few exceeding this. Their features are very handsome. They have a fine head of hair which they wear in various ways. Several also have a long beard which they rub as they do their hair with coconut oil.[2] The women are pretty and, something that is due to the climate, their food and the water, men and women and even old men have the finest teeth in the world.[3] These people breathe only rest and sensual pleasures. Venus is the goddess they worship. The mildness of the climate, the beauty of the scenery, the fertility of the soil everywhere watered by rivers and cascades, the pure air unspoiled by even those legions of insects that are the curse of hot countries, everything inspires sensual pleasure. And so I have named it New Cythera [Nouvelle Cythère] and the protection of Minerva is as necessary here as in the ancient Cythera to defend one against the influence both of the climate and of the people's morals.[4]

[1] 'O Venus, for they say it is you who grant rights to those who seek hospitality, may it be your pleasure to make this a happy day for those who set out from Troy and one our descendants will remember:' Virgil, *Aeneid*, I, 731-3.

[2] This is the Tahitian *monoi* consisting of coconut oil to which is added finely grated sandalwood, pollen from the pandanus flower and aromatic herbs.

[3] 'This comment is dated. No traveller could endorse such a comment today. In two hundred years the Tahitian woman has become a great consumer of "gold teeth". Inadequate nutrition, an over dependence on tinned and processed foods and, unbelievable though it may sound, a lack of vitamins have all contributed to dental decay in what is now French Polynesia, and numerous dentists have set up in Papeete. See P. Auget and P. Moorgat, 'Études dentaires dans les E.F.O.' in *Actualités odontostomatologiques*, 37 (Paris, 1957), pp 111-25' (ET).

[4] Minerva, with whom one can associate Athena, was the patroness of craftsmen and artists. Athena, the goddess of wisdom, was often represented as a warrior, with shield and helmet. Cythera was believed to have been the birthplace of Aphrodite, but it was also a centre of purple fishery and known as Porphyry.

I cannot as yet describe their form of government, their differences in rank and their distinctive marks. The cacique, we see this constantly, rules despotically, drives them away with a stick when they bother us, even saw to the return of items stolen from us even though he himself is a great thief, but he wants to be the only one to steal in his kingdom. As for indications of social differences, I believe (and this is not a joke) that the first one, the one that distinguishes free men from slaves, is that free men have their buttocks painted. Then the amount of paint on the buttocks and other parts of the body, the beard and moustaches, the length of the nails, hair hanging down or gathered up over the head, these nuances distinguish, I believe, the various degrees.[1] I shall write things down as I discover them.

Friday 8.

Today I settled into a great hangar on the bank of the river where we obtain our water, the soldiers, the sick from both sips, numbering around fifteen, and a few workmen. The hangar is immense, well built, and well covered by a kind of matting. I left only one exit with a barrier and a sentry. Only the cacique, his family and a few women are allowed in. These people are docile and one of our men, a stick in one hand, can drive them away without difficulty. But one has to keep an eye on everything because they are great thieves, although honest in trade. We are continuing to barter cocks, hens, pigs, fruit, fish, shells, lengths of cloth and other trifles they bring in exchange for nails, artificial pearls, buttons, and other baubles. The crowd around the hangar is enormous. They work willingly in exchange for nails, filling our casks, rolling them, carrying them. So far everything is going fine.

In the afternoon I asked the cacique for firewood. He led me quite close to the river in a plain similar to the Champs-Elysées[2] and gave me 4 large trees. Once these have been cut down I shall ask for more. I also made him understand that I wanted to sow some seeds that were unknown to him. This proposal seemed to please him greatly. I selected a piece of land and began to get it cleared, the cacique himself and his people helping to remove the weeds and stones. They are full of admiration for our gardening tools. I showed him that the land would need to be enclosed and shortly after the Indians brought a fence made of reeds that will simply have to be put up. I expect to sow in a couple of days.

There was a land breeze and a sea breeze; I had our cables checked. They are not scratched. Today we filled 45 water casks. We shall make all the haste we can, because with a strong sea breeze we would not be comfortable here. There is not room to drag for half a cable towards the land

Saturday 9.

We filled 45 water casks and obtained a few cords of wood. Once it was cut I had

---

[1] 'Generally speaking there were three social classes: the *hui arii*, the royal family and its relatives, the *bue raatira*, minor nobles and landowners, and the *manahune*, the ordinary people, with at the lower end of the scale the *teuteu*, the servants, and the *titi*, slaves or former prisoners of war' (ET).

[2] Readers who have visited Paris may find this comparison somewhat strange, but the area was little more than scrubland and marshes until the 1670s when Le Nôtre began to develop the avenue. Even in Bougainville's days, there were fewer than half a dozen buildings in what was to become the fashionable haunt of Paris society.

4. Pastel drawing by an unknown artist, representing Bougainville and his officers with islanders in Tahiti.
Held in the Rex Nan Kivell collection, Canberra, reprinted with the permission of the National Library of Australia, Canberra.

the Indians carry it to the hangar in exchange for a few nails. For the same price, they also help us to fill and roll the barrels. That is the good side. The bad is that they are the cleverest scoundrels in the universe, and that our camp is like the fair at Besons[1] where there is an immense crowd and where there are a few dealers, some sightseers and some rogues. The sentries are busy driving the crowd away and in spite of rounds, guards and precautions, they rob with long hooked sticks through the reeds that form the fence of the hangar, right into the tents erected inside. I am afraid that we might in the end be forced to kill a few as examples to others.

Anyhow many of the women and children are retreating into the mountains with their belongings. The cacique's house is quite empty. Today we saw his packages being taken away with many others. We cannot be too careful. I spoke to the cacique about this removal, and tried to make him understand that we wanted to live on good terms with him and his people. I think he told me that he wanted us to come back and that he would bring us some hens, pigs, wood and water. He marked 8 more trees to be brought down, but he carefully marked them so that their removal would not spoil the plain. He even showed us which side they should fall on.

We found some excellent cress and in large quantities, as well as some alleluia.[2] These two plants are excellent against scurvy. One also finds [here] ginger, indigo and sugar cane. This country is finer and could be wealthier than any of our colonies.

Sunday 10. Last quarter at 8.15.

At midday I observed latitude 17°33′.

Night ashore was somewhat disrupted. The Indians attempted to steal. Some even threw stones. A corporal, struck by one, saw the man, chased him and would have slashed him with his sword had it not been for a banana tree that parried the blow and had its trunk split. I ordered the officer in charge to fire on those who might be throwing stones.

This morning three lords from a neighbouring district came up with a present of cloth.[3] Their size and girth are gigantic. I gave them an axe, some nails and some gimlets. We fired a pistol in their presence, the shot piercing a plank. Their heavy highnesses fell backwards and it took us some time to persuade them to get up again. We had to repay their visit in the afternoon. The chief offered me one of his wives[4], young and fairly pretty, and the whole gathering sang the wedding anthem. What a country! What a people!

An Indian was killed today. I ordered the surgeons to examine the body to see whether he had been killed with a firearm, as the Indians are claiming.[5] A woman, seated by the body, was weeping bitterly and anointing his feet with coconut oil.

---

[1] Bezons is a small town near Paris, famous in the 18th century for its fair.

[2] The alleluia is the *Oxalis acetosella*. It got its name from the fact that in Europe it flowered around Easter.

[3] Bougainville provides additional details on this visit in his *Voyage*, p. 199, giving the name of the most important of these visitors as 'Toutaa'. This enables us to identify him with Tuteha, the chief of Paea, a well-connected personage of considerable influence. He was probably the brother of Queen Purea who had met Wallis the previous year and who would meet Cook in 1769.

[4] The French term 'femme' applying equally to a woman or a wife, only the context can guide the translator. Even then, one is often left guessing in such a case as this one.

[5] He was, as a certificate by the surgeon Laporte testifies. This certificate is inserted in the journal.

About 50 fusiliers sleep ashore and the two longboats come to anchor each night in front of the hangar, armed with swivel guns and blunderbusses. We saw today a superb red dye.[1]

I shall try to find out which plant they obtain it from. These people are very industrious and would soon reach the level of European nations if we brought them our crafts. The longboat collected 72 barrels of water and the Indians are continuing to bring wood as we cut it. We pay them with nails.

Monday 11.

I observed latitude at midday like yesterday: 17°33′.

The night was very quiet on land. A few Indians had remained in the evening hidden in the grasses, the constant sentry-rounds sent them away and only one came towards the middle of the night with a pig he was bringing as a gift and which he took great care to cause to squeal, shouting endlessly Taio which means friend. We are continuing our work on water and wood gathering. Our fat friends from yesterday came back on board at daybreak to bring me some pigs. I gave them an axe, a mirror and a small bell.

The surgeons' report states that the Indian was killed by a shot. There were several people around the body in a hut where they took him and where he lies covered with mats of the kind made locally. I am making and shall continue to make the most careful enquiries to discover the author of this murder. It must be a cowardly murder, since the murderer did not come to tell us: 'I have killed an Indian, I was forced to.' These good people have made no complaint to me about it, they no doubt believe that their compatriot deserved his fate.[2] They are continuing to visit us and to barter. They noticed we liked shells, they are collecting them and bringing them in large numbers. They have seen us pick certain plants, and they bring us similar ones, all for nails and other trinkets. I went for a fairly long walk. Everywhere one comes across groups of men, women and children peacefully seated in the shade of the trees. They seem happy and content. They greet us in a friendly fashion, offer us fruit and those we meet step off the path to let us pass always calling out Taio, Taio. They do not seem to rob each other, nothing can be locked in their huts, everything is hung up or on the ground without locks or guards. Being curious for novelties leads a few of them to steal, anyhow there are scoundrels everywhere and the chiefs tell us to fire on thieves.

Today a great ceremony has been held over a deceased man. The body is almost entirely decayed but I believe that they only abandon bodies after they have turned into skeletons, I do not know whether they then burn them. At least, one does not have to be afraid of being buried alive here as is the case in Paris.[3] Mr Duclos having

---

[1] The Tahitian *mati* was a bright red dye used to colour tapas and other material.

[2] Bougainville contradicts himself in his *Voyage*, p. 199. 'On the contrary he writes: "The locals came to complain about this murder", but adds later: "The islanders believed no doubt that their compatriot was in the wrong, because they continued to come into our quarters with their usual good faith"' (ET).

[3] The body seen by Bougainville presumably belonged to a local chief or a member of his family as ordinary people were normally given a simple burial. The fear of being buried alive in days when medical science was far less advanced than it is today was quite common, and it gave rise to the popular term for an undertaker, a *croque-mort*, from the belief held by many, probably with little justification, that the undertaker would bite (*croquer*, to bite or nibble) a body to ensure that life was indeed extinct.

warned me that the anchorage here is very bad, the bottom being covered with corals and rock, I went on board the *Étoile* in the evening to warn them that I planned to leave in two days' time. We shall have completed our water by them. Coming back, I heard a shot on land, I went over at once. A few rascals had thrown stones at the sentries. A pistol fired into the air drove them all away and no one came during the night.

Mr de Commerçon began collecting botanical specimens this afternoon. He is finding Indian botanical specimens here. Mr Verron has also settled down in the camp with a quadrant, a glass and a pendulum. I hope that time will allow him to take at least one good observation of the longitude.

Tuesday 12 to Thursday 14.

Observed latitude 17° 34′. What we feared has happened. Our kedge anchor cablet has been cut against the bottom, the cable of one of our bower anchors has similarly been cut near the clinch and at dawn we drifted onto the *Étoile* without doing her any more damage than an end of a studding-sail boom broken and on our part the breaking of our main brace.[1] We lowered our large anchor, passed a cable over to the *Étoile*, and took on board the cablet of her kedge anchor which she had on the sea bed and laid out another kedge on a cable of two hawsers secured end to end. We then raised our large anchor which was caught in the second kedge cable; and restored our fore topmast. At the same time I sent off [a boat] to take soundings in the W and see if there was a pass. The *Étoile* also sent us her sheet anchor which we lowered.

As it never rains but it pours, they came to tell me at the same time that because some soldiers had wounded some Indians, alarm had spread everywhere, that old men, women and children were fleeing with their belongings and that we might possibly have to face an army of these furious men. Such was our situation, we had to fear having to fight on land at the same time as our two ships were in danger of drifting ashore. I had myself rowed over to the shore and in front of the cacique and a few other Indians, I had four soldiers, suspected of being responsible for this misdeed, placed in irons. This action seemed to satisfy them.

Night was about to fall. I spent it ashore with 60 fusiliers making constant rounds. A few Indians approached but 7 or 8 shots fired in the air drove them off. One came along at about 9 o'clock to spy on us under the appearances of peaceful intentions and with the gift of a pig. I had him kept back until midnight and then led back beyond the line of the main guards, otherwise he would never have dared to leave.

The night was stormy, rain and wind. At around two a.m. a squall blew up which brought the *Boudeuse* to within one frigate's length of the coast.

I went on board immediately. After the squall, the wind began to blow from the land and this saved us for a moment from the coast. Mr Duclos was on board and I could depend on him for all the manoeuvres needed to save the vessel.

Wednesday 13.

Daylight revealed the extent of our misfortune. The cablet the *Étoile* had given us and that was holding us on her kedge anchor was cut, and so was a new cable

---

[1] A purchase attached to the main yard for trimming it.

secured to the anchor of 2800 she had given us. We had to drop our main anchor without a warping rope and remain with a single cablet and a kedge anchor as sole protection against the coast towards which a very strong breeze was driving us.

The *Étoile* sent us the end of the cablet of her second kedge anchor, which held us a little. We then sent her our longboat to help her to moor solidly, her anchors being fortunately on a bottom that was less bristling with coral heads than the bottom on which ours were. Once at least one of the vessels was safe in this manner, we saved the 2500 anchor of the *Étoile*, secured it to a cablet and sent it away.[1] It was a cruel day in that throughout all this time there was not a single moment when [we] were not afraid of drifting ashore.

However, some representatives I had sent to the Indians, the punishment inflicted on the soldiers in the presence of several of them, possibly fear itself, [which is] the beginning of wisdom, had brought them back to us. From the ship, I saw a crowd of people coming into our quarters; hens, pigs, bunches of bananas adorned this procession and promised peace. Mr de Nassau and a few officers were leading this crowd that they had appeased. I went to land at once with presents of material and iron tools. I distributed these to the chiefs, I expressed to them all my sorrow at the disaster that had occurred and made them understand that it would be punished. These kindly islanders overwhelmed me with signs of friendliness, the people applauded at this reunion and soon the ordinary crowd and the scoundrels were back. The occasion was useful. I saw two Frenchmen chasing Indians, with a stick in their hand. I tore the stick away from them and gave them a few blows: great acclamation on the part of the people whose chiefs ran up to save them from the effects of my resentment.[2] They also drove away with blows of their sticks the crowd that was hindering us and was robbing us although we were sick of it.

Thursday 14.

The night was peaceful ashore and at sea where it was quite calm. The boat I had sent to sound the W pass reported that there was one, I instructed the *Étoile* whose stock of water is complete to get under way and cruise outside. She was in the way of our manoeuvring. At 11 a.m., she was under sail, having wound ship[3] with a mooring rope to our ship. She cut at the splice point the cable of one of her anchors that had been safely lowered and we took it on board, which puts us in a safe position for the night which the lack of light and our various tasks are forcing us to spend here. I also took her longboat with one of her kedge anchors that I had lowered ahead of us on a 260 fathoms warping rope. Her boat came to join us when the ship was out at sea, with the officer I had send to sound the pass. I now hope to save the frigate. We collected three longboat loads of water, cleared all the belongings that were ashore where only the armed troops remained, and raised our main anchor. At night, I went

---

[1] I.e. it was paid out well away from the ship, by boat.

[2] The episode was mostly an act put on by Bougainville to impress the crowd, as it was unlikely that the sailors would have pursued the islanders with sticks unless there had been a theft or an attempted theft. The crowd's appreciation of the scene was probably more merriment at the spectacle than applause at justice being meted out.

[3] I.e. to turn a vessel at rest using lines either to anchors, the land or other stationary objects.

ashore and buried 108 paces from the river where we obtained our water and 13 paces from the edge of the sea the following inscription engraved on a piece of oak:

'In the year 1768, the 12 April, we, L.A. de Bougainville, infantry colonel, captain of the King's ships, commanding his frigates the *Boudeuse* and the *Étoile*, by order and in the name of S.M.T.C. Louis XV,[1] under the ministry of Mr de Choiseul, duc de Praslin, we have taken possession of an archipelago of islands that we have named the Bourbon Archipelago, stretching from the 18th to the 16th degree of latitude south and from the 149th to the 152nd degree of longitude W of the Paris meridian, in witness thereof, in one of the islands of this archipelago, situated in 17°34' of southern latitude and approximately 151° of longitude W, which we have named New Cythera, we have left the present inscription corresponding to the Act of Possession, signed by our officers and chief warrant officers whose names are appended in a bottle.'

This bottle containing the list of the two ships' staff officers and senior warrant officers, well corked and sealed with Spanish wax with the King's arms was buried above the inscription.[2]

Observation: Mr Verron observed ashore in this very place with the quadrant 17°35'3" of latitude.

This ceremony completed, I raised camp and brought everyone back on board. It was midnight and no Indian had approached our quarters since sunset. Our little yawl we used for this embarkation on account of the bar that breaks at full tide was stove in and we had to bring it back in a waterlogged state. Night was quite stormy.

Bearings of the anchorage after our dragging: the N point of the bay NW¼N4°N; the most N islet N¼NW1°W; the middle one N¼NW5°N, the nearest and most S one N¼NE4°E; the point of the starboard reef as you enter ENE; the S point of the bay SE3°E; the point of the portside reef as you enter SE4°S; the back of the bay S2°E; the r[iver] where we obtained water WNW. Such are the bearings according to which it will not be difficult to find the inscription by including the indications mentioned above.

Today the Indians asked us to fire a few shots in their presence. We complied with their request and caused a great deal of fear, all the animals shot at having been killed stone dead. They brought more refreshments than ever and after that the scoundrels had a great time. Apart from thefts, these are the best people in the world. All the women are weeping over our departure and the people are very eager for us to return. We sowed in the presence of the cacique and to the great satisfaction of the people: wheat, maize, beans, peas, lentils and the seeds of various vegetables.[3] Ever since we have been here, the winds have been almost always SE to NE through E. Sometimes there is a noticeable land breeze during the night and early in the day.

Friday 15.

At 5 a.m., we raised our anchor, cut our SE cablet, slipped the others and the

---

[1] The letters stand for 'Sa Majesté Très Chrétienne', 'His Most Christian Majesty', the formal title of the Kings of France.

[2] The parchment copy of this Act of Possession is held at AN, ref. 115 AP 3. There are 37 signatories. The hole in which the bottle was buried was five feet deep, according to Fesche (Martin-Allanic, *Bougainville*, I, p. 680).

[3] There is no evidence that any of these survived beyond germination.

Étoile's anchor cable in order to take immediate advantage of the land breeze that enabled us to emerge by the same pass as the one we entered by. I left my boats and those of the *Étoile* to raise her anchor and the kedge anchors. We could feel satisfied that we had at last got out of this nasty situation, but, no we still had to undergo the hardest trial.

Hardly were we half a league from the reefs when the breeze failed and a perfidious calm came to hand us over bound hand and foot to a violent current driving us towards the breakers. '*In brevia et syrtes urget miserabile visu.*'[1]

We spent almost an hour in this critical situation that was all the more cruel in that if at least the frigate had been wrecked on the shore, we would have had the hope of saving some of the belongings, whereas here even this possibility was missing and one's life would have been saved only [for us] to be condemned to new sufferings, the end of which would have faded in the distance.[2]

Our boats that I had recalled had come back speedily to tow us and they were beginning to gain against the current when a zephyr, a thousand times more charming that all those one celebrates in the opera, came to our rescue, filled our sails and at last we got away from these frightful reefs from which we were no more than a cable's length away. I then sent our boats back to look for our anchors. We tacked to and fro off the coast and joined up with the *Étoile*. At 5 p.m., our boats returned, bringing back the *Étoile*'s anchor and cable, one of our kedge anchors with its cablet and a 3rd cablet. The bad weather, the approach of night and the near exhaustion of the sailors did not allow them to raise up any others.

And so this detestable anchorage has cost us two large anchors, one of which belonged to the *Étoile* and four kedge anchors of which three are the *Étoile*'s but which we were using to moor with. Because I owe it to Mr de La Giraudais to say that without his help we would undeniably have drifted ashore and that his zeal saved us. We have three large anchors left and one kedge anchor, and the *Étoile* has three large anchors of which one is not very reliable.

Furthermore if we had delayed our departure by a mere quarter of an hour, we would have been lost without any resources whether on the reef, to which we discovered ourselves to be much closer when the breeze failed, or on the coast to which a violent squall from the S would have driven us, which forced us to let go of our topsails and would have caught us without moorings since all our anchors were unstocked.

Reflections on the moorings to be chosen here. We lost here 2 large anchors and 4 kedge anchors. If ships were to return here, they should avoid the place where we anchored. I have no doubt that with a survey one would find a good ground at the back of the large bay. Anyhow, by entering here through the pass the *Étoile* used to leave, Mr de La Giraudais assured me that between the two most N islands there is a

---

[1] 'Into shallow water and swirling sandbanks he drives us, a pitiable sight to see:' Virgil, *Aeneid*, I, 227. The 'he' in question is the god of winds.

[2] It is interesting to compare this thought with the actual disaster that befell La Pérouse in the New Hebrides in 1788, when both his ships were wrecked on the outer reef of Vanikoro, very few of the men and their belongings reaching the shore.

very safe anchorage for 30 vessels in from 23 to 12 and 10 fathoms,[1] ooze and grey sand, no tide, one league of searoom and several small rivers facing. To get out of the place where he had moored, this is the route followed by the *Étoile*: sailing N and N¼NW until abreast of the most E island, within a gun shot of it, steered N until abreast of the second island, then steered N¼NW and NNW until the most N island bears SE and SE¼S distant one league; one then steers NE and NE¼N in the offing. From the time of departure until then, one always has a depth of 17 to 30 fathoms, never less than 17.

Shortly after we sailed, the cacique came on board to make his farewells and a present of a large sail, and he brought us one of his Indians, begging us to take him with us and bring him back to him. I accepted this Indian, who may become of the utmost value to the people, and called him Louis.[2] I expressed my thanks to the good cacique and heaped gifts on him. A moment later, one of his canoes came alongside laden with bananas for us. When he left us, he hugged each one of us, weeping bitterly, as did the men and women who were in the canoes around the ship. There was a very pretty girl whom Louis went to embrace, he gave her three pearls he was wearing in his ears, kissed her once more and, in spite of the tears of this young lover, tore himself from her arms and went back on board.

I cannot leave this fortunate island without praising it once more. Nature has placed it in the finest climate in the world, embellished it with the most attractive scenery, enriched it with all her gifts, filled it with handsome, tall and well-built inhabitants. She herself has dictated its laws, they follow them in peace and make up what may be the happiest society on this globe. Lawmakers and philosophers, come and see here all that your imagination has not been able even to dream up. A large population, made up of handsome men and pretty women, living together in abundance and good health, with every indication of the greatest amity, sufficiently aware of what belongs to the one and the other for there to be that degree of difference in rank that is necessary for good order, not knowing enough about it for there to be any rogues or poor people, maintaining good order and works of public needs, i.e. public paths (do people who go about only on foot need any other?), markers for the reefs, fires at night on the islets used for fishing and navigation, plantations of fruit trees, etc. ... Having an elementary knowledge of those crafts that are adequate for men who still live in a state close to nature, working but little, enjoying all the pleasures of society, of dance, music, conversation, indeed of love, the only God to which I believe these people offer any sacrifices.[3] Here blood does not run on the altars or if

---

[1] 'The anchorage to which La Giraudais drew attention is Taipahia, sheltered by the land bridge linking Nansouty and Matupuuru islands. The *Étoile* had used Mahaena Pass, north of this reef' (ET).

[2] This name did not last. The Tahitian, Ahutoru, would take further names by the process, common in Tahiti, of adopting the name of his patrons: Poutavéry (Bougainville), Mayoa (Marion du Fresne). On Ahutoru, see Introduction: 'The Voyage'.

[3] Bougainville's knowledge of Tahitian society and customs was of course quite elementary at this stage, and together with the reactions of his companions his comments will make a large contribution to the Myth of Tahiti and the South Seas. In fact, sacrifices were often made to other gods than Venus: Mars received his tribute in time of war, the corpses of the defeated warriors being paraded around, and there were numerous *marae* or sacred places and altars where human sacrifices were carried out, as James Cook would be able to see for himself; one celebrated drawing by John Webber, 'A Human Sacrifice, in a Morai, Otaheite', is held at the British Museum, Add. MS 15513:16.

sometimes it reddens the altar the young victim is the first to rejoice at having spilt it.[1] There is no question of mysteries or secret ceremonies in its worship: it is celebrated in public and one cannot describe the happiness of these people every time they witness the raptures of an intertwined couple whose sighs are the only offering that is pleasing to the god. Each moment of enjoyment is a festive occasion for this nation.

To describe correctly what we witnessed, one would need Fénelon's pen, to depict it, Albani's or Boucher's brush.[2] This people possesses the gaiety of happiness and this tendency to that light jocularity that rest and joyfulness bring about. Men have several wives and girls all the men they want. We have seen children enjoying equally the care of the father or the mother. This nation's customs are enhanced by the greatest cleanliness. Everyone is endlessly washing and they never eat or drink without cleaning out their mouths before and after. I do not know whether they know war with strangers. I have seen a few with scars that our surgeons admired and which resulted from wounds caused by stones. Their weapons consists of bows, slings and lances made with a hard wood. I saw only one cripple, and even he seemed to have been the victim of a fall, no doubt from a coconut tree. Our senior surgeon assured me that he had seen on several of them traces of smallpox; I took every precaution to ensure that we did not pass on to them the other [pox].[3]

Do they have a religion? Or not? I saw no temple, no external sign of worship, those that we made in their presence neither impressed them nor interested them. In the leaders' houses one finds two large wooden figures, one for each gender. In order to find out whether these were idols, we knelt before them, then spat on them, stepped on them, these actions, each on so different, attracted in equal degree the laughter of the watching Indians.

As far as the dead are concerned, there is no doubt that they have a rite, to which they solemnly adhere and for which there are public ministers. I do not know whether they bury the dead. They expose them on stands under hangars almost to the point where they have turned into skeletons. They then remove them into their house and the *eatua* or minister of the dead comes to carry out his revolting task.[4] Up to this point, I am only stating what I saw, I put forward as probable or doubtful what seems to be and I do not forget Cicero's maxim: '*ne incognita pro cognitis habeamus his que temere assentiamus*'.[5]

I have spoken of their canoes, of their fishhooks, of their nets and seines absolutely similar to ours, of their ropes, of the materials of different colours, of their mats, of their sails. In general, what they make is very artistically fashioned and reveals

[1] A reference to the deflowering of a virgin.

[2] François de Salignac de la Mothe-Fénelon (1651–1715), theologian, author of elegant works on education and social questions; Francesco Albani (1578–1660), painter of frescoes on mythological subjects; François Boucher (1703–70), court painter since 1765.

[3] On the subject of venereal diseases in the islands, see the Introduction: 'The Voyage'.

[4] A *tahua*, or priest, of which there were several categories.

[5] 'Let us not treat as knowledge things about which we are ignorant and lend our assent to them.' From *De Officiis*, 1.6.18: the wording of the original is actually '*ne incognita pro cognitis habeamus iisque temere assentiamur*'. Cicero's *De Officiis* was known as 'Tully's Offices' and widely read in the eighteenth century as a textbook on ethics.

the greatest aptitude for all the crafts. What I believe to be their most valuable manufacture is a red dye that seemed to us to be superior even to that of the Gobelins.[1] I saw no sign of any minerals. Our chaplain assured me that he had examined a piece of some red mineral, but I do not know what has become of it. One would certainly find some pearls, as we obtained some pearl oysters.[2] Indeed it is to be wished for the sake of the inhabitants that Nature had refused them items that attract the cupidity of Europeans. All they need are the fruits which the soil liberally grants them without any cultivation, anything else, which would attract us, would bring upon them all the evils of the iron age. Farewell happy and wise people, may you always remain what you are. I shall never recall without a sense of delight the brief time I spent among you and, as long as I live, I shall celebrate the happy island of Cythera. It is the true Utopia.

Copy of the Act taking possession of the Island of Cythera.

In the year 1768, we L.A. de Bougainville etc... having, with the two above-mentioned vessels, travelled at the end of March in the above year through an archipelago of low islands almost drowned in their middle, that stretches from 18°50′ and 17°24′ of southern latitude and 140°25′ and 144°40′ of longitude west of the Paris meridian, we did not stop to survey these islands, because although they are inhabited they did not seem to us to be likely to be cultivatable. After sailing away, we discovered on 1 April and the following days an archipelago of high islands we named Bourbon Archipelago which lies between 18 and 16° of southern latitude and 149° to 152° of longitude west of the Paris meridian. We stayed to examine them and drop anchor in the largest of these islands at the end of a wide bay in the following bearings (bearings as shown 3 pages earlier). We spent several days establishing links with the local people, who are numerous, very tall, strong, industrious and gentle. We even, with their consent, set up a camp ashore where we obtained the water and wood the ships needed and bartered by means of nails, tools, glass items and other trinkets obtaining hens, pigs and various fruits. This island, the shore of which rises in an amphitheatre with large gullies and tall mountains, seems partly cultivated, the rest is wooded. Along the sea, at the foot of the high country, extends a stretch of low land covered with trees and habitations, everywhere watered by waterfalls and rivers. The products we saw consist of coconuts, bananas, yams, sugar cane, various other fruits and vegetables, a kind of wild indigo, various other fruits and vegetables [sic] and black, yellow and red dye. With the approval of the local cacique, we cleared a small patch of land, in which, to the great satisfaction of these people, we sowed wheat, maize, beans, peas, lentils and the seeds of various other French vegetables. The soil, [which is] everywhere remarkable, would grow in abundance everything that is entrusted to it and cultivation would be all the easier in that we saw none of the species of insects that plague the other colonies. On 12 April we took possession of the archipelago in the name of S.M.T.C. and, in the presence of several of our officers, we buried in this island the following inscription engraved on an oak plank (the inscription as is appears 4 pages back). This inscription together with the bottle have been buried 13

---

[1] The Gobelins were a family of dyers established in Paris, whose founder Jehan (d. 1476) discovered a new scarlet dyestuff. In 1662, Louis XIV's minister Colbert regrouped the various Paris dyers and tapestry-makers and purchased the Gobelins' premises. The factory has acquired a worldwide reputation.

[2] The pearl industry has become a major export earner in present-day Tahiti.

paces from the edge of the sea and 108 paces from the river where we obtained our water. This information, together with the above-mentioned bearings, is enough to enable them to be found, if the need arises. On 15 April, we sailed from this island, the inhabitants expressing a great regret at seeing us leave and the cacique came aboard to make his farewells and entrust us with one of his Indians whom he made us understand we should bring back.

In witness thereof, we have drawn up the present Act, which we have signed and had our officers and warrant officers from both ships to sign, as is required by our instructions, to be used as may be thought proper.

Made and signed on board the royal frigate the *Boudeuse*, the 15 day of April 1768.

At midday the S point bore from us SE4°S, the N point WNW5°W, the back of the bay S¼SW.

At 5.30 p.m., the islet off the N point of the bay bore NW5°W, the N point of the bay NW¼W, the middle of the back S¼SW, the S point of the bay SE2°S.

During the day the weather was stormy, a few squalls, rain and calms.

From 7 p.m. to 11 o'clock we covered 5 miles NE¼N2°E, and from 11 to midnight 3 miles NE¼NE. At midnight we hove to in order to hoist our boats. This operation was lengthy, the entire crew being stressed by the enforced labour of these last four days. We then tacked variously until 8 a.m. when we sighted the N point of the bay bearing SW¼S distant 10 leagues, from where I am taking my point of departure in latitude 17°6'30" and longitude 150°18'.

From 8 until midday steered NNW and at midday [...].

Estimated latitude: 16°52', observed: 16°54'. Calculated longitude: 150°25'.

Sighted other islands. Saw land No. 3 in latitude 17°4', longitude 150°40' by its centre. The Islands of the Allies [Isles des Alliés].[1]

At 10 a.m., sighted land to leeward which had the appearance of three islands: the most N bore W¼NW3°N 6 to 7 leagues, the most S, W¼SW3°W, 6 leagues, the middle one W2°S, the most W point of Cythera Island SSW3°S, the most E bearing S¼SE.

At midday these three islands seemed to form only one, the most S point of which bore SW¼S 5 to 6 leagues, the most N bore WSW5°S 4 to 5 leagues, on could see above it a more distant land. As soon as our Indian saw this island, he made us understand by the clearest of gestures that he had been there, that he had had mistresses there and that we would find as many refreshments as at Cythera which he calls *Enoua Taiti*, enoua meaning land.

The centre of this island would lie in latitude 17°4' and longitude 150°40'. Louis names it Oumaitia.[2]

---

[1] These words introduce the next islands seen, part of the Society Islands (Isles de la Société) whose inhabitants were allied to those of Tahiti, information the French obtained from Ahutoru.

[2] The three islands sighted earlier have now become 'this island'. Bougainville indicates that after leaving Hitiaa he sailed largely north, whereupon he sighted the island bearing west. The latitude given corresponds to that of Tetiaroa in 17°7', an atoll encompassing several low wooded islets. It was sparsely and indeed infrequently populated. Ahutoru's naming is not helpful: 'Oumaitia' has led to confusion over the years. Thus Martin-Allanic (*Bougainville*, I, p. 726) identified it with Ulieta or James Cook's Raiatea, which he believed Bougainville could see in the distance. Sharp did not believe that Bougainville ever saw Tetiaroa (*Discovery*, p. 116). Taillemite has no doubt that Bougainville is referring to Tetiaroa, which is described in his printed narrative as 'a modest height and wooded' (*Voyage* I, p. 233), a description confirmed by the Prince de Nassau: 'We saw on 17 April a low land covered with trees',

Note: Between what has been used during the stay by the two ships and what has been taken on board in the way of refreshments, we obtained through barter 800 heads of poultry and close on 150 pigs. Without the troublesome tasks of the last few days, we could have obtained far more because the Indians were bringing more than ever. The sick found much relief on land during their brief stay there. The air in this island is most salubrious and, once the good anchorage was identified, it would be one of the best places of call in the world. I forgot to mention that we obtained some excellent pigeons and that we received several pumpkins.

Saturday 16 to Sunday 17. New moon on the 17th at 0h 33′ a.m.

I am setting a course so as to fall between 15° and 14° of latitude in order to avoid the region of the dangerous islands and labyrinth of Roggewin.[1] We spent the night hove to, crowded on sail during daylight. At 5 p.m. the N point of the last land bore SSW 5°S, distance 7 to 8 leagues.

At midday, run corrected gives NNW 2°N, distance 84′.

Estimated latitude: 15°26′, observed: 15°34′.

Corrected longitude: 151°4′, 151°9′.

Sunday 17 to Monday 18.

Observed variation by azimuth: 7°3′ N.

Fine weather, fair sea, fresh ESE to E breeze. Steered NW¼W until nightfall then lay to during the night N¼NE to N, at daylight steered W¼SW.

At midday the corrected run is WNW 1°15′W, distance 82′30″.

Estimated latitude: 15°1′, observed 15°4′.

Calculated longitude: 152°25′, 152°35′.

Monday 18 to Tuesday 19.

Fine weather, fair sea, fresh E to SE breeze, steered W sailing at night under the topsails only and in daytime with all sails out. When night came, the Indian observed the stars, named a dozen of them for us, showed us by the stars in which point of the compass his country lay and that by steering NNW we would within two days reach a nation allied to his, which had all the refreshments we might need and especially women at our service. He gave us to understand that he was born there or had begotten a child there, his gestures indicated equally both these possibilities. Seeing that we were continuing on the same course and no doubt believing that we had not understood him clearly, he threw himself at the helm and, seizing the wheel, he wanted to turn the vessel towards the route he indicated. It is evident that these people navigate out of sight of land and set their courses by the stars. They have the basis of all the arts.

Journal, f. 24. That Ulieta is an unlikely rendition of Tetiaroa is confirmed by Joseph Banks who wrote in his journal of 'Thethuroa, a neighbouring Island', Beaglehole, *The Endeavour Journal of Joseph Banks*, I, p. 290. One could hypothesize that Ahutoru was in fact referring to the island of Moorea ('o-motea') further to the south-west, lying behind Tetiaroa.

[1] After failing to find Davis Land and coming upon Easter Island, Roggeveen sailed on towards the Tuamotus, discovering a number of them but losing his *Afrikaansche Galey* on Takapoto (which he named Disastrous Island). The entire group was given the not inappropriate name of the Labyrinth. He then proceeded to the northern Society Islands, including Bora Bora and Maupiti. See Sharp (ed.) *The Journal of Jacob Roggeveen*, pp. 116-40. Takapoto is in 14°30′.

At midday the run corrected is W¼NW 5°15′W, distance 112′30″.
Estimated latitude: 14°50′, observed: 14°53′.
Calculated longitude: 154°21′, 154°36′.

Tuesday 19 to Wednesday 20.
Variation observed occiduous: 7°10′, ortive: 7° NE.
Fine weather, fair sea, E to ESE winds. Sailed W with all sails during the day and at night under the topsails. The Indian still wants us to make for the land he is pointing out to us and is much distressed by our refusal.
At midday corrected run is W¼NW 3°N, distance 97′30″.
Estimated latitude: 14°40′, observed: 14°39′.
Calculated longitude: 156°1′, 156°21′.

Wednesday 20 to Thursday 21.
Observed variation occiduous: 7°27′.
Continuation of the same weather, same route on our part except for steering 8 hours W¼SW and for 6 W.
Run at midday W 2°30′N, distance 90′.
Estimated latitude: 14°34′, observed: 14°35′.
Calculated longitude: 157°35′, 158°.

Thursday 21 to Friday 22.
Observed variation occiduous: 7°45′.
Fine weather, fair sea, light E breeze; steered W¼SW all sails out during the day and under the topsails at night.
Since our departure I have suffered from a fairly high fever that finally left me last night.[1] This morning I spoke to the *Étoile*. All is well aboard her and her sick are in a good condition … Our Indian thinks of nothing but his women, he talks endlessly about them, it is his only thought or at least all the others he has relate to this one. He made us understand that if there are no women for him where we are going, we shall have to cut his head off.
At midday the corrected run is W 5°30′S, distance 84′.
Estimated latitude: 14°41′, observed: 14°45′.
Calculated longitude: 159°2′, 159°32′.

Thursday 22 to Saturday 23.
Observed variation ortive: 7°12′, by the azimuth: 7°33′.
Fair weather, light E breeze, course W¼SW.
At midday the run corrected is W¼SW 2°W, distance 75′.
Estimated latitude: 14°51′, observed: 14°57′.
Calculated longitude: 160°18′, 160°53′.

Saturday 23 to Sunday 24. Full moon on the 23rd at 9h25′ p.m.[2]
Observed variation ortive: 7°16′.

---

[1] According to Fesche's journal, Commerson was brought over from the *Étoile* on the 18th to attend to Bougainville; he remained on the *Boudeuse* until the 19th. Fesche, 'Journal', f. 296.
[2] This must be a mistake for the first quarter as the new moon is recorded on the 17th and the full moon again on 2 May.

Fair weather, very little wind, course W¼SW.[1]
At midday the compass point is W4°S, distance 55'30".
Estimated latitude: 15°1', observed doubtful 15°.
Calculated longitude: 161°15', 161°56'.
We have caught 4 or 5 bonitos.

Sunday 24 to Monday 25.
Observed variation by azimuth evening 7°10', morning: 7°2' NE.
Fine weather, fair sea, light breeze, course W¼SW.
At midday run gives W, distance 80'30".
Estimated latitude: 15°5', observed: 15°.
Calculated longitude: 162°37', 163°23'.

Monday 25 to Tuesday 26.
Observed variation occiduous: 8°15', ortive: 7°56'.
Fair weather, little wind, steered W¼SW until 6 a.m. when I altered course to W.
At midday the run is W45'S, distance 74'.
Estimated latitude: 15°1'.
Calculated longitude: 163°53', 164°44'.

Tuesday 26 to Wednesday 27.
Fine weather but very little wind NNE to ESE, courses W and W¼SW.
At midday the run corrected gives me W1°N, distance 58'.
Estimated latitude: 14°58', observed: 15°.
Calculated longitude: 164°53'. 165°49'.
Isaac Le Roi, from Noirmoutier, coastal pilot, died in the night of a stroke.[2]

Wednesday 27 to Thursday 28.
Observed variation ortive: 8°, by the azimuth: 8°6'.
Fine weather, almost no wind and even then we are forced to use almost no sails
on account of the *Étoile*. Steered W¼SW then W.
Estimated run W30'N, corrected W¼SW3°25'W, distance 44'.
Estimated latitude: 14°59'35", observed: 15°6'.
Calculated longitude: 165°37', 166°40'.
Observation: Mr Verron by the resultant of several observations, is, on the 27th at
midday, in 166°6'37" which gives me a difference with him of 1°13'37"E.[3]

Thursday 28 to Friday 29.
Observed variation occiduous: 8°30', ortive: 8°27'.
Fair weather, almost calm, E¼SE to E and E¼NE breeze, heavy seas from the S,
steered W.

[1] 'The 15th parallel having now been reached, the average route will now be westerly. Average speed
during the previous 24 hours: 5 knots' (ET).
[2] Further details on this man's condition can be obtained from St Germain: 'He had been complaining for several days of a sore throat, of feeling hot, and had asked to be bled. Yesterday he was in the sick
room mess and was given an infusion. This preliminary care did not prevent him from dying suddenly
in the night without anyone noticing': see La Roncière, 'Routier inédit', p. 235.
[3] This is relative to the noon position on 26th–27th, the first value given, which appears to be that by
account.

Estimated run: W¼NW3°W, distance 41′30″, observed: W1°N, distance 41′.
Estimated latitude: 15°, observed: 15°5′.
Calculated longitude: 166°19′, 167°23′.

Friday 29 to Saturday 30.
Fair weather, variable winds from E to SE, S then ESE, a few squalls but little wind, steered W.
Estimated run: W8°N, distance 53′, corrected: W1°N, distance 53′.
Estimated latitude: 14°58′, observed: 15°4′.
Calculated longitude: 167°14′, 168°24′.

May

Saturday 30 to Sunday 1 May.
Observed variation occiduous: 9°34′.
Almost calm, sailed W.
Run corrected: W5°N, distance 23′.
Estimated latitude: 15°1′, observed: 15°2′.
Calculated longitude: 167°38′, 168°53′.

Sunday 1 to Monday 2. Full moon on the 2ⁿᵈ at 0h33′ a.m.
Observed variation occiduous and ortive: 9°23′ NE.
Fair weather, the seas still high, winds E, very little breeze, sailed W. At midday the courses give me W9°15′N, distance 47′20″.
Calculated longitude: 168°26′, 169°46′.

Monday 2 to Tuesday 3. New land. Archipelago of the Small Cyclades [Petites Cyclades].[1]
Sight of land No. 1, 2, 3, 4. Approximately in latitude 14°20′, longitude 169°30′ at its middle, corrected 170°56′ at its middle.
Observed variation occiduous: 9°50′.
Fair weather, little wind from E to SE, then NE, sailed W until 6.30 a.m. Then sighted a high island the bearings of which were as follows: the S point NW¼W4°N, the N point NW3°W distance about 20 leagues. I then steered NW and NW¼N to haul along the island to windward.
Run at midday estimated for the 24 hours is WNW3°30′N, distance 46′, corrected run WNW1°45′, distance 45′.
Estimated latitude: 14°35′, observed: 14°37′.
Calculated longitude: 169°9′, 170°35′.
Bearings at midday: the N point NW5°W, the S one NW¼W, a headland that seems detached S of this point NW¼W5°W, another land fairly high that seems also to be detached and is the most S WNW5°N distance about 10 leagues.
According to these bearings, the middle of this island would be in latitude 14°20′, longitude 169°34, corrected 171°.[2]

---

[1] Bougainville will not retain this name, later calling the islands the Navigators. They are the Samoan archipelago.
[2] The islands coming into view are the Manua Islands, consisting of Tau in the east and Ofu and Olosega northwest of Tau.

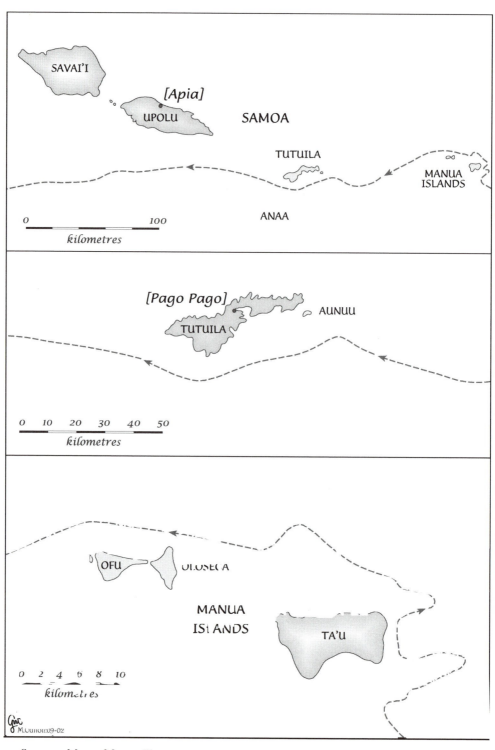

4. Samoa, 3 May–5 May 1768.

Note: combining this estimate with that of the following midday, the island would be 6 leagues from N to S.

Tuesday 3 to Wednesday 4.

In the afternoon the weather was fair, little wind variable from NE to NNE. At 5 o'clock as we were falling to leeward of the island, I altered course, we went about and steered E and E¼SE. The calm caught us during the night and lasted until 3 o'clock. Then, the winds coming from SSE, variable as far as S and SSW, we steered to pass N of the island. In the morning, there were squalls and rain. At 10 o'clock we made for the land which we ranged along a league and a half offshore. The headlands we sighted yesterday and thought to be detached really are and make up two fairly high islets, although they are not as high as the large island.[1]

A midday, as several canoes left the shore and seemed to want to approach us, I hove to in order to wait for them. Estimated run for the 24 hours is NNW5°30'W [distance] 37', corrected run NW¼N4°N 35'.

Estimated latitude: 14°4'30", observed: 14°6'30".

Calculated longitude: 169°27', 170°53'.

Note: We took various bearings during these 24 hours that I am not recording here. These bearings, necessary for the running of the ships, are not of any interest for geography. I shall henceforth content myself with recording the bearings according to which we determine the positions in latitude and longitude of the various lands.

Midday bearings: the N point of the large island SSE3°S, 1½ leagues, the W point of the same island SW5°S, 2 to 3 leagues. The N point of the middle island W¼SW2°W, 3 leagues, the S one WSW. From these bearings, the N point of the main island is in latitude 14°11' and its W point in longitude 169°35', corrected 170°59', which agrees with yesterday's estimate.

Wednesday 4 to Thursday 5.

Variation observed at sunrise by the azimuth: 7°22' by Mr Verron, 7°30' ditto by azimuth by the Chevalier du Bouchage. It is not the first time that navigators have noticed that, in the neighbourhood of land, the variation of the needle changed from that found out at sea.

At midday, I hove to wind aport. 8 or 10 canoes with outriggers came up, several with a sail, built like those of Cythera except that none was double and that they had neither stem nor stern raised, but they were all well decked. Each canoe can take about 8 people. The Indians first of all spent some time looking at the vessels, without daring to approach. Finally they became emboldened and came alongside. They are fairly tall, though smaller and less handsome than those of Cythera. A woman who had come in one of the canoes was hideous. None had a long beard. Their hair is black and tied up in various ways. One had very long red hair. They are naked, apart from a belt that covers their natural parts. Their thighs and stomachs are painted a bluish purple, as in Cythera, none had painted buttocks. We saw them with necklaces

---

[1] Mt Lata on Tau rises to over 3,000 ft (to 930 m). Ofu and Olosega, separated from each other by a narrow strait, are hilly but not mountainous. The *Boudeuse* originally approached the group from the south; it will now coast along their northern shores. All three islands are inhabited, with a population in excess of 3,000.

of flowers and others with painted grains like beads and hanging necklace-like in branches.

Louis took off his clothes and spoke to them, but their tongue is not the same, an interesting point.[1] Our Indian seemed to hold them in contempt and showed clearly that he had no wish to go with them. These islanders are used to trade and even skilled at it. They bartered for handkerchiefs and pieces of cloth some loincloths similar to those of Cythera, coconuts, yams, a few lances, some necklaces of painted seed and a fine blue bird with a red beak and red between the eyes, similar to a water hen. They had brought a cock, but nothing offered to its master appealed to him, and he took it back. The blue bird was in a cage and tied by one foot, the tie being fixed to the bars by a very fine piece of shell. There are therefore some *carret* in these islands or in the neighbourhood and some turtles.[2] Also these Indians have no knowledge of iron and appeared not to prize it in any way. Several examined some knives offered for barter and did not want them. At Cythera the first request of the Cytherans was for *aouri*.[3]

These islands are very high and steep sided, there is no low land on the seashore in the three parts we visited. We saw among the coconut trees a few huts with a round roof. The population did not seem very large. There is a very fine channel a league and a half wide between the large island and the second one and I think there could be an anchorage between the second and the third. I do not believe these islanders to be as gentle as our Cytherans. Their features are more savage and they displayed a great deal of mistrust. I saw one whose face was all scarred with smallpox.

SE to ESE winds. At one thirty I set the sail and steered WNW to sail by the third islet, then we sailed SW and finally SW¼S throughout the night because they saw from aloft three hillocks to the WSW.

We stayed under the topsails throughout the night to wait for the *Étoile*. At 2 a.m., the weather being very pleasant and the moon very fine, we sighted land ahead bearing W2°S and yesterday's bore NE4°E. At 5 thirty we let out all sails and sailed NW to approach the shore and range along the land, the E point of which bore NNW3°N and the W one WNW5°N distance 6 to 7 leagues. This island is long and high and of a pleasing appearance. The mountains slope down fairly gently and end in attractive plains on the seashore. The coast is steep and waves break almost everywhere along it. There is even a reef extending a fair distance out about two leagues from land, we sailed W to range along it. At the E point of this island there is an islet high in its

---

[1] The Austronesian migration from west to east across the Pacific passed by way of Samoa and consequently Bougainville was meeting Polynesians. Samoa was settled at least 1,500 years before Tahiti and the linguistic evolution which had taken place in the interim was sufficient to make Samoan unintelligible to a Tahitian.

[2] Strictly 'caret', or loggerhead turtle (*Caretta caretta*). However, the term 'caret' was fairly widely used at the time for sea turtle, including the hawksbill which produced commercial shell.

[3] The word is probably derived from 'iron', which the Tahitians had obtained from Wallis during the previous year: 'they seemed very fond of everything which they saw made of iron' (Carrington, *The Discovery of Tahiti*, p. 137). Whether their keenness was in fact due to an early awareness of the value of iron which they had, directly or indirectly, obtained from Spanish sources (the Spanish for 'iron' is *hierro*) is arguable. Banks reported that some Tahitians had obtained 'an Iron tool made in the shape of the Indian adzes' from '2 Spanish ships [that] had come here', but there is no doubt from the context that these two ships were Bougainville's. Banks, *Journal*, I, pp. 286–7.

centre that ends in a gentle slope by the sea. At its W point there is similarly a large round islet steep on all sides and slightly detached from the coast.[1]

The coast is divided approximately at the two-thirds point by a cape that is closer to the W than to the E. Its E point bears NE¼E and SW¼W from the middle cape and its W point bears from this same cape W¼NW 3°45'W, all corrected for the variation. It may have twelve leagues in length.

From 7 a.m. they have seen another land from the mast-head bearing WNW to NW¼W, high land.[2]

Numerous canoes appeared along the shore, a few approached the vessels but without coming alongside. They are built exactly like those of the islands seen yesterday, decked at both ends. I saw one that had in the middle of its two decks two rows of wooden pegs, like those of yesterday, but topped with fine sparkling white *limas*.[3] The Indians seem similar to those seen yesterday, naked except for the natural parts, the thighs painted. From the topmasts we saw 2 canoes with a sail to the S of us about 6 leagues off.

Le Maire, in approximately the same waters, saw a double one with a cabin on the stern, like those of Cythera, out of sight of all land.[4]

At midday the estimated run gives me WNW 5°W, distance 75', the corrected run is W¼NW 45'W, distance 71'.

Estimated latitude: 14°28', observed: 14°20'.

Calculated longitude: 170°39', corrected: 172°6'.

Bearings at sunset: the S point of the most W island bearing ESE 5°E about 4 leagues, the N one of the middle island E, 5 leagues, the most S head of the land ahead WSW 2°S about 15 leagues.

Sight of land No 5 approximately in latitude 14°18' and longitude 170°10'W by its centre, corrected 171°36'.

Bearings at midday: the cape situated at the one-third point along the island E 3°N distance 6 leagues, the W point of the island ENE 2°E. It is on these bearings that the estimated position of the island given above is based.

Sight of land No 6 in about 14°3' of latitude and 171°4' of longitude W, corrected 172°30', the prominent point to WNW of us approximately.

Correction. *Note*: this point, which is a small detached islet and forms the E point of this other island sighted at 4 p.m. bearing N¼NW 3°30'W, 7 to 8 leagues distant, will be in latitude 14°2'30" and longitude W 171°3', corrected 172°29'.

Bearings at sunset on the 5th: the NE point of the island N 2°E, the most W land NW of the compass. These bearings place this W point in latitude 14°8' and longitude

<hr/>

[1] The French are sailing along the south coast of the island of Tutuila. Off its south-west point lies the island of Aunuu which is roughly oval-shaped. There are several mostly rocky islets off its eastern extremity, Cape Taputapu. Bougainville will sail past the wide opening that forms Pago Pago Harbour and as he approaches the western part of Tutuila will note the prominent Steps Point.

[2] This is the island of Upolu. Today Tutuila forms part of American Samoa while Upolu belongs to independent Western Samoa.

[3] 'These are no doubt burgos, a type of sea snail, or trocas [trochids], another type of sea shell' (ET).

[4] Jacob Le Maire sailed with Willem Schouten from Holland in 1615 with the *Eendracht* and the *Hoorn*, pioneering the Cape Horn route into the Pacific, and sailed on through the Tuamotus and various island groups further west. They passed well south of the Samoan archipelago.

171°28'W, corrected 172°55' and give the island an extent of 8 to 10 leagues from E to W in its southern part.

Thursday 5 to Friday 6.

Corrected variation: 8° NE.

Cloudy squally weather with rain, SE to ESE winds altering E and ENE. In general, the approach of land, in these seas, especially during the waning of the moon, brings storms, and this squally weather with heavy fixed clouds on the horizon is an almost sure sign of the presence of some land and a warning to be on one's guard. We steered various courses during these 24 hours, compelled to alter course when the horizon ahead of us was too black. One cannot imagine the care and the anxiety with which one sails in these unknown seas, threatened on all sides by an unexpected encounter with land and reefs, an anxiety even more intense during the long nights of the torrid zone, the need for supplies, the shortage of water, the necessity to take advantage of the wind when it deigns to blow, not allowing one to spend the darkness hove to.

At night, I spoke with the *Étoile.* everyone there is in good health. At midday Mr de La Giraudais was in longitude 175° and had observed 14°21'. He complains of the slow progress of his vessel one cause of which is certainly the amount by which she is too high in the water, her draught being a foot and a half less than it should be.

An Indian from one of the canoes approached the *Étoile* this morning when he saw a piece of red cloth being offered, [he] dived into the water to come aboard, but having missed the ropes that were thrown for him, he had to go back.

Estimated run for the 24 hours WSW 3°45'W, distance 107'45".

Estimated latitude: 14°55'. Calculated longitude: 172°24', 173°51'.

Observation of longitude: from the results of five observations of the distance from the moon to Antares, the 5th at 4 a.m., Mr Verron concluded the longitude of the vessel on the 5th at midday to be 174°31'15".

From the results of four other observations of the distance of the moon to the same made at 10 in the evening, he found the longitude of the vessel for the same midday to be 174°45'21", so by taking the mean, the longitude of the vessel on the 5th at midday would be 174°38'18", which is a difference from my estimate of 4°00'42". We must have been carried W by strong currents since our departure from Cythera, I believe this more readily because all the navigators who have sailed across this sea reach New Guinea much sooner than they should. Moreover I would point out that during the monsoon when the sun is on this side of the Line, our estimates were W of the observations and that, since the change in the equinox, our differences are E of the observations.

Friday 6 to Saturday 7.

Observed variation ortive: 9°30' NE.

Cloudy stormy weather with intervals of rain, winds variable WNW, NW, then S to ESE and ENE. Sailed W when we could. In the evening, the topmen thought they could see land from the mast-head.[1] Never was night so black, until the moon rose we

---

[1] Although the French eventually concluded this was merely a cloud, it may have been a sighting of the high island of Savai'i, about 25 km northwest of Upolu.

were fumbling our way along with only the foretopsail at half-mast and the wind in the mizzen top. At 11 p.m., the moon cleared part of the clouds and allowed us to increase our canvas. In the morning, we saw no land, what we had thought we saw was merely a large cloud that had the shape of land. Two fairly long sea snakes passed alongside yesterday. At midday the run gives W¼NW4°W, distance 56½'.

Estimated latitude: 14°49', observed: 14°48'.

Calculated longitude: 173°22', corrected: 174°50'.

Saturday 7 to Sunday 8.

Corrected variation: 10°.

NNW winds. Fair weather but almost calm in the afternoon. I went on board the *Étoile* where everything is in good condition. During the morning, we had squalls and rain. Sailed W¼SW.

Estimated latitude: 14°49', observed doubtful: 14°44'.

Calculated longitude: 174°26', corrected: 175°55'.

Sunday 8 to Monday 9.

Squally weather, rain at intervals, calms and variable winds E to N and NNW, sailed W when the weather allowed it. During the night, we thought we saw land in the NW. We altered course then tacked back half an hour later, the weather having cleared and this land of clouds having disappeared.

At midday the estimated run is W¼SW4°45'W, distance 63', corrected W¼SW4°S, distance 65'.

Estimated latitude: 14°51', observed doubtful: 15°1'.

Calculated longitude: 175°31', corrected: 177°1'.

Observation: By 8 observations of the distance of the sun's rim to the moon carried out this morning at 8 o'clock, Mr Verron finds the longitude of the vessel at midday to be 180°0'15". Following my estimate of the route since his observation of the 5th, it would be 179°30'. There would have been less current running E and W during these last 4 days.

Monday 9 to Tuesday 10. Last quarter on the 9th at 6h10 p.m.

Observed variation azimuth: 9°46', ditto in the morning: 9°20'.

Fair weather, fairly good breeze but head wind NNW to W, SSW, SW and SSW. How much patience, O God, this navigation requires!

The run for the 24 hours gives NW¼W3°50'W, distance 56'.

Estimated latitude: 14°28', observed: 14°29'.

Calculated longitude: 176°30', 178°1'.

Observation: by 5 observations of the distance from the moon to Antares, the longitude on the 10th at 4 a.m. would be 180°14'30", which would make the one of the 9th at midday 179°37', a difference of only 23'15" with the previous one. The same day the 10th at around 10 a.m., 5 observations of the distance of the sun to the moon gave for that time 180°58'15" which, reduced to the midday of the previous day, give for the ship's longitude at midday on the 9th: 180°6'15".

3 observations: 180°0'15", 179°37', 180°6'15". Mean of the 3: 179°54'30" for the longitude on the 9th at midday. Difference with my estimate: 4°23'30".

Tuesday 10 to Wednesday 11.

Observed variation ortive: 10°13′ NE.

Sight of land No. 1 The Lost Child [l'Enfant Perdue (sic)]. Its estimated position: its NE point approximately in latitude 14°15′40″, longitude 177°24′ corrected: 179°. Its middle approximately in latitude 14°17′45″, longitude 177°25′ corrected 179°1′.

Continuation of the SSW, SW and S winds, small breeze, sailed close-hauled port tack.

At daybreak we saw a fairly high land that seems cut into two islands bearing WSW distant approximately 8 leagues. I have named it the Solitaire.[1]

At 8 o'clock bearings the S point WSW3°S, the N point W¼SW3°W. Since morning it has been calm, the ship hardly responding.[2]

At midday the S point of the island bearing WSW2°W, the middle WSW5°W to W¼SW1°S, its N point W¼SW5°W.

At midday run gives me WNW2°40′W, distance 38½'.

Estimated and observed latitude: 14°16′.

Estimated longitude: 177°6′, 178°42′.

Wednesday 11 to Thursday 12.

Observed variation ortive: 10°15′.

Note: I took possession in the King's name of the archipelago of the Small Cyclades and of the Lost Child.[3]

Calm and fair weather in the afternoon. At 7 o'clock I steered on the starboard tack, the winds being WSW, they then veered to WNW and NW, sailed S, SW¼S, SW, SW¼W. This morning we saw that this land forms only one island in which case it must enclose a very fine bay.

At midday the run gives me SW2°20′W, distance 51′.

Estimated latitude: 14°50′, observed: 14°52′40″.

Calculated longitude: 177°45′, corrected: 179°21′.

At 10 o'clock the middle of the island bore N and S of the compass distance 10 to 11 leagues, which would place its middle in latitude 14°15′ and longitude 155°30′, corrected 178°44′, an estimate that does not differ from yesterday.

Our Indian would like us to stop at all these islands, solely to make a sacrifice to Venus. It is his only preoccupation, all his thoughts are centred on that. This Indian is very intelligent and very shrewd. He examines and describes everything that is done in the ship, he knows what each one's task is. Above all he knows the helmsmen of both watches and is unwilling for anyone else to take up the helm. However he has not yet learnt one word of French, 1° because he is very lazy, 2° because our tongue is almost impossible to pronounce for a man whose language consists solely of vowel

---

[1] The French have now reached the Hoorn Islands, discovered in 1616 by Jacob Le Maire and Willem Schouten. They consist of two islands, Futuna, the 'high island' that rises to 2,590 ft (790m), and Alofi, a mile or so away. There is a Mt Bougainville on Alofi. The names Lost Child and Solitaire did not last.

[2] I.e. not answering the helm.

[3] Copies of the Act of Possession are held in BN, NAF 9407, 137 and 141. In it, Bougainville refers to a group of five islands situated in latitude 14 to 15° N and between the 169th and 172nd degrees of longitude west of Paris, with the Lost Child in latitude 14°17′ and longitude 179° W. He names the entire group the Archipel des Navigateurs, a term which will remain in use for many years.

sounds. The same reasons that allow it to be said that ours is not very musical make it inaccessible to his organs. In addition he has a deadly fear of any kind of pain and the slightest indisposition. He is timid by nature, mild, he despairs when he merely sees pain, the punishments that one cannot avoid inflicting are torture to him. Being accustomed to a soft life inevitably creates this character, what makes him indolent makes him gentle: '*Voluptates, blandissimae dominae, majores partes animi a virtute detorquent et dolorum cum admoventur favo praeter modum plerique exterrentur.*'[1]

Thursday 12 to Friday 13.

Observed variation in the morning by azimuth: 10°53′ by the pilots, 10°49′ by Mr Verron.

Deplorable weather, rain, squalls and constant head winds, variable from SW to NW. We tacked about grumpily all the more because the *Étoile* that sails much worse than we do (which is saying a great deal) is falling horribly to leeward and prevents us from maintaining enough sail to lessen the leeway.

At midday run gives me W¼SW 3°30′S, distance 55′.

Estimated latitude: 15°6′, observed: 15°7′.

Calculated longitude: 178°40′, corrected 179°40′ E of Paris

The Paris small mail service goes faster than we do.[2]

Observation: From 6 observations of the distance from the moon to Venus, Mr Verron has concluded that the longitude of the ship at midday on the 13th is 177°41′45″ E of Paris, which would be 45′15″E of our estimate since his observations of the 9th. I would have a difference with this observation of only 3°38′15″.

Friday 13 to Saturday 14.

Frightful weather in the afternoon and night, rain, squall, good breeze but head wind. Meanwhile after our six months at sea, we have refreshments left only for the sick. Wood is about to run out, several officers have a somewhat inflamed mouth and traces of scurvy in their gums. As for [me] I cannot master my impatience. Who could, Great God? Fresh gale when it is against us, calm when it is favourable.

At midday the run is WSW 1°30′W, distance 16′.

Estimated latitude: 15°12′40″, observed: 15°13′.

Calculated longitude: 178°55′, corrected: 179°21′ E of Paris.

Saturday 14 to Sunday 15.

Observed variation occiduous: 10°59′, ortive: 11°4′ NE.

Still W winds, SW to NW, squally weather, rain, slight calms, tacked according to the wind.

At midday the estimated run is W¼SW 3°45′S, distance 16′30″.

Estimated latitude: 15°15′35″.

[1] The quotation is from Cicero, *De Officiis*, II, 37: 'Pleasures, most ingratiating mistresses, distract the major parts of the mind from virtue; and, when the torches of pain and suffering are applied, most people are terrified beyond measure.'

[2] Although a postal service of sorts existed as far back as the sixteenth century, the Paris 'Petite Poste' was set up in 1653 for local mail only, with three deliveries a day. It was reorganized in 1760 by Piarron de Chamousset with a system of posting boxes in selected shopkeepers' premises. A postal service for provincial towns was first set up in 1672.

Calculated longitude: 179°10'20", corrected: 179°21' E of Paris.

Sunday 15 to Monday 16. New moon on the 16th at 8 a.m.
Observed variation ortive doubtful: 10°23'.
Continuation of head wind, squalls, rain and calms. This is not living, it is dying a thousand deaths each day.
At midday the estimated run gives WNW 3°45'N, corrected NW¼W 2°N, distance 21' 30".
Estimated latitude: 15°7', observed: 15°3'.
Calculated longitude: 179°28'30", corrected: 178°40' E of Paris.

Monday 16 to Tuesday 17.
Observed variation occiduous: 11°11', ortive: 10°25'.
Calm in the afternoon, at sunset a breeze came up from the ESE, we were unable to benefit from it, forced on account of the *Étoile* to remain hove to until 3 a.m. At dawn let out all sails and sailed W and W¼SW, the winds ESE to SE.[1]
Estimated run at midday W 5°N, distance 45', corrected route W¼NW 1°15', distance 51'.
Estimated latitude: 14°59', observed: 14°54'.
Calculated longitude: 179°41', corrected: 177°45' E of the Paris meridian.

Tuesday 17 to Wednesday 18.
Observed variation occiduous: 11°12' NE. Fair weather, good ESE to SE breeze, sailed W¼SW and hove to from one in the morning to 5 o'clock because the night was very black and the weather squally.
Run according to the observation: W 1°30'N, distance 119' 20".
Observed latitude: 14°51'.
Calculated longitude: 177°40', corrected: 175°40'.

Wednesday 18 to Thursday 19.
Observed variation ortive doubtful: 9°57'.
Fair weather, good SE to ESE breeze, sailed WSW under full canvass. Stayed hove to port tack during 5 hours at night.
At midday run gives me W¼SW 2°45'S, distance 117'.
Estimated and observed latitude: 15°19'.
Calculated longitude: 175°37', corrected: 173°33'.
Note: Warning to travellers. Several venereal diseases caught in Cythera have appeared recently on board both ships. They are of all the types known in Europe. I had Louis examined, who is riddled with them and he is being treated. It seems that his compatriots are not greatly concerned about them. Columbus brought it back from America. And there it is in an island isolated in the middle of the sea. I believe that this doctor who bet that by locking up a healthy woman with 4 equally healthy men the pox would appear after a while as the result of their activities, would win his

---

[1] 'Slowed by the *Étoile*'s poor progress, contrary winds or weak unfavourable winds, the *Boudeuse*, between 17 April and 17 May, averaged about two and a half knots (4.6 km), which explains the commander's impatience' (ET).

wager. A wise move was that decision I had taken of preventing the Frenchmen affected with this sickness from going ashore in Cythera.[1]

Thursday 19 to Friday 20.

Fair weather, good ESE to SE breeze. Sailed W¼SW and remained hove to from 2.30 a.m. until 4. We sail far better than the *Étoile*. The run for the 24 hours by observation is W2°15'N, distance 130½'.

Estimated latitude: 15°20' observed: 15°15'.

Calculated longitude: 173°22', corrected: 171°14'.

Friday 20 to Saturday 21.

Continuation of good ESE breeze, steered W¼SW. Hove to from 6 to 7.30 p.m. to wait for the *Étoile*, then from midnight to 4 a.m., wind aport.

Corrected run for the 24 hours: W2°40'S, distance 129'.

Estimated latitude: 15°17', observed: 15°21'40".

Calculated longitude: 171°, corrected: 168°54'.

Saturday 21 to Sunday 22.

Variation by azimuth observed in the evening: 10°50' occiduous, 10°34' ortive. 4th division. Sight of land. The Great Cyclades [Grandes Cyclades] approximately in 15° of latitude and 169° of longitude, corrected 166°45'.[2]

Fair weather, good ESE breeze, steered W¼SW all the afternoon then W during the night under the topsails double-reefed and the mizzen top. From 3 a. m. to 6 o'clock, we sailed W¼SW to W¼NW according to the points of the compass where the horizon was the clearest. At dawn saw the land ahead of us, a high land 7 or 8 leagues distant, presenting the appearance of two islands with a passage between the two of about 2 leagues width, which could be seen from NW5°W to S4°E. We kept close hauled, but unable to reach the channel between the two islands, I bore away to sail to leeward of the most N and I named this island Aurora and the S one that is separated from it by a channel Pentecost Island [Isle de la Pentecoste].[3]

At 10 o'clock we saw ahead of us a large isolated sugarloaf which I have named

---

[1] This last sentence is important in view of the subsequent debate on the presence of venereal disease in Tahiti, whether it was brought by the British (as indicated by the Tahiti expression *apa na peritane*, 'the British sickness') or whether it was endemic in the islands or brought in by earlier unidentified navigators or even was confused with yaws. See Introduction, 'The Voyage'.

[2] The Cyclades are a group of Greek islands in the Aegean sea situated around the island of Syros. Bougainville had already used-briefly-the term Small Cyclades to describe the Samoan islands. At this point in his voyage, he has reached what James Cook will later name the New Hebrides, present-day Vanuatu.

[3] The ships were in view of Maéwo (Aurora) and Arag-Arag, although the name Pentecost has remained (the alternative English name of Whitsuntide is sometimes found). The pass between the two is called Patteson Passage, after the missionary bishop John Coleridge Patteson. Aurora had already been seen by Pedro de Quiros on 25 April 1606 and named by him La Margaritana. He may have seen part of Pentecost in the distance, but this is doubtful and Bougainville has a good claim to being its discoverer. See Sharp, *The Discovery of the Pacific Islands*, pp. 63-4, Kelly, *La Austrialia del Espíritu Santo*, I, pp. 48, 195-6. The name Pentecost was given after the day on which this discovery was made.

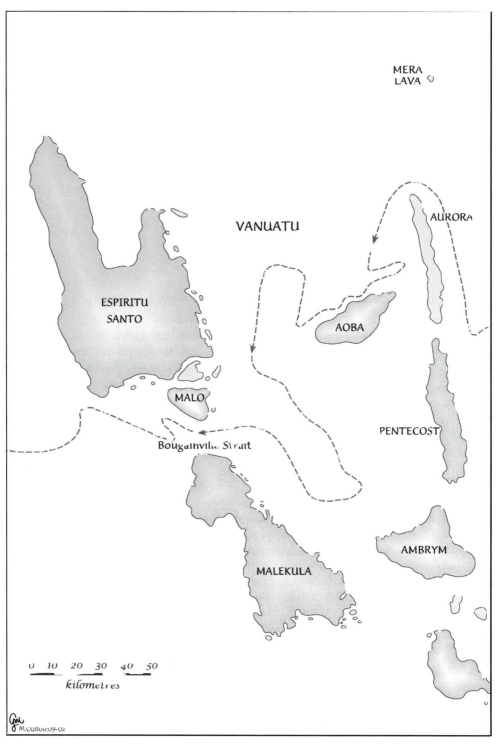

MERA
LAVA

VANUATU

AURORA

ESPIRITU
SANTO

AOBA

MALO

PENTECOST

Bougainville Strait

AMBRYM

MALEKULA

0 10 20 30 40 50
kilometres

5. Vanuatu, 22 May–28 May 1768.

90

Étoile Peak [Pic de l'Étoile] bearing NW and NW¼N.[1] Estimated run for the 24 hours: W2°N, corrected route: W4°N, distance 109′40″.

Estimated latitude: 15°18′, corrected: 15°11′.

Calculated longitude: 169°15′, corrected: 167°.

Sighted at midday the S point of Aurora Island [Isle Aurore] bearing S¼SW3°S, the land of Pentecost Island visible at the most S bearing S¼SE4°E, Étoile Peak which was what one could see the furthest N bearing NNW4°W.

Observation: by three observations made at three different times during the 24 hours and run back to midday on the 22nd, Mr Verron, from an averaged result, concluded that the ship's longitude on that morning was 165°40′E.

Sunday 22 to Monday 23.

SE winds, fresh breeze, we continued to range along Aurora Island, sailing NW¼N3°N until 3 o'clock then WNW for half an hour, then SSW until 8 p.m.

At 4 p.m. we again saw the channel separating Aurora Island from Pentecost Island, we had then to the SW another island that was covering up Pentecost of which only the N point was visible. I named this third island Lepers Island [Isle des Lépreux], the inhabitants we saw on it the next day being ridden with leprosy.[2]

It seems to me that Aurora and Pentecost islands lie together SE and NW and that the channel that separates the Pentecost and Lepers islands must run NE and SW.

From 4 o'clock yesterday we have seen in the distance other land stretching WSW as far as NW¼N.

During the night we tacked between Aurora and Lepers islands.

At 7 a.m. I bore away to range along Lepers Island, steering SW and sailing along the coast two leagues away.

At 9.30 I signalled the *Étoile* to send me an armed boat, I lowered my boats with a detachment of soldiers and sent them to land to get wood and look for refreshments, always keeping the ships within range to protect the boats ashore.

At midday the estimated run for the 24 hours is SW1°30′S26′15″, the corrected route is W¼SW1°45′S, distance 17′30″.

Estimated latitude: 15°30′, observed: 15°14′.

Calculated longitude: 168°57, corrected: 166°42′.

The run has been very incorrectly estimated on account of the violent currents that bear N here as we clearly observed when abreast the shore of Lepers Island while our boats were ashore. The last two latitudes, together with the bearings, are enough to straighten out our track.

---

[1] This is the island of Méré Lava (Mera Lava) discovered on 25 April 1606 by Quiros and named by him San Marcos. The routes taken by Quiros and Bougainville diverge at this point: Quiros had come from the north and now veered west and north around San Marcos, whereas Bougainville was coming from the west and after rounding the north of Aurora sailed south and then west. As a result, Quiros reached the wide bay at the north of Espíritu Santo, which he would name San Felipe y Santiágo, while Bougainville sailed south of the same island and sought Quiros's famous bay in vain.

[2] Aoba (also found as 'Omba' and 'Oba') lies between Espíritu Santo and Maéwo. It was probably seen by Quiros but merged with other lands and not clearly distinguished, and Bougainville can lay a claim to being its first real discoverer. The disease mentioned may have been leprosy, but was probably one or more variety of tropical skin disease.

During the morning, we saw a few canoes along the shore. None had sails and none tried to come alongside. One can see no huts on the coast, which is very steep and covered with trees up to the water's edge. Smoke rose up from several places in the middle of the forest from the top of the mountains down to the coastline.[1]

Quite close to land, we let out 50 fathoms of cable without finding bottom. Bearings at 3 p.m.: Etoile Peak by the N point of Aurora Island N¼NW and S¼SE, the same N point by the SW one of Lepers Island NNE and SSW.

At 4 p.m. bearing of the middle of Etoile Peak N4°W, 10 to 12 leagues distant, the N point of Aurora Island E¼SE5°E, 1½ leagues. These bearings and our position at 4 o'clock following our midday observations places this N point of Aurora Island in latitude 14°54'50" and longitude 169°11' E Paris meridian, corrected 167°30'.[2]

At 6 a.m. the SW point of Lepers Island bore SW¼S, the NE point of the same island SE¼S3°E distant approximately 2 leagues, the S point of Aurora Island in view SE5°E approximately 5 leagues, the N point of the same N approximately 6 leagues.

At midday bearing of the NE point of Lepers Island E5°N, 2 to 3 leagues, the SW one of the same island SW¼S about 6 leagues, the most N land of Aurora Island NNE about 8 leagues.

By these bearings combined with the altitude observed at midday, this N point of Aurora Island, found by the bearings of 4 p.m. and the altitude of midday yesterday to lie in latitude 14°54'50" and longitude 169°11', corrected 166°56', now appears to be in latitude 14°54' and longitude 169°11', corrected 166°56'. These two estimates agree perfectly and prove that the currents run only N, it has not yet appeared to us that they flow back S.

The horizon was too hazy throughout the morning for us to see the land lying to the W.

Note: from the midday bearings, the SW point of Lepers Island is in latitude 15°27' and longitude 168°44'E, corrected 166°2'.

Monday 23 to Tuesday 24.

Almost calm in the afternoon, the little breeze there was being variable from all sides. After dinner I went ashore in the island with three of the gentlemen. When I arrived I found the wood cut and already in the boats as well as some refreshments consisting in bananas, coconuts and *evis*.[3] Mr de Kervé who was in charge of the operation on land told me that a large number of islanders, their bows and arrows at the ready had wanted to oppose the landing. He carried it out in spite of their threats and took up a position at the edge of the forest. Gradually the Indians came closer with a more amiable appearance. He gave them a few presents and, with no further delay, set his workers to cut wood under the protection of the armed troop. Our boats

---

[1] The island is well populated, with several lakes in the interior. It is mountainous, rising in places to over 1,200 m (4,000 ft).

[2] The latitude given for this headland is accurate: it lies in 14°55'. The longitude shows the substantial error that has developed in the calculations: it should be 165°45' (168°05' East of Greenwich).

[3] This was the *Spondias dulcis*, known as the Vi apple or Tahitian apple. See infra, Journal of Fesche, note 20.

were facing, afloat on their grapnels.[1] In exchange for pieces of red cloth, the Indians helped to carry wood to the boats. A detachment also went to look for fruit. The Indians were never willing to barter their bows, nor their clubs, which are short and of some iron-wood.[2] They traded a few arrows. They remained suspicious, did not let go of their weapons and even those who did not have any bows were holding stones ready to be thrown. They endeavoured to make us understand that they were at war with the inhabitants of a neighbouring settlement. And indeed an armed group appeared advancing in good order. The others made ready to receive them, but there was no attack. When we had all returned to our boats, the islanders threw us a hail of stones and shot a few arrows. One sailor only was lightly hurt by a stone. A few gun-shots made these scoundrels flee and wounded or killed a couple. These people are ugly, short in stature, covered in leprosy. They are naked except for the natural parts. A few have painted chests. They are of two colours, black and mulatto. They have frizzy hair and thick lips, they pierce their nostrils to hang a few ornaments. I did not see any with a beard. They wear bracelets of pig or boar's teeth and pieces of turtle shell. We saw no huts, merely the remains of a couple and they must be rough low shanties. However they have cloth as in Cythera and material made of small palms. There is a kind of fig that is special to this country about which I have no other details.[3]

At 5 p.m. hoisted in our boats, it was almost calm during the night and the morning, the slight breeze being variable.

At midday the estimated run according to the bearings is W¼SW1°45'S, distance 17'20".[4]

Observed latitude: 15°18'. Calculated longitude: 168°39'30", corrected: 166°24'30".

In the morning we saw fairly clearly the land to the W.

Louis who had gone ashore with us expressed a great deal of contempt for these islanders.

Bearings at midday: the SW point of Lepers Island SE¼S4°E 3 leagues, the most NE point of the same island ENE3°N, the most N land of Aurora Island NE¼N. Land in sight to the W from WNW to SW¼S.

These bearings, joined to the altitude observed at midday places the SW point already determined yesterday in latitude 15°26' which agrees with yesterday's estimate and fixes our position at midday.

Tuesday 24 to Wednesday 25.
Observed variation by azimuth in the morning: 10°11' NE.

---

[1] A grapnel is a small anchor. Thus the boats were lying adjacent to the beach, bows to seaward, held in that direction by their grapnels, and ready for immediate embarkation and departure.

[2] This is probably the *Casuaria equistefolia*, one of the iron-woods found in these waters, about which Joseph Banks wrote that it was used for 'Clubbs of 6 or 7 feet long made of a very heavy and hard wood': Beaglehole, *The Endeavour Journal of Joseph Banks*, I, p. 386. However, Bougainville was unsure of what species this might have been, as his wording 'of some iron-wood' indicates, and it could have been some other hard wood, several species of which are found in this part of the world.

[3] This was presumably a fig, but with 800 species of the Ficus and the sketchy details provided, it is difficult to pinpoint the one seen by Bougainville.

[4] From Lepers Island, the French sailed on a westerly course towards Espiritu Santo, veering south once they reached a point of the coast slightly north of the present-day capital Santo, or Lunganville.

The calm continued during the afternoon. In the night, there was a slight breeze variable from SSE to SSW, SW then NE up to E. We took advantage of it to get out from the shelter of this large Lepers Island. At around 5 a.m., the ESE wind freshened and we sailed close-hauled port tack. The *Étoile* was still becalmed in the lee of the land. We made all sail to examine the SW land until 10 o'clock when I altered course to make for the flute which was hardly visible from the mast-head. Furthermore we had a vast extent of land all around us. I am not sure whether they form a single coast or whether it is a considerable number of islands. Various bights can be seen, either passages or bays. In the distance one can see the crests of mountains rising in levels and losing themselves among the clouds. The mountainsides, barren and red in certain places, seem to indicate that they contain minerals, elsewhere one sees charming plains, forested, and the sweeping glance suggests a rich country. Several canoes with sails cross from one land to another and seem to be good sailers.

At midday the estimated run give SW¼S5°W, distance 25'.

Estimated latitude: 15°34', observed: 15°36'.

Calculated longitude: 168°24', corrected: 166°9'.

Bearings from E to N through S and W: a point of Aurora Island ENE5°N, gap up to another point ENE4°E, continuation from there up to ESE3°E, gap up to a point E¼SE4°E, continuation from there to SE¼E, gap up to a point SSE2°S, continuation from there to S5°W, gap up to a point S¼SW, continuation from there to SW¼S, gap up to a point SW5°W, continuation from there to WSW5°W, gap up to a point W¼SW5°S. The land continues from there towards the WNW where it is lost from sight.

Note: the ship was in the position of the above bearings on the 25th at 3.30 p.m. in latitude 15°29' and longitude E 168°30', corrected 166°15'. This stretch of land gives the appearance of a great gulf.

Wednesday 25 to Thursday 26.

We spent the afternoon rejoining the *Étoile*. She joined us at sunset and we tacked variously throughout the night with variable ENE to ESE and E winds, fair weather, superb moonlight. At daybreak, it appeared from the bearings of the various lands that the currents had carried us a considerable distance to the SSW. I sailed rounding from NW¼W to W to examine the wide opening shown in yesterday's 3 p.m. bearings as stretching from WSW5°W to W¼SW5°S. If this is not an immense bay, it is a passage between these extensive lands.[1]

At midday I observed latitude 15°35' which I consider to be doubtful although it agrees within 10" with the estimate by account 15°35'10". Three other observers obtained 15° from 40' to 49'.

Calculated longitude: 168°15', corrected 166°.

Bearings for the 26 at midday: the starboard point as you enter NW¼N approximately 3 leagues. The port point as you enter S5°W approximately 5 leagues, the middle of Lepers Island NE.

---

[1] It is the latter, a strait some 15 km (9 miles) wide that would henceforth be known as Bougainville Strait, between Malo, an island just south of Espiritu Santo, and Malakula (or Mallicolo).

THE JOURNAL, MAY 1768

Thursday 26 to Friday 27.

Fair weather, fairly good E to ENE breeze. We sailed along the coast of the starboard land as you enter, at a distance of about 4 miles, sailed WNW. At 2 o'clock, I hove to and sent three armed boats to examine some rather pretty inlets that showed promises of good anchorages. Along the shore several Indians were gesturing at us. They are almost black. The *Étoile*'s boat fired on some of their canoes one of which had shot an arrow at it. The canoes soon disappeared. At 6 p.m. I hoisted in our boats and stood out to sea for the night. The officer in charge told me that he had found different depths from 40 to 15 fathoms some of which were rocky and others of sand and ooze. The inlets are formed by the various contours of the islands that overlap each other. During the night we followed various tacks off and on; at dawn we came back towards the coast and continued to range along it 1 league off, the weather being cloudy and the winds variable from N to NW and NE.

We see no land ahead of us and I am beginning to believe that this is not the Bay of St James and St Philip [Baye St-Jacques et St-Philippe], although the aspects are similar to the description given by Quiros and we are exactly in the latitude and the longitude indicated in the extract from his relation given in the History of navigations to the austral lands. The doubts will clear as we continue on a W course.[1]

I shall give tomorrow the estimated run for the two days if I can take the altitude and I shall endeavour in this way to correct by observation both the estimate and the observations. We are continuing to notice that the currents run strongly to the S.

Bearings. At sunset on the 26th. The most E land starboard on entering bearing ESE4°E 7 to 8 leagues, the most W point of the same W5°S about 6 leagues, the low lands between the two NNE5°E. The E point of the port land SE3°E 9 to 10 leagues, its W point SSE2°S about 10 leagues.

At midday on the 27th the N and the starboard land bearing NW to NW¼N4°N 8 to 9 leagues, the most E bearing E¼NE ditto. As the weather was hazy and tending to rain, one could not see the port or S lands.

The 27th at 6 p.m. The most S point of the starboard or N land NE¼E, the N point of the same NE3°E, a large islet named Tablebay NE¼N2°E, the most NW land N¼NW3°W. The port or S land from E¼NE to E¼SE.[2]

Friday 27 to Saturday 28.

Observed variation occiduous: 9°30'.

In the afternoon, cloudy weather, rain, calms, slight breezes from W to SSW and S. During a break we again had a view of the S land. Several times we thought we could see land ahead of us, land of clouds that disappeared when they did. We sailed W¼NW until 2 a.m. when we hove to, the weather being black with rain; in the morning, it was almost calm and I went on board the *Étoile* for various matters. Taking

---

[1] As Bougainville had sailed to the south of Espiritu Santo, he had no chance of seeing Quiros's bay of San Felipe y Santiágo. He will continue for some time to endeavour reconciling the Spaniard's landfall with what he could see around him. The History referred to in this passage is the seminal work by Charles de Brosses, *Histoire des navigations aux terres australes*, 2 vols, Paris, 1756.

[2] The appearance of an English name in an area not visited by any English navigator has proved puzzling. There is however in the approximate location indicated a Tutuba Island off the south-east coast of Espiritu Santo and one can only assume that the French obtained this name from the islanders.

my latitude observed at midday on the 26th of 15°35' and today's the run for the 48 hours would be W4°S 70' and I think it to be accurate.

Observed latitude: 15° 40'. Calculated longitude: 167°2', corrected: 164°47'. The currents have again been running S.

At midday on the 28th the most NW bearing NE¼N4°N about 15 leagues, the most SE bearing E¼NE about the same distance. The highest land NE¼E 9 to 10 leagues.

Note: These lands we have seen from yesterday are of an immense height and form a fairly unbroken chain above the clouds. I do not know how far they extend to the N. I would think that they go as far as 11°, that between 12 and 14° they can present a fine bay whose position was concealed by Quiros and that in 1722 Roggewin saw the northern end of these lands that he named Groningue.[1]

Saturday 28 to Sunday 29.

Observed variation occiduous: 9°47'.

Fine weather, fair sea, good S to SSE breeze, sailed WNW until 8 a.m. then W¼NW almost all night. We had courses brailed up and the main topsail shivering[2] to wait for the *Étoile* which sails worse than ever. Nevertheless we are compelled not to lose one instant. The state and condition of our food stocks require us to reach some European establishment. We have been for a long time on the same rations as the crew. We have only a few poultry left that are reserved for the sick and three turkeys that will do for three Sundays. We are eating one at midday today and it is a great cause of rejoicing for us. These large beans, called horse-beans, bacon and salt beef almost three years old make up such sad meals!

This morning we no longer saw land. I have named the archipelago we have sailed through the Great Cyclades and I have taken possession of it in the King's name.[3] Now Quiros's longitude and latitude are left behind. Where then is his great land? Would Bellin have been writing prose without knowing and would the Land of the Holy Ghost [Terre du Saint-Esprit] link up with New Guinea [Nouvelle Guinée]?[4] For my part, *Davus sum, non Oedipus*.[5] With constancy in keeping to the parallel we have been following, we shall soon know what the answer is.

---

[1] Bougainville is still misled by an appearance of concordance between Quiros's reported position and his own. Quiros himself sailed north from Espiritu Santo and did not sight the western side of the island. However, Torres who became separated from Quiros did sail south a couple of weeks later. Jacob Roggeveen's Groeningen was Upolu, seen in the distance on 15 June 1522, but nowhere near the area where Bougainville is now sailing: Sharp, *The Discovery of the Pacific Islands*, p. 100; see also Mulert, *De Reis van Mr Jacob Roggeveen*. The chain of mountains visible from the French ships is a range than runs roughly north and south down the western side of Espiritu Santo, with its most prominent feature being Mt Tabwemasana at 1,800 m (6,170 ft).

[2] i.e. with the wind blowing on the strengthened edge of the sail making it flutter in the wind.

[3] The Act of Possession is dated 30 May 1768. A copy is held at BN, NAF 9407, 142.

[4] Bougainville tends to lay the blame on J. N. Bellin whose necessarily broad-brush charts could only draw upon the scanty information available at the time. The reference to writing prose without realizing it comes from Molière's *Le Bourgeois gentilhomme*, in which the main character is delighted to discover that what is not verse is prose: 'I have been speaking prose for forty years without realizing, and I am most obliged to you to have told me this:' II, iv.

[5] 'I am Davus, not Oedipus:' Terence, *Andria*, 194. Davus (Davos in Greek) was a stock character in classical comedies, usually a slave attending as a sort of Jeeves to the household's eldest son. Hence Bougainville's comment: 'I am a humble servant, not a solver of riddles'.

At midday the corrected run for the 24 hours is NW¼W2°15′W, distance 82′.
Estimated latitude: 15°1′, observed: 14°57′.
Calculated longitude: 165°50′, corrected: 163°35′.

Bearings at 6 p.m.: the most NW land bearing NNE; the most E bearing E, land of an extraordinary height, one could see the tops of mountains above the clouds.

This archipelago of the Great Cyclades, judging its size from the area we have covered and what we have seen in the distance, may cover 4° of latitude and 5° of longitude.[1]

Note: Yesterday I checked a rather peculiar event on board the *Étoile*. For some time, a rumour had been circulating on the two ships that Mr de Commerçon's servant, named Baré, was a woman. His build, his caution in never changing his clothes or carrying out any natural function in the presence of anyone else, the sound of his voice, his beardless chin, and several other indications had given rise to this suspicion and reinforced it. It seemed to have been changed into a certainty by a scene that took place on the island of Cythera. Mr de Commerçon had gone ashore with Baré who followed him in all his botanizng, [and] carried weapons, food, plant notebooks with a courage and a strength which had earned for him from our botanist the title of his beast of burden. Hardly had the servant landed than the Cytherans surround him, shout that it is a woman and offer to pay her the honours of the island. The officer in charge had to come and free her. I was therefore obliged, in accordance with the King's ordinances, to verify whether the suspicion was correct. Baré, with tears in her eyes, admitted that she was a girl, that she had misled her master by appearing before him in men's clothing at Rochefort at the time of boarding, that she had already worked for a Genevan as a valet, that, born in Burgundy and orphaned, the loss of a lawsuit had reduced her to penury and that she had decided to disguise her sex, that moreover, she knew when she came on board that it was a question of circumnavigating the world and that this voyage had excited her curiosity. She will be the only one of her sex [to have done this] and I admire her determination all the more because she has always behaved with the most scrupulous correctness. I have taken steps to ensure that she suffers no unpleasantness. The Court will, I think, forgive her for this infraction of the ordinances. Her example will hardly be contagious. She is neither ugly nor pretty and is not yet 25.[2]

Sunday 29 to Monday 30.

Fine weather, fair sea, good ESE to SE breeze, we sailed W¼SW all sails out. At night we took in only the studding sails, as the moon clearly lit up the horizon. For dinner we ate rats and we considered them quite good.

---

[1] The New Hebrides (Vanuatu), including the Banks and Torres Islands to the north, stretch from 13°10′ down to 20°10′S and from 166°30′ to 169°53′E. Bougainville therefore was unaware of the existence of island groups to the north and south, but overestimates the longitude.

[2] Slightly different versions of this episode are found in the other journals. Vivez, for instance, in a rather lengthy account, names Ahutoru as the first Tahitian to realize she was a woman. Vivez of course was sailing in the *Étoile* with Commerson and Baret, whereas Bougainville, in the frigate, learnt about these events later. How much the commander knew about the Baret affair and when he really discovered it is open to speculation. See Introduction, 'The Participants'.

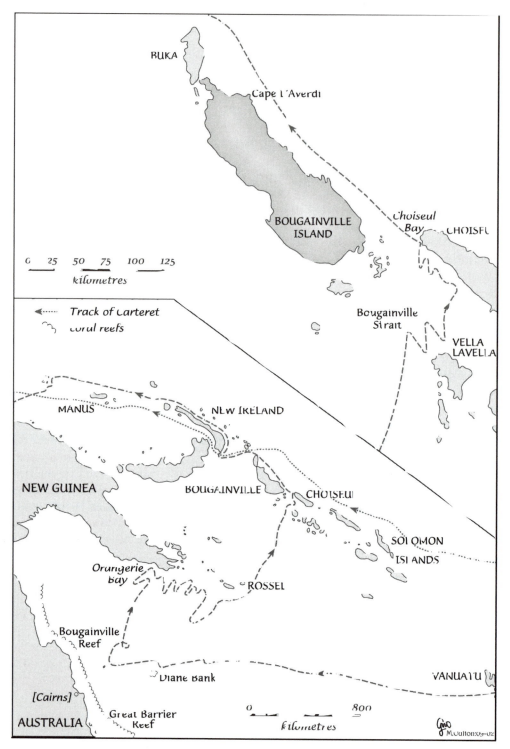

6. Vanuatu to New Guinea, 28 May–8 August 1768.

The run corrected for the 24 hours is W2°15′N, distance: 135′.
Estimated latitude: 14°56′, observed: 14°52′.
Calculated longitude: 163°31′, corrected: 161°13′.

Monday 30 to Tuesday 31.

Fine weather, fair sea, good SE to ESE breeze, we sailed W¼SW.
The run corrected by the altitude is W2°40′N, distance: 130′30″.
Estimated latitude: 14°52′, corrected: 14°46′.
Calculated longitude: 161°16′ E, corrected: 158°55′.

June

Tuesday 31 May to Wednesday 1 June.

Observed variation by azimuth: 10°10′ NE.
Fine weather, fair sea, light ESE to E breeze. We sailed WSW. At 7 p.m., sounded with 200 fathoms of cable, no bottom.
The estimated run for the 24 hours is W¼SW1°S, distance: 104′.
Estimated latitude: 15°8′.
Longitude Paris meridian east: 159°32′, corrected: 157°8′.

Wednesday 1 to Thursday 2.

Observed variation occiduous: 9°52′.
Not much breeze in the afternoon, at 11.30 p.m., the weather being very black, I hove to until 4 a.m. Good SE breeze. Steered W¼SW and WSW.
At midday the run was W¼SW5°15′W, distance: 86′.
Estimated latitude: 15°17′30″, observed: 15°18′.
Calculated longitude: 158°4′, corrected: 155°37′.

Thursday 2 to Friday 3.

Stormy ESE to SE gale, fairly rough sea, sailed during the day all sails set, heading WSW. At night, hove to during the night [*sic*] to wait for the *Étoile*, then sailed W¼SW under double-reefed topsails reefed mizzen topgallant sail. Hove to from 4.30 a.m. to 6 on account of a very black squall.
At midday the run is W5°15′S, distance: 148′.
Estimated and observed latitude: 15°32′.
Calculated longitude: 155°28′, corrected: 152°58′.

Friday 3 to Saturday 4.

Observed variation ortive: 7°4′ NE.
Stormy ESE to SE gale. Sailed W¼SW. All sails set during the day, the topsails reefed at night.
At midday the run is W3°S, distance: 152′.
Estimated latitude: 15°42′, observed: 15°40′.
Calculated longitude: 152°52′, corrected: 150°19′.

Saturday 4 to Sunday 5.

Observed variation: 8° NE.

Fair weather, continuation of the fresh ESE breeze. Steered W¼SW all sails set until night time when I steered W under reefed topsails.

At 11 p.m., we discovered some breakers to the S at approximately half a league distance (lst reef in latitude 15°41′ and longitude 151°35′, corrected 148°59′) we at once sailed close-hauled starboard tack steering NE and NE¼E until 4 a.m. when I hove to wind aport, drifting SW. At daybreak, I filled the sails and steered WSW to come closer to the breakers and examine them. At 8.15 they were sighted from the mast-head bearing SSE about 2 leagues, it is a small sandy island, quite flat against which the sea breaks. It extends from E to W. At 8.30 I steered W.

One can see from this encounter how risky this navigation is.[1] At midday the route is W4°15′N, distance 106′.

Estimated latitude: 15°33′30″, observed: 15°31′.

Calculated longitude: 151°2′, corrected: 148°26′.

Sunday 5 to Monday 6.

Observed variation occiduous: 6° 55′ NE.

Fine weather, good ESE to SE breeze. We sailed W¼SW, all sails set until night when I hove to. At midnight I set course for NE¼E until 4 a.m. when we bore back SW¼W. At 6 o'clock reverted to W¼SW. Sounded at 6 p.m., let out 240 fathoms without finding ground.

At midday the run is W3°S, distance 84′.

Estimated latitude: 15°34′, observed: 15°33′.

Calculated longitude: 149°34′, corrected: 146°55′.

For the last 24 hours numerous pieces of wood and a few fruit have been passing alongside. The sea has also considerably moderated. I have no doubt that there is land SSE of us.

Monday 6 to Tuesday 7.

Continuation of the fresh ESE to SE winds. I was steering W¼SW when at half past one, we became aware of breakers ahead, approximately 2 miles from the W¼SW to WNW (2ⁿᵈ reef in latitude 15°34′30″ and longitude 149°22′E, corrected 146°40′). We changed tack immediately and we steered NNE altering round until we returned to W and then W¼SW.

These breakers covered a wide extent with several rocks visible above the water. Some people thought they could see a low land SW of the breakers.[2] From 4 o'clock we had reverted to the W¼SW course when at 5.30, they shouted from aloft that they could see new breakers about 5 miles away from WSW to NNW with a fair number

[1] Bougainville has now reached the approaches of the Great Barrier Reef. He was fortunate in that he came upon these shoals and breakers before striking the dangerous coral barrier that stretches for 2,000 km (over 1,200 miles) off the east coast of Australia. The names Diane Bank and Bougainville Reef mark these discoveries on present-day maps. Caro, seeing a large number of birds over Diane Bank, called it Gannet Island, Caro, *Journal*, 4–5 June 1768.

[2] It is just possible that the Australian continent was sighted at this point. The French had reached the dangerous cluster of reefs and shoals now known as Bougainville Reef in lat. 15°30′ and long. 147°06′ E., almost due east of present-day Cooktown. At this point, Bougainville wisely decided to alter course and sail north and north-east, a decision that saved his expedition from coming to grief on the Great Barrier Reef.

of uncovered rocks, a third warning to me not to insist on seeking land along this parallel (3rd reef in latitude 15°17' and longitude 149°13' corrected 146°31'), these three reefs which we saw accompanied by several other rocks equally visible, and maybe by invisible shoals, seem to make up a chain running approximately SE and NW.[1]

At 6 o'clock I altered to the starboard tack and sailed close hauled. At midnight we changed course and returned to our track until 4 a.m. when we resumed the starboard tack sailing close to the wind. At 6.30 I set course to NNE, the winds being ESE stormy gale. We lowered our topsails.

Estimated run at midday were NNE30'E 56', corrected run NNE3°N 65'.

Estimated latitude: 14°42', observed: 14°20'.

Calculated longitude: 149°57', corrected: 147°15'.

Note: We are continuing to come across numerous pieces of wood, a few fruits and some seaweed, most of the pieces of wood are covered in barnacles, evidence that they have been in the sea for some time.

Observation. I had planned to approach land along the parallel of 15 to 16°, not that I was convinced that the southern land of the Holy Ghost as anything other than the archipelago of the Great Cyclades and that Quiros falsified either his discoveries or his narrative. However as Bellin places this land, I do not know on what basis, in longitude 148° East, I wanted to sail as far as that, which would have placed me in a position either of paying homage to that geographer's knowledge, or of visiting a coast unknown until now. However, if I go back to the longitude determined by Mr Verron's latest observations, I have passed the limit given by Bellin since yesterday at midday, I would have been in longitude 146°40', but according to my estimated longitude I was still 30 leagues from sight of land. The encounter with this succession of breakers does not allow me to continue to seek Quiros's southern continent here. These approaches do not bear any resemblance to the indications given by that navigator. Several other reasons compel me to come closer to the Line, [and] even to beat up towards the E.

Such extensive breakers announce a low coast and when I see that Dampierre, in this same latitude of 15°35' abandoned the western coast of this barren land, where there is not even any drinking water and whose approaches bristle with shoals and reefs, I conclude that the eastern coast is no better and I am inclined to think, like him, that all this land is nothing more than a mass of islands and banks.[2] Moreover these pieces of wood, these fruits, this seaweed that we are finding, the quietness of the sea, the currents, everything tells us that we have had land to the SE for several days and I am led to believe that it trends trend SE and NW, as do the reefs that defend it.

Other reasons peculiar to our situation urge me to sail towards known countries. I have only enough bread for three months at most, wood and water for hardly

---

[1] This is a clear sight of the Great Barrier Reef.

[2] Bougainville is comparing the voyages of Dampier with what he has seen off the east coast of the Australian continent, which leads him to this ungenerous interpretation. Dampier, the first Englishman to visit mainland Australia, reached the north coast of western Australia on 4 January 1688; several weeks were spent in what was undeniably a bleak and inhospitable land. He returned in the *Roebuck* in July 1699, sailing along the west coast to Shark Bay and what is now Dampier Archipelago. Nothing of what he saw suggested that the country held much promise.

a month, we have been 7 months at sea: my crews are truly in a state of health that is close to a miracle. But if I become embayed along a dangerous coast that may lack everything we are short of, especially water (Dampierre did not find any on his western area), how could I free myself from it with E and possibly NE winds these being prevalent in this season. I shall need an infinity of time: discouragement, shortages, sickness their inseparable companion, will soon have crushed our wretched crews. Furthermore I have no anchors left to sacrifice to save the ships in such dangerous waters. I therefore beg Mr Bellin's forgiveness, but I shall not verify whether he is right or wrong. I shall be satisfied with the fairly justifiable belief that we are very close to land, that this land is not that of the Holy Ghost, [and that] judging from its approaches it offers no promise of any facilities for the settlement of a colony that would be useful to its mother country.

Tuesday 7 to Wednesday 8.

Corrected variation: 7°.

Fairly cloudy weather with E to ESE squalls, sailed NNE until 6 p.m., NE until 8, close-hauled port tack from 8 p.m. to 2 a.m. when we resumed the starboard tack and sailed NE¼N to N¼NE, the winds becoming closer.[1]

At midday the estimated run is NN, distance 69'30", corrected NNE2°30'N, distance 76'.

Estimated latitude: 13°16', observed: 13°9'.

Calculated longitude: 150°24'25", corrected: 147°39'.

Wednesday 8 to Thursday 9.

Observed variation occiduous: 5°20', ortive: 5°10'.

Stormy E¼NE to E¼SE gale. We steered N until 10 p.m., from 10 to 2 a.m., kept close-hauled port tack, then resumed our course.

At midday the run corrected is N1°5'E, distance 107'.

Estimated latitude: 11°35', observed: 11°22'.

Calculated longitude: 150°26'30", corrected: 147°38'.

We are continuing to see numerous birds and to come across pieces of wood and various fruits.

After taking stock with the supply officer, I have bread left for two and a half months, vegetables for 50 days, meat is more abundant but it is extremely old. Generally speaking we have insufficient food left. '*O passi graviora dabit deus his quoque finem.*'[2]

Thursday 9 to Friday 10.

E¼SE to E and NNE winds. Steered N until 8 p.m. and changed tack until midnight, then remained close-hauled on the starboard back, little breeze, almost calm. At dawn, we saw land stretching from E to NW, an extremely high land in the hinterland with magnificent low lands along the seashore.[3]

---

[1] Meaning not entirely clear, but perhaps that the wind began to come round to head the ship so that she could no longer hold her course.

[2] 'You who have been through worse, God will grant an end to these misfortunes also.' Virgil, *Aeneid*, I, 198. This is an extract from Aeneas's address to his men upon reaching the shore after a storm.

[3] The expedition has now reached the south-east coast of New Guinea, east of present-day Port Moresby. The mountains in view are the Owen Stanley Range and straight ahead the Keveri Hills. The

Estimated run: N6°W, corrected N5°W, distance 65′.

Estimated latitude: 10°25′, observed: 10°17′.

Calculated longitude: 150°20′, corrected: 147°29′.

Bearings of land 5ᵗʰ division. At midday land from E to W and W¼SW. The current is taking us inshore.

Friday 10 to Saturday 11.

During these 24 hours, we have had calms, the current was carrying us towards the land and at 6 p.m. we were not one league from a small island at the eastern end of which is attached a reef that extends more than 2 or 3 leagues E. We spent the night in this worrying situation, making every effort with the help of long moments of light breeze to keep away from the shore. At midday we may have been some 4 leagues from land.

Note: while we were in the lee of the shore, a very sweet odour wafted towards us from the shore.

Sounded at 2 leagues from land without [finding] ground.

Throughout the day great fires were lit on the shore and we saw several canoes with sails. The low land is superb and the mountains whose crests are lost in the clouds, seem to announce a rich country. One can distinguish three levels from the plain up to the highest range which is in the hinterland more than 25 leagues away.

The run was S distance approximately 13′.

Observed latitude: 10°29′. Calculated longitude: 150°20′, corrected: 147°29′.

Sighted at midday the most E land in view bearing NE¼E, the most W to the NW.

Saturday 11 to Sunday 12 and Monday 13.

Observed variation occiduous: 5°41′.

Bearings on the 11ᵗʰ at sunset the most E land E¼NE2°E, the most W WNW, all at a distance of approximately 12 leagues.

The weather has been frightful these 48 hours, high wind, a great deal of rain, the sea rough, a thick fog that forces us to have resort to guns to keep in touch with the *Étoile* and the winds driving us towards the coast, we cannot make progress by tacking being compelled to wear,[1] having only one reef free in each topsail and the heavy seas causing us to make considerable leeway, we cover 18 points of the compass from one tack to another,[2] yesterday at 9 a.m., the wind having turned very fresh from the ESE to SSE, I sailed on the starboard tack, then at 1.30 we changed to the port tack for the night, the winds wild and squally. At 6 a.m. I altered tack for the day and we had no sight of land during some fairly fine breaks in the weather. During the

---

land the French could see probably extended from Keppel Point in the west to the Brumer Islands off the eastern extremity of Papua New Guinea.

[1] I.e. alter tack by turning away from the wind and allowing it to pass through the stern. A considerable amount of ground was lost doing this as opposed to tacking when the vessel altered tack with the bow passing through the wind. The latter was difficult to achieve in a square-rigged vessel when there was a heavy sea running as the waves beat the bow back onto the former tack and it is then necessary to back the sails and gain sternway in order to get the bow to pay off on the new tack.

[2] I.e. the tracks made good on the different tacks were 202½° apart, thus although tacking into the wind they were actually losing ground.

night tack, the winds altered aft a little and this morning they altered forward very conveniently as we were going to change tack. Praise God that we had sighted land before these days of fog and great winds. Indeed the winds being ESE to SE, I thought that by steering NE I was being excessively prudent because of the darkness, however we would have run the risk of being wrecked since on Friday morning we saw land stretching along the E. May we manage to round it and cursed be anyone who traces a shoreline on a chart with no other basis that his own capricious ignorance! I wish him a share of the cruel nights we are going through. No food and no sleep, such a diet puts one in ill-humour.

The estimated run for the 48 hours is SSW2°30'S, distance 22', the corrected route is S¼SW2°45'W, distance 31'.

Estimated latitude: 10°49', observed: 11°.

Calculated longitude: 150°12', corrected 147°20'.

Note: we have seen several sharks, a *bequune*[1] and, at nightfall, 7 or 8 of these fish named *cornets* jumped onto the waist of the ship where they were caught, a visit that did not fail to worry me.[2]

Monday 13 to Tuesday 14.

Continuation of the same weather, we tacked variously and made little progress by means of this dreary work, we saw land clearly as far as NNE4°N, some thought they could see it up to ENE. What could be seen distinctly consisted of two detached islands 8 or 9 leagues from us and I believe it is the land sighted to the E on Friday morning in which case, we would have progressed only by 6 compass points towards it in 4 days, it is a long time to be on tenterhooks.

I have made a precise list of all the food supplies on board the *Étoile* in order to make a cruel but inevitable reduction of the crews' rations.

The estimated run for the 24 hours is E¼SE3°S, distance 14', corrected SE4°30', distance 21'.

Estimated latitude: 11°3', observed: 11°16'.

Calculated longitude: 150°25', corrected: 147°33'.

Bearings at 6 p.m. the 2 islands at the end of the land, the most N N¼NE, the most E NNE4°N, about 8 to 9 leagues.

Tuesday 14 to Wednesday 15.

Observed variation occiduous: 6°1'NE.

The weather returned to fair and the winds ESE to E¼SE and E. We sailed S¼SW to S, S¼SE and SSE from midday until 6 a.m. when I altered tack and we sailed NE¼N and NE, all sails set, the sea fairly calm.

If the land, starting with the coast seen on Monday night, is only trending N¼NE, even NNE, we could hope to round it as long as the winds veer aft. We are in the greatest need of their kindness.

---

[1] The *bécune* belongs to the barracuda family (*Sphyraenidae*). Common in tropical waters is the *Sphyraena novaehollandiae*.

[2] Not strictly fish but cephalopod molluscs, known by the more popular name of calamary, these are common in most warm waters and some varieties, such as the *Omnatrephes* and *Todarodes*, are noted for their ability to fly above the waves and in this case land on deck.

The estimated run for the 24 hrs is S¼SW 3°15′W, distance 34′, corrected S¼SW 1°W, distance 41′.

Estimated latitude: 11°50′, observed: 11°57′.

Calculated longitude: 150°16′, corrected: 147°20′.

The run from Friday morning when we first sighted this land has been S2°15′W, distance 100′. At that time the nearest land bore from us N 2½ or 3 leagues.

Wednesday 15 to Thursday 16.

Observed variation occiduous: 6°10′, ortive: 6°6′.

Fair weather, the winds variable ESE to E. We sailed close-hauled starboard tack until 6 a.m. when at daybreak we saw land from the N to NE¼E 4°E. This latter seemed to form several islets separated from the mainland, that is the one that some people, on Monday night, thought they could see bearing ENE. I at once veered to the port tack. Our situation is most critical, should the winds continue to be so unfavourable.

The estimated run for the 24 hours is NE2°10′E, distance 55′, corrected run NE3°50′E, distance 61′30″.

Estimated latitude: 11°20′, observed: 11°29′30″.

Estimated longitude: 150°57′, corrected 148°.

Note: It appears from our bearings taken upon different sightings of this land that the currents here run only N and S. The chart I am drawing of our route proves it.

Thursdy 16 to Friday 17.

Observed variation occiduous: 6°16′.

Fair weather, fair smooth sea, variable wind from ESE to SE¼E. We steered S¼SE until 1 after midnight when we sailed close-hauled port tack for three hours. During the morning we saw several small islands from the NW to NE¼E. At 11.15 I altered course and we then saw the most W islet that I have named Ouessant on account of its great resemblance to the island of that name bearing NE¼N [dist.] 2½ leagues,[1] the middle island bearing NW¼N 2 to 3 leagues, at the same time a low land bearing ENE 5°N about 7 [leagues] and some breakers up to ENE, which seemed to link up with Ouessant. Other breakers extend to the NW of Ouessant for about half a league. The middle island is also almost surrounded by 2 reefs, the SE of one of which extends out for 1 league and the NW of the other 2 leagues.

Corrected run: ENE 1°54′E, distance 50′.

Estimated latitude: 11°18′, observed: 11°17′.

Calculated longitude: 151°47′, corrected: 148°50′.

Note: by sailing close along the islands, we clearly noticed that the current was bearing us towards the E.

Bearings at midday the 3 islands, the most W NW 7 to 8 leagues, the middle one N¼NW 4°W 3 to 4 leagues, Ouessant NNE 3°E 3½ leagues.

---

[1] Ouessant (Ushant) is an island situated off St Mathieu peninsula, separated from the western extremity of Brittany by the Chenal du Four. It is surrounded by several smaller islands, and has two central valleys running south-west to north-east. The island which Bougainville named after it is Tariwerwi in the Louisiade Archipelago, in latitude 11°10′S and longitude 151°13′ E. of Greenwich. The name Ouessant is still found on some maps today.

Friday 17 to Saturday 18.

The weather fairly fine in the afternoon, the winds ESE to SE. Thanks to a fairly favourable current we raised Ouessant and already were confident of finding the open sea when, at about 4 o'clock they called out from aloft that they could see the chain of breakers continuing from Ouessant to the E with an appearance of low land behind it. We therefore had to change course at 4 and then had abeam, to leeward, some breakers that were no more than a couple of miles away. The wind rose violent at sunset and compelled us to keep a minimum of canvass because our two topmasts, in quite poor condition, require careful handling. We spent the night tacking, spending longer on the SW tack than on the NE until 5 a.m. when with SSE winds and heavy seas we sailed E. At midday we could see nothing from aloft and the breakers seen last night must be lying to the NNE. Hope is reborn: it is certainly an occasion when one can say: 'Beautiful Iris, one is in despair even though one is still hopeful'.[1]

At midday the run gave me SE¼E 29'.

Estimated latitude: 11°33', observed doubtful: 11°29'.

Calculated longitude: 152°12', corrected: 149°15'.

Bearings at 4 in the evening Ouessant NW¼W 9 or 10 miles, the most E breakers in view E¼NE.

Note: for the last three days we have reduced by a 5th the quantity of vegetables added to the soup and by almost a third the allocation of biscuit or fresh bread. Soup is no longer served on fast days at midday. It is replaced by salt meat, this is the food we are the least short of. We have reached the point of resorting to expedients for firewood. The officers are similarly placed in respect of bread, vegetables and salt provisions which have been making up our food for a considerable time. The heavy seas have not allowed us to fetch a list of food supplies available on the *Étoile,* this I need to draw up an overall regulation for the rations.

Saturday 18 to Sunday 19.

Ortive variation observed: 6°.

And once again our hopes have been confounded. At 1 p.m. we saw one island bearing NE¼N distant about 4 leagues, then 4 or 5 others from N¼NW to NE¼E at various distances of 4, 5, 6 and 7 leagues. This chain of islands seemed to be trending ESE. At 2 o'clock we discovered 5 or 6 other islands the most E of which bore from us ENE5°N.

At 3.30 we sighted behind the islands a large high land bearing from us ENE distant about 9 to 10 leagues.

At 4 I altered course and sailed on the port tack close to the wind which blew SE to SE¼S. We kept to this tack until midnight when we sailed close-hauled on the other tack.

We saw the land at 6 a.m. and at about 9 we could count from aloft close on 20 islands or headlands that seemed detached from each other. Several were defended by breakers. At 10 o'clock I altered course.

At midday the run gave me ENE4°E 18'30".

[1] The line is a variation of a sonnet entitled 'Hope' by the pompous poet Oronte in Molière's *Le Misanthrope*: *'Belle Phillis, on désespère, alors qu'on espère toujours'* (Act I, scene ii).

Estimated latitude: 11°24′.

Calculated longitude: 152°30′, corrected: 149°33′.

Bearings at 4 p.m.: the most E islands NE¼E 4 to 5 leagues, the middle of the island called Round [Isle Ronde] NE¼N 5 leagues, the first island sighted N¼NW 3 leagues, a small one between the latter and Round Island N 3 leagues.

At 8 p.m.: Round Island bore NE 7 leagues approximately.

At 6 a.m.: sighted Round Island NNE approximately 7 leagues.

At 8 o'clock: Round Island N¼NE 6 leagues. At 10 the most E land ENE and E¼NE 8 to 9 leagues, Round Island NNW4°N 2 to 3 leagues, the most W island which is the one sighted first yesterday WNW.

Note: We had one poor goat left, a faithful companion of our adventures from the Malouine Island where I had taken her. She was Almalthea who daily gave us a little milk. Hungry stomachs called out for her death. I could only feel sorry for her and the butcher shed tears over the victim he was sacrificing to our hunger. She fed us for two days. I have heard it said that at the siege of Candia goats were put to another, less cruel, use.[1] May God preserve me from being reduced to either action. There are still two dogs left who were caught in the strait of Magellan. They too will have the honour of sustaining their masters. I think that they will vainly claim the sacred rights of hospitality. Even the eloquent Rousseau would not save them.

Sunday 19 to Monday 20.

Occiduous variation observed: 6°28′ NE.

The winds remained constant from SE to SE¼S, the sea very rough. During the 24 hours we tacked on the one hand more E than N and on the other more S than W. At 7 a.m. we sighted a stretch of land trending to the NE¼N and at 10 o'clock we had to tack to the SW.

Estimated latitude: 11°31′, observed 11°38′.

Calculated longitude: 152°44′, corrected: 149°47′.

Estimated rhumb line[2]: SE¼E3°15′E 14′, corrected: SE19′30″.

Sighted at midday the most E land in view bearing ENE5°N distant approximately 12 leagues.

Monday 20 to Tuesday 21.

Corrected variation: 6°.

Continuation of the invariable SE. We kept to our SSW board until one a.m. under the four main sails, the topsails double-reefed, stormy gale, squalls, the sea very rough. This is positively what we require in order to make no progress. From one a.m. we have kept to the starboard tack close-hauled.

The run for these 24 hours gives us SSW1°30′S, distance 16′.

Estimated latitude: 11°51′, observed: 11°53′.

---

[1] The reference is to the siege of Candia (former Heraklion, in Crete) by the Turks in 1666-9. Amalthea in Greek mythology is the goat who suckled Zeus; one of her horns was the origin of the term 'horn of plenty'. The 'less cruel use' alludes to a story (in Montaigne, *Essays*, II, ch. XII) that the Greek defenders noticed that goats struck by arrows ate a particular herb to cure their wounds, and followed their example.

[2] Another variant for 'run'.

Calculated longitude: 152°38′, corrected: 149°41′.

Tuesday 21 to Wednesday 22.

Corrected variation: 6°.

At 2 o'clock we saw land from the mast-head bearing NE¼E approximately 12 leagues distant. We then changed tack and sailed close-hauled starboard tack until 11 o'clock on the Wednesday. Stormy gale, very rough sea, forced to reef all the topsails. Our situation worsens daily. We are reduced from today to 3 quarts of water.

At midday the run was SSW 5°30′W, distance 68′.

Estimated latitude: 12°50′, observed: 12°52′.

Calculated longitude: 152°5′, corrected: 149°8′.

Wednesday 22 to Thursday 23.

Corrected variation: 6°.

Stormy gale SE to E and ESE, frightful sea, the most contrary weather we could be cursed with. We kept close hauled for 13 hours on one tack and 11 hours on the other with the four major sails, the topsails reefed. To add to our problems, the *Étoile* is sailing worse than ever.

At midday the run gave me ENE 2°30′E, distance 8′30″.

Estimated latitude: 12°52′, observed: 12°49′.

Calculated longitude: 152°13′, corrected: 149°16′.

For 10 days it has been impossible to lower a boat, so I cannot make any definite arrangements about the rations, not having any report on the *Étoile*'s supplies. I made the vessel come within hailing distance and instructed Mr de La Giraudais to cut down to our level in respect of bread, vegetables and water. O Bellin how much you are costing us! The Indian is very bored at not being able to land. However, he retains his cheerfulness, with a few exceptional moments when he becomes ill-humoured. His nature tends to mockery. He imitates with grace and lightness all those whose faces or manners lend themselves to ridicule. He notices everything and ably sorts out the whys and wherefores. He feels he has to improvise what seem like recitatives on everything he sees.[1] He forgets neither kindness nor harm done to him, but he is grateful and has no vindictiveness. He credits the moon, whose phases are well known to him, with a great influence on winds and weather. He even explained to us one night when it was extremely hazy, that in his country when the moon has that aspect (he calls this state of the moon *malama tamai* and we cannot as yet understand what that is) they sacrificed one or two men, either slaves or common people, *tata einou*, never any women, and that it was the *eatoua* who sacrificed them by means of a club.[2]

---

[1] Bougainville's wording suggests a form of song or chant: 'il improvise des récitatives obligés', giving an impression of some operatic recitative with obligato. It is evident that Ahutoru on a number of occasions resorted to Polynesian chants or *waiata*, used to mark events or situations considered significant. A Maori *waiata whakapai* is a chant expressing some strong emotion or praise.

[2] Sacrifices took place to placate the gods or on ceremonial occasions. A foggy night of an unusual appearance or, more often, an eclipse was interpreted as a sign that the gods were displeased, Henry, *Tahiti aux temps anciens*, pp. 234, 417–18. 'Marama' is a term used for the moon; the 'tata einou' mentioned were the 'tata'a fenua', or 'people of the soil'; victims were normally selected from the lower class of *manahune* or from prisoners of war, *titi*.

Thursday 23 to Friday 24.

Observed ortive variation: 6°48′.

Weather unchanged, the sea still as bad, the winds SE to ESE, We sailed S with the wind ESE and NE with a SE breeze, making little progress with ships that are such bad sailers [and] are no longer in condition and the rough sea forcing us to let out little canvas in order to preserve our damaged masting and to carry out manoeuvres that yield little result. Meanwhile we are using up our supplies of food and concern is already evident on almost every countenance. As for my part I remember what Aeneas, who was a hundred times more pious than I am, used to do in a case that was a hundred times less serious: '*spem vultu simulat, premit altum corde dolorem*'.[1]

Compass point, corrected: SSE1°S, distance 28′.

Estimated latitude: 13°12′ observed: 13°16′.

Calculated longitude: 152°23′, corrected: 149°26′.

Friday 24 to Saturday 25.

Observed variation occiduous: 7°4′, ortive: 7°2′.

The winds have been SE during the 24 hours and we sailed on the E tack, steering NE¼E and ENE, the sea still very rough. When the sun rose we saw land from the mast-head in the N, then we saw it continue almost to the NNE, a very high land whose extremity seems to form a large cape.[2] We hope that we shall at last emerge from this gulf of tribulations and that the land will trend NE and possibly even N. Anyhow, it is evident from today's observations combined with the previous ones and compared with our estimated run that the currents have borne us E by about 14 or 15 leagues.

Estimated run for the 24 hours NE4°E, distance 88′.

Estimated latitude: 12°12′30″, observed: 12°7′.

Estimated longitude: 153°31′, corrected: 154°14′, 151°17′.

Estimated run since observations of Monday midday: SE¼E40′S, distance: 57′. Corrected run since the same: ESE3°30′E, distance 95′.

Observation of the longitude: by observations of the distance from the moon to Antares and Spica, Mr Verron has concluded that the longitude of the ship at midday on the 25th is 150°5′30″ E of the Paris meridian, placing us 4°8′30″ further E than my own reckoning.

Bearings at midday the most W land NW distant 15 to 18 leagues, the most E N¼NE4°E 12 leagues approximately.

Saturday 25 to Sunday 26.

We have sailed NE¼E and ENE from midday to midnight without seeing any other land further E than the large cape that I have named Cape Deliverance [Cap de la Délivrance] and that we are rounding with an indescribable satisfaction. At midnight I changed to the other tack close-hauled until 3 a.m. when we reverted to the starboard tack, with the wind always SE, stormy gale and rough sea. At 8 a.m., seeing

---

[1] 'He feigns hope with his face, suppresses the deep grief in his heart:' Virgil, *Aeneid*, I, 209.

[2] The expedition is at last reaching the end of the Louisiade Archipelago. The final island along this chain is now named Rossel Island, after Paul de Rossel who sailed with D'Entrecasteaux and took over command of that expedition after the latter died in July 1793.

no other land to the E than the Cape Deliverance which was well to leeward I set course for the NNE.

At midday the run gave me NE¼N1°N, distance 71'.[1]
Estimated latitude: 11°15', observed: 11°7'.
Calculated longitude: 154°33', corrected: 151°36'.
Bearing at 6 p.m. Cape Deliverance bearing NNW4°N 8 to 10 leagues.
At 7 a.m. the said cape NW¼W1°W 10 to 12 leagues.
At 10 a.m. the said cape W3°N 12 to 15 leagues.

Sunday 26 to Monday 27.

Observed ortive variation: 6°42'.

Stormy gale from SE, the sea very rough. We continued sailing NNE without seeing land, the horizon was in truth very hazy. At 9.30 p.m. I almost hove to under the two courses starboard tack throughout the night. The winds were violent, the sea frightful, the weather squally and very dark. Such nights are frightening in these unknown seas. At 5 we filled the sails, steering NNE and N at 8 o'clock.

At midday the run is worth N¼NE5°E, distance 132'.
Estimated latitude: 9°5', observed: 9°.
Calculated longitude: 155°30', corrected: 152°33'.

Monday 27 to Tuesday 28.

Fair weather in the afternoon, the winds ESE, we ran to the NNW, studding sails out until 6 p.m. when I steered close hauled between ENE and NE¼E throughout the night which was spoilt by frequent squalls, rain and fog.

At 7 a.m. we saw land from NW to NW4°N about 10 [leagues] distant.

At 8 we saw other lands from ESE to ENE, approximately 12 to 15 leagues distant. We steered NNE.

At midday the run gave me NNE1°15'N, distance 94'.
Estimated latitude: 7°37', observed: 7°38'.
Calculated longitude: 156°1', corrected: 153°4'.

Sounded at 6 p.m. 200 fathoms no ground, the sea's appearance being extraordinarily altered.

Sight of land 6th division.[2] Bearings at midday: the E land from ENE to SE¼E2°S distant 8 to 9 leagues; 2 islands in the W, one bearing NW¼W, the other N5°E [sic] about 9 leagues, some very distant land NNE. These bearings place the most S of these islands in latitude 7° 56' and their most E part in eastern longitude 156°21', corrected 153°24'.

Tuesday 28 to Wednesday 29.

Observed variation ortive: 7°34'.

The winds have varied from SE to E, little breeze with squalls. During the day we kept close hauled on the starboard tack and during the night sailed on short tacks.

---

[1] 'On this day Bougainville took possession in the King's name of the lands he named Louisiades "situated between 12° and 9° of latitude S and 151° to 148° of longitude E". See BN, NAF 9407, f° 143' (ET).
[2] This sixth group of islands are the Solomons.

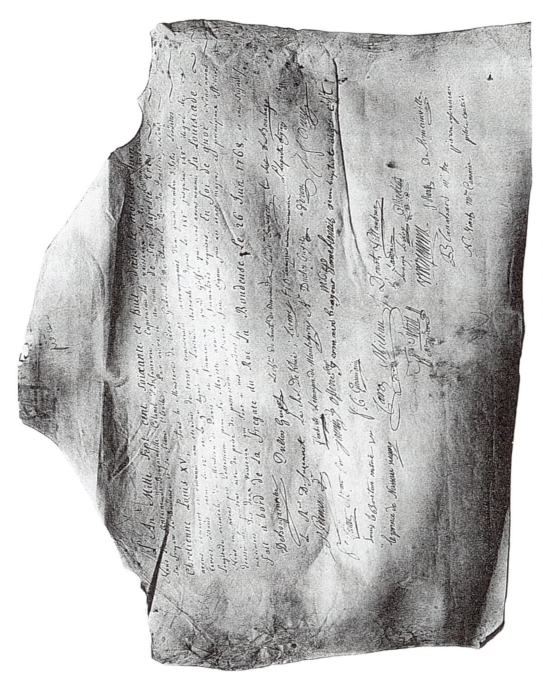

5. Document taking possession of the Louisiades Archipelago on behalf of the King of France. *Reprinted by permission of the Archives de la Marine (Colonies), Paris.*

The land is surrounding us, the major part stretching from SE¼S up to N by E and a few islands in the W from W¼NW up to N¼NW.[1]

At midday the run gives me NE¼N2°N, distance 35".

Estimated latitude: 7°9', observed: 7°8'.

Calculated longitude: 156°19'47", corrected: 153°22'47".

Bearings at sunset the E lands or islands from SE¼S2°S up to N3°E through E. The W ones from NNW as far as NW.

At midday the SE islands from S¼SE as far as E¼SE. The W ones from NW¼N as far as W¼NW5°W.

Wednesday 29 to Thursday 30.

Squally weather, fairly changeable and light breeze in the afternoon. At 3.30 I sounded being approximately 3 leagues from land, we found ground at 48 fathoms, white sand and pieces of broken shell. We tacked during the night to find ourselves at about the same distance from the land in the morning. At 6 a.m. I had the boats lowered with an armed detachment and sent them to examine a few coves that seemed to promise an anchorage, the depth found out at sea appearing favourable.

At about 10 o'clock, we saw a dozen canoes of various sizes coming fairly close to the ships without however wanting to come alongside. The largest of these canoes contained 22 men, the others 8 or 10, the smallest 3 and 2. These canoes are the first we have found in these seas that do not have an outrigger. The Indians are quite black, they have frizzy hair but long, a few of a reddish colour. They wore bracelets, and plaques on their foreheads and around their necks, I do not know of what material these are made, they seemed to me to be white. They were armed with bows, arrows and spears, they uttered great cries and it appeared to me that their attitude was not peaceful. I am very much afraid that we shall have to make ourselves feared. Where can one find men like the good Cythereans!

At midday the run gave me NW distance 8'.

Observed latitude: 7°.

Calculated longitude: 156°14'47", corrected: 153°17'47". Soundings: 48 fathoms, white sand, pieces of broken shell 3 leagues from land.

Bearings at 6 a.m. The SE land S5°E: the E point of the bay in which we are ESE4°S 4 leagues, the closest land NNE 2 leagues, the W point NW 4 leagues, the large island to the W W5°S.

At midday. The most E land ESE5°S, the most W bore NW5°W, the S point of the W land W5°S.

Soundings: 49, 48, 45 fathoms, same bottom.

Note: There flows here a strong current that bore us towards the ESE.

July

Thursday 30 to Friday 1 July.

During the afternoon, the weather was fairly fine, the winds ESE and SE. At 3 I

---

[1] After rounding Cape Deliverance, Bougainville sailed north and slightly east towards what is now called the New Georgia Group. He passed west of the islands of Vella Lavella and Ranonnga and, sailing

recalled the boats, at 4 they were back on board. Mr de Bournand who was in charge reported to me that he has everywhere found fairly good anchoring ground from 30, 25, 20 to 11 fathoms muddy sand but along the open shore and without any river. He saw only one small stream along the whole of the shore he surveyed. Broadly this coast is almost unapproachable, waves breaking everywhere along it. The mountains come right down to the shore and the land is completely covered with forests. There are a few huts in some small coves, but not many, the Indians residing in the mountains. The small boat was followed for a time by 2 or 3 canoes. An Indian stood up to throw a spear. Mr Lemoine, one of the volunteers, took aim but the shot misfired, the Indian sat down in the bottom of his canoe without throwing his weapon and the canoes paddled away from the boat. To be sure our situation is fairly critical. We have land on one side from S to NNW through E and N, on the other from W¼SW to NW and unhappily the weather is so foggy from the NW to the NNW that one cannot see beyond a couple of leagues. It is nevertheless in that gap that I am planning to find a passage. I am too far forward to draw back. A strong tide running from the N and bearing SE leads me to hope that I shall be able to sail through. What then are these lands that extend almost to 8° of latitude? The most southerly part of New Britain [Nouvelle Bretagne] and Dampierre's Pass are at the most in 6° 30'.[1] I hope that we shall soon be in a position to solve this geographical problem.

The tide made itself felt at its strongest from 4 to 5.30 p.m., the ships although driven by a good breeze found it hard to manoeuvre. It weakened at 6 o'clock. During the night we tacked from S to SSW on one tack, from ENE to N on the other, the weather squally with a great deal of rain.

At 6 a.m. we were at about the same point as yesterday at 5.30 p.m., sighting like yesterday the point of the coast we had sailed along bearing NNW and the W one NW, proof that there is an ebb and flow. I steered NW and NW¼N.

At 10 a.m. we passed through an extremely violent tidal race situated in the narrowest part of the channel which may be 5 to 6 leagues long, I have named it Raz Denys after my boatswain, a good old servant of the King.[2] The *Etoile* which crossed this race to our W found ground in 7, 6 and even 5 fathoms. The sea there was frightful.

Having then sighted at the end of the E what seemed like a fine bay, I kept to in order to send a survey party. A deluge of rain coming upon us at half past eleven forced me to delay until midday.[3]

At midday the estimated run is NW, distance 22'.

---

east of the Treasury Islands, found ahead the large landmasses which would become known as the islands of Choiseul and Bougainville.

[1] Bougainville's charts tended to be more confusing than helpful. Dampier Strait, between New Guinea and western New Britain, extends down to 5°50'S approximately and lies much further west than the *Boudeuse*'s position at the time.

[2] The expedition is making its way through present-day Bougainville Strait between the island of Choiseul and the islands of Fauro and Ovau offshore from the island of Bougainville. Oddly enough, although Bougainville Strait is deep at its centre and used by ships today plying between Australia and Japan, the present frontier dividing Papua New Guinea from the Solomon Islands snakes its way between the island of Bougainville and its offshore island group, the Shortlands, and Fauro-Ovau.

[3] 'This is Choiseul Bay, on the northern and western side of the island of that name' (ET).

Estimated latitude: 6°45′.

Calculated longitude: 155°59′, corrected: 152°57′.

Soundings: 1st cove: 30 fathoms fine muddy sand, 2nd cove: 29 fathoms ditto with shells, 3rd cove: 7 fathoms rocky bottom, 4th cove: 21 fathoms, muddy sand, 5th cove: 11 fathoms heavy red sand, very fine coral and shells.

Bearings at 5.30 p.m.: the SE land from SSE5°E up to NNW by E and N, the W land from W¼SW up to NW. Sounded at 4 p.m.: 45 fathoms yellow sand.

Bearings at 8 a.m.: the most S point of the W land WSW3°S 9 to 10 leagues, the most W point of the N land N¼NW2°W 5 leagues, the most E of the same land ESE.

Raz Denys: sounded at midday: 44 fathoms small coral, shells and gravel.

Friday 1 to Saturday 2.

Soundings at 2 o'clock: 20 fathoms, coral bottom.

At half past midday I sent the boats to sound and examine the bay and during this operation we endeavoured to remain in a position to follow their signals. The weather was fine, but almost calm. At 3 we saw the bottom below us in 10 and 8 fathoms, rocky bottom. We endeavoured to keep away from this shelf. At 4 our boats signalled a good anchorage and we at once manoeuvred, all sails set, to reach it. There was little wind and the tide, coming from the N, was against us.

Sighted while over the shelf, the small island off and S of Choiseul Bay [Baye Choiseul] bearing NW5°W approximately 4 leagues.

At 5 o'clock we again passed over the shelf with 10, 9, 8, 7, 6 fathoms rocky bottom. We even saw in the SSE a cable away an eddy that seemed to indicate that the bank could be touched at that point. Steering NW and NW¼W we came to increased depth. I signalled the *Étoile* to bear away so as to avoid this shelf and sent back her boat to guide her to the anchoring place. However we were making no progress, the wind being too weak for us to sail against the tide and night was hastening upon us. In two hours we have not advanced a half league and we had to give up this anchorage, it being impracticable to fumble our way there, surrounded as we were by shallows and reefs and at the mercy of rapid and irregular currents. I therefore steered W¼NW and WNW to return to the open sea, with frequent soundings. When the N point of the bay bore from us NE, I steered NW, then NNW and N. Bearings at 6 p.m.: the N head of the bay bore NW¼N, the SE head closest to us SE¼E, the S head of the W land SW, a fairly high islet in the channel approximately 3 leagues distant from the W land bore by its middle W3°S.

Details on the expedition by the boats

When our boats were about to come out of the bay, having finished their task, our men saw coming from the back of Warriors Cove [l'ance aux Guerriers] where a small river whose banks are lined by huts emerges into the sea, and advancing towards them in good order 10 canoes containing approximately 150 men armed with lances, bows, arrows and shields; when about 100 paces away, upon a general shout, the canoes divided to surround our 2 boats and the Indians, forsaking their paddles, took up a posture in readiness for hurling their arrows and lances. Our men anticipated them by firing a volley of muskets and 4 blunderbuss guns mounted on the boats. The

Indians kept their composure for some time but soon the rout was complete, several throwing themselves in the sea to escape into the woods. We captured 2 very long canoes, well carved and with the fore and aft parts rising about 5 feet, which shields them from the arrows when they present the prow. Their shields are oval-shaped, made of interwoven reeds excellently tied. They must be impenetrable to arrows. On the fore part of one of the captured canoes there was a carved man's head with a long beard. The eyes were made of mother-of-pearl, the ears of turtle shell and the face was like a mask. The lips were dyed bright red. These islanders are black, have frizzy hair they colour white, yellow and red. Their boldness, their custom of carrying offensive and defensive weapons, their skill at using them, prove that they are almost always in a state of war. It therefore belongs to the nature of man to be born an enemy of his fellow man.[1] Furthermore we note that blacks are much nastier than the Indians whose colour is closer to white.[2] We found in one of their canoes a partly cooked man's jaw.

Herewith the chart of Choiseul Bay and the soundings.

We saw some coconuts, a kind of prune and another fruit similar to a large hazelnut and we even found several in baskets taken in the canoes, which proves that the Savages eat them. We found in their canoes some *areka*, a fruit with a tart taste used in India by mixing it with betel and lime.[3] There were some in the same baskets as well as yams and extremely tightly and artistically woven nets with a fine mesh. I think they use it to catch birds, possibly paradise birds. We saw some fine *haras* in the woods.[4]

During the night the weather was squally with at times heavy rain.

At daybreak, I sailed N¼NW, not much breeze almost calm.

In the morning, a man fell overboard while working on the foresail shrouds, he escaped with a fright, having been saved immediately.

At midday the run was NNW 30'W, distance 30'30".

Estimated latitude: 6°19', observed: 6°18'.

Estimated longitude: 155°42'. Longitude corrected by the bearings: 155°23', finally corrected: 152°43'.

Note: we can see no land to the N and the pass through which we have emerged corresponds in no way with the details Dampierre gives on his [pass] I am still suspending judgment.

Bearings at 6 a.m.: the S point of the W land S¼SW 5°S, the N point of the same land W¼NW 5°N, the E land from E¼SE 5°S to SE¼S within sight.

Bearings at midday: the land we coasted along yesterday from ESE 2°S 9 leagues to SE 2°S about 7 leagues. The most S point of the W land S 2°E about 9 to 10 leagues, the most S small islands in view S 5°E 7 to 8 leagues, the N of a pass on the W land SSW 4°S 9 leagues, the S ones S¼SW 10 leagues, the most N high land N WNW 3°W 15 to 16 leagues, a kind of high peak above the clouds WNW 25 to 30 leagues.

---

[1] 'This comment was omitted from the published journal' (ET).

[2] Bougainville now distinguishes between black Melanesians and brown-skinned Polynesians.

[3] The nut of the *Areca catechu*, a genus of palm, which is chewed wrapped in a betel-leaf.

[4] The *ara*, a large parrot found in South America. More probably, the French saw some cockatoos, especially the sulphur-crested cockatoo (*Kakatoe galerita*) that is found from Australia to New Guinea and New Britain.

Saturday 2 to Sunday 3.

Observed variation in the evening by azimuth: 6°24'.

Fair weather but calm, the light breeze variable from SE to E. We steered E then NNW.

Run for the 24 hours gives me N¼NW3°45'W, distance 47'.

Observed latitude: 5°33'.

Calculated longitude: 155°10', corrected: 152°7'.

Bearings at midday: the most N point of the W land that I have named Cape dell'Averdi W5°N 12 to 15 leagues, the peak sighted yesterday SW¼W about 25 leagues, the most S point in view S¼SE.

Bearings at 6 p.m.: the W point of Warriors Island SE1°S, the middle of the Passage Islands [Isles du Passage] S¼SE2°S, the most S point of the W land S¼SW, the most N point in sight W¼NW2°N.[1]

Caught a dolphin, a dorado and a fish that we do not know.

Sunday 3 to Monday 4.

Observed variation: 7°15' NE.

Almost calm during the afternoon and the night, the weather fair, the breeze coming from the SE and ESE, we sailed WNW. In the morning the weather became squally and the winds varied from W to SW and S to S¼SE. I set a WSW course to land on an attractive and level island where I very much hope to find an anchorage. I have named it Archers Island [Isles des Archers].[2]

At midday the estimated run is NW¼W5°30'W, distance 42', corrected NW¼W5°15'N, distance 58'.

Estimated latitude: 5°13', observed: 4°57'.

Calculated longitude: 154°25', corrected 151°42'.

Note: for the last 2 days the currents have been running strongly NW.

Bearings at 6 p.m.: Cape Laverdi WSW 10 leagues, the peak SW¼S, the most E land in sight SSE 15 leagues.

6 a.m.: Cape Laverdi SSW3°W, other detached lands from the SW to W.

Midday: the most E land SE¼S4°S 12 to 15 leagues, Cape dell'Averdi S¼SE5°S 12 leagues, the most W point of Archers Island SW¼W 5 leagues.

Sounded without finding ground.

---

[1] Having sailed through what is now Bougainville Strait between the islands of Choiseul and Bougainville, the French are sailing off the northern coast of Bougainville Island. The somewhat Italianate 'Cap dell'Averdi', usually written simply 'Laverdy' by Bougainville, is named after Clément-Charles-François Laverdy (1723-93), Contrôleur Général des Finances in 1763-8; he was executed during the French Revolution. The cape is now usually shown as Cap L'Averdy. The high peak is Mt Balby (3,100 m, 10,200 ft), the highest point in the Emperor Range; the south point of Bougainville Island then in view is probably Kieta or Koromira Point, the Warriors Island is Choiseul, the islands in Bougainville Strait the Ovau Islands, and the most northerly point then in view some part of Buka Island.

[2] This name, chosen because islanders came up to barter bows and arrows, was soon replaced by Bouka (and Boucka) from what Bougainville understood them to be saying. The island of Buka is not truly level: there is a range of mountains, rising to 400 m (1,300 ft) in the west; however, Bougainville was approaching its eastern side which is mostly level and undulating. Buka had previously been discovered by Philip Carteret on 25 August 1767 and named by him Earl of Winchelsea's Island and subsequently Lord Anson's Island.

Monday 4 to Tuesday 5.

Light SSE to SSW breeze in the afternoon. We sailed close-hauled on the port tack. The island where I would like to anchor looks charming. A plain halfway along the coast seems to be partly cleared and has a number of habitations. Several canoes came out from the shore and stayed off examining us. Finally, after a number of gestures inviting them, three canoes each with 5 or 6 men, came up to the vessel and bartered a bow, some arrows and a coconut for some trifles. As they left, one of them shot an arrow that did not reach the ship. These Indians are black, with short frizzy hair, [they are] totally naked, with pierced and noticeably lengthened ears. The bows are longer than any we have yet seen in these seas. Their canoes are large enough for two men to row side by side. They have no outrigger and their fore and aft parts are only slightly raised. These blacks chew betel, having red teeth. They often repeated the word *Bouka*, I have given that name to their island. Those of Choiseul Bay also use betel. We found sacks where there were leaves with areca and lime.

I do not believe that we shall succeed in reaching this island, we almost have a head wind and a fast current, running NW, is clearly bearing us away. During the night, we sailed S¼SW and SSW and this morning the island was quite distant from us bearing E and SE. We shall have to seek an anchorage in New Britain. However we are in the greatest need of land, water is going to run out, scurvy is beginning to extend its evil empire, 5 or 6 officers are affected by it. Boredom and bad food would bring about a number of other ailments.

Yesterday evening we saw some very low land in the NW¼W and towards 10 a.m. we sighted a high land from WNW almost to N. As it was calm this morning I sent to the *Étoile* for some flour, brandy, salted meats and a little firewood

Estimated run for the 24 hrs is WSW2°15′S, distance 31′, corrected by the altitude and the bearings it is W distance 55′.

Estimated latitude: 5°10′, observed: 4°57′.

Estimated longitude: 153°56′, corrected: 153°30′, 150°33′.

Sighted at 6 p.m.: Boucka Island [Isle Boucka] from SE5°S distant about 6 leagues to S4°E 6 leagues. At the same time we saw from the mast-head some low lands from NW to NW¼W.[1]

Sounded without finding bottom.

Bearings at 6 a.m.: Bouka Island E5°N to SE5°E about 7 to 8 leagues, the low land seen last night from N¼NW to NNW.

At midday: the land of Bouka bearing from SE¼E to ESE 10 to 12 leagues. The low land NNE 7 to 8 leagues.

At eleven o'clock the *Étoile* reported land between W and N. At midday we saw that this was two islands. One bore N about 10 to 12 leagues, we lost sight of it.[2]

---

[1] Bougainville has now sailed around the northern coast of Buka and is making for the islands of New Britain and New Ireland. The low land seen is Nissan Island, named Sir Charles Hardy's Island by Carteret, but discovered by Jacob Le Maire in 1616, part of a small group which the Dutch named Groene Eylanden. For Carteret's track ahead of Bougainville, see Wallis, *Carteret's Voyage*, I, pp. 177-87.

[2] New Britain and New Ireland are now coming into sight.

Tuesday 5 to Wednesday 6.

Fair weather, all sails set, sailed WSW to reach a high shore. At about 5 p.m. it seemed to us to include several bays and I decided to seek an anchorage there. During the night I tacked about to keep it to leeward.

At daybreak, we were 5 or 6 leagues from land. We also sighted a new land high and of a fine aspect from SW¼W4°W to NW¼W2°W distant from 18 to 12 and 10 leagues. We sailed NW and NW¼N to reach the land. At 9.30 I lowered the boats to examine a bay possibly 3 or 4 leagues from us and stayed within reach of their signals.

At midday the estimated run for the 24 hours is WNW4°15′N, distance 46′ 30″, corrected the route is W3°30′N, distance 50′.

Estimated latitude: 4°34′, observed: 4°53′.

Calculated longitude: 152°40′, corrected: 149°40′.

Note: the currents having been running S.

Continuation of Wednesday 6. Anchorage moored E and W.

At 1 p.m., the boats signalled an anchorage and we sailed at once for it. At 3 we dropped anchor in 33 fathoms fine white and oozy sand.

The run from midday to the anchorage has been NNE4°15′E, distance 8′. Bearings of the anchorage: Duclos Island [Isle Duclos] W¼SW1°30′W, Fishermen's Point [Pointe des Pêcheurs] E of the entrance W¼SW1°S, the W point W¼NW, the back of the harbour SE¼E.

Its estimated position: latitude: 4°46′, longitude: 152°43′.[1]

Corrected position: latitude observed on land by means of a quadrant: 4°49′27″, observed longitude: 149°44′15″.[2]

Thursday 7.

It rained all night. The day was also almost always wet. I sent our water casks ashore, I set up a small camp there and they began to collect water and wood and do the washing, all matters of major importance. The landing is superb on fine sand and the back of the harbour has 3 streams, within a distance of 200 paces, useful the first one to get our water, the second for the *Étoile*'s, the 3rd for the washing. The wood is on the seashore and there is some of several kinds, all very good for burning, several excellent for working with. The two vessels are within hailing distance of each other and of the shore. Those are the advantages of this port, we could not wish for anything better to obtain water, wood and carry out all the repairs the two ships most urgently need. Furthermore this port and its surrounding area are uninhabited, which allows us a precious tranquillity and freedom of action. However one must keep on one's guard, because on the bank of a small river, a third of a league from our camp, we found a canoe as though stored there and 2 huts. This canoe has an outrigger, [is] very light and in good condition. We also found several remains of fires and meals, such as

---

[1] This line is crossed out in the manuscript.

[2] The ships have entered Carteret's harbour, which he named Gower's Harbour and which Bougainville would call Port Praslin after the Minister of Marine César-Gabriel de Choiseul, duc de Praslin. Its present name is Kambotorosch. Nearby Lambom Island Bougainville would name Duclos Island, but it is also referred to as Isle aux Marteaux; Carteret had called it Wallis's Island. See on this Wallis, *Carteret's Voyage*, I, pp. 179–81.

heaps of large shells and heads of animals that Mr de Commerçon told us were boars. We even thought we heard the shouts of men in the mountains.[1]

The inconvenience of this place is that one finds neither coconuts nor bananas nor any of the resources one might have obtained, willingly or unwillingly, from an inhabited country. If fishing does not prove to be abundant, we shall have found here only safety and the bare necessities. I am very much afraid that our sick will not recover their health as they did in Cythera. In truth, we do not have any who are seriously affected, but several are sick to some extent and if they do not improve here, the progress of the sickness will be rapid.

This morning I went for a walk to the small river where the canoe was. A sailor from my boat, looking for shells along the seashore, found a piece of a lead plaque on which was an inscription in English. This is what the piece contained:

(anc) HORD HERE ICK MAJESTYS[2]

The rest is missing, there are traces of the nails that fixed the plaque to the tree from which the Savages will have torn it and broken it up. This encounter led me to search in the neighbourhood of this anchorage and in the afternoon, about two leagues N of this cove, in another one fairly deep on the same shore, I found traces of the English camp, wood cut with the axe and the harpoon, a fairly large tree on which the lead plaque was attached about 10 to 12 feet up. The traces are well marked, and still fresh, some of the nails remain in place and the nails are not rusted. On the tree itself there are notches made by the English or by the Savages and none of all this is old.[3] I do not believe that more than three months have elapsed since our good neighbours passed this way.[4] The spun-yarn that presumably held the tents down is fresh, the shoots on the cut trees would in France be coming from a single drive of the sap and I should think that the English vessels that put in here were the same whose traces and inscriptions we found in Port Galant.

I consider our encounter on this coast at about the same spot as the English had put in as one of the most singular effects of chance.

Friday 8.

E or SE. Deplorable weather, almost constant rain night and day, we are collecting wood and water and have begun to examine our ropes. One finds here some reeds similar to those of Malacca but I am told they are not ripe.[5] I have had our sheet anchor secured to the cat-head,[6] having only 2 since Cythera.

---

[1] In the published narrative, Bougainville comments that these were in fact the calls of 'large crested wood-pigeons with a blue plumage known in the Moluccas as the crowned bird.'

[2] Bougainville rightly surmised that 'HORD' was part of 'anchor[e]d', but did not guess that 'ICK' was part of 'Britannick'.

[3] 'We nailed a piece of board on a high Tree on which were engraved the Engl. Colours, Capt & Ships Name, time of coming & sailing from and Name of the Cove', Wallis, *Carteret's Voyage*, I, pp. 183–4. The notches seen by Bougainville may have been taboo marks, indicating that the tree was probably earmarked for a feast or other purpose, ibid, p. 182n.

[4] In fact almost a year had gone by: Carteret put into Gower's Harbour at the end of August 1767.

[5] The reference is to the rattan *Calamus rotang* of Malaya.

[6] This would normally be the largest anchor on board. It may have been stowed in board and hence not available for use in emergency. The cat-heads are two large pieces of timber extending from the bow to which the anchors are secured and from which they are let go. The sheet anchor was therefore got ready for use.

Saturday 9.

Fairly good weather in the afternoon but it had rained during the morning. We are obtaining wood and water. Our fishing is not productive. We have underrun our cables[1] and even raised one anchor to check the quality of the bottom. It is good.

Sunday 10.

Pouring rain all day, unhealthy weather. A sailor has died on board the *Étoile*. His illness was complicated and was in no way related to scurvy. We buried him on the shore.[2]

Monday 11.

Today we have the finest weather in the world. We continued with the necessary work. At midday Mr Verron observed on land with the quadrant a latitude of $4°49'27''$.

This evening, while fishing with the seine opposite our encampment, a young sailor was bitten in the water by an animal he told us is about a foot long, as thick as a thumb, with a black body and a white patch on its head. This snake bit him on the thumb, the lad grabbed it and flung it far away, the animal hurled itself at him a second time and bit him on the side.[3] They did not pay any notice to it, although he complained and right away had some difficulty in walking. He came back on board a quarter of an hour later and, after eating his soup, uttered loud cries. He was brought up to the quarterdeck after being given some theriac and rubbed his side. He was made to walk by two men taking it in relays in order to make him sweat. The surgeon then scarified and rubbed the bitten side with theriac and he was made to drink *eau de luce*.[4] The walk was continued until he could no longer stand. As soon as he was laid down he entered into convulsions, losing consciousness whenever they reached their peak. He was complaining of a fierce pain in his stomach. Mr de Commerçon, called in for a consultation, approved what had been done until his arrival and continued the treatment until 10 o'clock when the patient began to improve,[5] sweating regularly and sleep already sending down its beneficial poppies.[6] The patient's colour did not change from the time of the bite. The wound at his side had turned blue almost at

---

[1] This operation is carried out from a boat and means that the cable was lifted into the boat and passed over it while the boat moved out from the ship until it was over the anchor so that the cable could be inspected to see if there was any damage caused by chafing on rocks or similar obstructions on the seabed.

[2] The sailor was Jean Loran, of St Malo, who was suffering from 'an inflammation of the lower intestine', according to Vivez.

[3] There are some 55 species of sea snakes or *Hydrophidae*, all of them venomous but they seldom attack unless, as in this case, they feel threatened or are coming ashore to lay eggs. The *Homalopsinae* are another family found in northern Australian and Papuan waters, but they normally keep to rivers or estuaries, and it is less likely that one of these was the attacker.

[4] *Luce* is a dialectical low-Breton word for the bilberry or whortleberry (*Vaccinium Myrtillus*).

[5] Commerson will claim the credit for this cure in a letter to his brother-in-law of 16 January 1770: 'Who but I administered the specific for the bites of snakes, even unknown snakes?' Montessus, *Martyrologe et biographie de Commerson*, p. 129.

[6] A quasi-literary reference to the poppy, i.e. opium, used in sleeping-draughts. Bougainville adapted it from a 1725 prose poem, *Le Temple de Gnide*, by Charles de Secondat de Montesquieu (1689-1755), a work popular enough in its day but one that the famous political philosopher would probably wish to see forgotten.

once and the blood drawn out by scarifications was already dissolving. His eyelids were also swollen, but this swelling did not last.

Tuesday 12.

Fine weather day and night. We are speeding up our work, this place providing no refreshments for our sick. There are a few cabbage-palms but few in number and hard to obtain.[1] The young sailor bitten yesterday has spent a very good night and day. He is out of danger, but may feel the after-effects of this accident for a long time.

I went ashore yesterday to set up a pendulum clock with seconds, a quadrant and the stands necessary for telescopes. There will be a minor eclipse of the sun tomorrow, visible only in the Southern Hemisphere. It is interesting for us and for geography to observe this well.

Note: the Cytheran has made us understand that in his country, there are snakes along the shore that bite men at sea and that all those who are bitten die. Their medical science is, I think, not advanced. He is amazed to see the sailor walking. Very often, examining the results of our crafts, and all the means by which they increase our faculties and multiply our strengths, this islander is filled with admiration at what he sees and he blushes for his own country: *avuaou* Tahiti, he tells us sadly.[2] He will be even more surprised when he is able to understand that this very ignorance is what protects his compatriots' happiness.

Wednesday 13. Eclipse of the sun observed.

The night was very fine and the day very fair, with however a few showers but which did not adversely affect the observations. Mr Verron took corresponding altitudes to determine the operation of the clock and he was observing with a telescope of 9 feet. Mr du Bouchage was observing with my achromatic Dollond telescope, which is over 3 feet, my post being at the clock.[3]

The eclipse began at 10h 56′ 26″, it ended at 0h 33′ 31″, the immersion was doubtful by about 1′, the emersion certain. Its size was about 3′. Since from the corresponding heights, we found that the clock was fast by 5′ 10″, this correction will have to be made to the above times.

---

[1] This term is to some extent a generic one, making identification difficult. Carteret collected some 'cabbages' that were the tops of felled coconut trees: 'you must cutt down a tree for each Cabbach which makes a great distruction of these usefull trees', Wallis, *Carteret's Voyage*, I, p. 184. Bougainville may well have followed suit as he later refers to 'trees already felled'. Alternatively, there may have been a few true cabbage-palms, such as the *Areca oleracea* or the *Cordyline australis*.

[2] The most likely translation here is 'Alas, poor Tahiti'. The following sentence, expressing Bougainville's belief that Tahiti's isolation and lack of sophistication was the islanders' good fortune was omitted from the printed *Voyage*, and Bougainville there stressed how inferior and sad Ahutoru felt when faced by the Europeans' superior knowledge: 'He did not like to admit that he was affected by our superiority over his nation. One cannot believe how proud he is' (*Voyage*, p. 281). From this, Bougainville concluded that Atuhoru was high-born and that clear social distinctions existed in New Cythera.

[3] The various observations taken at different times were made to check the accuracy of the calculations. The Dollond achromatic telescope was demonstrated before the London Royal Society on 8 June 1758; it was over 5 feet in length; see Dollond, J, 'An Account of some Experiments concerning the Different Refrangibility of Light', *Phil. Trans.*, 1758. The Dollonds were a family of instrument makers of French Huguenot origin who had settled in London. John Dollond (1706-61) was a member of the Royal Society and was assisted by his son Peter (1730-1820).

This observation is all the more necessary because by these means, we shall at last be able to determine the width of this vast ocean, until now so uncertain.

I had a base measured and all the necessary bearings taken to draw an exact chart of this harbour that I have named Aquarius Harbour [Port du Verseau].[1]

Note: as my estimate for the position of Cythera was within 12 to 13′ of the observations Mr Verron made there, I have allocated the difference of 3° I find here between by estimate and the longitude determined by the eclipse, I have, I said, allocated it to the daily estimate from Cythera to here.

Thursday 14.

Very fine weather, the winds still being SE to E and ENE, fairly good breeze. I had a drawing made of a waterfall that supplies the water to the *Étoile*'s stream. Art would struggle in vain to produce at Versailles or Brunoy what Nature has cast here in an uninhabited spot.[2] What hand would dare to build the leaping platforms held back by invisible links whose graduations, almost regular, promote and vary the spilling flood, which artist would have dared to create these storeyed masses that make up basins to receive those sheets of crystal water, coloured by enormous trees some of which rise up from the basins themselves? It is quite enough that there are privileged men whose bold brush can trace for us the picture of the inimitable beauties. This waterfall would deserve the greatest painter.

Friday 15.

S, SSW. Cloudy weather, intervals of rain. I went to survey the back of this bay and the various anchorages it contains. We have finished collecting our water and our wood and we are going to work on finishing repairing the *Étoile*. This vessel had a great many repairs and work to be done and not many men.

The few refreshments we can get here and possibly the quality of the land are not helping our scurvy cases to recover. Every day new cases are found on both ships.

At 10 p.m. there was an eclipse of a satellite which the bad weather did not allow us to observe.

Saturday 16.

S. SSW. Frightful weather, pouring rain night and day. This abominable weather holds up the work and increases cases of sickness. We killed today snakes, scorpions and a kind of animal the length of a finger with a hard shell on its body with prominent prickles along the sides, with a fairly long tail and six legs.[3] We have also seen since the first days some boars in various parts of the forest but we did not kill any.

---

[1] This name, based less on a zodiac sign (July is not the month of Aquarius) than on its alternative meaning of 'The Water Carrier' in view of the heavy rain the French had to endure for much of the time, will not be retained. The harbour would soon become known as Port Praslin.

[2] The fountains and cascades of the château at Versailles are famous. Brunoy, near the town of Sénart, was the home of a wealthy financier Jean Pâris de Montmartel (1690–1766), once renowned for its stepped cascades; see R. Dubois-Corneau, *Jean Pâris de Montmartel, banquier de la Cour*, Paris, 1917. Bougainville asked Charles de Romainville to draw the waterfall: his drawing depicts a wide waterfall cascading down in a series of nine steps.

[3] This somewhat fearsome insect is a *eurychantha horrida*, of the *Phasmidae* family.

Sunday 17.

S. SSW. It rained all night. The day passed with neither rain nor sun. We have been ready to leave for two days but the *Étoile* is holding us up. We are working to assist her. We have taken from her the quantity of food that is appropriate for our number and this vessel, our supply ship, is almost light. Unhappily there are no stones here suitable for ballast. We have to make up for it with wood that is very difficult to cut; our crews are exhausted, having been at sea without respite since Montevideo.

Monday 18, Tuesday 19.

S, SSW. Shall I always have to write bad weather, strong gale, storm, constant rain? That remains the weather of these 48 hours. Our crews are worn out, this land is providing them with nothing more than an insalubrious air and extra labour. Tomorrow, if weather permits, we shall be in a position to sail.

For the last 3 days, I have cut down by more than a third the dried beans that make up our soup, I say our because everything is now distributed equally: officers and crews are on the same footing, our situation, like death, makes all men equal.

The only refreshments the forest gives us are *lataniers*[1] and a kind of cabbage palm: even then they are few in number and have to be fought over with the swarms of enormous ants that have forced us to abandon several stumps of trees that were already felled.

Wednesday 20.

S. SSW. We were unable to sail today. The weather having been frightful, strong gale, rough sea and torrential rain. For the last 3 days seabirds have been in the bay, proof that it is bad weather outside. Our men went to get palm cabbages and lataniers.

They brought me a very extraordinary animal. It is an insect about 3 inches in length of the mantis family. Almost all its body parts are made up of a material that, even on close examination, one could mistake for leaves: each of its wings is half a leaf which when the wings are together forms a whole one. The underside of its body is a whole leaf of a colour that is more dead than the upper one. The animal has 2 antennae and 6 legs the upper parts of which are also pieces of leaves. Mr de Commerçon has described this strange animal and I am keeping it in alcohol.[2]

One finds in the forest a great multitude of infinitely varied insects, some very fine birds, parrots, and especially several types of large turtledoves, all very pretty. The flora in respect of plants and trees is that of the Indies. The reefs that line the coasts contain treasures of conchology. I am persuaded that in this respect we have collected here a great number of new species. I have a shell that I believe is of an absolutely new type. It is a most elegant heart in its shape and material. We have found only one.

Note: I was mistaken, this shell is known but rare. It is called *diconcha cordiformis*

---

[1] The *Corypha umbraculifera* is found in Mauritius, other parts of the Indian Ocean and in Asia.

[2] The French had come upon a leaf-insect, a phasmid of the *Phylliidae* family. There are about 20 species of these insects that range from the Seychelles eastward to the Pacific as far as Fiji. The only type known in New Britain is the *chitoniscus feedjeanus*.

and described as '*quae duplicis valvae curvatura proecipua, verticibus nutantibus, cordis figuram exprimit, sutura perpendicularis divisa*'. Klein, p. 137.[1]

The hammers, a rare shell, are fairly common here. We collected 10 in a very small space.[2]

I buried at the foot of the tree on which we had fixed the clock used for the astronomical observations the following inscription engraved on an oak plank:

'In the year 1768, we L.A. de Bougainville, infantry colonel, captain of the King's ships, commanding the frigates the Boudeuse and the Étoile, in the name of and by order of H.M.C.M. Louis XV, under the ministry of de Stainville, duc de Praslin, we took possession of this land and of the adjacent islands that we have named the Archipelago of […]. In witness thereof, in the harbour called Aquarius Harbour and situated in southern latitude 4° 49′ 27″ and longitude 149° 44′ 15″ E of Paris, astronomically observed. We have deposited at the precise location of these observations the present inscription corresponding to the Act of Possession that we are taking to France.'[3]

Thursday 21.

S. SSW. The night was frightful, a deluge of rain, thunder, wild squalls, the darkness of chaos. The day was less bad, nevertheless it was not possible for us to sail. The sea outside is frightful and would not allow us to hoist in our boats, because we would have to do so outside, seeing that one could sail from here only by means of a stern-fast on an anchor that the longboat would raise.[4] The rain continued almost without a break, the wind was less violent and from the S. We went to fetch some remedies and food in the woods. Some of us went to Turtledove Island [Isle des Tourterelles] to hunt and cut down some palm trees and *lataniers*. Another squad of which I was part went to the mainland to look for an acid fruit similar to the *evi* of Taiti or the hog-plum of our colonies[5] that we discovered yesterday. We found a few of these fruit to which the boars seem to be very partial, almost all those we picked up having felt these animals' teeth, the traces being still fresh. The search parties were practically unsuccessful. All we got out of it was being soaked to the skin.

There is here a kind of aromatic ivy similar to the betel which we believe has an antiscorbutic property. Several of our men soak the leaves in wine, wash their swollen

---

[1] 'Which obtains the shape of a heart from the noble curve of its double valve and its curling vertices, being divided by a perpendicular suture.' Bougainville is quoting from Jacobi Theodori Klein (1685–1759), author of *Tentamen methodi ostraeologicae*, but especially of *Naturalis dispositio echinodermatum*. There was however a French translation available since 1754, the *Ordre naturel des oursins de mer et fossiles*.

[2] The *Malleus* or hammer-shell/hammer-oyster, has the shape of an inverted T. It belongs to the same family as the pearl oyster and is fairly widespread in the shallow waters of the Indo-Pacific region. The island facing the anchorage on Romainville's chart, named Hammer Island, has roughly the shape of this shell and the name in smaller lettering, 'Hammer Cove', suggests that this is where the French found the ten shells mentioned.

[3] This document has apparently been lost: 'It is not included among those copied by Margry (BN, NAF 9407)' (Martin-Allanic, *Bougainville*, I, p. 766n.).

[4] The ship would have to use a hawser from its stern, to an anchor, to point her towards the entrance of the bay. When pointing in the right direction the hawser would be slipped and the ship would sail out leaving the boat to weigh the anchor and then rejoin outside the bay.

[5] Bougainville is referring to the *Spondias mombin* of the West Indies, the fruit that Carteret calls the 'Jamaica plumb' (Wallis, *Carteret's Voyage*, I, p. 186). The variety found in the Pacific region is the *Spondias cytherea* Sonnerat, sometimes called the Tahiti plum.

legs with this infusion and several have already seen the swelling diminish. One also finds a few mangoes, excellent fruit common in the Indies.

Friday 22. Earthquake.

S. SSW. These 24 hours have been 24 hours of rain and bad weather. Although every kind of need urges us to go and seek lands where we could at least find necessities, we have been unable to sail, the winds being S and the sea very rough.

At 11 a.m. there was an earthquake which we felt on both the 2 vessels. It lasted approximately 2'. On the shore, the sea rose and fell by about 4 feet in height and during the tremors several of our men who were on the reefs, busy picking shells, went to seek a refuge in the boats. The earthquake was not extensive and caused no appreciable damage on land.

Saturday 23.

Continuation of the S to SSW wind. The night and the day gave us the picture of the deluge when the sluice gates opened. We were unable to go out and our situation worsens with every day that is lost without making headway. I went to examine the exit by following the channel between Turtledove Island and the mainland. This way out will be of use to us in case the winds remain S.

Sunday 24 to Monday 25 midday.

S. NE, WSW, SSW. It rained all night almost without interruption. Dawn brought fine weather and calm, we raised our mooring anchor, I sent men to land to fix a rope to some trees, I had the kedge anchor fixed stern-fast then hove short on the outer anchor. I made several attempts to sail during the day, sending out to check on the weather outside. Finally, at 5 p.m., I was beginning to redo the moorings when a fresh breeze rose from the back of the harbour: we were under sail in less than ¾ of an hour and finally came out of this wet and unpleasant hole by the pass we had used to enter. I left the kedge anchor to the *Étoile* for her to work herself out with and my small boat to help to tow her in case of need. At 8 o'clock we had our anchor in place and 3 of our boats hoisted in. We stayed hove to wind aport to wait for the *Étoile*, who appeared at around 7.30. Our boat, which had remained with the *Étoile*'s longboat to raise the kedge anchor, rejoined us at 1 after midnight and we hoisted it in without delay. During the night it rained heavily with little wind from the WSW to SSW, S and SSE and we remained close to the wind starboard side.

Draught on leaving: 13 feet 10 inches astern, 13 feet 4 inches bow, the anchors and the boats excluded.[1]

At daybreak I steered from E¼SE to NNE, rounding as does the land that I believe to be New Britain and that I want to round by the N. I suspect that we are coming out of the bay Dampierre names St George Bay [Baye St-Georges]. The English will have come here looking for Port Montaigu.[2]

---

[1] I.e. the draught was taken while the ship was in harbour so that the anchors were still down and the boats not yet hoisted.

[2] On leaving Port Praslin, Bougainville sailed ESE, which brought him back into the open waters of the Pacific Ocean, then following the trend of the east coast of New Ireland, sailing NNE and N to NNW. Unaware that Carteret had recently discovered that there was a passage north of Port Praslin and that what he thought was the east coast of New Britain formed a separate island, Carteret's 'Nova

Run to midday gives me by the courses E6°30′S, distance 17′30″.
Estimated latitude: 4°47′.
Estimated longitude: 152°55′, observed: 149°56′15″.
Bearings at midday: the most N land that we are following N4°W 8 to 10
leagues, the most S of the same coast W 5 to 6 leagues.

Monday 25 to Tuesday 26.
   Observed variation doubtful: 7°4′.
   Fair weather, fresh breeze, the winds variable from SE to S and SSE. We sailed
along the coast and I am becoming more convinced that this is New Britain.[1] The land
is very high and forested, we have seen signs of only a few habitations along the area we
have travelled these 24 hours; I am estimating this from the small number of fires and
the little area of cleared land. We saw nothing that had the appearance of a bay. During
the night we carried little canvas and even remained lying to for close on two hours in
order to await the *Étoile*. At daybreak the wind freshened and we sailed N to NW¼N,
following the trend of the land. I have allocated 35 tents to make into large trousers for
both crews. Without the various distributions of clothing that have already been made,
it would have been impossible for these poor people to be clothed for such a lengthy
campaign, especially during the downpours we have passed through. For the rest I have
nothing to give them and anyhow it is high time we reached some places inhabited by
Europeans where we could find a range of resources. I have been compelled once
more to cut back the bread ration by one ounce, the little food we have left is partly
spoilt. However one must eat the bad with the good. For two months we have banned
butter from our vegetables as uneatable, we now eat it on bread as if it was the best
Vanvres butter.[2] What a measuring rod! And when will it end? Such is our situation
that we are suffering at the same time from the past that has weakened us, the present
whose sad features are repeated time and again and the future whose indeterminacy is
the most cruel of our woes. My personal sufferings are multiplied by those of the
others. Nevertheless I must make it known that no one is allowing himself to be
defeated and that courage outclasses misfortunes.
   The currents have carried us towards the SSE.
   At midday the estimated run for the 24 hours is N5°E, distance 64′ 30″, the corrected route NNE1°30′N, distance 43′.
   Estimated latitude: 3°43′, observed: 4°10′.
   Estimated corrected longitude: 153°, observed corrected: 150°11′45″.
   Bearings at 6 p.m.: the most N point NNW5°N approximately 6 leagues, the

---

Hibernia', Bougainville sailed south to avoid becoming embayed in Dampier's St George's Bay. To the
French, Dampier's reports seemed accurate enough, as the passage through what is now St George
Channel seen from the south gave the impression of being barred off by the west-trending coast of
New Ireland and the islands of Mioko and Makada in the centre of the channel. Port Montagu was further west, on the coast of New Britain.
   [1] Bougainville will naturally continue to use the term New Britain for what is in reality New Ireland.
   [2] 'According to the *Dictionnaire sentencieux* (1778), "there are only two kinds of butter that good society dares to mention: the butter from Vanvre [sic] and the butter from La Prévalais". Cf F. Braudel, *Civil-
isation matérielle et capitalisme*, I, p. 155' (ET). The area is Vanves, at the time a small village outside Paris.

most S point SW¼W4°S ditto, Bournand Island [Isle Bournand] from NNE5°E to NE¼N 9 to 10 leagues.

Note: I am giving to the islands offshore from this large land the names of officers from both ships.[1]

Bearings at midday: The most NW land of New Britain bearing WNW3°N, 6 to 7 leagues, the most SE bearing S5°E, 5 to 6 leagues, Bournand Island from E¼NE2°E to ENE2°E about 8 leagues. The middle of Oraison Island [Isle d'Oraison] N¼NW3°N 10 leagues.

Tuesday 26 to Wednesday 27.

Observed variation occiduous: 7°41′ NE.

SE calm. Fair weather, fresh SE breezes until two hours past midnight then calm until midday. The part of the land we are sailing along is far more beautiful, more level than the one we followed after our departure, accordingly it is inhabited.

We have seen numerous plantations and much cleared land. No canoe appeared on the shore but smoke was seen rising in a great many places.

The current caused us to be set towards the SSE.

The estimated run for the 24 hours is NW distance 63′30″, corrected NW3°W, distance 58′40″.

Estimated latitude: 3°26′, observed: 3°31′.

Estimated longitude: 152°16′, observed 149°28′.

Bearings at 6 p.m.: the most E land of New Britain bearing SE¼S4°S 8 leagues, the most W bearing WNW 10 to 12 leagues.

At midday: the most W point of New Britain bore WNW5°W 9 leagues, the most E bore SE5°E 7 to 8 leagues, Oraison Island by its centre ENE5°30′E about 12 leagues, Du Bouchage Island [Isle du Bouchard] by its centre N¼NW5°W.

Wednesday 27 to Thursday 28.

Calm S, SSE. The calm continued for the remainder of the day and the following night. As we were almost in the centre of the pass between New Britain and Du Bouchage Island, I sounded at 6 p.m., they let out 230 fathoms of line without finding bottom. There was very little wind in the morning. At about 10 a.m., a local canoe came up quite close to the *Étoile*. I do not know what happened on this occasion.[2] The country spread out before us is very beautiful and seems to be inhabited in all areas as well as in the small islands.

At midday the run gives me NNW4°N, distance 20′.

---

[1] There are a number of substantial island groups off the east coast of New Ireland, from the Feni Islands in the south to the Tabars in the north. Identification of Bougainville's sightings was made by Martin-Allanic, *Bougainville*, I, pp. 791-3. Bournand Island was first sighted by the expedition of Schouten and Le Maire in July 1616 and named S. Jans; it is one of the Feni Islands, or both seen as one. Oraison is Kaan Island, now the Tanga Islands, and Du Bouchage is Geritte Denis, part of the Lihir group. Suzannet Island was known for a while as Gardner's Island; it is a sighting of the Tabar Islands.

[2] A journal, probably written by Commerson, tells us that 'at about 10 o'clock we saw a canoe coming towards us, containing 9 bearded blacks. They came within hailing distance and gestured for us to go ashore. They came very close and there threw a hail of stones at us while uttering great shouts. We fired 4 or 5 musket shots at them, they fled. Some were wounded and they returned to land.' *Journal*, in BN, MS 2214:5, entry of 27-28 July.

Observed latitude: 3°22'.

Estimated longitude: 151°58', observed: 149°10'.

Bearings at 6 p.m. the most W point of New Britain bearing W¼NW5°N, 9 to 10 leagues, the most E bearing SE, Du Bouchage Island from NNW5°N as far as N5°W 7 to 8 leagues.

Sighted at midday: the most E land of New Britain bearing SE, the most W bearing W¼NW, Du Bouchage Island from NNE4°E to ENE5°N, 5 to 6 leagues, Suzannet Island [Isle Suzannet] from NW1°N to NW¼W4°N.

Thursday 28 to Friday 29.

Observed variation occiduous: 7°15',

Calm, SSE. Almost calm during the 24 hours and the slight progress that a weak breeze might allow us to make is almost entirely negated by the opposing current. In truth we are being cruelly treated and I cannot express the extent of my suffering.

Several canoes came up to the 2 ships and came very close to us, in one there were 6 men, 5 in another. They are black and naked, their hair short, woolly and in some cases powdered white. They wear their beards fairly long. The canoes are long, narrow, with an outrigger, the bow and stern a little more raised up than the body of the canoe are partly carved and fairly delicately so. The blacks were inviting us by gesturing to go ashore, we were inviting them to come aboard, but our invitations, and even the gift of a few trifles thrown into the sea, did not inspire in them sufficient confidence for them to come alongside. They were armed, one even threw by means of a sling a stone that did not reach the deck, I believe this was to test whether they were out of reach of our shots. However, if they had tried this experiment once more, we were ready to punish them. After going round the ship and picking up what we had thrown, they struck their canoes in unison with great shouts and returned to land. These men are tall and seem strong and agile. They showed us several times a kind of bread. Almost all wore bracelets around their arms. Tree leaves cover their enormous and pendulous nudity.

Estimated run for the 24 hours NW¼W5°W, distance 28'30", corrected run WNW3°W, distance 15'.

Estimated latitude: 3°9' observed: 3°17'.

Estimated longitude: 151°44', observed: 148°56'.

Bearings at 6 p.m.: The W point of New Britain W¼NW2°W, the E one SE2°E, Du Bouchage Island from NE¼N3°E up to E¼NE, its islet NNE4°E 9 leagues, Suzannet Island from NW3°N to NW3°W.

At midday: the NW land of New Britain W¼NW 12 to 15 leagues, the SE one SE 8 to 9 leagues, Du Bouchage Island NE3°E to ENE5°E 6 to 7 leagues, its islet NE 8 leagues, Suzannet Islet NW to NW¼W1°W 7 to 8 leagues

Note: two waterspouts formed in the NE that lasted 8 or 10'.

Distance covered in a straight line from the harbour: NW¼N1°N 109', 14 leagues [sic] in 15 days.

Friday 29 to Saturday 30.

Almost dead calm during these 24 hours and still the contrary current. Estimated run: NW¼W45'W, distance 31', corrected: NW¼W2°N 17'.

Estimated latitude: 3°, observed: 3°7′.

Estimated longitude: 151°30′, observed: 148°41′.

The Indian canoes surrounded us all the morning. The blacks coveted everything and offered nothing in return. They seemed to display bad faith in their bartering, they have decided today almost without ceremony to come alongside. A few shots fired from the *Étoile* on the canoes surrounding her caused the entire fleet to depart.

Bearings at midday: The SE point of New Britain bearing SE2°S, Du Bouchage Island E to E¼NE 10 to 12 leagues, Suzannet Island from N4°E to NNW 4 to 5 leagues, the NW point of New Britain W¼NW2°N 10 to 12 leagues.

Saturday 30 to Sunday 31.

Still calm with contrary current.

Estimated run NW¼W2°30′W 32′, corrected: NW¼W4°W 24′.

Estimated latitude: 2°50′, observed: 2°55′.

Longitude: 151°9′, observed: 148°20′.

This morning 15 or 20 canoes surrounded the *Étoile* and the blacks no doubt thought they would capture it by hurling stones. A few shots soon dispersed them. They left behind one canoe for their pains. Suzannet Island is very beautiful and well populated.

Bearings at midday: the SE point of New Britain bearing SE¼E3°S 6 to 7 leagues, the NW point W¼NW3°W 8 to 9 leagues, Suzannet Island from E4°S to NE¼N2°E 7 to 8 leagues.

August

Sunday 31 July to Monday 1 August.

Observed variation occiduous: 6°43′.

Very light breeze from E to SE, we continued to haul along New Britain, this part is very low and even the extremity is a semi-drowned point that seems to be 6 or 7 leagues in length.[1] A few canoes came along to observe us but they kept a fair distance away. At night we gave them a spectacle of star and serpent rockets.

Estimated run: NW¼W3°W 45′, corrected: WNW5°N 43′.

Estimated latitude: 2°33′, observed: 2°35′.

Estimated longitude: 150°31′, observed 147°41′.

Bearings at midday: The most W in view W¼SW, the drowned land seen from the mast-head W and W¼NW, the most E land SE¼S, Suzannet Island from E¼SE4°S to NE¼E2°E 15 leagues.

Monday 1 to Tuesday 2.

Calm and rain, a few gusts of squally wind during the storms. What a trial and

---

[1] The expedition has now reached the northern extremity of New Ireland, off present-day Kavieng. A number of low islands, the Tsalui Islands, separate it from New Hanover (present-day Lavongai), a substantial and high island that can easily give the impression that it is a mere continuation of New Ireland. Bougainville will recognize the insularity of New Hanover which appears in his journal as Kérué or Kervé Island.

how long it is! I have steered NW to avoid a labyrinth of islands lying off the end of New Britain.

Estimated run: NW3°W 46', corrected: NW¼W5°30'W distance 42'.

Estimated latitude: 2°3', observed: 2°15'.

Estimated longitude: 150°, corrected: 147°3'.

Bearings at midday: the most SE point of New Britain SE4°S 9 to 10 leagues, the most NW S¼SW5°W 7 leagues, Kerué Island [l'Isle Kerué] from SW¼W5°W to W¼SW5°W 10 to 12 leagues.

Tuesday 2 to Wednesday 3

Calm and pouring rain. What manoeuvres should one adopt? One should accept misfortunes patiently, so they say.

Estimated run: NW4°30'N, distance 26'. Run corrected from the bearings: NW¼W5°15'N, distance 20'30".

Estimated latitude: 1°55'40", corrected: 2°3'30".

Estimated longitude: 149°44', observed: 146°47'.

Bearings at midday: Kérué Island from SW 2°W to SW¼S2°W, the land of New Britain in view S¼SE 8 to 9 leagues.

Sighted at 5.30 on Wednesday evening from the mast-head Le Corre Island [Isle le Corre] bearing N.[1]

Wednesday 3 to Thursday 4.

Storm, thunder, rain but at least there was a wind, a fresh ESE to SE breeze. I set a course for the NW¼W.

Estimated run NW5°W, distance 59'30".

Run corrected from the altitude and the bearings NW¼W1°30'W, distance 62'.

Estimated latitude: 1°26', observed: 1°30'.

Calculated longitude: 148°51', observed: 145°53'.

Bearings at midday: Dampierre's Squally Island [Isle Orageuse][2] from SW4°W to WSW5°S 6 to 7 leagues, its islet W¼SW3°W, Matthias Island from W3°S to W4°N 12 to 15 leagues.

Thursday 4 to Friday 5. Last quarter.

SE, SSE. It rained in the afternoon and almost all night, that is the sickness, but there was a breeze, that is the cure. I steered NW¼W until 10 p.m., then WNW. At 1 a.m. we thought we could see land to the W¼SW and I steered NW then W at 8 a.m. At 10 o'clock the weather having become fine, we again saw the land glimpsed during the night, it is a small island. I believe that the two others we saw yesterday are the ones Dampier called Matthias Island and Squally Island. I sounded at 2 a.m. we did not find ground.

Run for the 24 hours NW¼W3°45'W, distance 90'.

Estimated and observed latitude: 00°45'S.

---

[1] This is probably is a distant view of lonely Tench Island. The expedition will now make its way through the wide Ysabel Channel, separating New Hanover from the St Matthias Group.

[2] Dampier came upon the Matthias Group from the opposite direction. Strictly his Matthias Island is the large Mussau Island, and his Squally Island where he encountered 'many hard Squalls and Tornadoes' (Dampier, *A Voyage to New Holland*, ed. J. Spencer, p. 205) is Emirau off its south-eastern point.

Longitude: 147°33', 144°36'.

Bearings at 6 p.m.: Matthias Island from SSW 3°S to WSW 4°W 6 to 8 leagues.

At midday: the island out at sea SE¼S 3°E 10 to 12 leagues.

Friday 5 to Saturday 6.

Occiduous variation observed by azimuth: 6°10' NE, ortive by azimuth: 5°40'.

SSE. Fair weather, light gale, we have not seen any more land and I have tried to sail away from it in order to have less rain and more wind. I steered W¼SW. The *Étoile* compelled us to remain without canvas and even hove to for part of the night.

Estimated run: W 5°30'S, distance 53', corrected: W 1°N, distance 58'.

Estimated latitude: 00°50', observed: 00°44'.

Longitude: 146°35', 143°37'.

Saturday 6 to Sunday 7.

Observed variation in the evening by azimuth: 6°12', in the morning ditto: 6° 3'.

Fair weather, fresh ESE to S breeze. I steered W¼SW and WSW when we could.

Estimated run: W¼SW 4°15'W [distance] 78'. Corrected: W 3°30'S, distance 80'.

Estimated latitude: 00°53'50", observed: 00°49'.

Longitude: 145°15', 142°16'.

Sunday 7 to Monday 8.

Observed variation by azimuth ortive: 5°50'N.

SE to SSE. Fair weather, fresh SE to SSE gale. I steered W¼SW and WSW. At 8 p.m., I decided to sail SW¼W, the horizon having a very hazy appearance at sunset from W to WSW. At dawn we steered W¼SW. The *Étoile* continues to prevent us from making any headway and this delay is cruel in our situation.

Estimated run: W¼SW 45'S [distance] 80', corrected: W¼SW 3°45'W 84'.

Estimated latitude S 1°5'45", observed: 1°.

Longitude: 143°52', corrected: 140°52'.

At 11.30 a.m. we saw a low land bearing W 5 or 6 leagues.

Monday 8 to Tuesday 9.

Fairly good weather, light ESE gale, we steered W¼SW and WSW, ranging approximately 4 miles off the island seen this morning, it is lowlying, forested and broken into several islets linked by sandbanks and reefs. There are a great number of coconut trees and habitations on this island. The huts are high, almost square and well roofed. Several canoes were fishing, none came up to examine us. These isolated men seem to be happy anchorites.[1] This morning we saw a line of low islands, wooded and surrounded by breakers, a disastrous encounter. We shall endeavour to get out of it. I have named this chain of islands the Chessboard [Isles l'Echiquier].[2]

---

[1] The island is Kaniet and it can be credited as one of Bougainville's discoveries, although it may have been sighted by Alvaro de Saavedra in 1558 and by Ynigo de Retes in 1545: Sharp, *Discovery*, pp. 19-20, 30-32, 119. South of Kaniet lies an island group, which is now known as the Hermit Islands. It was so named (Los Ermitaños) by Francisco Maurelle in January 1781, who believed them to be Bougainville's Anchorites. Bougainville did have a sighting of these to the south, but he had already given the name of Anchorites to the northern group.

[2] This name evokes primarily the image of islands displayed like chess pieces on a board, but the term *Echiquier* was also in use to describe a line of ships in sawtooth formation, as well as a tactical disposition of troops preparatory to engaging in battle.

Estimated run: W¼SW3°15'S, distance 70', corrected: W¼SW45'W [distance] 81'30".

Estimated latitude: 1°17'40", observed: 1°15'.

Longitude: 142°42', observed: 139°40'.[1]

Bearings at 5 p.m. Anchorites Island [Isle des Anchorites] N5°30'W, another from aloft WNW.

At 5.30 a.m.: two small islands fairly high bearing SE¼S3°S approximately 7 leagues.

At 8 a.m.: seen an archipelago of low islands bearing NW to SW.

At 11.30, sighted the low islands bearing NW¼N1°W 1 league and a half as far as SW 4 or 5 leagues, another low one also bearing S5°W.

Tuesday 9 to Wednesday 10.

Calm, SE. The weather was squally with rain and a fresh breeze. We steered to round these low islands. Towards 5 o'clock, the calm delivered us over to a violent current that was driving us towards the coast. The night was worrying. The danger of being carried to the land was only too real. I had two anchors made ready to go and readied some cable on the deck. We sounded 5 or 6 times without finding bottom: for such is one of the dangers of these shores: almost two ship's lengths from the reefs that line them one is unable to lower an anchor. Fortunately the weather was not spoilt by any storm: there came even a light N breeze around midnight that enabled us to progress a little towards the SE. The wind freshened at 8 a.m. and led us away from these low islands that I believe to be uninhabited. During these two days, we saw neither fires, nor canoes, nor huts. Some people thought they had seen one but not clearly.

At midday the estimated run according to the bearings is SW5°S25'45", corrected it is SW1°S27'20".

Estimated latitude: 1°34'40".

Longitude: 142°24', 139°21'.

Bearings at 6 p.m.: the line of islands named the Chessboard from N¼NW to SW, approximately through its centre and 1 and a half leagues from land, the SW island or Boudeuse Island from E1°S to E¼SE1°E, about 3 leagues.

At 6 a.m.: the Chessboard from NE1°E to W¼NW2°W, the Boudeuse [island] from ENE about 3½ leagues.

At midday: the most N of Chessboard Islands bore N¼NW1°W 4 to 5 leagues, the most S bore WSW3°W same distance.

Wednesday 10 to Thursday 11.

Observed variation occiduous by azimuth doubtful: 3°55'.

---

[1] 'There was a serious error in the longitude on that day, of some 2°, the consequence no doubt of the strong currents that prevail in this region. The route followed by the *Boudeuse* led her to the largest and most dangerous atoll in these seas, Ninigo Archipelago, called *L'Echiquier* [the Chessboard] by Bougainville. It consists of six atolls, quite close together, around 150 islands or very low islets that stretch over close on 100 kilometres, some of which are wooded. Worried by the presence of this dangerous area, Bougainville went round the archipelago by the south, sailing between it and the island of *La Boudeuse* (Liot Island)' (ET).

Fair weather, fresh SE breeze. At 3.30 we had rounded the low islands. Since at sunset we had been able to see open water for some 8 leagues to the SW¼W, we followed this course with all sails set until midnight, then we sailed on under topsails and the mizzentop.

Run for the 24 hours WSW4°30'S 95'.

Estimated latitude: 2°15', observed 2°17'.

Longitude: 141°, observed: 137°56'.

Bearings at 6 p.m.: the last two islands one after the other NE approximately 4 leagues.

Thursday 11 to Friday 12.

Observed variation by azimuth ortive: 4°3', 2nd: 3°51'.

Calm, SE. Fair weather, little breeze, at midday we saw land from S¼SE up to S¼SW, an extremely high land with high mountains; soon we saw it extending towards the W, which proves that it is New Guinea.[1] Practically calm throughout the night and in the morning; steered W¼SW and W.

At midday the run is W¼SW1°20'W [distance] 41', corrected: W¼SW2°30' [distance] 46'30".

Observed latitude: 2°24', calculated longitude: 140°20 , 137°10'.

Bearings at midday: the land from SE¼S approximately 12 leagues to W¼SW2°W 8 to 9 leagues.

Note: the currents were running S and I think W.

Friday 12 to Saturday 13.

Observed occiduous variation by azimuth: 4°4', by amplitude: 4°30' ortive by amplitude: 4°24'.

E, calm, SW to W. Fair weather, fresh E breeze in the afternoon and part of the night. Then the weather turned stormy with rain, lightning and a notable burst of thunder. Calm followed and a light variable SW to W breeze. Meanwhile the ravages of scurvy are spreading. There are new cases daily. What food, Good Lord, is our lot! Rotting bread and small quantities of it and meat the smell of which the most intrepid cannot bear when the salt is removed. In any other circumstances our salted provisions would be thrown overboard.

Estimated run for the 24 hours WNW2°45'W, 48'30", corrected NW¼W4°15'W, 69'.

Estimated latitude: 2°7', observed: 1°58'.

Longitude: 139°24', corrected: 136°10'.

The land we are following is uneven, some high parts, some low parts with double folds and enormous mountains. The coast seems to offer several anchorages.

Bearings at 6 p.m. land stretching from SE5°E to W¼SW approximately 12 leagues.

---

[1] 'These were indeed the mountains of New Guinea, the highest of which rise to 5,000 metres. The *Boudeuse* was some 100 miles east of the great Geelvink Bay' (ET). Geelvink Bay is now Teluk Cenderawasih

At 8 a.m.: land from W¼SW to SE¼E3°E; the middle of the 2 Cyclops bearing SE¼S4°S.[1]

At midday: the land bore from W5°S to SE¼E.

Saturday 13 to Sunday 14.

Observed doubtful ortive variation: 3°6′.

Calm, SSW to SW. Light breeze in the afternoon, almost calm since midnight and a breeze varying from SSW to SW. We saw the continent in the distance. There is something like a vast gulf here in which several islands are situated.

Estimated run for the 24 hours: WNW1°15′N, 52′ 30″, run corrected by the altitude and the sight of land: WNW3°N 71′.

Estimated latitude: 1°38′, observed: 1°28′.

Estimated longitude: 138°24′, observed: 135°6′.

Bearings at 6 p.m.: land from SE¼E to WSW4°W.

At midday: Oger Island [Isle Oger] bore WSW5°S to SW5°W 4 to 5 leagues, Aiming Corner [Coin de Mire] SW¼S4°W 4 leagues, Cliff Island [Isle des Falaises] from SW¼S4°S to SSW2°W same distance, the main land from SSE5°E to SW5°W 12 to 15 leagues.[2]

Sunday 14 to Monday 15.

Observed variation by azimuth at 5 p.m.: 3°20′, by amplitude: 2°55′.

Little periods of calm. There was almost no breeze during the day and the little that we got from time to time was most variable. The *Étoile* is still preventing us from benefiting from a breeze when there is one. We have had to stay under the topsails and even remain hove to during part of the night. That vessel is not working any more. Thunder and rain for part of the night.

At 10 a.m. we fell in with a very noticeable tidal race. It was running N and NNE and carrying a prodigious quantity of trees, trunks, leaves and fruit. It seems to indicate the presence either of some great rivers in the mainland or a channel around here running N and S. We had abreast of us a very low and level land that I believe to be an island I have named Alie Island [Isle de Alie].[3] The continent mainland appeared in the distance, a high land that seemed to end in a cape.[4] The

---

[1] Bougainville is in view of the high mountain ranges on either side of Humboldt Bay, so named by Dumont d'Urville in 1827 (now Teluk Yos Sydarso). The Cyclops are relatively low at just over 300 metres; nearby is an isolated peak, 155 m high, that became known as Bougainville Peak. These names survived for many years, but all these geographical features now have Indonesian names. Thus the Dutch settlement of Hollandia became known as Sukarnopura and subsequently Jayapura. The *Boudeuse* and the *Étoile* sailed down from the Ninigo group and the islands south-west of it to reach the coast of West Irian not far from the present-day frontier between Indonesia and Papua New Guinea. They were now to sail offshore in a westerly and north-north-westerly direction.

[2] The expedition is reaching Walckenaer Bay in which there are three small islands, known as the Podenas (Pulau-pulau Podena). Bougainville sighted two of these, *Ile des Falaises* (Cliff Island) and *Coin de Mire* (Aiming Corner). The island that he name after the *Boudeuse*'s second pilot Carles Oger is probably the one known as Jamna.

[3] The expedition is now off the great Mamberamo River. Nearby are the Koumamba Islands, of which Alie Island is one, though the name has not been retained.

[4] The expedition is reaching the most northerly feature of New Guinea approaching Geelvinck Bay; it was known for many years as Cape D'Urville, again following that navigator's visit of 1827 (now Tanjong Narwaku). Near the mouth of the Mamberamo lies the town of Teba.

water was white and so changed that we sounded twice but without finding bottom.

Estimated run: W¼NW2°N 36′, corrected: W¼NW2°45′N, 45′20″.

Estimated latitude: 1°20′, observed: 1°17′.

Estimated longitude: 137°40′, corrected observed: 134°22′.

Oger Island seemed to us to be inhabited, we saw fires rising from it. This morning I gave the name of Moulineau to an enormous mountain on the mainland.[1]

Bearings at 6 p.m.: The W point of Oger Island bearing S3°W 2 leagues and a half to 3 leagues, Cliff Island by its middle SSE1°S 3½ leagues, the most W of the continent SW 12 to 15 leagues.

At midday: the most W point of Alie Island SW 4 to 5 leagues.

The horizon being very hazy, we could not see any other land.

Monday 15 to Tuesday 16.

Observed variation by azimuth at 5 p.m.: 3°2′ NE, amplitude occiduous: 3°15′.

Variable, calm. Light breeze variable from N to E then calm, heavy storm during the night, rain, we have been the toy of the tide that very often was running N, the ship would not steer. It was calm during most of the morning.

Estimated run: W¼NW2°N 21′30″, corrected: NW¼W4°30′W 28′30″.

Estimated latitude: 1°12′, observed: 1°3′.

Estimated longitude: 137°19′, observed: 133°57′.

Sighted at midday from the mast-head a high land bearing SW.

The currents are continuing to run N.

Tuesday 16 to Wednesday 17.

Observed variation by azimuth at 5 p.m.: 3°24′.

W, calm, SW. As the *Étoile* was very distant from us and the winds W, I decided to tack towards her and join her. At 5.30 p.m. we had reached her. That vessel is finding it very difficult to make way in this tide and is weary. Such is our misfortune that her sluggishness again prevents us from taking advantage of the light wind that would enable us to progress. There have been many arguments about where Hell is situated, truly we have found it.

Calm during the night, followed by horrible thunder, and heavy rain, all that followed by a fairly fresh head wind.

Estimated run: W¼NW5°N 14′, corrected: W¼NW30′N, 18′30″.

Estimated latitude: 00°59′ S.

Estimated longitude: 137°6′, observed: 133°39′.

No sight of land.

Wednesday 17 to Thursday 18.

Calm, SW to W. Almost calm, rain and head wind for what there was of it. No, a man's life is not relative to the length and the kind of test to which we are being subjected so forcibly.

Estimated run: NW¼N2°30′N 36′, corrected: NW1°15′W 44′40″.

Estimated latitude: 00°30′S, observed: 00°25′S.

---

[1] 'Moulineau is the crest of the Gauthier Range rising to approximately 2,000 m' (ET).

Longitude: 136°44′, 133°14′.

Sight of land at 4 p.m. stretching from SSW5°W to SW5°W 15 to 20 leagues distant, high land.[1]

At 9 a.m. sighted an island bearing SW¼W.

At midday it bore SW¼W1°W approximately 10 leagues.

Thursday 18 to Friday 19.

W to SW. Head wind from W to SW. We tacked still having fairly high seas coming from the W. We saw land from SSW to SW, high land. The *Étoile* signalled there was another bearing W¼NW to WNW. We have been unable to see it at any time during the day.

Estimated run: W¼SW 15′, corrected: NW¼N3°W 26′30″.

Estimated latitude: 00°28′S, observed: 00°4′S.

Longitude: 136°30′, observed: 132°58′.

Sighted at 5 p.m. the land bearing SW2°W 8 to 10 leagues.

At midday from SSW to SW 15 to 18 leagues.

The currents are still running N.

Friday 19 to Saturday 20.

WNW to SW. Head wind from WNW to SW, heavy seas from the W. We sailed close to the wind on both tacks.

All we had on board was a dog from the Strait of Magellan, young and plump. We killed it, ate it and found it excellent. He fed us for two days. Our scurvy cases are getting worse and their number is increasing.

Estimated run: W4°20′S, corrected: W¼NW00°30′N, distance: 39′.

Estimated latitude: 00°7′S, observed: 00°4′N.

Longitude: 135°36′, observed: 132°20′.

Bearings at 6 p.m land stretching from SSW5°S to SW 12 to 15 leagues.

No sight of land [in the morning], the currents having carried us towards the N. Crossing of the Line.

Saturday 20 to Sunday 21.

Observed occiduous variation by azimuth: 2°41′, ortive by azimuth: 2°6′.

Head wind from S to SSW. We steered close-hauled on the port tack.

Estimated run: W5°S, corrected: W¼NW4°58′W 80′20″.

Estimated latitude: 00°2S, observed: 00°12′N.

Longitude: 134°44′, observed: 131°.

The currents are still running N.

Sunday 21 to Monday 22.

Observed ortive variation by azimuth: 2°2′.

Calm, W, NW, SSW. Calm in the afternoon. At 8 p.m. the winds came up from

---

[1] The ships are now sailing to the north of the great Geelvinck Bay and the great Schouten Islands (now Pulau-pulau Briak) have come into view. They cover an area of over 660 sq. miles and were discovered in July 1616 by Willem Schouten and Jacob Le Maire. The high land that first came into view would have been the island of Biak, which is dominated by Mt Sombunen (244 m).

the W then the NW. I steered SW until 7 a.m. when we saw land ahead of us. I then set course for the W.

Run: SW¼W 4°40′S distance 39′, corrected: SW¼W 2°S 40′40″.

Estimated and observed latitude: 00°12′S.

Longitude: 134°14′, 130°27′.

Sighted land from SSE to SW, extremely high land.

Second crossing of the Line.

Monday 22 to Tuesday 23.

Observed occiduous variation amplitude: 2°15′, ortive azimuth: 2°5′.

During the 24 hours the winds have been variable having practically boxed the compass.[1] We steered according to their fancy. A violent current bears NW here. We are convinced of this by the various bearings we have taken, having not lost sight of the coast. We sounded several times without finding ground.

The run for the 24 hours is W 42′, corrected 45′. Estimated latitude: 00°14′S, observed: 00°12′S.

Longitude: 133°32′, 129°42′.

Bearings at 6 p.m. the land stretching from SE¼S to WSW 5°W, the closest land bearing WSW.

At 6 a.m. the SE¼E land bore SW 4°W.

At midday New Guinea bore from ESE 4°E to SW 3°W 12 to 13 leagues, 2 islands one bore SW¼S 1°30′S the other SW 1°30′W 4 leagues.[2]

Tuesday 23 to Wednesday 24.

Observed occiduous variation by amplitude: 1°55′ NE.

S. SSW. At 2 p.m. I lowered two boats and sent them to the outer island where there seemed to be a village in order to endeavour to obtain some refreshments. A shoal that runs along this island and even extends quite a distance towards the E forced our men to make a wide turn to round it and land on the island. Our men found neither habitations nor refreshments. What from a distance had seemed to be huts were merely rocks hollowed out by the sea and eroded into caves. It was 9.30 p.m. when the boats returned to the ship. The *Étoile* had had time to catch up. She was over 4 leagues behind us. She is causing us enormous delays.

At around 9 p.m. the winds freshened from the SSW to S. I kept close-hauled on the port tack until we had run 5 leagues then ran 3 leagues on the other tack, then we reverted to the port tack with variable S¼SE to S and SSW winds.

The mainland was trending SW and SSW and a great pass opened out between that coast and an island further to the NW.[3] I wondered whether I should manoeuvre to make my way through the open passage or whether I should pass to leeward of the NW island. I decided to follow the latter course, having no hope of making any

---

[1] To 'box the compass' is to make a complete revolution; the French equivalent is more literal, '*faire le tour du compas*'.

[2] 'These two islands are the Mios Saya' (ET). Now Pulau Milos Su.

[3] Bougainville has reached the end of the great island of New Guinea and is in view of Waigeo which is separated from it by Dampier Strait. He will now have to find his way through the labyrinth of islands large and small that will eventually lead him to a Dutch-administered outpost.

progress against the wind and tide. The wind SSW stormy gale and the tide running NW. We had two consecutive alerts during the morning. First of all they shouted from above that they could see a line of breakers ahead of us and we immediately changed tack, then having examined more carefully these breakers we found them to be the eddies of a violent tide. We resumed our course and an hour later they called out from the forecastle that they could see the bottom under us. The matter was urgent, but the alert was as short as it had been serious, the water promptly reverting to its normal colour, I cannot decide whether those who called out that they could see the bottom actually saw it. 5 or 6 people claim they did.

My poor boatswain Denys Couture died of scurvy at 2 p.m. today. The King has lost a good servant.[1]

Estimated run: W3°45'S 23', corrected: W2°S 59'.

Estimated latitude: 00°13'39", observed 00°14'S.

Longitude: 132°37', 128°43'.

Bearings at 6 p.m. the land of New Guinea stretching from E¼SE to SW5°W distant 14 to 15 leagues. A land further N bore W4°30"N distant 9 to 10 leagues. The W head of Cave Island [Isles des Cavernes] SSE3°E distant about 2 leagues.

At 6 a.m.: the land of New Guinea from E4°S to SW¼S and SSW, the land seen yesterday at sunset to the W bore from WNW3°N distant from 7 to 8 leagues to WSW4°W stretching to WSW4°W same distance..

At midday: the land of New Guinea from ESE to SSW about 15 leagues, the N point of the island we are coasting along bears NW¼W4°N, a cape stretching out from the the most S point bore SW5°S, the middle of an island further S SW5°S approximately 6 leagues.

Wednesday 24 to Thursday 25.

S, SSW. We sailed along the land with S to SSW winds, good breeze and a strong current running NW. This land is high, forested and the shore is steep almost everywhere, nevertheless there are a few sandy coves in which we saw some huts. We also saw some canoes but few in number as were the huts. This coast has very few inhabitants. Around 3.30 we saw some low land in the N and at sunset we discovered more bearing W and W¼NW. Here we are once again hemmed in.[2]

We spent the night tacking and at dawn we had land to the E, SSE, S¼SE, SW¼S, SW, W¼SW, W and W¼NW, with two passes between the various islands one of which lies almost NE and SW, the other E and W. With the winds still S and SSW, fresh breeze, we tacked throughout the morning. At 10 a.m. I signalled to the *Étoile* to carry out some soundings. They found sandy ground at 45 fathoms. Mr de la Giraudais told me that yesterday at around midday he sailed over a shoal and saw the bottom for about 1'30".[3] The *Étoile* was following our track and his report confirms that of our men.

---

[1] Bougainville had a high opinion of Couture and had named a tidal race in Bougainville Strait after him.

[2] 'The land seen to the south is part of Waigeo and the northern one the low atoll of Aiu [Ayu] which extends over some fifty kilometres. Other islands appeared in the west with, behind them, Gilolo Passage [Selat Gilolo] that links the Banda Sea to the Pacific Ocean, a route Bougainville could have followed' (ET).

[3] This confirms similar reports made the previous day and noted in Bougainville's journal.

Estimated run for the 24 hours: W4°N 43′, corrected by the altitude and the observations: WNW4°45′W 59′.

Estimated latitude: 00°11′S, observed: 00°4′N. 3rd crossing of the Line.

Calculated longitude corrected by the observations: 131°45′, 127°47′.

I believe that we have entered the Papuan Islands. We shall have to get away from them, but the exits will not be easy. Furthermore we have an eternal headwind or calms and soon we shall have nothing to eat. The vegetables ran out last night. The number of scurvy cases is increasing daily. There are now 45 on the sick list.

Bearings at 4 p.m.: sighted a low land from N¼NW3°N up to N5°W, approximately 8 leagues distant; one and ½ hours later we saw a reef attached to this island stretching from N¼NW to NE¼N.

At 6 p.m.: the most N island bore NE¼N2°E, its centre being 6 to 7 leagues distant, the island we are sailing along from ESE4°S up to WSW 10 to 11 leagues, 2 islands further W than this point the one bearing W¼SW2°S, the other W¼SW5°W in view.

During the morning: the most E land in sight bearing E5°S, the most W bearing SW¼S, 9 to 10 leagues, the middle of the small islands further W than this point SW5°S 7 to 8 leagues, a great island from SW up to W¼SW2°S, a more N island from W3°N to W¼NW4°W 9 to 10 leagues, a more N island by its centre W¼NW4°W 5 leagues, the islets closest to us S¼SE 4 to 5 leagues, other islets further E SSE, the most E island through its centre E5°S 10 to 12 leagues.

Thursday 25 to Friday 26.

Observed variation doubtful: 1° NE.

S, SSW, SSE. I have decided to tack in order to reach the SW passage.[1] This is because on that side we can see the open sea to the S for a distance of at least ten or twelve leagues. The main thing, in our situation, is to sail into higher southern latitudes to be sure that we have rounded both New Guinea and the mass of islands spread between it and the Moluccas.

During the entire day, we made hardly 3 leagues, the winds remaining S to SSW with a barbarous constancy. The night was more favourable towards us, thanks to Latona's daughter whose silvery light enabled us to tacks between the rocks and the islets.[2] Furthermore the current that had been contrary for as long as we were abreast the W pass became helpful when we reached the SW channel. At half past 4 in the morning we rounded the most NW islets and with the SSE breeze set a course for the SW.

I sounded several times at 55 and 65 fathoms, grey fine muddy sand. The bottom is shallow as one approaches the N land and increases when one nears the S one. The *Étoile* which was closer by two miles to the latter than we were found 80 fathoms.

---

[1] Two straits were available to Bougainville as he sailed along the northern coast of Waigeo (he had already passed Dampier Strait between Waigeo and the New Guinea mainland), one almost due west that would have taken him towards Halmahera, the other south-west and much more likely to bring him to a Dutch outpost. This latter passage became known as Bougainville Strait.

[2] Latona in Roman mythology was the mother of Diana who was a moon goddess and whose name was used in poetry to represent the moon.

At 6.30 a.m., when we were well out of the channel, we sounded without finding ground.

The run for the 24 hours by the altitude and the bearings is SW¼W1°30'W distance 51'.

Observed latitude: 00°23'S. 4th crossing of the Line.

Calculated longitude: 131°4', 127°4'.

Bearings at 6 p.m.: the large island from E to SW¼S4°S distance [of] this last point about 7 leagues, the most E islet at the E point of the SW channel SW5°S, the NW islands from WNW to NW¼N, the NW point of the W channel W¼NW4°W approximately 5 to 6 leagues.

At 8 a.m.: the middle of several islands bore N¼NE 7 leagues, the middle of the channel from which we are emerging ENE2°N 6 leagues, the SW point of a more S island ESE 4 leagues, the NW point of another island SSW2°S 3 leagues, the S point of an island N of the latter W approximately 6 leagues.

At midday: the port island from E3°S to SE¼S 6 to 7 leagues, the starboard island from NW2°W to NW¼N2°N 7 to 8 leagues

Friday 26 to Saturday 27.

Observed occiduous variation by azimuth: 1: 00°59' NE, 2°: 00°36' NE, ortive by azimuth: 00°30' NE.

S, SSW. The winds have prevailed from SSW to S, even up to SE and calm. I steered SW, even SW¼S when we could. We saw several islands and the altitude proves to me that we are finally in the Indies Sea.

Note: these islands are 5 in number, there were 7, two disappeared in an earthquake some years ago. The Dutch name them the 5 Islands and send [ships] there; neighbouring people go fishing there.

Estimated run: SW¼S1°15'S 52', corrected: SW¼S4°30'W 61'.

Estimated latitude: S1°7', observed: 1°11'.

Calculated longitude: 130°32', 126°26'.

Bearings at 3 p.m.: sighted a small island bearing SE, at 4 o'clock it bore SE¼E4°S 8 leagues.

At 6 o'clock, this island bore ESE3°S 7 to 8 leagues, the E island from E¼NE4°N to NNE2°N, the W island N to N¼NE2°N 10 to 11 leagues.

At midday: the most E island bore E¼SE 6 to 7 leagues, a large round islet named Big Thomas [Gros Thomas] sighted this morning at 10 o'clock bore SSW4°S 4 to 5 leagues.[1]

Saturday 27 to Sunday 28.

Observed ortive variation by amplitude: 00°6' NE.

S, SSW. Unendingly S to SSW breeze together with calms. We steered W in the afternoon and tacked during the night to round the large islet. I sent instructions to the *Étoile* to lower her boat and have the nearby islands examined. The calm weather will not be totally wasted. There is a spring tide.

Estimated run by the altitude and the observations is SW1°S 23'20".

---

[1] '"Gros Thomas" is an isolated islet named Pisang, 480 metres high' (ET).

Observed latitude: 1°28'.

Calculated longitude: 130°17', 126°10'.

Bearings at 6 p.m.: Big Thomas SSE2°E, 3 or 4 islets in the W from W¼SW3°W to W¼NW2°W.

At midday: the middle of Big Thomas ENE2°N about 5 leagues, the W islands from W¼SW4°S to SSW1°W distant 2, 3, 4 and 5 leagues.

Sunday 28 to Monday 29

Occiduous variation by azimuth: 30' NE, amplitude occiduous: 40' NE, ortive by azimuth: 42'.

S, SSW. Fine weather if it was good, but the head wind remains constant. The *Étoile*'s boat came back at around 9 p.m. Mr de la Giraudais advised me that no refreshments were found on the island they visited, nor any kind of spices. A black in a canoe came on board of his own volition. He was armed with two spears and wore a gold earring. He seemed in no way surprised to see men who were so different from him. These gentlemen believe that he is an escaped slave from some inhabited island, those they visited not having any inhabitants. Also there is a fierce current between these islands with no anchorage. Their trees and plants are similar to those of New Britain.

Run according to the observations and the latitude: SW1°S 33'.

Observed latitude: 1°52'.

Calculated longitude: 130°2', 125°53'.

Bearings at midday: Big Thomas NE¼N3°E 11 leagues, the islands visited yesterday from NW¼W2°W to NNE1°E 5 to 8 leagues.

Note: During the 24 hours we have passed several considerable tidal races.

Sounded no bottom.[1]

Monday 29 to Tuesday 30.

Observed variation by azimuth at 4.30 p.m.: 00°2' NW, amplitude occiduous: 00°5' NW, ortive amplitude: 00°00'.

S, SSW, SW. The weather was fine but calms or head wind. What trade winds these are! The current is running strongly SW. We passed several very strong tidal races. I sounded and no bottom was found.

The run estimated by the latitude and the observations is SW distance 37', corrected: SW2°W 38'.

Observed latitude: 2°18'.

Calculated longitude: 129°36', 125°25'.

Bearings: at 2 o'clock another island bearing W 10 to 12 leagues.

At 6 p.m.: the closest island to the NW bore NW¼N2°W 5 leagues, the one sighted this afternoon WNW 7 to 8 leagues.

At 9 a.m.: the most N island N5°30'W 10 leagues, that seen at 2 o'clock at NW¼N 10 to 12 leagues.

At midday no land in sight.

---

[1] 'On 29 August, Bougainville took possession of a "vast stretch of land and islands uninhabited by any Europeans, situated between 5° of latitude S and 2° of latitude N from longitude 150° to 126° E of Paris." He does not name these lands. (BN, NAF 9407, f° 145)' (ET).

Tuesday 30 to Wednesday 31.

Occiduous observed variation: 00°25′ NE.

SSE, S, SSW, calm. Almost calm and headwind. At 4.30 we sighted land to the S, a high land and at the time very fogbound. I believed it to be Ceiram.[1] During the night the current caused us to come closer to land. Without this favourable current, I doubt that we would ever get out of this area.

At midday the run by the observations and the latitude is S2°W, distance 25′.

Observed latitude: 2°43′.

Calculated longitude: 129°36′, 125°24′.

Bearings at 6 p.m.: the land from S¼SW to SSE at a great distance.

At 6 a.m.: the land from SE5°E to S¼SW5°30′W approximately 8 to 9 leagues.

At midday: the land from ESE5°E up to SW, this latter at a great distance.

Note: 1° We have seen day and night some great fires on the shore. It is forested and enormous mountains rise up from place to place. We took soundings without finding bottom.

2° Comments on the winds in these parts: in the Moluccas [Moluques], they call N monsoon the W one and S monsoon the E one because during the former, the winds blow more usually from the NNW than from the W and during the 2nd, they come most often from the SSE. These S winds dominate in the same way during this monsoon in the Papuan islands [Isles des Papous] and on the New Guinea coast. We know this from sad experience.

September

Wednesday 31 to Thursday 1 September.

Calm in the afternoon, fresh breeze during the night. We ran SW¼W until midnight, the wind and the current enabling us to make great headway. From midnight until 3 a.m. we sailed E¼SE, then for an hour SW then until 6 o'clock E¼SE still fresh breeze. We were in a bay, several fires were lit along the shore and at dawn we saw two boats with a sail of the shape used by the Malays. I hoisted the ensign and the Dutch flag and fired one gun. The boats went ashore without coming near us, perhaps they do not like their masters' colours, and perhaps the Dutch do not allow the locals to go on board any ship… The land at the back of the bay is low and level, surrounded by high mountains and the bay is sprinkled with several islands. I steered WNW to round a fairly large one that bore from us W at the point of which lies an islet and an uncovered sandbank with a reef that appears to stretch for one league out to sea. With a quite fresh S breeze, we sailed W¼SW until midday.

Run calculated by the latitude and the observations: WSW1°W 49′30″.

Observed latitude: 3°1′.

Calculated longitude: 128°51′, 124°28′.

---

[1] The expedition has sailed south through the Molucca Seas and has now come in view of Ceram (or Seram), a large island of some 17,000 sq. m that rises up to 3,000 m. The French will now veer west towards the island of Buru where the Dutch outpost of Caieli (present-day Kayeli) was situated in a large and sheltered bay on the north-east coast.

Note: the bay into which we sailed is a bay sprinkled with islands, in the NW of the small island of Ceram. The island one then rounds is Bonao, which is cut in two by a channel and makes up two islands. Valentin gives some excellent details on this part of the Moluccas, that I recognize as being correct.[1]

Bearings at 6 p.m.: the most E land bearing ESE, the most W bearing SW 5°W. We can see approximately 15 to 18 leagues of coastline.

At midday the N point of Bonao Island bore E¼NE, its S point E¼SE 4°S 5 leagues, the S point of Kilaing Island S¼SE 3°S 7 leagues. The land of Manipo Island from S 5°E to S¼SW 5°W in view.

Note: The Dutch have endeavoured to destroy cloves in Bonao, Kelang and Manipo.[2] They have an outpost on the latter island. Kelang supplies good building timber and even has pine trees of which the inhabitants sell the resin.

Thursday 1 to Friday 2

Observed variation occiduous: 00°41′ NE.

Fresh SSW to S and SSE breeze. We tacked between Bonoa, Kilaing and Manipo. At 10 p.m., we sighted the land of Buru [Bouru] and as my plan was to call there rather than at Amboina, we tacked variously during the night to stay within reach of that island.[3]. At daybreak, I recognized Cajeli Bay [Baye de Caieli] and we tacked in order to enter it. The Dutch have a small post here and not knowing what the situation is in Europe, it is wise to risk discovering what it is only in a place where we have superior forces. I rounded a bank surrounding the port headland of the bay as one enters. I lowered my flag to half-mast and fired one gun to request a pilot as I could see a number of boats in the bay. None came and so I sent mine to take soundings and let me know the depths. At 1.30 p.m., I dropped anchor in 27 fathoms sand and mud, excellent holding ground. I let go a second anchor. The *Étoile* anchored close to us.

At midday the run for the 24 hours is SW¼W 1°5′W, distance 37′.

Observed latitude: 3°21′.

Calculated longitude: 128°20′, 124°7′.

From midday to half past one we covered 4 miles SW, which places the anchorage in latitude 3°24′40″ and [longitude] 124°3′. We have observed since then latitude 3°25′.[4]

Bearings at 6 p.m.: the E point of Kelang Island [Isle de Kelang] E¼SE, its W point SW 5°S, the E point of Manipo Island [Isle de Manipo] SSE 5°E, its W point S 5°W, a small islet further W S¼SW 2°S.

Bearings at the anchorage: The N point of the bay N 4°E approximately 2 leagues, the E point which is on the port hand as one enters NE 2°E ½ leagues. A

---

[1] Fr. Valentijn was the author of *Oud en Niew Oost-Indien*, published in Dordrecht in five parts in 1724-6.

[2] Dutch successes in maintaining their clove monopoly were waning by the late 1760s and soon the French would set up their own plantations in Mauritius and other French colonies. See Ricklefs, *A History of Modern Indonesia*, pp. 104-5.

[3] Amboina lies south of Seram. Off the north-western extremity of Ceram lie several islands. The distance separating the westernmost point of these islands from Caieli (Kayeli) Bay is approximately 40 km (25 m.).

[4] Kayeli is situated in latitude 3°23′ South and in longitude 127°06′ East which is equivalent to 124°46′ East of Paris. Bougainville's reckoning accordingly reveals an error of 43′ West, which corresponds to his error at his time of departure from New Britain.

peninsula W¼NW1°W approximately ¾ league. The end of a breaker that extends more than a league to seaward from the peninsula NW¼W. The flag of the Dutch Residence S¼SW5°W.

Continuation of the Friday.

A boat came alongside as soon as I had anchored, with two unarmed soldiers one of whom who speaks French asked me on behalf of the head of this factory why I was coming here.[1]

Note: the Company's[2] instructions require residents of these factories, upon their first contact with foreign ships, to send two unarmed soldiers.

I sent him an officer to tell him that hunger was bringing me to this port, that by giving me food and refreshments he would enable me to leave it and that the sooner the better.[3] Towards 5 p.m., I went ashore with almost all the officers and we were perfectly well received by the Resident. He was most generous with his offers of beer and gave us a good supper that our appetite caused us to prize more greatly than the best meals we had ever had.

Note: The Dutch were amazed at the voracity we displayed when eating. For my part, I declare that this supper was one of the most delicious moments of my life.

This Resident's grade is that of an under-merchant, he has under his orders one sergeant and 25 men. His fort is a weak wooden palisade with a battery of 5 or 6 small-calibre guns, for what they are worth.

Note: they had originally built a stone fort here, an accident caused it to blow up in 1689 and since then they have been satisfied with a surrounding palisade. It is named the Defence and I believe that to be a nickname. 14 habitations of Indians, almost all Moslems, previously scattered through the island, are today clustered around the Residence and make up the township of Cajeli, on the banks of a small crocodile-infested river. There are a few black quarters scattered in the island where they cultivate rice. Unhusked rice is called *padi*.[4]

The Company obtains from this post only black and white ebony wood and a few other varieties of hardwood valued for works of joinery. There are at present 3 Dutch vessels in this port. The largest is a snow of 14 guns named the *Draak* whose function is to cruise among the Moluccas, particularly against the Papuans and the Ceramese. The Dutch are at war with these nations.[5]

---

[1] According to reports in the Dutch archives, the first contact with the French was made by a port official, Georg Frederic Teylinger, who had been sent by the resident: General Archives, the Hague, VOC archives, vol. 3140, f. 25. See also Martin-Allanic, *Bougainville*, I, p. 795n.

[2] I.e. the Dutch East India Company, founded in 1602, which held a monopoly of Dutch trade in the east. A good source on its background is Bruijn, *Dutch-Asiatic Shipping in the Seventeenth and Eighteenth Centuries*.

[3] 'Bougainville sent the resident a letter outlining the purpose of his call, which is held in the Archives of The Hague (Gen. Arch, Ven. Oost-In. Compagnie, 3140, f° 28). The officer in question was Kerhué' (ET).

[4] Bougainville is correct: the Malayan word *padi* means unhusked rice or rice in the straw. From this, the English obtained the term 'paddy fields'.

[5] In addition to the *Draak* ('Dragon') there were the sloop *Nagelbooem* ('Clove') and a light local schooner or *pantjaling* the *Liefde* ('Love'). A fourth is also mentioned in journals, the schooner *Goede Frouw* ('Good Woman').

Note: Captain Kop Le Clerc, from Saxony,[1] who commands the *Draak*, told me that, over the previous two years, they had attacked the people of Ceram on several occasions. He had even captured one of their forts. It should be said that these people defend themselves vigorously, they are armed with spears and poisoned arrows and have a fairly high number of muskets. There was even a 2-pounder gun in the fort taken by Mr Le Clerc, no doubt the English supply them with these weapons.

They do not even have a post on Ceram. They discovered on that island two years ago some new cloves. They travel in strength there to burn them out. In general by adopting an excellent policy they keep cloves only on Amboina, nutmeg at Banda and cinnamon at Cerlan.[2] These important commercial centres are strongly fortified and well supplied with men. The Company maintains an infinity of other outposts the purpose of which is to maintain its authority over all these countries, to see that all the spices are destroyed except in those places where it is sure of keeping them under control and to prevent contraband.[3] The English have been ceaselessly roaming about these islands in recent years. They leave from Bankoul and visit these parts from which they sometimes obtain contraband nutmeg and cloves.[4] Two years ago in a factory neighbouring this one (Savaï N of Ceram) an English frigate, the *Kimberg*, 26 guns, commanded by Mr Watson, used musket shots to obtain a pilot so as to enter there and caused many vexations. Byron put into Batavia. Another English frigate, a year later also went into Batavia, obtained an enormous quantity of provisions and in November, during the W monsoon, left towards the E. I believe that they are setting up close to the Moluccas an establishment linked to the vessels of which we found traces in New Britain.[5]

Saturday 3 to Wednesday 7.

I made a deal with the Resident who will supply me with rice, a few hundredweight of dried venison and fish, a few cattle and some venison for our daily use. They are plentiful on the island as are boars. Poultry is not abundant, however we shall obtain a few. The price is not a consideration. Anyhow everything here belongs to the Company directly or indirectly, it alone buys and sells. The Resident is a despot. He has 100 slaves to attend to his house who are not natives of this region. The local

[1] Martin-Allanic speculates that this officer was one Kristophe Le Clerc, possibly originally spelt Leclerc, and the son of an exiled French Protestant. See Martin-Allanic, *Bougainville*, I, p. 796n.

[2] Ceylon (Sri Lanka) was at the time a Dutch possession.

[3] The Dutch Company did indeed take great pains to maintain its monopoly among the spice-producing islands, but not all observers shared Bougainville's attitude towards this policy of exclusiveness. Carteret, who was not well received in the Dutch establishments, made a number of unfavourable comments, see Wallis, *Carteret's Voyage*, I, pp. 197-8, 208-9, 236-7; II, 36-7, 402-3.

[4] Bankoul is Bengkulu on the south coast of Sumatra (Sumatera), the English's 'only remaining permanent post in Indonesia'. Ricklefs, *A History of Modern Indonesia*, pp. 79, 117.

[5] At the time, Bougainville was unaware of Carteret's voyage and his surmise that Britain was planning to encroach on the Dutch possessions was natural and based on earlier clashes between the two powers in this region – the Dutch had driven the English from the Spice Islands and the Malay archipelago between 1613 and 1632 but by the middle of the eighteenth century the English had begun to recover from these earlier setbacks. John Byron put into Batavia with the *Dolphin* and the *Tamar* in late 1764 on his way home to England. See Gallagher, *Byron's Journal of his Circumnavigation*. Samuel Wallis in the *Dolphin* called at Batavia at the end of November 1767, and Carteret followed in early June 1768, remaining there for a number of weeks. Wallis, *Carteret's Voyage*, I, pp. 228, 250-60.

natives are not slaves. Slaves come either from Macassar or Ceiram, the inhabitants of those two islands selling each other. He is a man born in Batavia, his name is Hendrick Ouman, he married a girl born in Amboina. His grade in under-merchant, the 3rd in the Company's service: this is the hierarchy: assistant, bookkeeper, under-merchant, merchant, grand merchant, governor. Military men rarely achieve any command. These political rankings have a linkage with the military. An ensign is an under-merchant, a lieutenant a merchant, a captain a grand merchant, but to obtain a command or to govern a post, it is necessary to forsake the military condition and adopt the mercantile one. Moreover the general in Batavia has the final word in all military appointments; he puts forward nominations for political grades and positions but his nominations have to be ratified by the *stahtouder*, head of the Company.[1] The local chiefs or *Orancayes*[2] stay close to the Resident. He pays them some attention and through them controls the people. Finally this nation is very unattractive. I do not believe that there are 50 whites here.

Landing is very difficult, the country is flat and especially at low tide boats have to remain at quite a distance from the shore. The sea here is very bad during the N monsoon. Besides we immediately raised our second anchor, the quality of the bottom allowing us to remain with a single anchor.

Stay at Boero

I sent my sick to land where they slept during our stay here. The good air one breathes here[3] and the fresh food will save them, I hope. I have obtained water through the company's slaves. Our men have no other occupation than walking and relaxing. The two crews are receiving fresh meat and their joy is indescribable. We went several times on battue stag hunts.[4] The country is charming, broken up by groves, plains, hills with vales watered by little streams. There are vast quantities of deer and boars and neither mosquitoes nor sand flies. The Dutch brought in the first deer. The woods are inhabited by a great number of well plumaged birds. One can find there some most attractive parrots, the type of wild cat that carries his young in a pouch situated against its lower abdomen, the bat whose wings are of enormous width, monstrous snakes that can swallow a sheep, and another variety that clings to the top of trees and throws itself at passers-by (we killed two of them).[5]

The river of Way Abbo whose banks are almost entirely covered by dense trees is

---

[1] I.e. the Stadholder of the Netherlands.

[2] *Orang kaya*, literally 'the powerful men'.

[3] Scurvy, in the days before vitamins were known, was generally put down to 'bad air', as well as to a shortage of fresh food. It had been noted that foul air was breathed in prisons, where scurvy cases among long-term prisoners were fairly common, and below decks on long sea voyages – hence the practice of 'perfuming' sailors' quarters with vinegar or some other form of disinfectant. Sending men affected by scurvy to walk along the shore during a brief call was another common practice.

[4] I.e. hunting expeditions which were organized, presumably by the Dutch, with a number of beaters in attendance.

[5] The 'wild cat' is a marsupial of the Didelphidae family, related to the opposum; the bat is the Javanese megabat or *Pteropus vampirus* that achieves a wingspan of 1.4 metres or close on 5 feet; the sheep-swallowing snake is a python which, like the boa constrictor, crushes its prey and swallows it whole; and the arboreal snake was probably related to the green python of New Guinea, *Chondropython viridis*

infested by enormous crocodiles that devour beast and men. There have even been cases of men in canoes carried away at night by these animals that are kept at bay by means of lighted fires.

Describe fort, garrison, by 50 whites, reinforcement of the buildings and the engineer.[1]

Note: the island of Bouro or Boero has an estimated circumference of 64 leagues, i.e. approximately 18 from E to W and 13 from N to S. It was formerly a dependency of the King of Ternate. Cajeli, the main settlement, is situated in a marshy plain extending for almost 4 miles between the rivers Way Soneweil and Way Abbo, this latter is the largest in the island, its waters are often very muddy. This township consists of the Residence and the homes.

Note: there is a fine pepper plantation, we have seen it and this sight told us that this tree is similar to New Britain's. I did not know it at the time.

The country's natives are divided into two classes. The Moors and the Alfurians.[2] The former are entirely subject to the Dutch and are gathered under the rule of the Residence. They are very lazy, but good people, furthermore observing the law of Mohamed, that is to say they wash frequently, do not eat pork and have as many wives as they can support. They live on fruit, sago and fish. On feast days, they eat rice the Company sells them. It is for themselves that they take as many wives as they can feed, but these are not seen. Each village has its mosque built of reeds and planks. These people are not very warlike.[3] They greatly fear the Papuans who come sometimes as many as 2 and 300 to burn the houses, carry away everything they can and take slaves. Their last visit took place three years ago. The *orencaies* or chiefs stay close to the Resident and European policy has been able, throughout these islands, to sow among the various chiefs a yeast of reciprocal jealousy that ensures a general state of slavery. Should a chief plan some plot, another will discover it and warn the Dutch. The Alfurians, without being the Company's enemies, are to some extent independent. They live scattered here and there in the inaccessible mountains that fill the hinterland of this island. They live on fruit, sago and from hunting. Their religion is not known.

---

[1] This is a note made by Bougainville to remind him to prepare a report for his superiors or for subsequent publication. His published narrative incorporates information on the fort and the locality, following a rearrangement of his notes; he describes the countryside, the fauna and the islanders, then gives a brief outline of the economy and the political situation, concluding with a description of the Residence and the governor's lifestyle: *Voyage*, pp. 399-407.

[2] The Alfurs, tribes inhabiting parts of the eastern Indonesian archipelago, probably of Proto-Malay and Indonesian descent with an admixture of Papuan elements. In the Seram-Buru region there was also some affinity with the people of Oceania.

[3] There is an interesting footnote to Bougainville's visit and his evidently amicable relations with both the Dutch and the native islanders. On 3-15 September 1793, the expedition of D'Entrecasteaux called at Kayeli, 'the little town of Cajeli, with its mosques and its houses quite numerous enough to form a kind of city.' The French received two callers, 'elderly men who had known Mr de Bougainville during his stay in that port, and who after so considerable a lapse of time wept with joy at meeting compatriots of that kindly navigator:' Rossel, *Voyage de Dentrecasteaux envoyé à la recherche de La Pérouse*, I, pp. 470-71 and 477. 'Rossel gave each of these veterans a medal commemorating his own expedition's visit': F. Horner, *Looking for La Pérouse*, p. 196. Rossel was in command at the time because D'Entrecasteaux had died and his second-in-command D'Auribeau was seriously ill; in addition, there were over 60 cases of scurvy on board the expedition's two ships, evidence of the hardships that expeditions in these distant seas had to undergo.

They are not Moslems because they raise and eat pigs. From time to time, the Alfurian chiefs come to call on the Resident who makes them some small gifts.

The Moors sold us poultry, goats, coconuts, bananas, fish, eggs and parrots. They gained from the situation but this money will not stay in their hands. The good Resident will find ways to take it away from them.

Note: The rate of exchange here is one gourde piastre[1] for 2 rupees. The loss is 9 s[ols].

All the heavy cattle belong to the Company, as do the deer hunting rights. The Resident has a few hunters he sends out. He gives them lead and powder for 3 shots against which they have to bring back two animals for which they then are paid 6 sols each. If they bring back only one, the cost of one shot of powder and lead is deducted from what is owed them.

The Moors are very mild people and seem to be quite attached to the Dutch who teach them to be very fearful of foreign nations.

Describe the house and policy of the Resident.[2]

The Company's policy towards the local natives is the same in almost all its outposts. However there are differences for large populations. It gives substantial annual subsidies to the kings of Ternate, Tidore, etc… to be sure that they will destroy the spice plantations in their dominions and will not engage in any contraband.

I do not know whether I am mistaken in my conjectures, but I believe that this trade, so precious to Holland, is about to suffer some mortal blows. The English are constantly manoeuvring to appropriate a part of it. Every year, they appear in these countries and make various attempts in the various regions. The Dutch themselves are convinced that the English have taken away some charts[3] and there is almost no doubt left that they have begun an establishment on some island in these climes. The attempt that Watson made on the Papuans in 1764 was unsuccessful. These Indians, pretending to be peaceful, attracted his longboat into a river. They were lying in ambush in great number on the banks and captured the longboat without a single man being able to escape. Almost the entire crew were Bengalese who were kept prisoner. The Europeans were attached to stakes, forcibly circumcised and died under torture. A midshipman, a son of My Lord Sandwich, who was in this longboat, was a victim of this unhappy fate.[4] Watson, whose task he said had been to hand over the Philippines to the Spanish, had obtained permission from the General to sail there by the Moluccas route under the pretext of the monsoon winds. It appears since then that this General realized his error, at least he sent instructions to all the posts to refuse him water and landing rights.[5]

---

[1] The *piastre-gourde* was originally a Spanish coin worth approximately 105 French *sols* and was the main European currency used in the East and eventually in Mauritius: see Pingré, *Voyage à Rodrigue 1761–1762*, III.

[2] See Note 1, p. 147 above.

[3] This is a literal translation of Bougainville's '*plans*', but it is possible that the word should be '*plants*', i.e. seedlings or plants.

[4] John Montagu, fourth Earl of Sandwich (1718-92), First Lord of the Admiralty in 1748 and again from 1771 to 1782. James Cook named the Sandwich Islands (Hawaii) after him.

[5] The *Gentleman's Magazine* for March 1772 denies that this episode took place, and points out that the Navy had no frigate named *Klinsberg*, as reported by Bougainville in his 1771 *Voyage*: see Bougainville, *Voyage autour du monde* (ed. Bideaux and Faessel, 2001), p. 341.

Note: Graaf's voyage to the East Indies contains numerous details on the Moluccas.[1]

There is no doubt that the administration of the Company, to retain the exclusivity of the spice trade, is very wise and that navigation through these seas is very difficult and known only to the Dutch. Nevertheless one finds spice plantations in Ceram, the Celebes, Bouton and certainly even in other neighbouring islands. The monsoons make these enterprises easier in that by going to the Moluccas at the end of the W monsoon, one disposes of the whole E one to act without Batavia being able to send any assistance to the posts that are under attack. In truth the Dutch keep a watch over the fortifications of Amboina, Banda, Macassar and Ternate, but the frequent earthquakes to which these islands are subject often totally destroy all the works. Those of Banda and Amboina were destroyed in this way just a few years ago. Banda has been rebuilt, Amboina has not yet been completed. In addition, however strong the garrisons may be that they place in these posts, sickness is continually waging war against them and the mortality rate is enormous in those islands. I would add that the Dutch have few naval forces in these regions. One frigate at Ternate, a 14-gun snow here, a few sloops, are those obstacles to be overcome? In truth the governors of Amboina and Banda visit their administrative settlements annually with a fair number of *coracores* but the date of this cruise is known; furthermore an armada of these *coracores* would scarcely impress warships.[2]

Note: The Resident was due to leave almost immediately after us to go to Amboina and attend the festivities and banquets given by the governor in mid-September to the main *orencaies* of the islands that are dependencies of his administration. All the Residents of the subordinate factories are to be present and after the feasts, the Governor, with these Residents and *orencaies*, goes on a tour of his territories. When they return, there are more festivities and rejoicings. The same practice is followed in Banda.

I could not praise too highly Resident Hendrik Ouman's kindnesses towards us. Undeniably our arrival created a delicate situation for him, but he behaved as a gentleman. He sent us [a copy of] his superiors' instructions forbidding him to receive any foreigner and asked to state the reasons for my call in writing. Our urgent needs provided all too evident a motive, moreover I was the stronger of the two. He therefore carried out with good grace what he could not avoid doing and added to his actions the manners of a frank and honest man. His home was ours, food and drink were available at any time, a form of courtesy that was precious indeed to someone who still felt the effects of famine. He came to dine on board and gave us 2 formal meals the second of which was the farewell dinner required by etiquette. The cleanliness, the elegance and the good food of these meals astonished us in such a small outpost.[3] This good Dutchman's home is very attractive, entirely in the Chinese style.

---

[1] Nicolas de Graaf, author of *Voyage aux Indes orientales et en d'autres lieux de l'Asie*, Amsterdam, 1719.

[2] Caracore, a large reed vessel with oars and a sail in use in the Malay archipelago and the southern Philippines. It is often found with an outrigger on each side.

[3] This reception is in marked contrast with Carteret's unhappy experiences in the Dutch East Indies, which were partly his own fault and which led to a protracted and acrimonious correspondence between the various parties involved. See Wallis, *Carteret's Voyage*, I, pp. 76-93, II, pp. 365-437.

Everything is arranged to take advantage of fresh air. It is surrounded by gardens and a pretty river. An avenue leads to it from the seashore. His wife and daughters, dressed in the Chinese style, are excellent hostesses. They spend their time arranging flowers for distillations and for bouquets and treating betel. The air one breathes in his charming house is delightfully perfumed and I would gladly have made a lengthy stay there. What a contrast between this gentle and peaceful existence and the unnatural life we have been leading for the last 10 months.

The Resident does not speak French, but he had with him Captain Le Clerc who speaks it well and an engineer captain Mr Gédéon Dulez de Chateaumein who says he is Swiss and whom we would have taken for a Frenchman. He is back from Amboina and has been in these parts for 5 years where he is in charge of the fortifications of Banda. He has his wife with him, a young creole born in Batavia the daughter of sub-governor of Banda. This officer has been most useful to us here, as well as Mr Le Clerc. They gave us all the help they could. The day after our departure, Mr Dulez was due to sail with his wife on Mr Le Clerc's ship. He is due to carry out an inspection at Makassar from where he will go to Batavia.

Boero provides few attractive shells. Precious shells, an item of trade for the Dutch, are found on the coast of Ceram, at Amblau and at Banda and are sent to Batavia where interesting collections may be found. One also finds cockatoos at Amblau. The Resident gave us birds' nests, bats' ears and nutmeg mushrooms to eat.

I must say a word on the impression that this European establishment made on our Cytheran. One realizes that his surprise must have been great when he saw men of our own colour, and houses, gardens and domesticated animals in large numbers and in such a range, and at the hospitality that was shown in so open and knowledgeable a manner. He behaved intelligently towards the Dutch. He began by making them understand that he was a chief in his own country and that he was travelling with his friends for his own pleasure. He endeavoured to imitate us exactly during our visits, when out walking, or at table. I did not take him with me on my first visit ashore. He imagined that this was because he was knock-kneed and wanted sailors to climb on him to straighten them out. He often asked us whether Paris was as attractive as this factory.

Note: There have been 3 almost consecutive earthquakes in this island: one on 7 June, the second 12 July, the third the 27th. It was on 22 July that we felt one in New Britain. These earthquakes have terrible consequences for navigation, in that they often create new islands and shoals where none had been before and sometimes destroy existing islands. All the houses here are low and made of bamboo.

On Tuesday afternoon I loaded the animals and other refreshments supplied to me by the Resident. His memorandum of costs is very high, but I have been assured that the company sets the prices and that one cannot alter its tariff. It is true that the meat is better here than in any hot country I know. Venison is also very good, and the poultry excellent. There are few vegetables seeing that only the Company's garden has any available. Fruits are rare and of few varieties, coconuts, bananas, grapefruit, a few limes, bitter oranges, lemons and very few pineapples. A very good variety of barley grows here named *ottong* as does the Borneo sago used for a gruel that is strange to meet. Boreo butter has a reputation in this country that we did not consider justified.

On the Wednesday morning, I took on the sick and made all the arrangements required to sail. The breeze here blows from the sea and the land alternatively, the latter during the night, the former during the day.

Astronomical observations: Mr Verron determined our anchorage to lie in 3°23′35″ of southern latitude and 123°41′45″ of longitude E.

At 8.30 p.m., the land breeze having started, I despatched a boat to anchor with lights at the head of the bank and we set sail.[1] I had all the trouble in the world to raise the anchor. The messenger[2] of the capstan broke and we had to resort to pulleys with untarred ropes to raise it up from this sticky mud in which it was embedded. When we had rounded the head of the bank, I hoisted in all my boats; the *Étoile* did the same and during the night we endeavoured to make way towards the N.

Wednesday 7 to Thursday 8.

At dawn, we had almost calm and the land breeze. The current bore us W.

At midday, the run since our departure deduced from the bearings and the latitude is NNW4°W distance 20′.[3]

Observed latitude: 3°7′.

Calculated corrected longitude: 123°54′.

Note: the good air of this land and the consumption of refreshments that we have found there have made this stay, however brief it may have been, salutary to our sick. They are feeling appreciably better. This same fresh food also gave strength back to our crews. We are sharing the good effects with them and one needs to have undergone our sufferings to realise what satisfaction one feels at seeing the end of the cruellest part.

Bearings at 8 a.m.: the E point of Boeroo Island bearing SE¼S3°E 7 to 8 leagues, the most W bearing W approximately 3 leagues, Manipa Island from ESE3°E to ESE4°S 8 leagues, Kelaing Island E4°S to E¼SE 12 leagues.

At midday: the E point of Boero bearing SE4°S 8 to 9 leagues, the W one W 1½ leagues, the NE point of Cajeli Bay SE¼S1°E 4½ leagues.

Thursday 8 to Friday 9.

Observed occiduous variation amplitude: 00°33′ NE.

The winds have been as they were in the bay, variable land and sea. The calms also decided to join in. We kept to the most favourable tacks but the current did more for us than our sails. All this part of Boero is almost uninhabited, we only saw fires at its western extremity. The country is cut through by high mountains and covered in forests.

---

[1] A stowaway, Nicolas Chapelle, who had been sailing in Dutch vessels for four years, was found after the *Boudeuse* left port and was added to the muster roll.

[2] A messenger cable is a hawser with its two ends lashed together to form an endless loop. It is secured to the cable by nippers (short lengths of rope) and passes round the capstan which is then used to weigh the anchor. By releasing the nippers and resecuring the messenger to the cable further down an endless strain can be kept and the anchor hove in without stopping.

[3] Bougainville points out in his published *Voyage*, p. 317, that he decided he would not change his dates until after he had reached Mauritius but would continue to keep to the dates indicated by the sequence of his journal, 'warning that instead of Wednesday 7, in India they were using Thursday 8.'

At midday the run deduced from the latitude and the bearings is WNW 5°W 40′.

Observed latitude: 2°55′.

Calculated longitude: 123°16′.

Note: this island named Choulibesi (written Soele besje) is occupied by the Dutch, there is a bookkeeper, a sergeant and 25 men in a redoubt called Claverblad or cloverleaf.[1]

Bearings at 6 p.m.: the most E land of Boero in sight bore SE¼E4°E, the most W bore W¼WS3°S.

At 6 a.m.: the most E land of Boero ESE, the most W SW2°S 6 leagues, the S point of an island N of Boero NW¼N 5 to 6 leagues, its N point NNW4°W.

At midday the most E lands of Boero ESE5°E 10 leagues, the most W S¼SW4°W 4 leagues, the E point of the N island N5°W, its W point NNW5°N 6 leagues.

Friday 9 to Saturday 10.

Observed occiduous variation amplitude: 00°30′ NE.

The winds have been WSW to SSW along the land, S and SSE when we sailed out to sea, fairly fresh breeze, fair sea. We passed through some strong tidal races. I steered when we could SW¼W in order to find land between Wawoni and Button, wishing to pass by the strait of Button. In this season it is dangerous to pass E of this island, one would risk being driven onto this coast by the currents and the wind; in which case one could get away only when the W monsoon is well settled. This passage is also preferable at this time to the one S of Toukanbesi, all the more because the latter is dangerous even for Dutch pilots.[2] Furthermore I am aware that sailing from here to Java is full of great difficulties. *Custodiat Deus introitum et exitum.*[3]

At midday the estimated run is WSW4°S 85′, corrected SW¼W2°45′W, distance 75′.

Estimated latitude: 3°33′, observed: 3°34′.

Calculated longitude: 122°11′.

Bearings at 6 p.m.: the most E lands of Boero ESE in view, the most W SSE3°E 7 to 8 leagues.

At midday the *Étoile* signalled land between S and W but we did not see it, in fact there was none.

Julien Launay, tailor, died of scurvy at 9 o'clock this morning.[4]

We have seen several whales and dolphins.

Note: the currents here set E.

---

[1] This is a reference to the Celebes or Sulawesi.

[2] The expedition is sailing towards the islands at the south-eastern extremity of Sulawesi. The obvious route appeared to be due south into the Banda Sea, but the strait between the islands of Button (Pulau Butung) and Wawoni (Pulau Wawani) had been recommended to the French as quicker and safer than wending their way through the archipelago of Kepulauan Tukangbesi. Bougainville uses both spellings 'Button' and 'Bouton' for Butung.

[3] 'May God keep the entrance and the exit:' Vulgate, Psalms, 120, verse 8.

[4] 'He was beginning to convalesce, two wild sessions of brandy finished him off:' Bougainville, *Voyage*, p. 320.

Saturday 10 to Sunday 11.

Observed occiduous variation: 00°19′ NE

Fair weather, calm sea, fresh SSE to SE breeze. I sailed SW¼S as long as the wind would allow. At midday I changed tack until 2 o'clock as I wanted to reach land in daylight to identify properly the entrance to the strait. At 9.30 a.m. we saw the land of Wawoni, a high island especially in its centre. At 11 o'clock we sighted Button Island.

Run for the 24 hours: SW3°45′W 51′, corrected: SW1°45′W, 46′30″.

Estimated latitude: 4°10′, observed: 4°8′.

Corrected calculated longitude: 121°35′.

At 8 a.m. sighted the island of Wawoni bearing W¼SW to SW.

Bearings at midday: the N point of Wawoni W5°N 7 leagues, its S point SW¼W4°W 8 to 9 leagues, the NE point of Button Island SW¼W4°S 9 leagues.

Sunday 11 to Monday 12.

Observed ortive variation amplitude: 00°21′ NW.

The winds have been variable from SSE to SE and even ESE. We steered SW¼W until nightfall to come up to the land and adequately survey the entrance to the strait. We ranged along the entire coast of Wawoni and at 6.30 p.m. I put back out to sea. We spent the night tacking so as to come up to windward of the point of Bouton and be near it at daylight. In fact at 5.30 a.m., we were about 3 leagues to windward of it. I had my boat lowered, the *Étoile*'s came alongside and I took it in tow. At 9 we entered the strait with S fresh breeze that lasted until 10.30, then light airs until 11.30.

One needs to range along the shore of Bouton. When one sails from the E, the NE point of Bouton is of a modest height cut into several hillocks. The cape that forms the entrance on the port side of the strait is cut into a cliff with a few white stones forward. Behind this point on the E is a pretty little bay in which we saw a small Malay boat with a sail. The corresponding headland of Wawoni is low and extends to the W. The land of Celebes lies straight ahead. As you enter you see the pass open to the N between Wawoni and Celebes but the S one is almost closed. One sees in the distance a low land divided into some kinds of islets. On the land of Bouton, as one enters, one discovers large rounded capes and some attractive bays between these capes. There are two rocks off one of these capes that one inevitably takes for two ships in sail, the one fairly large, the other small.

Note: The Dutch name this strait Button Street [La rue de Button].[1]

At 10.30 I had sounding taken about ¼ of a league from the coast of Bouton and 1½ leagues E of the above-mentioned rocks, they found 45 fathoms sand and ooze.

At midday we had passed a little beyond the 2 rocks, they lie off an islet behind which appears an attractive passage. We saw there a Malay building shaped like a square box with a canoe in tow progressing slowly with sail and oar along the shore. A French sailor I took on at Boeroo and who has been sailing in the Moluccas with the Dutch for the last 4 years, told me it was a boat belonging to Indian bandits who

---

[1] The Dutch term is *straat*, which has both these meanings.

were endeavouring to take prisoners in order to sell them.[1] Meeting us seemed to unsettle them. They hauled up their sail and using poles fled behind the islet.

I will add that the land of the Celebes we are passing to starboard is everywhere green and presents a delightful picture through the variety of its low lands, hills and mountains. The width of the strait so far varies from 7, 8, 9 to 10 miles.

The run for these 24 hours is SW¼W2°W 40'.

Estimated latitude: 4°30', observed: 4°29'.

Calculated longitude: 121°3'.

Bearings at 5 p.m.: the most S point of Button in view S¼SW2°S 10 to 12 leagues, the N point of the same SW¼W 6 leagues, the S point of Wawoni WSW 2½ leagues, its most N point in view N 3 leagues.

At 6 a.m.: the most N point in sight of Wawoni N3°E 6 to 7 leagues, the S one NW¼W4°N 5 to 6 leagues, the N point of Button W¼NW2°W 2½ to 3 leagues, the most S of the same in view S¼SW4°S 9 leagues. The land of Celebes completed the picture.

Bearings at midday: Bouton from SW¼S4°W 1½ leagues to ENE by E and S; Wawoni from NE5°E to N. An islet off a headland of the Celebes N¼NW 6 leagues, the closest land of this island W¼NW 2 leagues.

Note: the island of Pancassani, the islets that lie N of it between it and Celebes and a point of the latter seemed to form but a single land.[2]

Monday 12 to Tuesday 13.

We continued our course into the strait, the winds changing in accordance with the channel and allowing us to proceed gradually from W to S, always with a leading wind until 5 a.m. when the breeze came from the ESE, whereupon we remained close-hauled on the port tack so as not to get away from the coast of Button. At 5 o'clock I had soundings taken and we found 27 fathoms excellent mud about 1 league from Button and 1½ leagues from Pangasani. At 6.30 p.m., the winds veering forward and the tide being against us, I dropped anchor in the same ground of 27 fathoms soft mud. The depth is quite level in this part of the channel.

First anchorage in this strait in 27 fathoms soft mud. Its bearings: a point of Button in line with one of the Celebes N¼NE4°E, the N point of Pangasani NNW1°W distant about 3 leagues. The E point of Panganasi that seems to make a first narrows (the land of Bouton appearing, behind this point, to leave no passage open) S¼SW1°W. A low point of Button behind which there seems to be a cove or bay S5°W 2 leagues. Here the channel is approximately 3 leagues wide.

I return to my account. At 2.30 we sailed past a superb harbour on the Celebes coast. This land is a delightful sight. At 3 o'clock, Pangasani Island and the islets N of

---

[1] 'Took on' is a generous expression. The ship's records indicate 'discovered on board'. The man was Nicolas Chapelle, from the town of Fécamp in Normandy.

[2] The strait into which the expedition has entered is complex and winding. Passing between Wawani to the north and Butung to the south, the French would have seen Sulawesi ahead of them seemingly closing what could have been at first sight merely an inlet. Several further openings then appeared: straits leading north, south and south-west. Bougainville selected the southern strait between the islands of Butung and Muna (his 'Pancassani'). This strait is quite long and gradually narrows, with a number of small islands scattered along the route.

it separated from each other and we saw the various openings they offer. The high land of Celebes is visible behind. At 5.30 we were hemmed in so that neither entrance nor exit was visible.

I believe that around 2 p.m., the tide began to run against us. The sea was at the time up to the base of the trees on the shore which leads me to believe that the tide here comes from the N at least at this time of year.

There are a few habitations on this part of Button, several fires could be seen at night. The land of Pangasani seemed to be highly populated judging from the quantity of smoke. The land is low, level, covered with fine trees and I believe it to be full of spices as well as Celebes, Button and Wawoni. The English often come to these parts and it is certainly not for a change of air. Here now are the first French vessels to appear in these important areas. O my country, wake up, it is time. Neptune has not sworn you an eternal hatred. Is it not his custom to assist Venus's favourites? But it is only by an assiduous devotion that one obtains the sympathy of the gods.

In the morning, a great number of outrigger canoes came around the ships and the islanders of Boero brought hens, bananas, cockatoos, and parrots. They asked for Dutch money, especially silver two-*sol* coins.[1] They also readily accepted knives with red handles. They told me that a large vessel belonging to the company had passed by here ten days earlier. These Indians belong to a considerable group living on the heights opposite our anchorage. The habitations are located on 5 or 6 mountain ridges. Everywhere the land is cleared, separated by ditches and well planted. They cultivate rice, maize, potatoes and yams. Their bananas have an exquisite taste. They seem to be skilled traders but display fairly good faith. The men are neither handsome nor ugly. They go about naked, except for a belt and a kind of turban made of a cotton cloth. They wanted to sell us some lengths of coloured cotton material, but they were very coarse and we did not buy any.

I showed them nutmeg and cloves and asked them for some. They replied that they had some dried ones at home and that when they wanted some, they went to get them in Ceram and in the neighbourhood of Banda when there is no doubt that their suppliers were not the Dutch; a good lesson for us.

At 10.30, I had the anchor raised and we tacked almost without making any progress, the breeze blowing feebly from the S to SW.

At midday the run for the 24 hours is SSW 5°W 22'.

Observed latitude: 4°48'.

Calculated longitude: 120°52'.

Tuesday 13 to Wednesday 14.

We continued tacking, the tide running in our favour and making little progress until 8 p.m. when, the wind dropping, I came to anchor in 36 fathoms soft mud. We had by then passed the first narrows. Tacking we coasted within a pistol shot range of 6 detached rocks in front of Panagasani. We sounded several times and found 27, 30, 35 fathoms, mud, quite close to six rocks, we let out 15 fathoms of line without finding

---

[1] Probably the *stuyver*, worth one-twentieth of the guilder, a coin widely used in Dutch India and the East Indies in the seventeenth and eighteenth centuries; but a positive identification is difficult, as other coins of lower value were also current.

bottom. We passed outside 4 fairly large and populated islands that lie closer to the coast of Button than to that of the NW. This coast rises in an amphitheatre with a low land at its base that I believe to be often water-covered. I deduce this from the fact that the islanders have their habitations on the mountain ridges. Maybe also, as they are almost always at war with their neighbours, they wish to have a line of forest between their homes and the enemies who might attempt a descent.

Note: I am not giving any bearings here, I am showing the anchorages on my chart.

In this part of the strait nearly all the land is cleared on both sides and there are numerous habitations.

At 7.45 a.m. I raised anchor with all sails set there being little wind and we tacked until midday when having seen a bank to the SSW, I dropped anchor in 20 fathoms sand and mud.

The run for the 24 hours is SSW 5°S 20′.

Estimated latitude 5°7′.

Calculated longitude: 120°46′.

Several canoes appeared this morning, one of which bore an unfurled Dutch flag at its stern. All the others moved away to make room for it when it approached. It was a transport for one of the Button *orencaies*. The Company is kind enough to grant them its flag and the right to display it.

Wednesday 14 and Thursday 15.

I sent my boat to take soundings around the shoal and I sailed at one in the afternoon to try, with the tide, to gain a few leagues. There was no way this could be done. The winds were too weak and too brief, we lost about a half league and at 3 thirty, I dropped anchor in 13 fathoms, sand, mud, shells and coral. We always used a kedge anchor, the slight breeze and the good holding ground allowing us to.

The officer sent in the boat to take soundings between the bank and the land reported that, close to the bank, there are 8 and 9 fathoms of water, the depth increases as one gets closer to the land of Bouton, a high and steep land, broadside to a superb bay. The depth increases until one no longer finds ground with 80 fathoms approximately mid-channel between the shoal and the land so that, if the calms were to dominate in this part, there is an anchorage only close to the bank. The quality of the bottom in the area is good. There are several banks between this one and the coast of Pangasani that extend from E to SW of our anchorage.

One could not therefore urge one too strongly to keep, throughout this strait, to the land of Button. It is along this coast that the good anchorages are found, it conceals no dangers and furthermore the winds blow from it the most frequently.

In this part of the strait almost up to the exit, the coast of Button gives the appearance of a mass of islands, but that is because it is cut by several bays that must make some superb harbours.

Observed ortive variation by azimuth: 2° NW.

The night was fine and windless. At 5 a.m. the breeze coming from the ENE, I set sail and steered for the land of Button. At 7.30 we had rounded the bank. The wind dropped. I lowered my longboat and small boat and signalled to the *Étoile* to do the same. The boats towed us until 4 o'clock, the tide being in our favour. We sailed in front of two magnificent bays, in front and quite close to the high land, we found no bottom. At 3.30, a good breeze came up from the ESE, I sailed to seek an anchorage within reach of the narrow pass through which one emerges from this strait.

However the more we progressed, the less we saw any exit. The starboard land and the port land form on each side a part of a circle that joins up and seems to leave no opening.

At 4.30, we were abreast a very open bay and I sighted a local canoe that seemed to be going into it towards the S. I set my boat on to it to try and bring it in and thus obtain a pilot. Meanwhile our other boats were taking soundings. A little offshore and almost off the N point of the bay, we found 25 fathoms sand and coral. Afterwards we lost the ground. I tacked on the other course, then hove to in order to give the boats time to seek an anchorage. They signalled 45, 35, 29, and 28 fathoms muddy ground along the land linked to the S point of the bay. We went to this anchorage, with the help of the longboats and at 5.30, we came to with one of our large anchors (the 2nd) in 35 fathoms, soft mud. The *Étoile* anchored S of us.

Just as we had finished anchoring, my boat came back with the Malay vessel, which provided us with a local pilot who asked us for 4 *ducatons*[1] to act as our guide. I shall pay them with very good grace. This gentleman is sleeping on board and his boat is going to wait for him on the other side of the pass. In fact one can also get there through the far end of a bay close to the one we are in now. All that is needed is a short portage of the canoes. Not long before raising anchor, I thought I could see the port extremity of the pass, but it has to be guessed at. From where we are it overhangs a two-level rock that makes up the starboard point. We shall see it tomorrow, with God's help.

At midday observed a latitude of 5°10′.

The run from yesterday midday to the anchorage is S¼SW2°W 12′20″.

Latitude: 5°19′.

Calculated longitude: 120°42′45″.

Friday 16.

The night was very fine. At 2 a.m., I tacked, it was 4 o'clock before we filled our sail. There was no wind, with our boats we reach the mouth of the pass. The sea was then very low along both banks and as we had found that the tide runs from the N in this strait, I expected that we would have a favourable current. The wind was astern and a good breeze. It was useless, it forced us to struggle for an hour and a half against the current. We had to go back. The *Étoile* anchored almost at the mouth of the pass off the coast of Button, in a kind of elbow where the tide flows back and therefore is not so noticeable. The breeze helping, I continued to fight for more than an hour against the flow, almost without losing ground but the wind having given up, I lost a good mile and was forced to anchor in 30 fathoms, sand and coral. I remained under sail steering to lessen the strain on my anchor which was only a very weak kedge anchor.

Note: near the pass by which one emerges from Button Strait, the flood comes

---

[1] 'Approximately fifteen francs, states the *Voyage*' (ET).

from the S, at least at this time of year. I do not know what the limits of the two races are, the one coming from the N and the one from the S.

The flood had started at around eleven o'clock. I dropped anchor at one o'clock; at 4 the wind having freshened a little and with almost a slack tide, I raised my anchor and with all the boats ahead of us, I sailed into the pass, followed by the *Étoile*.

At 5.30, the narrowest part was satisfactorily negotiated and at 6.30 I dropped anchor in the bay of Bouton below the Dutch post.

This post is a collection of 7 or 8 bamboo huts with a kind of palisade, adorned with a flagpole. There reside on behalf of the company a sergeant and 3 men. It must be said that this coast offers a most agreeable picture. It is cleared almost everywhere with a number of huts. Coconut plantations are widespread. The land rises in an amphitheatre and its appearance is improved by the cultivated plots. The seashore is devoted to fisheries. The coast that faces Button is no less pleasant and no less populated.

Throughout the day, canoes surrounded the ships. They came as though to a fair, laden with hens, parakeets, cockatoos, bananas, coconuts, eggs and pieces of cloth. Food is now as plentiful as it was short earlier. We are driving scurvy away but there are a number of stomach ailments: a dangerous indisposition in hot countries where they turn into dysentery and soon become a fatal illness.

Note: in all these parts of the Moluccas, on land as at sea, it is deadly to sleep in the open air, especially when there is moonlight.

Note: I say it again, I advise those who might come here to obtain the currency the Dutch use in the Moluccas. With this money, they will obtain refreshments quite cheaply. One must especially obtain those Dutch silver coins that are worth two *sols* and a half. Since the Indians do not know the currency we have, they set no value on our 24-*sol* coins nor on the Spanish *reals*.

Continuing the description of the pass. When one comes from the N, it begins to open out only when one is about a mile off. On the Bouton side lies a detached rock, which looks exacted like a galley with a tent on it. The shrubs covering it create the tent effect.[1] The land of Button is moderately high, covered with houses and the shore is enclosed by fisheries. At low tide, the galley is linked to the land on one side, when the tide is high, it is an islet. On the same side of Bouton, the ground rises as one continues along the pass; it is almost everywhere cleared into an amphitheatre and adorned by huts.

The other side of the pass is steep from the N entrance, almost sheer and it is everywhere bare rock covered nevertheless with trees, but that side is not inhabited like the other. There are only 5 or 6 huts along that length of the passage.

A mile and a half or 2 miles from the pass on the N side, nearer Button than Pangasani, one finds depths of 20, 18, 15, 12 and 10 fathoms muddy bottom, as one advances S in a boat, the ground changes, one finds sand and coral in various depths of 35 fathoms to 12, then one loses the ground.

Note: the current in these narrows is strong enough to enable one to pass

---

[1] Bougainville spells the word '*tante*', the French word for 'aunt'.

through in calm weather, even with a weak head wind. It is not sufficient to defeat a strong head breeze, which would then allow one to sail through backing and filling.[1] In any case, unless one has a fairly strong favourable breeze, it is desirable to have one's boats ahead in order to control one's steering in the tortuous windings of the channel.

The pass may be a good quarter of a league in length, its width varies from about 150 up to 400 *toises* (this is an estimate based on a quick look).[2] The channel snakes along and on the Pangasani side, at approximately two-thirds of the pass, I mean of its length, lies a fishery that warns one to keep away from that side and range along the coast of Button. In general one must endeavour to keep to the middle of the channel.

Emerging from the land of Button, several islands that lie to the SW and the land of Pangasani form a wide gulf and the best anchorage is opposite the Dutch Residence about 1 mile from land.

Our Buttonian pilot is not churlish. He helped us with his knowledge as much as any man could who knows the area and understands nothing about the way our ships manoeuvre. He took the greatest care to warn us of the dangers, shoals and anchorages. But he always wanted us to steer straight for where we were going and took no account of the way we hug the wind to cope with it and benefit from it. He also thought our draught was of eight or ten fathoms. In the morning, another Indian came aboard, a very knowledgeable old man, whom I believe to be our pilot's father. They stayed with us until evening and one of my boats took them home after we had anchored. They live near the Residence. Naturally I loyally paid the dear pilot who, with his much honoured father, drank copiously of our brandy, convinced no doubt that Muhammad could not see anything. However they refused to taste any of our meals; a few bananas and some betel, such was their food.

Note: coming into this strait, one must not fail to get such a local pilot. It is easy and useful.

Saturday 17.

The night was fine and almost windless. I had the sails readied at 4 a.m., we were under way at 5 o'clock. For as long as the tide was flowing in our favour there was little wind and even then it was a head wind; in spite of the wind we made a little progress. At 11 o'clock, as the tide was beginning to turn, the breeze freshened and we tacked. One has to beware of a bank that extends a fair distance offshore from an island that is W of Button. While tacking we took several soundings with 50 fathoms of line without finding ground.

At midday the run from the anchorage is SW¼W 15′.

Observed latitude: 5°31′30″.

Calculated longitude: 120°34′15″.

Note: we obtained here a monkey the Cytheran firmly believes to be a man. We

---

[1] In a narrow channel a vessel, carried along by the current, uses its sails, filled to make headway and backed to make sternway, to avoid being carried too close to the shore on either side.

[2] These estimates give in modern terms a length of some 1,500 yards (1.4 km) and a width varying between 320 and 850 yards (290 to 780 m).

shall have as much trouble convincing him that this is not so as we will that we have two substances one of which survives the other.[1]

From early morning, a multitude of canoes came around the ships and trading began. Everyone benefited from it. These Indians are selling their foodstuffs at a much better price than they get from the Company's vessels, but this price is low both in itself and relatively to our needs. The crews have an abundance of poultry and other refreshments. We have also bought some fairly attractive cotton goods.

Around 9 a.m., I received a visit from 5 *orencaies* from Button. Their canoe was flying the Dutch flag. They were well dressed and they all carried a stick with a silver knob on which was engraved VOC. The oldest had an M above the V.[2] These chiefs brought the gift of a goat which I gave to the sick and they bade me welcome with best wishes for a safe voyage. I made them a gift, in the King's name, of a length of silk cloth which we cut into 5 lots, one for each of them and I taught them to identify the national flag. They offered us all the assistance at their disposal: water, wood, refreshments. I offered some liquor, it was what they were waiting for and Muhammad allowed them to drink to the health of the Sultan of Button, to France, to the prosperity of the Company and to the success of our voyage. They told me that in the previous 3 years, at various times three English vessels had passed by, to which they had supplied water, wood, poultry and fruit, that they were their friends and that they could see that we too would become their friends. At that time, their glasses were full but they had recently emptied them. I believe that these people could be useful. They also told me that the Sultan of Button lived in this very district. He must be powerful if power can be measured by the population, because his island is large and well populated. The *orencaies*, after taking their departure, went to pay a visit to the *Étoile*. Undoubtedly, they will also have toasted the health of their new friends there.

Our pilot came back to see us this morning. He brought me a few coconuts, the best I had yet seen. He warned me that, once the sun was high, the SE breeze would be very strong. I gave him a large glass of brandy in exchange for this useful information. As a result all the canoes left us around eleven o'clock, not wishing to run any danger at sea when the strong breeze came up.

Saturday 17 to Sunday 18.

The pilot was right. As we were tacking towards the island to the W of Button, the SE breeze came up, fresh and strong. We altered course to WSW and made good progress in spite of the tide. Towards half past 3 we sighted the end of Pangasani. As soon as it is morning we sight the high mountains of Cambona Island on which is a peak visible above the clouds.[3] Towards half past 4, we sighted a corner of the land of Celebes. At sunset, we hoisted in our boats then with all sails set sailed WSW.

---

[1] Meaning that human beings, unlike animals, have a soul as well as a body. Bougainville deleted this paragraph when he was writing his *Voyage*. It was a very minor incident, which could be read as unflattering to the Tahitian, and there are underlying philosophical implications in the comments that he felt were best omitted.

[2] VOC is the monogram of the Dutch East India Company (Vereenigde Oostindische Compagnie), but it is not possible to guess what the 'M' might have stood for.

[3] The island is Pulau Kabaena, the highest point of which is Mt Sabanpolulu. The French are now following a westerly course to pass between Sulawesi and the island of Selayar that lies approximately 16 km south of it. The safe pass is only about 10 km wide.

I am setting my course to find the island of Saleyer 3 or 4 leagues to the S so as to then seek the strait that lies between that island and Celebes.

We had the studding-sails set top and bottom throughout the night. At 10 p.m. I steered W¼SW.

At dawn we did not see land, which proves that during the night we had currents that caused us to lose about 3 leagues. I had been warned that between Button and Saleyer the currents run E. I continued on the W¼SW course. At half past 9 we sighted the high land of Saleyer bearing WSW to W¼NW. As we advanced, we discovered a lower headland that seems to end this land to the N and then I steered W¼NW and WNW to find the islands that lie N of Saleyer and form the strait that bears that name.

At midday the run for the 24 hours is WSW1°15′W distance 93′.

Observed latitude: 5°55′. Calculated longitude: 119°7′.

Observed occiduous variation by azimuth: 1°7′NW, occiduous amplitude: 00°53′ ortive amplitude: 1°10′NW.

Bearings at 6 p.m.: Cambona Island [Isle de Cambona] from N¼NW4°W to NW¼W2°N distance 3 to 4 leagues, a small island further S than the E point of Cambona N2°W approximately 3 leagues, the most S point of the island the most S of Button bearing E¼SE5°S 7 to 8 leagues.

At 6 a.m.: the land most W yesterday evening bore NE5°E to NE¼E4°E 15 to 16 leagues.

Bearings at midday: the land of Saleyer from SW5°S to WNW approximately 5 leagues, one could see something like 2 islands N of it, namely the most S bearing WNW5°N 5 leagues (that is a headland of Saleyer itself), the most N bearing NW¼W 6 to 7 leagues.

Sunday 18 to Monday 19.

Observed variation occiduous amplitude: 00°45′, ortive amplitude: 00°40′. The winds NE and ENE, very light breeze, we steered successively NW¼W, then NW, then between NW and NW¼N. We recognised that what appears at first to be a first island N of Saleyer is a fairly high part of it, linked to the rest of the land by a tongue of land that is extremely low and that this high part ends in an almost drowned point. Then one sees two fairly long islands of moderate height the distance between them being approximately 4 to 5 leagues; then making way one discovers a quite low small island between the two larger ones first sighted to the N of Saleyer. The good pass lies close to this small island, either to leeward or windward. It seemed to be wider to the S of that small island than to the N of it and I decided to choose that one. I am naming this small island Passage Island, the one to the S of it South Island [Isle du Sud], the one to the N, North Island [Isle du Nord].

Note: These are the islands the Dutch call Bougerones and I have heard them pronouncing this pass as the Boutsaron.

The wind has been very weak all day and we had to heave to at nightfall to wait for the lazy *Étoile*. She caught up only at around 8 o'clock.

Note: at 7.30 MrVerron by the observation of Lyra's meridian passage, found that we were at that moment in 5°4′ of southern latitude.

Fortunately the moon was favouring us until midnight with her kindly rays, which decided me to make for the pass as soon as the *Étoile* was close to me. I kept to the middle of the pass the width of which may be six to seven miles. At 9.30 we were N and S of Passage Island, the middle of South or Port-side Island [Isle du Sud ou Basbord] bore from us between S and S¼SE. I steered W¼SW until 1 a.m. then broached to until 4 o'clock. Before and in the pass I had several soundings taken by hand and no ground was found. From this morning, we have been following the southern coast of Celebes and to reach it, I steered to NW¼W then, following round the shore with a fresh ENE breeze.

The route since yesterday midday is W5°N 66′.

Observed latitude: 5°49′. Calculated longitude: 118°1′.

Bearings at 6 p.m.: the most S land of Saleyer in view S¼SW3°S, the N point of the same W distance 1⅓ league, the S point of South Island W5°N 1½ league, its N point W¼NW5°N same distance, the S point of North Island NW¼W1°N about 5 leagues, the middle of Passage Island by the middle of North Island NW5°W about 3 leagues, the N point of N Island NW1°W about 5½ leagues, the land of Celebes in sight WNW3°W to NW.

Bearings at midday the coast of Celebes bearing ENE 5 to 6 leagues to WNW2°W approximately 1½ leagues, we were less than one league from the shore.

Monday 19 to Tuesday 20.

Observed occiduous variation amplitude: 1°10′ NW, ortive amplitude: 00°53′ NW.

We continued to range along the coast of Celebes with very fresh ESE to SE breeze. In truth it is hard to find a more beautiful country anywhere in the world. At the foot of the mountains an immense plain stretches out, everywhere cultivated and everywhere adorned with houses. The seashore is nothing but an endless plantation of coconut trees. The eye of a sailor who had only just got away from salt meats sees with delight herds of cattle roaming about these pleasant plains embellished at intervals by clumps of trees. The population of this island appears to be substantial. At half past midday, we were abreast a large township whose houses stretched out along the beach under the shade of the coconut trees.

This southern part of Celebes ends in three long points, level and lowlying with between them two fairly deep bays. It is in the most E one that Turate is situated. The night before, we passed the trading posts of Bompanga and Bomtin.

I had several soundings taken; a few miles E of the above-mentioned township, we began to find ground in 20 and 18 fathoms grey sand. We were approximately one and a half leagues from the land. We then lost the bottom then found it again in 20 fathoms. At 4 o'clock, we found only 9 and 8. At the time we were abreast the third point, the coast then trended NNW and we had ahead of us an island about 3 leagues long, low and fairly level, named Tanakeka. It was then a matter of rounding the dangerous shoals of Brill or Lunette and Bunker.[1] Two routes were available for this, one to the N the other to the S of Tanakeka Island. If one passes N the channel has a depth

---

[1] 'After passing before Bomtin (Bonthain) and rounding Tanakeka (Tanakeke) Island the *Boudeuse* is sailing towards Java, leaving to the north the banks of Laars and to the south those of Sabalana and Pater Noster' (ET).

of 10 to 6 fathoms: one can anchor to leeward behind a tongue of sand that stands out, one then sees three other small islands N of the first one, one emerges between the two to the N by keeping to the northern side. This route was indicated to me by Wood Rogers in the journal of his voyage around the world.[1] He followed it, guided by a local pilot. The passage to the S of Tanakeka seemed to me shorter and less complicated. I sent the *Étoile*'s boat to take soundings ahead of me and I followed it under the topsails, the *Étoile* keeping to my wake. I had given instructions to the officer commanding the boat to maintain a course that would allow him to pass a league and a half from the island.

From 5 p.m. when we set off behind the boat and gaining a little to the N, as we advanced, the sound line gave 8, 9 10, 11 and for a long time 12 fathoms; opposite the most northerly of the islands, going more W, we had 13, 14 and finally 15 and 16 fathoms.

Note: the bottom is of heavy grey sand and gravel.

I recalled the boat at 7 o'clock and set sail for the SW¼S close hauled on the port tack, sounding from hourglass to hourglass[2] and always finding 15 to 16 fathoms until 10 o'clock. The depth then increased. At 10 thirty, we found 70 fathoms, bottom of sand and coral, then we found none with a line of 120 fathoms.

At midnight, I signalled to the *Étoile* to hoist in her boat and let out all canvas. I then steered SW in order to pass in mid-channel between Brill and the shoal of Sararo, sounding hourly, still without finding ground.

At dawn, we saw no more land. The S¼SE wind caused us to steer SW¼W at 8 o'clock. At 10 o'clock, I set course for the WSW.

The run for the 24 hours is W¼SW 40′S, distance 192′. Observed latitude: 6°10′, calculated longitude: 116°21′.

Comment: when the breeze is not favourable and fresh, to set off to round the Lunette, one should anchor below Turate and wait for a steady wind. Otherwise, one runs the risk of being carried on by the currents, without being able to defend oneself from this dangerous shoal.

Bearings at 6 p.m.: the island of Tanakeka from NE¼N, 1 and a half leagues, to NNW5°N same distance, the middle of a small low island to the NW of Tanakeka NNW1°W one and a half leagues, a low point of Celebes E¼SE2°S approximately 4 leagues.

At 3 p.m., we sighted a local boat making for the SW. I chased it, wanting to obtain if possible a pilot, [or] at least some information on the route. I caught up with it before 4 o'clock and as it had hove to on the starboard tack giving the appearance of waiting for us, we took in sails to come up to it. Then it made for the land and in

---

[1] Woodes Rogers (1678-1732), after raiding Spanish settlements along the coast of South America, sailed from California in January 1710 with the *Duke* and the *Duchess* and two captured Spanish vessels. One of his officers was William Dampier. After taking on supplies in Guam, he went south towards the Celebes and the Moluccas and reached Batavia in June. His account of the expedition, *A Cruising Voyage round the World*, was published in London in 1712. See *Who's Who*, pp. 208-10.

[2] Equivalent to a half-hour, the hourglass used on board ship emptying in half an hour, at which point the bell was struck. When the first half-hour of a watch had gone by, the bell was struck once; it was struck twice when the next half-hour had elapsed and so on until the four-hour watch came to an end with the ringing of 'eight bells'.

spite of a few gun shots that came close to it, it got away from us. That boat had on board pirates from this coast whom the Dutch enslave when they catch them: and so, taking us for Company vessels, it preferred the risk of being sunk to a slavery it believed inevitable. I strongly praise these Indians' courage and certainly, had I not needed a pilot, I would not have wished to interrupt their voyage.

Tuesday 20 to Wednesday 21.

Observed occiduous variation amplitude: 2° 18' NW, ortive amplitude: 2° 14' NW. Banks of Saras, Sesten, the Hen, Pater Noster, Tangayang.

The winds have varied from SE to SSE, fresh breeze. I steered W until 7 p.m. when, rounding the bank of Saras, I set course for the W¼NW, sounding hourly without finding ground. We thus kept to the channel between the Sestenbank and the Hen to starboard, the Paternoster and the Tangayang to port. We had all sails set day and night, the advance I had gained over the *Étoile* giving me all the time I needed to take soundings.

I had been told that from the Sesten, the current ran S towards Tangayang and that consequently, when the sounding line gave 25 to 30 fathoms, one should immediately alter to the N, the depth lessening suddenly. The midday observation on the contrary placed us further N.

The run for the 24 hours gave me W¼NW2°N 113'.

Observed latitude: 5°52', observed: 5°45'.

Calculated longitude: 114°31'.

Wednesday 21 to Thursday 22.

Occiduous observed variation amplitude: 1°15'.

I continued to steer W to find the islands of Alambai in the latitude where I believed them to be. We had not seen them by sunset.[1] I steered W¼SW until we had covered the extent of the lookout's range. At 8 o'clock I sounded and they found 40 fathoms, bottom of soft mud and sand. I then set course for the SW¼W and WSW until 6 a.m., then for the W¼SW until midday. We sounded hourly until dawn and the reading stayed at 40 fathoms, soft mud until 4 o'clock when we obtained only 38 fathoms same ground.

At half past midnight, we met a local vessel making way towards us, it at once changed to the starboard tack close-hauled: we found a second one at 7 a.m., that proved to be no more anxious to assist us… It is clear that the local sailors do not care to deal with the Company's ships.

At midday the estimated run for the 24 hours is W¼SW4°45'S, distance covered 118' 30". The run corrected by the latitude and the midday sounding that gave 45 fathoms soft mud, is W¼SW45'S, distance covered 116'20".[2]

Estimated latitude: 6°17', observed: 6°9'.

Calculated longitude: 112°37'.

Note: I am navigating today on the large-scale chart of the island of Java by Mr

[1] The French are using D'Après de Mannevillette's *Neptune oriental* of 1745, including a somewhat unreliable atlas. 'Alambai' are probably the islands of Salembu.

[2] With the latitude as one position line and the 45 fathom depth contour and nature of the bottom, shown on the chart, as the other, it is possible to obtain a fix and hence correct the day's run.

d'Après de Mannevillete. The small-scale one he provided on the Moluccas and surroundings is not at all accurate, neither in respect of the lie of the coasts and islands nor even in respect of a few essential latitudes. The straits of Button and Saleyer are very incorrect. He has not even shown the 3 islands that lie in the latter pass. Admittedly this skilled navigator declares that he does not in the least feel answerable for this chart of the Moluccas and the Philippines, having been unable to find any suitable reports to construct it. One could wish for the safety of Neptune's children that compilers of charts might show the same delicacy as that illustrious sailor whose charts are so worthy of their reputation. The 2nd part of Mr Danville's map of Asia, published in 1752, is excellent from Ceram up to the Alambaï Islands.[1]

Note: I do not believe the latitude and longitude in which he situates them to be correct. These islands are 4 in number, 3 small and a larger one. Mr d'Après places the latter ESE of the other three, approximately 7 leagues distant. This is wrong because these 4 islands are together.

I used it during my passage and everywhere checked by my observations the accuracy of his positions and the lie he puts down for the interesting parts of this difficult navigation. I will even say that New Guinea and the Papuan islands are more credible on his map than on any other I have handled. It is with real gladness that I am taking this opportunity to give due credit to Mr Danville's work. I am well acquainted with him and he has given me the impression of being as good a citizen as he is an honourable man and an enlightened man of science.

Since we got away from the land, the currents have been running N and I believe also E. We shall verify this when we reach Java.

Thursday 22 to Friday 23.

Observed ortive variation amplitude: 1°28'.

Fine weather, fresh SE breeze until dawn when it veered S and S¼SW. We sailed W¼SW until 8 a.m. then WSW, we sounded several times and obtained 47, 45, 41, 42 fathoms always soft mud.

Note: the current continues to bear N.

At midday the estimated run is W¼SW 1°45'S, distance covered 108', corrected route W¼SW 2°30'W, distance 106'.

Estimated latitude: 6°33', observed: 6°25'.

Calculated longitude: 110°51'.

Friday 23 to Saturday 24.

Observed variation: 00°45'NW.

The winds have varied from SE then ESE, to E. They veered with rain to SW and returned to SSW and S, fair weather, fresh breeze. I made my course more S until 6 o'clock. We sighted the island of Java at one thirty from the top of the masts. From 6 o'clock we stayed close hauled on the port tack, sounding frequently. In the afternoon we obtained 40 fathoms, from 10 p.m. 28, 25, 20; from 9 a.m. as we neared the

---

[1] Jean-Baptiste Bourguigon d'Anville (1697-1782), cartographer and Royal Geographer, published an *Atlas de la Chine* in 1727; he eventually published more than 200 maps and charts.

coast 17 fathoms; at midday 10 fathoms and the bottom was always of soft mud. Yesterday in the afternoon, we saw a great number of fishing boats some of which were at anchor and had their nets out.

This morning we saw, in addition to a multitude of fishing boats, a ship leaving the land, then three others one of which was a three-master and two snows. One of the snows was making for the NE, the other two are following the same course as us and have their Dutch flag unfurled at the stern.

At midday the run for the 24 hours is W40'S 89'.

Estimated latitude: 6°25'50", observed: 6°25'30".

Calculated longitude: 109°22'.

Bearings at 6 p.m.: the most E land SE and SE¼S, the most W in view W¼SW 5°W. I am not estimating distances the horizon being too hazy.[1]

Bearings at midday: the large land of Alang Point [Pointe d'Alang] SE¼S approximately 2 leagues, Mandali Island [Isle Mandali] SW¼W2°S 2 miles, the most W land WSW approximately 4 leagues.[2]

Note: by locating my position at midday on Mr d'Après' chart in accordance with the bearings taken on the land, I find: 1st the coast of Java is placed 9' to 12' further S than the average of our meridian observations, 2nd the trend of Alang Point is not correct as he shows it trending WSW and SW¼W whereas in truth it trends from Mandali Island to the W¼SW for approximately 15 miles after which it reverts to an S trend, 3rd he does not show the true extent of the coast, 4th by following the indication of his chart, we would from one midday to another have covered 13 miles less to the W, either because that coast is more extensive, or because the current may have carried us to the E.

Saturday 24 to Sunday 25.

Observed ortive variation: 1°17'.

The winds came from the ESE to SE until 8 a.m. We sailed WSW until 4 p.m. to reach the ships that were ahead of us; then we spoke to the snow. They told us they were on their way to Samarang as was the large vessel. They were effectively keeping to that route. We saw another ship a 3-master at anchor between the islands of Dowa and Swalon. They are taking on rice and timber at the various factories the Dutch have on the Java coast.

At 4 p.m., I set course for the W¼NW until 4 a.m. so as to pass well off the large gulf shown on the charts, then W until midday, the breeze coming from the land and very weak.

We continued to take frequent soundings. From midday to 4 o'clock, while we were following along the coast at a distance of 2 or 3 miles, we obtained 9 and 10 fathoms muddy ground; at 7 o'clock 30 fathoms, at 9 o'clock 32, at 11 o'clock ditto, at one and at 3 a.m. 34 fathoms and always muddy bottom

---

[1] From a known height on the deck the distance to the horizon is known, i.e. with a height of eye of 5.8m (19 feet) the horizon is 5 miles off, and this can be used for estimating distances.

[2] Alang (Tanjung Bugel) is the eastern extremity of a mountainous peninsula on the north coast of Java in latitude 6°24' South and longitude 111° East of Greenwich. Nearby is the small island of Pulau Mandalika.

We continued to encounter a multitude of fishing boats.

The estimated run for the 24 hours is W2°N 80', corrected by the altitude W2°S 80'.

Estimated latitude: 6°23', observed: 6°28'.

Calculated longitude: 108°2'.

Bearings at 4 p.m.: the middle of Poulo Mandali ENE3°E 5 leagues, the most S land where a ship is anchored S¼SW 3 leagues, the middle of Poulo Dowa SSW3°S, same distance, the Carimon Java Islands [Isles Carimon Java] in view to the N¼NW4°W 8 to 9 leagues.[1]

At 5.30 p.m.: the most E land of Java E¼NE5°E approximately 5 leagues, the most W SE5°S approximately 6 leagues, the middle of Carimon Java N2°W 8 leagues.

Sunday 25 to Monday 26.

Observed occiduous variation: 1°15', ortive amplitude: 1°30'.

Unhappily it has been almost calm until 5 p.m., I say unhappily because we would have been interested to see the land before dark in order to set an appropriate course to pass between the point of Indermaye and the Rachit Islands [Isles Rachit], then to seaward of the covered rocks that lie W of them. We could see nothing at sunset.[2]

From midday we steered W and W¼SW. At 6 o'clock until midnight I set a W and W¼NW course, sounding hourly with 25, 24, 21, 20 and 19 fathoms, muddy bottom.

At one o'clock I sailed W¼NW and at 2 o'clock NW until 4 then NW¼W until 6. My intention, reckoning that at 1 a.m. I was mid-channel between the islands and the mainland of Java, was to make for the N of the rocks. The sounding line had given me 3 times 20 fathoms, then 22, then 23. I estimated that I was 3 or 4 leagues NNW of the Poulo Rachit.

At sunrise, I found I was far from my reckoning. We saw the coast of Java from the S¼SW up to W several degrees N and at 7.30 we saw from the mast-head Poulo Rachit bearing NNW and NW¼N about 7 leagues distant. This sighting produced an enormous and dangerous difference with Mr d'Après' chart. I suspended judgement until the midday altitude could attribute this difference to the currents or indict the chart. I set course for W¼NW and NWN in order to examine the coast properly. It is extremely low and level and one cannot make out any mountain in the hinterland. The winds were SSE to SE and E, fresh breeze.

At midday the run for the 24 hours are W¼NW 78'.

Estimated latitude: 6°12', observed: 6°13'.

---

[1] The peninsula off which the French have been sailing trends rapidly south and gives way to the long gulf of Pemalang. There are a number of small islands off the coast, those of Dowa and Swalon mentioned by Bougainville being shown on D'Après' *Neptune oriental*. The islands of Carimon Java Bougainville sees to the north are the Karimunjawa group.

[2] With charts that provided only a generalized outline of the coast, Bougainville was concerned to avoid possible dangers at this late stage of his voyage. He was reaching the end of Pemalang Gulf, after which the coast trends rapidly north towards Indramajo Point (Tanjung Indramayu) with Pulau Rakit some distance off shore.

Calculated longitude: 106°46'.

Bearings at 10.30 a.m.: Indermaye Point [Pointe Indermaye] S¼SE approximately 3½ leagues, the middle of Poulo Rachit NNE4°E approximately 4 leagues and a half.

At midday: the most S point of Indermaye E¼SE2°S about 4 leagues, the middle of Poulo Rachit NE about 5 leagues.

Let us now judge the case according to the altitude and the bearings at midday. On Mr d'Après chart the gulf between Poulo Mandali and Indermaye Point is 22' less extensive from E to W than it is in reality and the coast is placed 16' further S than it is situated according to our observations. The same correction has to be made for Poulo Rachit adding that the distance between these islands and the coast is greater by at least two leagues than the chart indicates. Otherwise the lie of the coast seems fairly accurate as far as one can tell from the estimations made by sight and while sailing. I will point out that the above-mentioned differences are very serious for anyone using that chart and sailing by night.

Sounded at 8 o'clock: 21 fathoms, at 9 o'clock: 23, at 11 o'clock: 19, at midday: 18, soft mud.

Monday 26 to Tuesday 27.
Observed occiduous variation amplitude: 1°23'.

With the ESE breeze we sailed W¼NW ranging along the coast at 3 or 4 miles to avoid those rocks hidden below the water. At 1 o'clock a boat that was at anchor ahead of us sailed on the port tack, which led me to think that the current was changing and becoming unfavourable for us. At 2 o'clock we spoke to the boat. A Dutchman who was in charge and seemed to be the only European with some coloured men told us he was on his way to Ambon and Ternate and was coming from Batavia which he reckoned was 26 leagues away.

At 5 o'clock the breeze came from the sea and as we could only sail WNW, a course that would have caused us to come upon some dangerous banks spread through the bay formed by the points of Indermaye and Sidari, that anyhow one can only work on short and unsure tacks between these banks on one side and the underwater rocks on the other, I decided to anchor at 6 45 in 13 fathoms, soft muddy bottom. Consequently I merely lowered my kedge anchor with 70 fathoms of cablet.

At 2 a.m., the breeze having changed to SSW and SW, I raised anchor and steered NW. The breeze was extremely weak. At midday the estimated run for the 24 hours is WNW4°30'N 47', the run corrected by the latitude is NW¼W2°30'W 49'.

Estimated latitude: 5°51'40", observed 5°49'.

Calculated longitude: 107°12', longitude corrected by the observations: 105°49'.

Note: the observations of the meridian altitude have given from 5°47' to 5°52'. The mean of this latitude compared with the midday bearings would give only 2' or 3' too far S for this part of the coast of Java on Mr D'Après' chart, which is equal to W because one should assume that the estimation of the bearings is absolutely correct.[1] The relative trending of the various parts of the coast seems fairly right. We shall be

---

[1] Meaning unclear, but it appears to mean that since the bearing of the land from the noon position was due west the latitude would be the same regardless of how far away the land might be.

able to verify its accuracy for the E to W stretch when we reach Batavia. Moreover all this land continues to be very low.

Bearings at 6 o'clock: the most W land in sight W 3°S 6 to 7 leagues, the most E SE¼E 3°S approximately 5 leagues, the most W point of Indermaye S 4 miles.

Soundings: at 1 o'clock: 19 fathoms, at 2: 15 fathoms, at 3: 14 fathoms, at 4: 10 fathoms, at 5: 10 fathoms, at 6: 13 fathoms ooze.

Soundings from the time of departure: at 4 o'clock: 16 fathoms, at 8: 20 fathoms ooze, at 10.30: 22 fathoms, at 11: 23 fathoms, at 12: 24 fathoms.

Note: at 10.30, we obtained coral with a first sounding, I had a second sounding taken a moment later, we had mud as usual.

Bearings at midday from the mast-head the coast stretching from S to SW¼W, 5 to 6 leagues distant.

Saw a ship to the WSW.

The currents as we can see from the bearings seem to have carried us to the W by about 13', they also bore us N.

Tuesday 27 to Wednesday 28.

Fine weather, winds variable from ENE to N, NNW, then SSE and SE. I altered course slightly N to avoid the shoals of Sidari Point. We sounded hourly. At midnight I set a course for the W¼SW and WSW. At 2.45 a.m., we saw an island to the NW 5°N, approximately 3 leagues, convinced then that I was further than I reckoned, I dropped anchor in 23 fathoms, muddy bottom, in order to wait for daylight after sailing for an hour to the SW. At daylight we recognised all the islands of Batavia Bay and noted the following bearings: Edam Island [Isle d'Edam] on which there is a flag SE¼S 4 leagues, Rotterdam S¼SE 2°S, the middle of Onrust roadstead SSW 4°S 5 leagues.

Soundings: at 1.30: 25 fathoms, at 3 o'clock: 26 fathoms, at 4 o'clock: 22 fathoms, at 5.30: 23 fathoms, mud. Sighted at that time the most W land bearing SW 4°S, the most E in view S.

Soundings: at 9 o'clock: 23 fathoms, at 11 and 12: 22 fathoms, at 12.30: 18 fathoms, at 1 o'clock: 19 fathoms, at 1.30: 23 fathoms, at 2 o'clock: 25 fathoms, at 2.30: 27 fathoms, at 3 o'clock: 26 fathoms.

Note: the depth lessens gradually after this up to the anchorage in front of the town.

I estimated myself by then to be in latitude 5°49'20" and longitude 106°25' whereas according to our bearings we were really in 5°50' of latitude and 105°58' of longitude, a difference no doubt caused by the currents.[1]

I first raised anchor at 10.30 but the wind having completely failed and the current being against us, I anchored under sail with a kedge anchor and sailed again at half past midday.

I set course for the island of Edam until I was about ¾ of a league from it. Then, the dome of the great church of Batavia bearing from us S, we sailed towards it until 5.45 when we dropped anchor in this town's roadstead in 6 fathoms, muddy bottom and bearings: the river mouth S 2°E approximately ½ league, the island of Edam

---

[1] The error in longitude is only 24', a small difference considering the charts he had at his disposal and the various currents he found between the islands.

N¼NE5°E 3 leagues, the starboard buoy as one enters NNW ½ league, Onrust NW¼W3°N.

The *Étoile* anchored close to us, moreover one anchors here with a single anchor with another ready to let go.

According to the estimate of our track, Batavia would be in southern latitude 6°11′ and longitude 104°52′ E of Paris.

After anchoring, I sent Mr de Bournan to notify the General of my arrival and make arrangements for the gun salute. There is in the middle of the roadstead a ship that serves as flagship. It is the one that returns the salute to merchant and Company ships. They came from that ship to ask who we were, [we replied] a ship of the King of France; to several other questions, no reply. The Chevalier de Bournan came back very late and having been unable to see the General who is now living in a country house 3 leagues from the town. He was taken to the Sabandar of the Christians who assured him that he would advise the General and that if I was willing to land on the morrow he would present me to His Excellency. Great![1]

On the 29th in the morning, I went ashore and accompanied by the Sabandar called on the General. He struck me as a simple and polite man. He will return our salute shot for shot, will see to all our needs and is allowing our sick to be taken into the Company's hospital on the same footing as his own sailors. I am to hand the Sabandar a list of our various requirements and everything is under his control. I believe that he will not suffer any loss with us nor with the suppliers.

There is in town a very fine furnished hotel where one obtains board and lodging for 2 rixdollars[2] a day. Foreigners, other than officers from warships, cannot lodge elsewhere. I did not use my privilege, as I found that house very convenient. We all took a room there. We also hired several coaches. They are two-seaters, drawn by 2 horses and the cost is 2 piastres a day.[3] Presented by the Sabandar and accompanied by my officers I made the customary courtesy calls, namely to the General, the Director-General, the Admiral of the Indies (he is a vice-admiral of the States General) and the President of Justice. One can see that the Sabandar is a man who deals with foreigners: there are two of them, one for Christians and one for pagans. These are fine lucrative posts.

September–October

Stay in Batavia[4]

Changes made to the crew at Batavia. Desertions:

French deserters found on board after departure:

---

[1] The Governor-General of the Dutch East Indies was Petrus Albertus van der Parra, born in Sri Lanka in 1711, died in Batavia in 1771; he had entered the service of the VOC at the age of twenty and been appointed to his present position in 1761. Reports and sundry papers relating to Bougainville's voyage through Indonesian waters are held at The Hague, General Archives, VOC volume 3140:32.

[2] A rixdaler or risdaler was worth approximately 53 French *livres*.

[3] Joseph Banks in October 1770 hired two of these conveyances, 'which are a kind of open Chaises made to hold two people and drove by a man setting on a Coachboax, for each of these I paid 2 Rxr 8s/ a day by the month'. Banks, *Journal*, II, p. 186.

[4] Bougainville's journal contains nine blank pages at this point. Similarly, the lists of changes made among the crew, desertions and men found on board are left blank; this information appears in the crew lists.

October

Departure from Batavia

Sunday 16th [October]

Everything being on board my frigate, I raised anchor at 8 a.m. with the land breeze and steered so as to anchor W of the buoy one finds to starboard as one enters.[1] We dropped anchor at 9.15 in 7½ fathoms soft mud and bearings: the said buoy E¼SE one half mile, Edam Island NNE4°E 3 leagues, Onrust NW¼W 2 leagues ⅓, Rotterdam N2°W 1½ leagues.

The *Étoile* that had been waiting for its supply of bread, having received it during the day, raised anchor at 3 a.m. and steering for the fires I had had lit, anchored near us at 4 o'clock.

Monday 17.

At 5 a.m., I raised anchor and steered N¼NE to pass E of Rotterdam about 1 mile and a half, then NW¼N to pass S of Horn and Harlem then at 8 o'clock W¼NW to W¼SW to haul along the N of the islands of Amsterdam and Middelburg (there is a flag on the latter); continued W leaving to starboard a buoy placed S of small Cambuis.

Bearings at 8 o'clock: Edam Island E3°S 5′, the centre of Batavia roadstead SSE4°E 4′, Onrust S4′, the W point of Middelburg SW3°S6′.

At midday: the NW point of Bantam Bay W approximately 24′, Onrust SE¼E 4°E 12′, the island of Middelburg E¼SE4°E 4′, the large Cambuys N¼NE to NNW.

Observed latitude: 5°55′.

At midday we were N and S of the SE point of Great Cambuis [Grande Cambuis] approximately 1 mile. From there I steered to pass between two buoys, one S of the NW point of Great Cambuis, the other E and W of Cannibal Island [Isles des Anthrophages] or Poulo Laki.

From there until the anchorage the course was W¼NW.

At 5.30 p.m., the current driving us towards the coast, I dropped anchor in 11 fathoms mud and the following bearings:

The NW point of Bantam Bay W¼NW2°W 5 leagues, the middle of Poulo Baby NW5°W 3 leagues, Poulo Lay E¼SE2°S. when I recognized St Nicholas Point [Pointe St Nicholas], it is so to speak the landmark for ships coming from the Strait of Banca [Détroit de Banca], I sent for Mr Caro, Lieutenant in the service of the India Company and taken as pilot for these waters. I handed the control of the track over to him until our departure from the Strait of Sunda.

Note: There is another route to come out from Batavia than the one I have followed. Upon leaving the roadstead, one sails along the coast of Java, leaving to port a barrel that serves as a buoy approximately 2 leagues and a half from the town, then one sails along Kepert Island to the S, then one follows the coast and one passes

---

[1] Bougainville's shipboard journal does not contain any details on the fortnight's stay in Batavia. These appear in his published narrative. He was made welcome by the Dutch, but had no wish to prolong his stay unnecessarily, especially as dysentery and other illnesses soon affected the men, including Ahutoru.

between two buoys one of which is situated S of Middleburg Island [Isle de Middle-bourg] and the other opposite on a bank at the point of the mainland. Then one finds once more the buoy that is S of small Cambuys [Petite Cambuys] and at that point the two routes converge.

Tuesday 18.

At one thirty in the morning, we began to raise the anchor, at 1.50 the cablet broke and having been unable to [secure it?] quickly enough, we lost the kedge anchor because it was not buoyed.[1] From the time of our departure until 8 o'clock, we sailed W¼NW and WNW. From 8 to midday, steered W to SW.

From midday steered WSW to SW¼S with variable winds. At 5 o'clock the winds being brief and the current unfavourable, we dropped anchor in 18 fathoms, muddy bottom and the following bearings: the point of Pepper Bay [Baye au Poivre] SW¼S2°W 6 leagues, the point of Negery SE¼E 1 league, the middle island N5°E 5 leagues, Sambourico Island [Isle de Sambourico] NW 9 leagues, Cracota WNW 7½ leagues.

Bearings at 8 a.m.: Point St Nicholas W5°S 3½ leagues, Poulo Baby from E¼NE to ENE3°E 3 leagues.

At midday: the small Toque bore E5°N 2′, the S point of Middle Island [Isle du Milieu] NW¼W 4′, the large Toque [Grande Toque] N5°E 2½ leagues.

At ½ past midday a Dutch soldier came aboard with a register on which he asked me to write the name of the two French vessels, [I did so] willingly. I saw on this register the *Swalow* [sic] passed on 18 September.[2] This soldier sold me for 16 piastres 10 turtles and 83 hens. His post is ahead of Bantam. At 6 p.m., after we had anchored, a soldier came from another Dutch post with a register on which I saw the passage of the *Utile*, a French vessel, dated June, going from Mauritius [Maurice] to the Philippines.[3] This soldier sold me 10 turtles for 10 piastres.

Wednesday 19.

The winds were very fresh SSE to SE throughout the night. At 4.30 a.m. we weighed anchor and sailed SW to SW¼W until 8 o'clock when we sighted the island of Cracata bearing NW¼N4°N 7 leagues, the point of Pepper Bay SE¼E5°S 2 leagues.

From 8 until 10.30, steered close-hauled then I decided to pass to leeward of Prince's Island being unable to round it. We therefore sailed W¼NW and WNW until midday. We sighted the island of Cracata bearing NE¼N3°N, the NW point of Prince's Island [Isle du Prince] SE¼E3°S. Observed latitude: 6°30′.[4]

---

[1] This was the ninth anchor lost during the voyage.

[2] Carteret in the *Swalow* had sailed on 15 September 1768 from the island of Onrust, 'not having returned, as is the common custom, into Batavia road; in the evening of the 20th, we anchored on the SE side of Prince's Island, in the Streight of Sunda'. *Carteret's Voyage round the World*, I, p. 261. He was therefore just over a month ahead of Bougainville.

[3] The *Utile* was a merchant ship, under the command of Yves Cornic, that had been sent by Pierre Poivre, the administrator of the Isle de France, to look for spices in the Moluccas. It was wrecked on Timor on 9 January 1769, only two men, an officer and a sailor, surviving the disaster. Various documents and reports concerning this expedition are listed in Taillemite, *Bougainville*, I, p. 432n.

[4] The French sailed to the western extremity of Java, Tanjun Pujut, and south-south-west into Sunda Strait between Java and Sumatra. Ahead lay Pulau Rakatau and the famous volcano of Krakatau (Kraka-toa). Bougainville kept close to the Javanese shore until he approached the wide south-western bay of

At 9 a.m. we saw to windward a ship bearing a Dutch ship that was entering the strait. It seemed to me to be a vessel of 50 guns.

Note: all the scurvy cases are cured, but there are numerous men affected by dysentery.

We continued to range along Prince's Island. At one o'clock I had my boats hoisted back in and waited for the *Étoile*'s in order to send Mr Caro back to her. At 3 o'clock sighted the E head of Casnaris Bay bearing S5°E, Cracata Island NE¼E, at 4 o'clock the earlier-mentioned point SE¼E. By these two sets of bearings, I find myself at 4 o'clock 13 miles from the NW point of Prince's Island from which I am taking my point of departure on Mr D'Après' chart in latitude 6°21′ and longitude 102° E of Paris.

Until Wednesday midday, the winds varied from SSE to SE¼S fresh breeze. We kept close to the wind all sails set except at night when we reduced to the main and mizzen topsails to wait for the *Étoile*.

At midday the run from my point of departure is SW2°45′W, distance 102′ 40″.

Note: I am taking here a second latitude with the declination of the true date, that is to say of the 20th and not the 19th which is our day on the ship, and leaving this declination as it is for Paris without making the change required by the difference in meridians.[1] Mr Caro told me that the latitudes were shown in this fashion on D'Après'. I find this hard to believe.[2]

Wednesday 19 to Thursday 20.

Observed variation ortive: 2°30′ NW.

Fair weather, SSE wind, the sea fairly rough, we steered close to the wind port tack, all sails high.

At midday the run gives me SW5°W, distance 103′.

Estimated latitude: 8°45′, observed: 8°37′.

Calculated longitude: 99°23′.

Thursday 20 to Friday 21.

Fair weather, SE wind, the sea fairly rough, steered close hauled on the port tack. At midday the run gives SW2°W 99′40″.

Estimated latitude: 10°9′, observed: 9°55′.

Calculated longitude: 98°11′.

Note: I have decided not to wait any longer for the *Étoile* which does not need me to reach the Isle de France. The health of the men who are entrusted to me and the King's interests require me now to shorten as much as I can this campaign that is already too long.

Casnaris (Pepper Bay); he decided not to attempt to negotiate Penaitan Strait, rounding instead the island of Penaitan (Prince's Island). At that point, he entered the Indian Ocean.

[1] 'As the change of date that should have been carried out when the *Boudeuse* crossed the 12-hour meridian had not been made, it was necessary to take this into account with all astronomical observations' (ET). The latitude is calculated from the meridian altitude of the sun, corrected for the observer's height of eye and refraction, and the declination of the sun (its distance north or south of the equator). The latter was tabulated for each day, hence the requirement for the correct date.

[2] This note was subsequently crossed out.

Friday 21 to Saturday 22.

Observed variation ortive: 1°28′ NW.

Fair weather, fresh SE and SSE breeze, fairly heavy seas, steered SW. Our main topmast is substantially split about the cap. I have had it double-woolded.[1]

The run for the 24 hours is SW3°30′W, distance: 131′.

Estimated latitude: 11°21′, observed: 11°22′.

Calculated longitude: 96°31′.

Saturday 22 to Sunday 23.

Observed occiduous variation: 1°50′, ortive: 2°5′.

Fine and good weather. I steered SW and SW¼W all sails set and keeping a careful lookout for Cocos Island, a low island with a reef on its SE point.[2] We did not see it but a good number of birds in the morning.

The run is SW2°S, distance 146′.

Estimated latitude: 12°50′, observed: 13°2′.

Calculated longitude: 94°35′.

Sunday 23 to Monday 24.

Observed variation occiduous: 2°20′.

Continuation of fine and good weather. I steered WSW, all sails set.

The run for the 24 hours was WSW1°30′S, 146′.

Estimated latitude: 14°5′, observed: 14°4′.

Calculated longitude: 92°17′.

Monday 24 to Tuesday 25.

Continuation of the same good weather: courses between W¼SW and WSW. Run for the 24 hours: WSW30′W, distance 148′.

Estimated latitude: 14°57′, observed: 15°.

Calculated longitude: 89°55′.

We have a great many sick, their number is growing daily. The convalescences are very lengthy. For the last two days several have been affected by high fevers.

Tuesday 25 to Wednesday 26.

Observed variation occiduous: 2°54′, ortive: 2°30′.

Even better than yesterday: same courses. They gave me at midday WSW2°30′W, 180′45″.

Estimated latitude: 16°1′, observed: 16°2′.

Calculated longitude: 86°56′.

Wednesday 26 to Thursday 27.

Better and better, stormy ESE to SE gale, steered W¼SW.

At midday the run was W¼SW4°45′S, distance 196′.

Estimated latitude: 16°53′, observed: 16°56′.

---

[1] To woold is to wind a rope around the mast or yard, in this case two ropes, in order to strengthen it or prevent further splitting.

[2] The Cocos or Keeling Islands are a group of 27 coral islands and reefs that make up two main atolls, West Island and Home Island, in latitude 11°48′ to 12°11′ and longitude 96°50′ East of Greenwich. Bougainville sailed just too far south to see them.

Calculated longitude: 83°40'.

Thursday 27 to Friday 28

Same weather, same course.

The run has given me WSW 5°W, distance 160'.

Estimated latitude: 17°54', observed: 17°49'.

Calculated longitude: 81°.

Friday 28 to Saturday 29.

Observed variation occiduous: 3°36', ortive: 3°55'.

Same weather, same course, a little less wind during the morning.

The run gives W¼SW 3°40'S, 175'.

Estimated and observed latitude: 18°34'.

Calculated longitude: 78°2'.

Saturday 29 to Sunday 30.

Observed variation occiduous: 4°22', ortive 3°55'.

Same weather, steered W and W¼SW all sails set.

The run gives give W¼SW 3°W.

Estimated latitude: 19°, observed: 18°59'.

Calculated longitude: 75°21'.

The weather has turned very cold. It has been necessary to revert to winter clothing. The dysentery cases are not improving. The fevers are a little less persistent.

Sunday 30 to Monday 31.

Same weather, same course, the sea very rough.

I have fished the main mast[1] whose top was arching to such an extent that one could fear it might snap 6 feet below the cat-harpings.[2] To all appearances the main piece[3] is broken. I also had the main topgallant sent down to relieve the strain on the sick mast and for the same reason we took 2 reefs in on main topsail.

Run: W¼SW 5°15'W, 170'.

Estimated latitude: 19°29', observed: 19°18'.

Calculated longitude: 72°21'.

Note: the man named Tostain, of St Malo, carpenter sailor, died during the night of a dysentery that had begun at Bouro.

November

Monday 31 October to Tuesday 1 November.

Same weather. I steered W and W¼SW until 6 p.m. then W wishing to sight Rodrigue Island [Isle Rodrigue].[4]

Run: W¼SW 4°W, 167'.

---

[1] To 'fish' is to add a length of timber along a mast or yard to strengthen it; the French call this 'to twin [*jumeller*]'.

[2] 'Cat-harping' is the bracing of the shrouds by a criss-crossing of short ropes or, as is the more common practice in modern times, by means of iron cramps.

[3] The mast of a warship was composed of several pieces; the main piece was the piece in the middle.

[4] Rodrigues Island lies approximately 560 km east of Mauritius in latitude 19°42' south and 63°25' east of Greenwich. It is a relatively high island surrounded by a coral reef.

Estimated latitude: 19°40', observed: 19°39'.
Calculated longitude: 69°25'.

Tuesday 1 to Wednesday 2.
Observed variation occiduous: 6°43', ortive: 6°50'.
Fair weather, rough sea, the wind less fresh. Steered W¼NW until 2 a.m., then W.
The run was W¼SW 5°W, distance 119'20".
Estimated latitude: 19°39', observed: 19°53'.
Longitude: 67°19'.

Wednesday 2 to Thursday 3.
Observed variation: 7°18'.
Light gale from E to E¼SE. I steered W¼NW. Run: W 1°S, 81'20".
Estimated latitude: 19°47', observed: 19°54'.
Calculated longitude: 65°52'.
By several different observations Mr Verron observed the ship's longitude today at midday as being 64°6'.

Thursday 3 to Friday 4.
Observed variation occiduous: 8°55'.
Light gale the same as yesterday; steered WNW until 6 p.m. then W¼NW.
Run: W4°45'N, distance 98'40".
Estimated latitude: 19°45', observed: 19°46'.
Calculated longitude: 64°8', longitude according to Mr Verron: 62°22'.

Friday 4 to Saturday 5.
Sighting of Rodrigue Island.
Observed variation occiduous: 9°42', ortive: 9°39'.
Fine weather, fair sea, fresh E to ESE gale. We sailed W¼NW until 7 o'clock then WNW until 4 a.m. so as to pass N of Rodrigue in case we were more W than our estimate, then W.
At 11 o'clock, we sighted Rodrigue Island distant approximately 10 to 11 leagues.
At midday run gives me W30'N, 126'30".
Estimated latitude: 19°44', observed: 19°45'.
Calculated longitude: 61°55'30".
Bearings at midday Rodrigue Island from W¼NW 1°20'N to W¼NW 2°W, distance 8 to 9 leagues. Its latitude observed on land is 19°40'.

Saturday 5 to Sunday 6.
Observed variation occiduous: 11°37'.
Good E to ESE breeze, we steered W to pass S of Rodrigue giving a wide berth to the chain of reefs that surrounds it on all sides except the N. At 4 o'clock we were N and S of the NE point, from which I deduce the following difference in our estimate from Prince's Island to Rodrigue. Mr Pingré[1] observed 60°52' of longitude E of

---

[1] This was the noted astronomer Alexandre-Guy Pingré (1711-96) who had sailed to Rodrigues Island in 1760 to observe the Transit of Venus.

Paris and at 4 o'clock, I find myself by my estimate in 61° 26′ of longitude W of Paris so supposing that the observation made on the island in a house was made 2′ W of the point that I had N and S at 4 o'clock, my difference over 1200 leagues of run shows a shortfall of 00°34′, and with the observations of Mr Verron of the 3rd shows an advance of 1°12′.[1]

At 7 p.m., I set a course W and W¼NW and at midnight for W¼NW.

At 4 p.m., we were in latitude 19°50′55″ and longitude 60°54′E.

From 4 o'clock to midnight the run gives me W5°S, distance 146′.

Estimated latitude: 20°00′27″, observed: 20°4′.

Corrected longitude: 58°24′.

Sunday 6 to Monday 7.

Observed variation ortive: 13°22′.

Fair weather, fairly rough sea. E to ESE winds; we steered W¼NW and WNW.

Run for the 24 hours: W5°N, 146′30″.

Estimated latitude: 19°58′, observed: 19°51′.

Calculated longitude: 55°49′.

My calculations at midday place me 13 and a half leagues from the town that the observations of Messrs d'Après and the Abbé de la Caille determined at longitude 55°7′31″ and latitude 20°9′44″ and 9 leagues from Round Island. This estimate could not be more accurate.[2]

Call at the Isle de France

From Monday 7 to Tuesday 8.

Fair weather, ESE wind, we sailed so as to pass between the islands. At 5 p.m. we were N and S of the middle of Round Island [Isle Ronde] At nightfall we fired a gun to get the fire of Gunners' Point [Pointe aux Canoniers] lit, but this light, mentioned by Mr d'Après in his instructions, is no longer lit, so that after rounding Coin de Mire, I was quite troubled to avoid the dangerous reef that stretches out ½ league off Gunners' Point. I steered W¼SW and WSW from 8 o'clock until 10. Then I tacked to endeavour to keep to windward of the port. From time to time I had a gun fired and finally between 11 o'clock and midnight, a boat came alongside with a royal pilot.

I thought my troubles were over and I had handed over the command of the ship to this pilot who told me we had to continue tacking. He was the master and I

---

[1] 'The correct longitude of Rodrigues Island is 61°05′ east of the Paris meridian; Pingré's calculation therefore gave a result that was 18′ too far west and Bougainville's was 34′ east of the latter; his longitude was therefore a mere 16′ out, which is excellent. Since his point of departure was affected by an error in the opposite direction, for which D'Après de Mannevillette's chart was solely responsible, one can say that there was a certain compensation of errors' (ET).

[2] Bougainville is now arriving at Port-Louis, the administrative capital of the Ile de France. The port is situated on the west coast; Bougainville approached the island from the north and had to make his way between a number of small islands that lie off the north coast, Serpent Island, Round Island, Flat Island, Gabriel Island, Gunners Quoin. The Abbé de la Caille is Nicolas-Louis La Caille (1713–62), mathematician and astronomer, who sailed to the Cape in 1751 to determine the position of stars in the southern skies. He was the author of the *Caelum australe astelliferum* (1763) and of a narrative of his voyage to the Cape. See on him Armitage, 'The Astronomical Work of Nicolas-Louis La Caille'.

agreed. At 3.30, he grounded us near the Bay of Tombs [Baye aux Tombeaux]. I immediately set all sails in order to make sternway and made arrangements to bale, lower the boats, etc. Fortunately the ship paid off on the right tack and we got out of this deadly situation. What a fate it would have been to come aground in port through the fault of some ignoramus paid by the King and for having followed the rules.

Note: 1. This pilot is called Mr de la Grange, a word to the wise. 2. I advise those who are making for the Isle de France, when they realize that they will be unable to reach the entrance of the harbour in daylight, to stay out at sea for the night and to windward of Round Island, not hove to but tacking with a good load of sails on account of the currents. In fact there is an anchorage between the islands where the bottom is of sand but one should drop anchor only if absolutely necessary.

We steered back out to sea and ran two tacks until 7 a.m. when the dear pilot anchored us SW of two flags called the Two Brothers [Deux Frères] that indicate the channel at approximately half a cablelength. I immediately sent a message ashore to advise of my arrival and requested the port's assistance. The port captain, Mr Mervin, came with several longboats and had us towed into harbour.[1]

The *Étoile* arrived at 6 p.m. and anchored outside the port.

We found here that we were one day behind and we made the required adjustments. Thus our Tuesday is a Wednesday.

I sent the sick to the hospital, drew up a list of my requirements in rigging and food supplies and began immediately to get the frigate ready for careening. I took all the port's workers and the sailors of the *Étoile*, having resolved to leave before her.

On the 16th and the 18th we breamed[2] the frigate. Her sheathing is wormeaten but the free-board is as good as when it left the shipyards. Some 42 feet of our outer keel were carried away when we grounded thanks to the worthy pilot: we were lucky to get off so lightly.

We left our mainmast, which had a split at the lower end and was likely to give there as much as at its top where the tie was broken. I had a main mast made of a single piece, which is 18 inches longer than the other. We had been given the first one at Brest being supposedly of 81 feet and it was only 77 feet 2 inches. I changed my two topmasts, my anchors, cables, cablets, etc … I handed over all my old food stocks to the royal stores and took others for 5 months. On 10 December everything was ready and I gave orders for our departure.

I handed over to the Legion 23 soldiers including one corporal and four soldiers from the detachment that left from France with me, the balance is made up of the deserters taken on at Batavia.[3] I also left ashore, at Mr Poivre's request,[4] the iron and

---

[1] More correctly, Merven. Joseph Étienne Merven was appointed port captain in July 1767. (AN, Colonies E130, Merven file).

[2] To bream is to burn off the weed, shells etc. that have accumulated on the ship's bottom during the voyage.

[3] These numbered eleven. They received a royal pardon for their desertion on condition that they served in the army for a period of eight years (AN, Colonies B 201, f. 312).

[4] Administrator or *Intendant*, Pierre Poivre (1719–86) served at the Isle de France from 1766 to 1772. He played a major role in developing the port and other facilities in the island and in building up a French spice trade, endeavouring to break the Dutch stranglehold. See on him, Ly-Tio-Fane, *Mauritius and the Spice Trade: The Odyssey of Pierre Poivre*.

the nails taken on board the *Étoile*, my cucurbit, my cupping-glass,[1] numerous medicines and a quantity of items that are of no use to us and which this colony greatly needs.

Messrs de Commerçon and Verron remained, the first to study the natural history of these islands and of Madagascar where we [the French] are today attempting to re-establish ourselves,[2] the second to be in a position to travel to Pondicherry to observe the transit of Venus on 9 June 1769.[3] I was also asked for the services of Mr de Romainville, the engineer, to help with works being carried out in the island and for a few volunteers and pilots for the India route. I left this engineer and Messrs Fetch, Duclos the elder, Ouri and Oger. It is fortunate that after such a long voyage one is able to enrich this colony with men and needed equipment. The satisfaction this gives me is lessened by the loss we sustained here of Mr Du Bouchage, *Enseigne de la Marine*, a person of outstanding merit, who died on 26 November of a dysentery he caught in Batavia. We also lost Mr Lemoyne's young son, who died of consumption on [15] November. He had been made *Garde de la Marine* at the last promotion.

[Bougainville then adds two pages of comments on the situation in Mauritius and the arrival and departures of several ships. His next section, entitled 'Departure from the Isle de France', lists in greater detail the changes made to the complement of the Boudeuse.]

Departure from the Isle de France

Changes to the crew made in this island.

I lost at the Isle de France the Chevalier du Bouchage, *enseigne de vaisseau*, and Lemoine de Montchevry, *Garde de la Marine*.

At the *intendant*'s request, I left behind for the King's service in the colony: the Rev. Fr Lavaisse, chaplain, Messrs Fetch, volunteer, Verron, pilot observer, Oury and Oger, first and second pilots, De Romainville, infantry lieutenant sailing in the *Étoile*, Pierre Duclos son, volunteer ditto, Commerçon des Humbert, naturalist ditto and his valet a girl [disguised] as a man.

I gave over to the Legion 23 soldiers including all those taken on at Batavia and a corporal and four soldiers from my detachment.

I left sick at the hospital: Mr Herpin, assistant pilot, 2 sailors and one boy. Mr de St Germain also remained ashore being seriously ill.

Took in as replacements: Mr des Longrais, naval clerk, Mr Hervel, senior pilot with the Company, a second boatswain, a carpenter, 3 English sailors and one boy.

---

[1] The French term is *ventouse*, the primary meaning of which is a cupping-glass, as used in bleeding as a medical treatment (as an alternative to leeches); and the reference to medicines as the next item makes this translation likely. However, a *ventouse* on a ship is a ventilating device, known in English as an air-scuttle, and this is an alternative possibility.

[2] In 1768 the Comte de Maudave set up a French outpost on Madagascar, where the French had been attempting to establish settlements since 1642. Maudave's colonizing attempt petered out after a couple of years and was taken over, partly as a private venture, by Maurice de Benyovski in 1774: see Dunmore, *French Explorers*, I, p.235.

[3] On the British side arrangements had already been made for James Cook to observe the Transit of Venus in the Pacific Ocean, and he was by then on his way to Tahiti in the *Endeavour*.

I also took on as passenger the Chevalier de Tremergat, sub-lieutenant in the navy.[1]

[Bougainville sailed from Port Louis on 12 December 1768. The voyage of exploration can be considered to have ended at the Isle de France, a view clearly reflected in Bougainville's journal which from this point contains little more than basic navigational details and ends with the entry of 14–15 February 1769.]

[1] The muster roll provides additional information on these changes. There was, for instance, one black servant named Aza belonging to the Prince of Orange, who was taken on as a passenger for the Cape of Good Hope. The three English sailors are named, Joseph Joansouh (presumably Johnson), Thomas Philisse (possibly Philips) and Filden (Fielden); the latter, a Londoner, was to desert at the Cape. Longrais was taken on to replace the ailing Saint Germain; he would return to Mauritius a few months after the *Boudeuse* reached France. Louis de Tremergat had had a remarkable career in the navy, losing a leg at the Battle of Quiberon in 1759; he became a member of the States of Brittany in 1787.

APPENDIXES

APPENDIX 1

The Muster Roll

The consolidated muster roll for the Boudeuse *was compiled from Bougainville's own journal by Étienne Taillemite for his edition of the voyage, where it appears on pp. 160–81. Further information was taken from two other rolls kept in AN Marine, one of which, C6 370 105ff, is countersigned by Bougainville, Duclos-Guyot and the Intendant of the Isle de France, Pierre Poivre. It covers the period 4 October 1766 to 25 March 1769. First names are not shown in the case of the officers. The men, as would be expected, are predominantly Bretons, St Malo and the nearby town of St Servan providing the great majority. Where known, their place of origin is indicated in the roll entry. Monthly pay is shown in French pounds (livres). The date on which the man concerned came on board then follows. Promotions and rises granted during the voyage are shown in italics. There were remarkably few deaths or desertions. Deaths are indicated by an asterisk next to the name of the man concerned, desertions by an asterisk with the addition of the letter d. Some did not complete the full circumnavigation in the* Boudeuse, *being taken on later in the voyage or transferred from or to the storeship, or left behind for a variety of reasons at the Isle de France; they are identified by mean of a small dagger. Short stays in hospital that did not lead to repatriation or to the man being left behind are indicated in entries in the muster roll. Duclos-Guyot and his sons, Le Corre and Fesche are listed from 4 October 1766, Bougainville from the 20ᵗʰ, the three enseignes and the gardes from the 26ᵗʰ, most of the petty officers and men from the 24ᵗʰ; the soldiers arrived on board on 2 November. The* Étoile's *muster roll has not been located.*

The English equivalent of rank or title is used, with the French term shown between square brackets.

Senior Officers [*Officiers Majors*]

L.-A. de BOUGAINVILLE, Capitaine de Vaisseau, 250 *l*. with a supplement of 120 *l*. as commanding officer. [See mess and other allowances below.]

N. DUCLOS-GUYOT, Fireship Captain [*Capitaine de Brûlot*], 125 *l*.

Chevalier de BOURNAND, Enseigne de Vaisseau, 66 *l*. 13 *s*. 4 *d*. *Lieutenant de Vaisseau from* 18 *August* 1767.

Chevalier d'ORAISON. Same rank, same rate of pay.

★Chevalier du BOUCHAGE, Same rank, same rate of pay. Died at the Isle de France 26 November 1768.

†Mr DUSSINE, Lieutenant de Frégate, 70 *l*. Joined 20 October but disembarked on 13 November before the departure from the Loire.

Chevalier de SUZANNET, Garde de la Marine, 30 *l*. with a supplement of 20 *l*. for carrying out duties of officer. *Enseigne de vaisseau from* 18 *August* 1767.

Chevalier de KERHUÉ, Garde du Pavillon, 36 *l* with supplement of 20 *l* for carrying out duties of officer. *Enseigne de vaisseau from* 15 *August* 1768.

Josselin LE CORRE, Officer of the Blue [*Officier bleu*], 50 (possibly 70) *l*.

†Mr de SAINT-GERMAIN, Naval Clerk [*Écrivain de la marine*], 100 *l*. Landed sick at the Isle de France 8 December 1768.

†Rev. Fr LAVAISSE, Chaplain [*Aumônier*], 50 *l*. Disembarked at the Isle de France 8 December 1768.

Mr LAPORTE, Senior Surgeon [*Chirurgien-major*]. 70 *l*. plus allowance for 190 men at the rate of 1 *s.* per man: 9 *l.* 10 *s.*

Mess and other allowances from 20 October 1766

To Mr de Bougainville, Capitaine de Vaisseau, commander, for his personal table costs, wages and subsistence of his servants and indemnity for his furnishings at 28 *livres* a day and for advance payment for 365 days from 20 October 1766 the sum of 10,220 *livres*.

To the same for 11 senior officers at 2 *l.* 10 *s.* each per day, 27 *livres* 10 *sols* and in respect of an advance for 180 days: 4,950 *livres*.

To the same for ditto at 4 *l.* each, 44 *l.* per day and for 180 other days: 7,920 *livres*. Total: 23,090 *livres*.

Passenger

The Prince of ORANGE NASSAU ZIEGUEN [= SIEGEN].

Volunteers [*Volontaires*]

†Pierre DUCLOS-GUYOT. 30 *l*. Landed at Rio Janeiro 12 July 1767 and transferred to the *Étoile*, replaced by Mr Lemoyne.

Alexandre DUCLOS-GUYOT. 30 *l*.

†Charles-Félix-Pierre FESCHE. 30 *l*. Disembarked at the Isle de France 8 December 1768.

Warrant Officers [*Officers Mariniers*]

★Denis COUTURE, from St Servan, Master [*Premier maître*]. 70 *l*. Died 24 August 1768.

Germain BONGOUR, from St Servan, Second Master [*Second maître*]. 50 *l*. *increased to* 55 *l. on* 24 *October* 1767. Sent to the hospital at the Isle de France 24 November 1768, released 9 December following

Boatswains [*Contres maîtres*]

Jean PERE (or PERRÉ), from St Malo, 36 *l*. *increased to* 40 *l. on* 24 *October* 1767. Sent to the hospital at the Isle de France 2 December 1768, discharged the 7th of the said month.

†Nicolas GILBERT, from St Servan. 36 *l*. Sent to the hospital at Paimboeuf 13 November 1766 and paid off.

Pierre DAVIAUD, called Chauvel, from St Malo. 36 *l*. *increased to* 40 *l. on* 24 *October* 1767. Sent to the hospital at Batavia 1 October 1768, discharged 15th of the said month, to the hospital at the Isle de France 9 November 1768, discharged 29th of the said month.

Pierre GESLIN, from St Servan. 36 *l. increased to* 40 *l. on 24 October* 1767. Sent to the hospital at the Isle de France 1 December 1768, discharged the 5th of the said month.

Boatswain's Mates [*Bossemans*][1]

Guillaume MENOU, from St Servan. 30 *l. increased to 36 l.* Sent to the hospital at the Isle de France 1 December 1768, discharged on the 5th of the said month.

Benoist RAIZE, from St Malo. 30 *l. increased to 36 l.* Sent to the hospital at Batavia 1 October 1768, discharged the 10th. Ditto at the Isle de France on 28 November 1768, discharged 5 December.

Laurent ROUSSAY (or RONÇAIS), from St Colomb. 30 *l. increased to 36 l. on 24 October 1767.*

Leading Seamen [*Quartiers maitres*]

Alain NOBLIER (or NOUBLIER), from St Malo. 24 *l. Rated boatswain's mate at 30 l. on 24 October* 1767. Sent to the hospital at the Isle de France 1 December 1768, discharged on the 5th of the said month.

Julien MÉHU, from St Servan. 24 *l. Rated boatswain's mate at 30 l. same day.* Sent to the hospital at the Isle de France on 16 November 1768, discharged the 9th December following.

†Jean HAMON, from St Servan. 24 *l.* Disembarked with permission at Montevideo 28 February 1767 on account of ill-health.

Nicolas CANTIN, from St Servan. 24 *l. Rated boatswain's mate at 30 l. on 24 October* 1767. Sent to the hospital at the Isle de France 1 December 1768, discharged on the 5th, returned on the 6th and discharged on the 9th.

Coxswains [*Patrons*]

★(d) Joseph LÉPINE, from St Malo. In charge of longboat [*chaloupe*]. 27 *l.* Deserted at Montevideo 27 August 1767.

Jean CROCHARD, from St Malo. In charge of cutter [*canot*]. 24 *l.* Rated boatswain at 30 *l.* on 24 October 1767.

Pilots [*Pilotes*]

†Jean RIODUJOUR, from St Nazaire. First pilot [*Premier pilote*]. 60 *l.* Disembarked sick at the Malouine Islands on 18 April 1767, transferred to the *Lièvre*.

†Jean-François OURY, from St Malo. Second Pilot [*Second pilote*]. 36 l. *Promoted First Pilot at 45 l. on 18 April 1767.* Disembarked at the Isle de France 9 December 1768.

†Charles OGER, from St Malo. Second Pilot. 33 *l.* Raised to 40 *l.* on 18 April 1767. Disembarked at the Isle de France 9 December 1768.

†Pierre Joseph HERPIN, from Paimboeuf. Second Pilot. 33 *l. Raised to 36 l. on 24 October 1767.* Sent to the hospital at the Isle de France on 7 December 1768 where he still was when the frigate sailed on the 12th of the same month.

---

[1] A *bosseman* was an assistant boatswain dealing especially with the cables and anchors. The term fell out of use in the early nineteenth century.

†Louis Guillaume BENOIST, from Montfort. Assistant Pilot [*Aide pilote*]. 30 *l.* disembarked at Montevideo 9 November 1767 and transferred to the *Étoile*, reimbarked at Batavia on 9 October 1768, sent to the hospital at the Isle de France 9 November 1768, discharged 10 December.

François Guillaume de VIENNE, from St Malo. 30 *l.*

Gunners [*Canoniers*]

Pierre FEUILLET, from St Malo. Master gunner [*Maitre canonnier*]. 45 *l. Raised to 50 l. from 24 October 1767.*

Laurent MICHEL, from St Servan. Assistant Master Gunner [*Second canonnier*]. 33 *l. Raised to 36 l. from the same date.*

Assistant Gunners [*Aides-canonniers*]

Jean-François FROUSSART, from St Servan. 27 *l. Rated Assistant Pilot at 30 l. on 24 October 1767.*

Guillaume HIGNARD, from St Malo. 27 *l. Raised to 30 l. from 24 October 1767*, sent to Batavia hospital 1 October 1768, discharged 10 of the same month.

Jacques MALOLOUET, from St Malo. 27 *l. Raised to 30 l. from the same date.* Sent to Batavia hospital 1 October 1768, discharged on the 9th, sent to the hospital at the Isle de France 25 November 1768, discharged on the 30th.

Jean-Baptiste LOCHET, from St Malo. 24 *l. Raised to 27 l. from the same date.* Sent to the hospital at the Isle de France 25 November 1768, discharged on the 30th.

Jean-Jacques Henry SABLON, from St Malo. 27 *l. Raised to 30 l. from the same date.* Sent to the hospital at the Isle de France 1 December 1768, discharged on the 5th.

†Jacques GAILLET, from St Malo. 27 *l.* Sent to the hospital at Brest on 21 November [1766], paid off on 5 December.

Jean René GIOT, from St Servan. 24 *l. increased to 27 l. from 24 October.* Sent to the hospital at the Isle de France 25 November 1768, discharged on the 30th.

Jacques POUPION, from St Malo. 24 *l. increased to 30 l. from 24 October.* Sent to the hospital at the Isle de France on 1 December 1768, discharged on the 5th.

René LOCQUET, from St Malo. 24 *l. increased to 33 l. from 24 October 1767.* Sent to the hospital at the Isle de France on 25 November 1768, discharged on the 30th.

Jean GUERRIER, from St Servan. 21 *l. increased to 24 l. from 24 October 1767.* Sent to the hospital at Batavia 1 October 1768, discharged on the 8th, to the hospital at the Isle de France 1 December 1768, discharged on the 5th.

Nicolas LESOURD, from Le Croisic. 21 *l. increased to 24 l. from the same date.* Sent to the hospital at the Isle de France 25 November 1768, discharged on the 30th of the same month.

Carpenters, Caulkers and Sailmakers

†Mathurin TAUPÉ (or TOUPÉ), from St Servan. Master Carpenter [*Maître charpentier*], 40 *l.* Landed sick at the Malouine Islands on 19 April 1767, transferred to the *Lièvre* at Montevideo, where he died.

Etienne LIONNOIS, from St Servan. Assistant Master Carpenter [*Second charpentier*] at 30 *l. Raised to 36 l. and appointed Master Carpenter from 20 April 1767.*

François FROGER, from St Servan. Assistant Carpenter [*Aide-charpentier*] at 24 *l.*
   Raised to 27 *l. from* 24 *October* 1767. Sent to the hospital at the Isle de France on
   11 November, discharged the 9 December following.

†Jean GIRARDEAU, from Chantenay. Master Caulker [*Maître calfat*]. 50 *l.* Sent to the
   hospital at Brest on 22 November [1766], paid off on the 23rd and replaced on
   the 24th.

Guillaume DANTONVILLE, from St Servan. Second Caulker [*Second calfat*] at 24 *l.*
   Raised to 27 *l. from* 24 *October* 1767.

Louis LE BRETON, from St Servan. Master Sailmaker [*Maître voilier*]. 36 *l. Raised to*
   40 *l. from the same date.*

★(d) Toussaint LEBRUN, from Agde. Second Sailmaker [*Second voilier*]. 27 *l.* Deserted
   at Montevideo 28 August 1767.

Henry MARAIS, from St Malo. Assistant Sailmaker [*Aide voilier*]. 24 *l. Raised to* 30 *l. on*
   24 *October* 1767.

Jean-Baptiste ALLEGOT, from Recouvrance. Master Caulker. Taken on [at 50 *l.*?] at
   Brest to replace Jean Girardeau.

Able Seamen and Sailors [*Matelots*]

Jean MARTIN, from St Malo. Helmsman at 20 *l. Rated Leading Seaman [Quartier*
   maître] at 24 *l. on* 24 *October* 1767. Sent to the hospital at the Isle de France on 1
   December 1768, discharged on the 5th following.

René BIZEUL, from Guérande. Helmsman, 20 *l.* Sent to the hospital at the Isle de
   France like the above.

Thomas CORDIER, from St Malo. Helmsman, 20 *l.* Sent to the hospital at the Isle
   de France 25 November 1768, discharged on the 30th of the same.

Georges RENAUD, from St Servan. Helmsman, 20 *l. Leading Seaman at* 24 *l. from* 24
   October 1767. Sent to the hospital at the Isle de France like the above.

★Julien LAUNAY, from Pontchâteau. Tailor. 20 *l.* Died at sea 10 September 1768.

★(d) Honoré LAMOUREUX, from Guérande. Pilot's boy. 18 *l.* Deserted at Montevideo 1 November 1767.

Jacques TALVART, from St Servan. 18 *l. Topman at* 20 *l. on* 24 *October* 1767. Sent to
   the hospital at the Isle de France 9 November 1768, discharged the 26th of the
   same.

Jean CUQUEMEL, from St Servan. 18 *l.* Sent to the hospital at the Isle de France 25
   November, discharged the 30th on the same.

Pierre LEROUX, from Le Croisic. 18 *l.* Ditto.

†Jacques BONNIER, from Le Croisic. 18 *l.* Sent to the hospital at Brest 26 November 1766, paid off on the 28th, reinstated the 29th, dismissed 6 December.

Pierre Charles ALAIN, from St Malo. 18 *l.*

François LEMAY, from Le Croisic. 18 *l.* Sent to the hospital at the Isle de France 25
   November 1768, discharged on the 30th.

†Jean CLOSSE, from St Servan. 18 *l.* Sent to the hospital at the Isle de France 9
   November 1768, stayed in the said hospital on 12 December following, date of
   the frigate's departure.

Cantin REFUVEILLE, from St Malo. 18 *l. Rated topman at* 20 *livres* 24 *October* 1767.

Alexis SAMSON, from St Malo. 18 *l*. Sent to the hospital at the Isle de France 24 November 1768, discharged 6 December following.

Pierre LE GAL, from St Servan. 18 *l*.

François DENORD, from St Malo. 18 *l*. Sent to the hospital at the Isle de France 11 November 1768, discharged 26th of the same.

Pierre LENORD, from St Malo. 18 *l*.

François LEMARCAN, from St Malo. 18 *l*. *Helmsman at 20 livres from 24 October 1767.* Sent to the hospital at the Isle de France 3 December 1768, discharged on the 5th of the same.

Augustin BLAIZE, from Paimpol. Surgeon's assistant,[1] 18 *l*.

François DIARDIÈRE, from St Malo, 18 *l*. *Topman at 20 l. on 24 November 1767.* Sent to the Batavia hospital 1 October 1768, discharged 6th of the same.

Louis RENAUD, from St Malo. 18 *l*. *Topman at 20 l. from the same day.* Sent to the hospital at the Isle de France 25 November 1768, discharged the 30th of the same.

Gilles DURAND, from St Malo. 18 *l*. *Ditto at 20 livres from the same day.* Sent to the hospital at the Isle de France 1 December 1768, discharged the 9th of the same.

Olivier LAUNAY, from St Coulomb. 18 *l*. *Topman at 20 l. from 24 October 1767.* Sent to the Batavia hospital 1 October 1768, discharged 15th of the same, ditto to the hospital at the Isle de France 9 November, discharged the 9 December following.

† Malo Julien GAUCHER, from St Malo. 18 *l*. Paid off at Paimboeuf 15 November 1766.

Pierre PINAUT, from St Servan. 18 *l*. *Topman at 20 l. from the said date.* Sent to the hospital at the Isle de France 25 November 1768, discharged the 30th of the same.

†Nicolas DURAND, from Nantes. 18 *l*. Sent to the hospital 21 November [1766], paid off on the 28th, replaced on the 29th at Brest.

†Jean THOMAS. From Nantes. Carpenter. 18 *l*. Disembarked at Montevideo 28 February 1767 on account of sickness.

Servan NABOURG, from St Servan. 18 *l*. Sent to the hospital at the Isle de France 1 December 1768, discharged the 5th of the same.

Jean GUILLON, from St Servan. 18 *l*. *Topman at 20 livres from 24 October 1767.* Sent to the hospital at the Isle de France like the previous one.

Jean LEFRANC, from St Servan. 18 *l*. *Assistant Carpenter at 24 l. from 19 April 1767.*

★Jean TOSTAIN, from St Malo. 18 *l*. Sent to the Batavia hospital 1 October 1768, discharged on the 15th of the same. Died on board 30 October 1768.

Jean MERLET, from St Malo. 18 *l*. *Topman at 20 livres from 24 October 1767.* Sent to the hospital at the Isle de France 1 December 1768, discharged on the 5th of the same.

†François LAURENS, from St Malo. 18 *l*. Sent to hospital and paid off at Paimboeuf 15 November 1766.

Alain BÉTUEL (or BETHUEIL), from St Malo. 18 *l*.

Jean-Pierre SCOLAND, from St Malo. 18 *l*. *Topman at 20 livres from 24 October 1767.*

---

[1] The French term was '*frater*', from the Latin for 'brother'. It was falling into disuse as fewer medical men were using Latin terms.

Sent to the hospital at the Isle de France 1 December 1768, discharged the 5ᵗʰ of the same.

†(d) Jean ESNAULT, from St Malo. 17 *l.* Deserted at Montevideo 12 November 1767.

François BUCAILLE, from Dinan. 17 *l. Made cook's mate at 21 l. from 22 August 1767 to replace Jean Salot.* Sent to the hospital at the Isle de France 17 November 1768, discharged the following 9th December.

Jean HENON, from St Malo. 17 *l. Rated topman at 20 l. on 24 October 1767.* Sent to the Batavia hospital 1 October 1768, discharged on the 5ᵗʰ.

Clement BAROST, from St Malo. 17 *l. Rated topman at 20 l. on 24 October 1767.* Sent to the hospital at the Isle de France 25 November 1768, discharged on the 30ᵗʰ of the same.

Thomas LECUYER, from St Malo. 17 *l. Increased to 18 l. from the above date.*

†(d) Jean PIERRE (or PERRÉ), from St Servan. 17 *l.* Deserted at Brest 5 December [1766] and placed in Pontaniou [gaol], released 13 December and dismissed the same day.

†(d)Jean GODIN, from St Servan. 16 *l.* Deserted at Brest 5 December [1766] and placed in Pontaniou, released 13 December and dismissed the same day.

Antoine HOTTE, from Le Croisic. 16 *l.*

Pierre SIMON, from St Servan. 16 *l. Raised to 18 l. from 24 October 1767.*

Jean-François LE CACHEUR. 16 *l. Raised to 18 l. from the same date.* Sent to the hospital at the Isle de France 1 December 1768, discharged the 5th of the same.

Louis RUÉLAN, from St Servan. 16 *l. Raised to 18 l. from the same date.* Sent to the hospital at the Isle de France 9 November 1768, discharged 7 December following.

†Laurent JOUBIN, from St Malo. 16 *l.* Sent to the Brest hospital 3 December 1766, discharged and paid off on the 5ᵗʰ of the same.

Etienne AUGER, from St Malo. 16 *l. Raised to 18 l. from the 24 October 1767.* Sent to the hospital at the Isle de France 1 December, discharged the 5ᵗʰ of the same.

†(d) François AUDOUARD, from St Malo. 16 *l.* Deserted at Montevideo 28 February 1767.

Alexandre LEGRAND, from St Servan. 16 *l. Raised to 18 l. from 24 October 1767.* Sent to the hospital at the Isle de France on 1 December 1768, released on the 5ᵗʰ of the same.

Vincent LANGLOIS, from St Malo. 15 *l. Raised to 18 l. on the same date.*

†Mathurin LEMONNIER, from St Malo. 15 *l.* Sent to the Brest hospital 29 November 1766, paid off 5 December.

Guillaume LE FAR, from Le Croisic. 15 *l.* Sent to the hospital at the Isle de France 25 November 1768, discharged on the 30ᵗʰ of the same.

†Julien BERNARD, from Paimboeuf. Caulker, 15 *l.* Sent to the Brest hospital 1 December 1766, paid off on the 5ᵗʰ of the same.

Suliac BOISSELIER, from St Malo. 15 *l. Raised to 17 l. from 24 October 1767.*

Joseph METAYER, from St Malo. 15 *l.* Sent to the Paimboeuf hospital 13 November 1766, taken back on at Brest the 26 November following.

Toussaint HEQUET, from St Malo. 15 *l. Raised to 18 l. from the same date.* Sent to the hospital at the Isle de France 1 December 1768, discharged the 5ᵗʰ of the same.

Pierre BALARD, from St Malo. 15 *l. Raised to 18 l. from the same date.* Sent to Batavia hospital 1 October 1768, discharged on the 8th of the same.

Laurent SOUQUET. 15 *l. Raised to 18 l. from the same date.* Sent to the hospital at the Isle de France 9 November 1768, discharged 5 December.

Jacques LE FRANÇOIS, from St Servan. 15 *l.*

François VALET, from St Servan. 15 *l. Raised to 16 l. from the same date.*

Jacques ROSSET, from St Malo. 15 *l. Raised to 16 l. from the same day.*

†François ADAM, from St Servan. 15 *l.* Sent to the Paimboeuf hospital 13 November 1766 and paid off the same day.

François LAUNAY, from Cancale. 15 *l.*

François LORRE, from St Servan. 15 *l. Raised to 18 l. from 24 October 1767.* Sent to the Batavia hospital 1 October 1768, discharged the 11th of the same.

†François PARIS, from Paimboeuf. 15 *l.* Sent to the hospital at Brest 24 November 1766, paid off on the 28th, replaced on the 29th.

†Jean BOISSERPE, 15 *l.* Sent to the hospital at Brest 1 December 1766, paid off 5th of the same.

Marc PINARD, from St Malo. 14 *l.* Sent to the Batavia hospital 1 October 1768, discharged the 11th of the same.

Louis BOURSAULT, from St Servan. 14 *l.*

†Jean-Marie BEAULIEU, from St Malo. 14 *l.* Sent to the hospital at Paimboeuf 13 November 1766 and paid off the same day.

Jean-Marie LÉA (or LÉAS), from Brest. 14 *l.* Sent to the Batavia hospital 1 October 1768, discharged the 9th of the same, ditto at the Isle de France 10 November 1768, discharged 1 December following.

Pierre-Jean JAGOT, from St Servan. 14 *l. Raised to 17 l. from 24 November 1767.* Was sent to the hospital at Brest 29 November 1768, discharged 2 December.

Jean-François LEMEAUX, from St Malo. 14 *l.* Sent to the hospital at Paimboeuf 13 November 1766, taken back on at Brest the 27 November following. Sent to the hospital at the Isle de France 10 November 1768, discharged the 7 December following.

★(d) François RENAUD, from St Malo. 14 *l.* Deserted at Montevideo 28 February 1767.

Guillaume HERCOUËT, from Cancale. 14*l. Raised to 16 l. from 24 October 1767.* Sent to the Batavia hospital 4 October 1768, discharged the 15th of the same, sent to the hospital at the Isle de France 9 November 1768, discharged the 6 December following.

Jean BOUAN, from Cancale. 14 *l. Raised to 16 l. from the said date.* Sent to the Batavia hospital 5 October 1768, discharged the 15th of the same, sent to the hospital at the Isle de France 11 November 1768, discharged the 1 December following.

Jacques HÉLAR, from Morlaix. 14 *l. Raised to 16 from the said date.* Sent to the hospital at the Isle de France 30 November 1768, discharged the 6 December following.

Claude (or Guillaume) DURAND, from Nantes. 14 *l.* Sent to the hospital at the Isle de France 6 December 1768, discharged the 10th of the same.

Pierre PLANCHET, from Redon. 14 *l.*

LaurentVENTAUBERGHEN, from St Malo. 14 *l. Raised to 17 l. from 24 October 1767.*

†Nicolas DUTUIL, from St Malo. 13 *l.* Sent to the Brest hospital 21 November [1766], paid off on the 28th, replaced on the 29th.

Pierre BESNARD (or BERNARD), from St Malo. 13 *L. Raised to 16 l. from 24 October 1767.* Sent to the Batavia hospital 4 October 1768, discharged on the 11th of the same.

Jean TURPIN, from St Malo. 13 *l. Raised to 16 l. from the same date.* Sent to the hospital at the Isle de France 9 November 1768, discharged 1 December following.

†François TUELOUP, from St Servan. 13 *l. Raised to 16 l. from the same date.* Sent to the hospital at the Isle de France 25 November 1768 where he remained when the frigate sailed on 12 December 1768.

François COSSE (or COSTE), from St Servan. 13 *l. Raised to 16 l. from the same date.*

Pierre-Joseph SARROT, from St Servan. 13 *l. Raised to 17 l. from the same date.* Sent to the hospital at the Isle de France 9 November 1768, discharged 6 December following.

†Charles POUCHAIN, from St Servan. 13 *l.* Sent to the hospital at Brest 22 November [1766], paid off 28th, replaced 29th, discharged from hospital and dismissed 20 December.

Louis JEAN, from Cancale. 13 *l. Raised to 17 l. on 24 October 1767.*

Jean PEDRON, from St Servan. 12 *l.*

★Jean CHENET, from St Servan. 12 *l.* Drowned 30 January 1768.

Etienne LAISNÉ, from St Servan. 12 *l.*

Noël BITEL, from St Servan. 12 *l. Raised to 15l. from 24 October 1767.*

Pierre DAVIAUD, from St Malo. 12 *l. Raised to 15 l. from 24 October 1767.* Sent to Batavia hospital 4 October 1768, discharged 15th of the same, sent to the hospital at the Isle de France 24 November 1768, discharged 7 December following.

Pierre SOULIVAN, from St Servan. 12 *l.. Raised to 15 from the same date.*

Guillaume NEVEDEC, from St Brieuc. 14 *l.*

Victor SOR (or FORT), from St Servan. 12 *l. Raised to 15 l. from 24 October 1767.* Sent to the hospital at the Isle de France 28 November 1768, discharged 9 December following.

Michel GABARIAU (or GABORIAU), 12 *l.* from Nantes.

Replacements taken on at Brest 29 November 1766

Hamon Joseph LARS, from Recouvrance. 15 *l.* Replacing Nicolas Dutuil.

Prigent KEREBEL, from Recouvrance. 15 *l.* Replacing Nicolas Durand.

Julien ABALEA, from Recouvrance. 13 *l.* Replacing Charles Pouchain.

Jacques PAGE, from Recouvrance. 12 *l.* Replacing François Paris.

Félix TRÉHORET, from Recouvrance. 12 *l.* Replacing Jacques Bonnier.

Jean TRÉGUIER, from Recouvrance. 12 *l.* Replacing [blank]

Soldiers from the Recruits Regiment

Brémont Company: André Philippe OSSERE, sergeant. Sent to the hospital at the Isle de France 17 November 1768, discharged 7 December following.

Beauregard Company: Jean-Marie DUPUY, corporal. Sent to the hospital at the Isle de France 9 November 1768, discharged 9 December following.

★(d) Virgine Company: Joseph PRUDENT, ditto. Deserted at Montevideo 6 November 1767.

Nesle Company: Louis RICARD, remunerated. Sent to the Brest hospital 21 November 1766, discharged 26th of the same.

†La Perelle Company: Louis HÉROUARD, ditto. Disembarked at the Isle de France 10 November 1768 to serve in the Legion.

Fusiliers

Nesle Company: Louis MERY. Worked aloft from 14 November 1767.

Beauregard Company: Jean MAHÉ. Worked aloft from 6 December 1766.

†Braquemont Company: François BATHINE. Sent to the hospital at Brest on 26 November 1766, where he stayed.

★(d) Fontaine Company: Jacques Nicolas LEROY. Sentenced 15 June 1767 to loss of one month's pay for theft on board ship. Worked aloft from 1 January 1768. Sent to the hospital at Batavia 1 October 1768, deserted on the 14th of the same.

★(d) La Perelle Company: Nicolas BLAIN. Deserted at Montevideo 19 October 1767.

★(d) La Perelle Company: Joseph GOTINEAU. Sent to the Brest hospital 21 December 1766, discharged 24th of the same, deserted at Montevideo 2 November 1767.

Beauregard Company: Louis LE GRAND. Worked aloft from 14 November 1767.

★(d) Nesle Company: Henry FÈVRE. Deserted at the Malouines 27 May 1767.

Nesle Company: Jean-Baptiste MONSAN. Sent to the Batavia hospital 1 October 1768, discharged the 10th of the same.

★(d) Braquemont Company: Jean-Baptiste GOGET. Deserted at Montevideo 3 November 1767.

Braquemont Company: Antoine PAQUET.

Fontaine Company: Jean CHAROT. Worked aloft from 14 November 1767. Sent to the Batavia hospital 1 October 1768, discharged the 10th of the same. Sent to the hospital at the Isle de France 10 November 1768, discharged the 16th of the same, sent back 17th of the same, discharged 7 December following.

Fontaine Company: Martin STEIN. Sent to the hospital at Batavia 1 October 1768, discharged the 10th of the same. Sent to the hospital at the Isle de France 17 November 1768, discharged 7 December following.

Beauregard Company: Connerat RHEIMS. Worked aloft from 14 November 1767. Sent to the hospital at the Isle de France 10 November 1768, discharged the 16th of the same, sent back 17th of the same, discharged 7 December following.

Beauregard Company: Louis-François METTANT.

Fontaine Company: Charles BIODET.

La Perelle Company: François LAMORY [and] François BINET.

Nesle Company: François RIBAUT.

†Vergines Company: Joseph Victoire RAPIN. Sent to the hospital at Brest 29 November 1766.

Braquemont Company: Charles ROSSIGNOL.

La Fontaine Company: Nicolas BELANGER. Sentenced to the loss of one month's pay on 15 June 1767 in accordance with the 1675 Ordnance Article 1039 for theft on board.

Vergines Company: Jean-Baptiste LAURENT; Nicolas VALON; Philippe PIVIER, the latter sentenced to the loss of one month's pay on 27 January 1767 for theft on board.

Supernumerary Warrant Officers [*Officiers Mariniers*]

Jean JAMAIN, from La Rochelle. Master Gunner. 50 *l. Raised to 55 l. from 24 October 1767.*

★Isaac LE ROY, from Noirmoutier. Coastal Pilot [*Pilote côtier*]. 50 *l.* Died 27 April 1768.

René LAISNÉ, *from Avranches. Second Surgeon.* 50 l.

François DOUARIN, from Lamballe. Assistant Surgeon. 36 *l. Raised to 40 l. from 24 October 1767.*

Sébastien BRISSON, from Marsilliac. Ditto. 32 *l. Raised to 36 l. from the same date.*

Gilles BRÉARD, from Valogne. Senior commissary clerk. 21 *l.*

Joseph AMELOT, from Nantes. Senior Valet. 21 *l.*

Louis AMONNEAU, from St Malo. Cooper [*Tonnelier*] 21 *l.*

†Jean PATOT (or PALOT), from Tours. Cook. 21 *l.* Disembarked at the Malouines 21 April 1767.

†Joseph NEAU, from Nantes. Master Armourer [*Maître armurier*]. 50 *l.* Disembarked at the Malouines 18 April 1767.

Butchers and Bakers

Etienne LEBOUL, from Nantes. Butcher. 21 *l.*

Michel MANGEARD (or MINGARD), from Alençon. Baker. 21 *l. Reduced to 18 l. from 24 April 1767.*

Pierre SAINT-MARC, from Alençon. Baker. 18 l. *Appointed Master Baker 24 April 1767 instead of Michel Mingard.*

Musicians

†Pierre KERMASSON (or KERMANON), from Mesquer. 30 *l.* Disembarked sick 11 November 1766.

†Martin FINET, from Orleans. 30 *l.* Disembarked ditto the same day.

Pierre LEGLAY, from Nantes. Sailor. Bagpipe player. 12 *l.*

†Guillaume DÉRICOURT, from Strasbourg. Hautboy. 50 *l.* Believed to have disembarked with permission at Montevideo in November 1767.

†Daniel DÉRICOURT, details and information as for the previous Déricourt.

Ship's Boys [*Mousses*]

★(d) Jean-Baptiste KERANGUIADER, from Guérande. 7 *l.* 10 *s.* Deserted at Brest 5 December 1766.

★(d) Jean-Jospeh DESCAZEAUX, from Nantes. Deserted at Brest 5 December 1766.

Alexis HECQUET, from Vannes.

Louis FRANQUELIN, from St Malo. 6 *l.*

Michel RODRIGUE, from St Malo.

†Pierre LEFÈVRE, from St Nazaire. Disembarked sick at Paimboeuf 14 November 1766.

Jean-Marie THERRU, from Lion.

Pierre HERPIN, from St Malo.

Gilles BRÉMONT, from St Servan.

Nicolas DUTERTRE, from St Servan. Rated sailor at 12 *l.* from 24 October 1767.

Jean BRIAND, from St Servan. Disembarked at the Malouines 19 April 1767, reimbarked at Montevideo 4 August.

Pierre COCHOIS, from Rennes. 7 *l.* 10 *s.*

Jean HERPIN, from Paimboeuf. 6 *l.*

Jacques MENÈS, from Guadeloupe. Taken on at Paimboeuf 15 November to replace Pierre Lefèvre. 7 *l.* 10 *s.*

Jean PINÉANT, from Brest. 7 *l.* 10 *s.*

Guillaume Etienne Marie THOMAS, from Brest.

[†Alexis DUJOUR, disembarked at the Malouines 18 April 1767. Another boy understood to have been on board for the voyage was Jean THIBEAUDEAU.]

Servants
For Mr de Bougainville
†Benoit FONGLAIR, known as Benoist de Juville. Remained with Bougainville's cousin at Buenos Aires.

Bernard DENIS, from Vire, Coutance diocese.

François-Nicolas LE GREC, from Neuilly.

Pierre LANGEVIN, from St Malo.

For Mr Duclos
Jean HONS, from Luxemburg. 12 *l.*

For the Chevalier de Bournand
Jacques FIÉ, known as Sully, from Touraine. 12 *l.*

For the Chevalier d'Oraison
Yves LOUÉDEC, from Tréguier. 12 *l.*

For Mr du Bouchage
Pierre LE CAM, known as Lapierre, from Guingamp. 12 *l.*

For Mr de Kerhué
†Jean BALAY (or BALLARD), known as Blain. 12 *l.* Deserted at Montevideo, then caught and transferred to the *Étoile.*

For Mr de Suzannet
Guillaume LE DANFF, from Comore, Vannes diocese. 12 *l.*

For the Prince of Orange Nassau-Ziegen
†Julien Guillaume BUARD, from Le Mans. Dismissed by his master at the Malouine Islands and transferred to a Spanish frigate.

Kitchen Servants

†Jean LEDOUX, from St Malo. Dismissed at the Malouine Islands.

Claude COLLET, from Le Mans.

†Michel ERARD, taken on at the Malouines instead of Ledoux, exchanged in November 1767 for Henry MESLAY, one of the *Étoile*'s cooks.

In addition I took on three Blacks for my service.

Replacements taken on at the Malouines
Warrant Officers

†Louis RIBOUCHON, from Plumels, Department of Vannes. Master Armourer at 50 *l.*

†Jean LÉCRIVAIN, from St Malo. Assistant Gunner at 24 *l.* Sent to the Batavia hospital 5 October 1768, discharged the 15th of the same month. Sent to the hospital at the Isle de France 23 November 1768, discharged the 30th of the same month.

†Pierre JUEL, from St Malo. Leading Seaman at 24 *l.* Sent to the hospital at the Isle de France 9 October 1768, discharged 7 December 1768.

†François DUVAL. Ditto at ditto. Sent to the hospital at the Isle de France 1 December 1768, discharged 5 December.

†Pierre GILET. Ditto at ditto. Sent to the hospital at the Isle de France 1 December 1768, discharged 5 December.

†Joseph SIRE, Ditto at ditto. Sent to the hospital at the Isle de France 9 November 1768, discharged 6 December.

Seamen at 20 *livres*

†Louis CADET, from St Servan.

†Etienne LE BLAS, from Triguet, department of St Brieux. Disembarked 13 July 1767 at Rio de Janeiro to go over to the *Étoile* where he has been working since the 14th.

†François COUSIN, from St Malo. Sent to the hospital at Batavia 1 October 1768, discharged on the 10th. Sent to the hospital at the Isle de France 9 November 1768, discharged 1 December.

Seamen

†Louis BOUCHER, from St Malo. 18 *l.* Sent to the hospital at the Isle de France 9 November 1768, discharged 10 December.

†Louis LE RAY, from Paramay [Paramé], department of St Malo. 15 *l.* Disembarked at Montevideo 28 September 1767 with permission.

†Jacques RODEC, from La Rochelle, apprentice seaman at 12 *l.* Sent to the hospital at the Isle de France 9 November, discharged 5 December 1768.

†Julien BEGUIN, from St Malo. Ditto.

†Jacques BATAILLE, from St Malo. 12 *l.*

Boys at 7 *l.* 10 *s.*

†Jean VEILLON, from Pleurtuit, department of Dinan. Sent to the hospital at Batavia 5 Octobre 1768, discharged 15th of the same month. Sent to the hospital at the Isle de France 1 November 1768, discharged 10 December following.

†Laurent BERDIER, from St Colomb.

†Pierre DUBAYE, from Landrecy in Picardy.

Passenger

†Michel-François de NERVILLE, formerly commander in the Malouines. Disembarked at Montevideo 10 November 1767.

Captain's Servant from 21 April

†Michel HEVARD, from Chalandray, Avranches diocese. Disembarked at Montevideo 12 November 1767 to transfer to the *Étoile*.

People embarked at Rio de Janeiro

★Mr Jean-Robert Suzanne LEMOINE MONTCHEVRI, from Rochefort, *volontaire* at 18 *l. Promoted Garde de la Marine 18 August* 1767. Died at the Isle de France 15 November 1768.

†Pierre-Antoine VÉRON, pilot observer [*Pilote observateur*], from Rouen, at 70 *l.* Disembarked at the Isle de France 30 November 1768.

★(d) Joseph RIVIÈRE, seaman, from Souillac in Quercy. Deserted at Montevideo 29 October 1767.

†Laurent LEFEBVRE, from Angoulême, boy at 7 *l.* 10 *s.*

Black Servants

†Jean RUBI, servant to Mr Le Corre. Disembarked at Montevideo 17 August 1767.

†Jean, servant to the Chevalier d'Oraison. Disembarked ditto 1 November 1767.

†FRANÇOIS-XAVIER, servant to the Chevalier de Kerhué, *Garde Pavillion de la Marine*, acting officer.

Others

★(d) Jean Vanture PARIS, from Marseille. Drummer. Deserted 10 October 1768 at Batavia.

†François COURTEIS, Spaniard, sailor from Cadiz. Found on board 13 July 1767. Disembarked at Montevideo 2 August 1767.

†François CARCIR, from Ferrol, sailor. Ditto 2 August 1767.

People taken on at Montevideo

†Joachim GAUGER, from Buenos Aires. Boy at 7 l. 10 s. *Rated seaman at 12 l. per month on 24 October 1767*

†Philippe CHARLES, *from St Servan. Boy. Taken on 11 September 1767. Disembarked at Montevideo 3 November 1767.*

†Jean-Baptiste COMMON, from Paris. Soldier. Disembarked at the Isle de France with permission on 10 November 1768 to serve in the Legion.

†François-Henry MELAY, from Pierrefont, diocese of Soissons. Captain's servant. Sent to the hospital at the Isle de France 18 November 1768.

Taken on at the Island of New Cythera

†LOUIS, Indian sent by his people.

Taken on at Buru

†Nicolas CHAPELLE, department of Fécamp, found on board. *Seaman at 18 l.*

People found on board of the said frigate on her departure from Batavia 16 October 1768

†Etienne MASSAR, from Wardin in Lorraine, aged 28, drummer from Batavia. Disembarked at the Isle de France 10 November 1768 to serve as a soldier in the Legion.

†Jean LESCAR, from Castres in Albigeois, aged 25, soldier at Batavia. Sent to the hospital at the Isle de France 10 November 1768.

†Jacques Nicolas COSSARD, from Paris, parish of St Marguerite, aged 23, carpenter.

†François-Juste MICHAUD, from Dôle in Franche-Comté, aged 22, grenadiers sergeant at Batavia.

†Jacques JUBLIN, known as CHEVALIER, from Arras in Artois, aged 36, soldier at Batavia. Sent to the hospital at the Isle de France 10 November 1768.

†Jacques LESCAR, from Castres in Albigeois, aged 27 soldier at Batavia. Disembarked at the Isle de France 10 November 1768 to serve in the Legion.

†Jacques MEUNIER, from Arlan, jurisdiction of Ossonne, aged 33, soldier at Batavia.

†Aubin SOLEAU, from Angers in Anjou, aged 28, soldier at Batavia.

†Jean-François GIRE, from Namur, aged 24, stonemason at Batavia. *Rated seaman on board 10 November 1768 at 12 l. a month.*

†Martin RAYMOND, from St Martin de Castillon in Provence, aged 33, carpenter at Batavia. *Rated seaman on board 10 November 1768 at 12 l. a month.* Sent to the hospital at the Isle de France 25 November 1768, discharged 7 December following.

† Grégoire MARTIN, from Georges in Dauphiné, aged 30, soldier at Batavia. Disembarked at the Isle de France 10 November 1768 to serve in the Legion.

†François PALLIAUD, from Cousance, diocese of Châlons in Champagne, age 30, soldier at Batavia. Sent to the hospital at the Isle de France 10 November 1768, discharged 9 December following to serve in the Legion.

†François LE ROY, from St Barbe in Lorraine, aged 25, sailor at Batavia. Sent to the hospital at the Isle de France 9 November 1768, discharged 26th of the same month, *reimbarked as seaman the said day 26 November at 18 l. per month.*

†Jean DUGAISE, from Oscot in Flanders, near Ghent, aged 22, house carpenter at Batavia. *Appointed sailor on board 10 November 1768 at 12 l. a month.*

†Florent MESSINO, from Langres in Champagne, aged 27, stonemason at Batavia. Disembarked at the Isle de France 10 November to serve in the Legion.

†Pierre DELATTE, from Argenteuil, 2 leagues from Paris, aged 17, soldier at Batavia. Disembarked at the Isle de France 10 November 1768 to serve in the Legion.

Replacements and increase in crew numbers at the Isle de France

†Jean LEVEILLE, servant to Mr Le Corre, instead of Jean Rubi. Taken on 19 November 1768.

†Louis LA BICHE, from Brest, helmsman, taken on 28 November 1768.

†Pierre STANDELET, from St Servan, Second master, taken on 8 December 1768.

†Oliver BALUET, from St Malo, assistant carpenter, taken on ditto.

†Joseph JOANSOUH [= JOHNSON?], from Caux, English sailor, taken on ditto.

†Thomas PHILISSE, from Sousbéré [= Sudbury?], ditto, taken on ditto.

†(d) Barnabé FILDEN [= FIELDEN?], from London, ditto, taken on ditto. Comes from the *Ambulante,* Deserted at the Cape of Good Hope 16 January 1769.

†Jean BOUTONS, from St Malo, boy, taken on as ditto.

†Mr des LONGRAIS, clerk, replacing Mr de St Germain, taken on 9 December 1768.

†Louis PORUS, boy, taken on as ditto.

†AZA, black for the Prince of Orange, taken on 9 December 1768, disembarked at the Cape of Good Hope 9 January 1769.

†PIERRE, creole black for Mr de Bougainville, taken on 9 December 1768.

†Joseph HERVELLE, from Port-Louis, senior pilot replacing Jean Riodujour. Taken on 10 December 1768, comes from the *Laverdy*.

†Nicolas MÉNAGE, from Dinan, boy, taken on 10 December 1768.

APPENDIX 2

The Journal of Caro

Jean-Louis Caro, second-in-command of the Étoile, *left a 'Journal du Voiage du Tour du monde par Caro' in twelve volumes or 'cahiers'. This is not a true shipboard log, but an account written up in France after Caro's return, which incorporates much of his daily log. There are long gaps in the account of the expedition, with on several occasions a summary of events; at other times simply omissions. Part of this manuscript has been lost; what remains is held in the Archives Nationales, Paris, under the reference Marine 4 JJ 1, No. 5.*

Included among his papers is a letter to an unnamed correspondent, dated 16 May 1767, summarizing the first part of the voyage, from the island of Aix to Montevideo. He also enclosed a letter to be forwarded to his wife, and requested his correspondent to report on his behalf to the Compagnie des Indes, thus ensuring that he retained contact with it, as his nominal employer, so that he would find it easier to resume service when the Bougainville expedition came to an end. He was in fact welcomed back and appointed as second-in-command of the Duc de Duras *in January 1770. It is something of an indication of how slow methods of communication could be in his day that the letter is recorded as having been received in Paris on 14 February 1768.*

His journal, starting with his departure from South America in July 1767, consists largely of navigational details, with a few details on meetings with the Patagonians and Fuegians in the Straits of Magellan. There is a gap between the point at which the ships emerged into the Pacific Ocean, with the singing of the 'tédéome', and the first sighting of the 'islands of Cuiros' on 22 March 1768.

His journal reflects his down-to-earth nature and shows someone who, as Étienne Taillemite commented in his Introduction, brings to mind Montaigne's 'simple and unsophistic-taed man whose background ensures the veracity of his testimony': in other words, being free from the extraneous comments added to their reports by men like Commerson and even Bougainville, Caro's journal is thus a straightforward and useful source of information on the progress of the expedition.

Through the Tuamotus

21 to 22 March [1768]
Sighted the islands of Cuiros [sic]

    ... At 6 a.m., the look-out of the main top who was on watch said that he could see land which we soon sighted from the poop-deck bearing from us SE¼S 4 leagues, which seemed to be several islands, the look-out counted 7 or 8 that looked like islands but very small, visible in fair weather from 6 or 7 leagues. Shortly after, the look-out said he could see another island ahead of us that we [also] detected from below, bearing W 5°S. We promptly signalled land in sight to the *Boudeuse*, and we steered W¼NW and WNW to pass N of the said island following the *Boudeuse* which was about a quarter of a league ahead of us.

At 8 a.m., the said island bore from us WSW about 8 leagues, which appeared to us to be very wooded, the whole shore of a very fine white sand, the island moderately high, visible from 6 or 7 leagues. The sea breaks a great deal at the SE point, there may be from the first islands we sighted to this one a distance of approximately 11 leagues SE and NW.

At 9 a.m., the *Boudeuse* signalled to us to take soundings, we ran out 200 fathoms of line, no bottom. Then we filled all sails, making for the WSW. At 9.30 we were a short half league from the island, enjoying greatly the greenery of the trees and how wooded this island was, we saw at the same moment two blacks running along the shore on very white sand, shortly afterwards 15 to 20 blacks came out of the forests who ran along the seashore then returned into the forest. The *Boudeuse* hove to for a moment, then she set her sails and steered W as we did. We saw two blacks who had stayed on the shore who each carried a small switch, who were gesturing and beating the sea with their little sticks. We saw 3 or 4 huts at the edge of the woods. Through the spyglass these blacks seemed to be quite naked. There are also some breakers off the most N point which is the one we ranged along the closest to. It seemed to us that the sea was breaking all around the island like on a bar. I think a landing would not be easy. I doubt that this island from what I saw of it is more than one good league round, at the most. It is very low, only the trees make it visible from the distance I have recorded.

They lit fire on the most N point; approaching this island it seemed to me to lie SE and NW. We were quite surprised when we saw those two men. Never would we have imagined that a small island like that one could be inhabited. Who the devil went and placed them on a small sandbank like this one and as far from the continent as they are and surely not many vessels would have passed this way since Cuiros passed here in 1603, we are the first in 165 years who have passed by…[1]

Wednesday 23.

… At 6.15 a.m., we sighted land bearing from us N and N¼NW 5 leagues. Shortly afterwards, the weather cleared a little, we saw it stretching from NW¼N to N¼NE. All this land seemed to us to be very low and drowned, we could only see small clumps of trees separated from each other. We did not expect to see land so soon, we reckoned that from the land which we saw yesterday to the nearest one there was 80 leagues according to what the French chart indicates and since yesterday's midday readings to 6 this morning we have covered only 21⅔ leagues W. This shows that when one is in places that are not known one must always go carefully during the night and carry little canvas. We continued on our NW route, the wind ESE light gale light rain until 8 a.m. when I sighted the most E land bearing N¼NE 4½ leagues, the most W extremity NW3°N 5 to 6 leagues….

[1] Quiros sailed near, but did not see Akiaki. Bougainville named this island 'The Lancers [l'isle des Lanciers]', but Caro's reference to switches and small sticks gives a less imposing image. Yet the island spear or tao made of coconut wood or pandanus usually measures 10 to 15 ft. As for Polynesian migrations, the French could not, at this stage, imagine the distances the islanders could cover in their canoes or the extensive pattern of island settlement.

23 to 24 March.

... We saw in this area several canoes sailing, we counted up to 6 and 8 blacks in each canoe, their sails seemed to be a long and dun-coloured square. They landed on different islands, we also saw two kinds of small boats at anchor near one island, all these boats were on the other side of the bar, we saw none on this side ... The entire coast is full of terrible reefs and breakers, the sea seemed quite low along the shore, there are large stones and large corals that protect these islands from the sea, without this I believe there would have been no sign of these islands for a long time. It is here according to Cuiros's memoir, that he saw land in 18° 43′ and according to his description of it, it corresponds to it without any possible error. He saw the inhabitants of these islands drink only water from coconuts, not having any drinking water anywhere on this land.[1] We agree, it is not possible in a land that, from what we saw, was not more than ½ a cable-length in its widest part, that one could find water fit to drink. One can also see the sea beyond the most N bar. We continued our WN¼W course under full sail, the look-out said he could still see land extending to the NW, this worried us, we reckoned that it was a continent and that we would find it quite difficult to extricate ourselves. Finally at 3 o'clock, the look-out said he could no longer see the land continuing, that he could only see the same part of it, but that the most N point of the bar extended further on the N side, this put our minds at rest and made us hope that we would find a pass to get out of this sad place that has nothing but sand like the land we saw yesterday... [At 6 o'clock] spoke to the *Boudeuse* which was hove to and told us that we were going to try to stay under light canvas overnight until tomorrow at daybreak and that we would make for the land and that he planned to send a boat to shore if it could land to try to obtain some information about the inhabitants of this island and about the land itself. We hove to starboard to the wind, double reefed the topsails lying to N to NNE until 7 p.m. when the *Boudeuse* and we made for the S¼SE with the two topsails and the mizzen topsail, light E breeze, calm sea. There was a great deal of lightning and a little thunder during the night. We continued on the S¼SE tack until 3.30 a.m. when we altered tack and sailed N¼NE with the same rig. At 5.15 we let out the reefs, added canvas, steering N¼NW. At dawn we saw the land bearing from us N to NNE approximately 5 leagues which is the same land as yesterday evening. Kept to a N¼NW course to try to reach the land, the winds ESE and E fresh breeze, fair weather, fair sea. At 7 a.m. the *Boudeuse* which was a little ahead of us bore away to W¼NW and immediately set her studding sails, we did likewise delighted that she had done that. If we had sent our boats ashore, we would have lost a day and maybe without being able to step on land whereas now until nightfall we shall progress W which will enable us to see if we come upon any other land and endeavour to find a good land where we might obtain water and a few supplies...

I believe following the memoirs of Mr Cuiros that this island we have just seen is the same as he saw and coasted along the N part. But the islands we saw the day before yesterday, I do not believe that he saw them, there is too great a difference in

---

[1] This belief was expressed by other travellers because of the lack of watercourses on atolls. However, wells could be dug where water, brackish at first, then drinkable, could be obtained.

latitude and he saw land only in 18° 40′ and they are marked in 20° 00′.[1] This island seemed to me according to the track we followed from midday to 6 p.m. to run or lie ESE and WNW approximately 11 or 12 leagues long. From what I saw there is no danger in ranging quite close along it, all the dangers are no further from land that ½ a league. We saw no shallows or any danger. As for a landing it seemed very difficult and I believe that we would not find one on the side we sailed along where we could go ashore. According to the French chart it shows a small island 20 leagues ENE of the most E point of the land that Cuiros shows in 18° 30 to 35′ which bears from me at midday W 5°S distance 50 leagues, which could be the island we saw the day before yesterday which is situated fairly close to the route we covered yesterday. I sighted at 9 a.m. the most W land bearing NE¼E 5 leagues and we lost sight of it at 9.30....

Friday 25 March

At 5.30 a.m., we saw land from WNW to S distant 5 to 6 leagues, all this land low and drowned. We shook out the reefs of the topsails, set more sail and steered NW and NW¼W, from 6 to 8 a.m. we had both calms and rain. The wind freshened from the NNE to E at 8, the weather cleared, we steered NW and NW¼W all sails set. This land, seen from the topmasts, looked the same as the one sighted yesterday, lined by reefs all along the shore and also submerged with isolated clumps of trees. We are weary of seeing those [islands] so often and so close to each other.

This last one seems larger than the one we left yesterday morning, we consider ourselves very lucky that all these islands and stretches of land are as safe as they are and that we are not finding the shoals or other dangers one fairly frequently comes across near land as low as this.

According to the latitude I observed at midday, Mr Cuiros did not sail among these islands nor sight them, he says that he saw several islands in latitude 18°43 to 48′. At midday I estimated the run these last 24 hours WNW at 18 leagues. Observed latitude S:17°54′, estimated S: 17°58′. Calculated longitude 142°00′.

At midday I sighted land to the S distant 4 leagues from the deck, but from aloft they can see land stretching further to the SSW as far as the eye can see, I do not know whether this is an island or more extensive land as is shown on the French chart. All I know is that this is not the land seen by Cuiros, as far as a large stretch of land is concerned that cannot be, because Cuiros coasted along several islands in 18°45′ to 48′ and others he sailed past further south.

Sunday 27 March

Note: Since the time we sighted the first islands when he saw that one yesterday evening, we have covered a distance of 80 leagues always within sight of land.

The Call at Tahiti

Sunday, 3 April

... This island seems very large and very high to us, I believe it to be as high as

---

[1] Caro's deductions are broadly correct in spite of his own erroneous latitude. The French sighted Vahitahi and Akiaki, neither of which was seen by Quiros. Vahitahi is in latitude 18′30′ and Akiaki in 18°30′ and Akiaki in 18°05′.

the island of Bourbon or Mascarin,[1] the S point forms a large cape, the N one gradually falls away and appeared to be low. We believe this land forms two islands; as far as we can tell from Round Island at the most S extremity of this island, I reckon it can be 20 leagues, they lie according to our dead reckoning from midday yesterday to today midday in ESE and WNW.

Continuation of Tuesday 5.

Various comments [on what] we saw yesterday afternoon, namely: At 2 p.m., we saw a small canoe crossing ahead of us in which there were two men paddling. When they were a little to windward, they stopped to examine the vessel. Some time later, another canoe with one man in it also stopped to examine us. The *Boudeuse* which was a short league ahead of us seemed to be surrounded by several and some came up alongside. The closer we got to land the more canoes we saw leaving the land and coming towards us and the *Boudeuse*, most of them paddling and few with a sail. At 4 o'clock, the sea was covered with canoes. At 5 p.m., we had at least twenty alongside. They came with the utmost confidence, they had in their canoes coconuts and bananas and some fruit like large pears that had a slightly bitterish taste.[2] The others had also some fruits that we believe are those Mr Anson indicated in his memoirs.[3] Several ate some of them raw and others cooked, this fruit has a taste close to that of the [sweet] potato which is very doughy and could serve as bread in case of necessity.[4] All these Savages appeared to us to be most affable and in no way embarrassed. They are all robust, of average height, well built in appearance, some are mulatto coloured, some whitish, others reddish and the rest black. They all have frizzy hair, they usually wear a long beard like the capuchins. They had children of 7 or 8 in their canoes. Their canoes all have outriggers, those that do not [sic] are two tied together. These people are all naked having only small pieces of cloth to cover their nudity.

Note: There was in two canoes tied together a Savage who gestured with his hands, his feet and his entire body that he greatly wished to come aboard. The canoe came alongside, even though we were making more than a league an hour, and this man caught the rigging chains and climbed on board and pushed the canoe out to sea. He wore 3 or 4 loincloths of white cotton with a hole for his head to go through, it was like a great shirt sleeveless and opened on both sides. This man did not seem surprised to be among us, he began to walk around the ship and sometimes looked with an expression of astonishment. It seems that he is above the others, when he talks to them, they do right away everything he has told those who are in the canoes.

[1] Bourbon is now known as Réunion, situated in the Indian Ocean close to Mauritius, part of the Mascarenes group (Caro's 'Mascarin'). Its highest peak is the Piton des Neiges at 3,069 m (10,072 ft); the highest point of Tahiti is Mt Orohena at 2,241 m (7,355 ft).

[2] Identification is fraught with risks. Martin-Allanic believed it to be the avocado (*Bougainville*, I, p. 659n), but this is unlikely. It may have been the Yellow Apple or Vi (*Spondias dulcis*). Banks describes it as 'not unlike an apple which when ripe is very pleasant' (Banks, *Journal*, p. 342). Its flavour is somewhat akin to that of a quince, which seems to correspond to Caro's 'slightly bitterish taste'.

[3] George Anson, in 1742–3, did not sail anywhere near the Society Islands, but kept to the Northern Pacific. An account of his expedition, drawn from his papers, was compiled by Richard Walter and contains details of fruits and other supplies found on Tinian and other northwestern Pacific islands, which explain's Caro's comment.

[4] The breadfruit (*Artocarpus altilis*), locally known as the *uru*.

We ranged along the land a short league away, which seemed to be an excellent land very cultivated and seemed highly populated all along the seashore. Their houses struck us as very well made and very pretty, the prospect is charming but anchorages are not many. The coast seemed to be very safe and quite steep to all along the shore, close to land there is a small bar where the sea breaks.

Note: At 6 p.m., we signalled to our Savage boarder to call a canoe over to take him and go ashore. He gestured that he wanted to stay on board and told the canoes to bear away. We made him understand that we were going to make for the open sea, he made us understand that he would come with us and that he wanted to stay on board.

In the night, all the canoes went back to land and this morning at daybreak left the shore and came alongside with coconuts, bananas, small pigs, a little fish and a few shells. In one canoe there were two young girls of 13 or 14, fairly passable and a little whiter than the most whitish men; they appeared fairly affable and I believe that they are not jealous, all the men who are in the canoes all ask us for women. Our boarder seems to desire one greatly, he looks at everyone to see if he could not find a woman, he likes being on board, he does not seem to want to return to land, I believe that he knew about muskets, when he saw some in the main room, he began to say: poux, poux and gestured that it made people die.

We saw 3 or 4 canoes that had at the stem a figure rising by about 3 feet, made of wood that showed only a head and the remainder being carved in open work. All the canoes had on board a small one, I mean a small banana tree they gave upon coming alongside that we take for a sign of peace.

None of the canoes had weapons of any kind whatsoever. It seems that they live in a great state of peace[1] and do not seem harmful and if we can find an anchorage, we shall have nothing to fear from these people. Their canoes are sewn with coir like the Pondicherry *chelingues*,[2] they make a great deal of water so they are forced to bail them out continually. These people swim like fish. I have seen several canoes filled with water, this does not trouble them.

5 to 6 April.

... At 6 a.m. the *Boudeuse* and ourselves made for the land, we sent boats ahead of the vessels and followed the shore to find an anchorage between the reefs, lacking everything, especially water. Fortunately at 11 a.m., our boats signalled that they had found an anchorage between the reefs in 12 and 15 fathoms, bottom of sand and coral a little ooze.

The *Boudeuse* was first to enter and anchored between the land and the reefs where the sea breaks a great deal and we ranged along close to the starboard side till we could see the ground under us but we had 4½ fathoms of water, all the reefs were clear of the water it being low tide. We went further in than the *Boudeuse* and dropped anchor at 2 p.m. We are 2 cable lengths from land and a similar distance from the reefs.

---

[1] Caro was not to know that Wallis's visit in the *Dolphin* in the previous year had sufficiently impressed the Tahitians with the superiority of European weapons to ensure a peaceful reception for their next visitors. Ahutoru, the Tahitian 'boarder', had obtained his knowledge of guns from the same visit.

[2] These small boats, used on the Coromandel coast, were constructed of small planks tied with coconut fibre or similar binding material.

We moored with the bower anchor having the cable veered out and the end secured to the bits, the second bower anchor is ready with another kedge anchor,[1] our starboard anchor is in 22 fathoms, ground of grey oozy sand, the port anchor is in 10 fathoms same ground. Moored SE and NW like the coast when we are at anchor. We are moored opposite 3 small rivers two cables from us, the water seems very easy to obtain. I hope that our crews will recover soon, if we can give them some refreshments for they are in great need of them. We have several cases of scurvy. There are 3 little sandy islets bearing N and NNE on the reefs, one of them is very wooded and the two others less so, they are inhabited by a few Savages.

From 7 to 8 April.

In the afternoon Mr de Bouguinville[2] sent his soldiers ashore with tents and baggage to settle on land. When everything was landed, the king of the locality with the chiefs made some gestures and made Mr de Bouguinville understand that they could not remain ashore at night and that in daylight we could do what we liked and they took themselves the baggage to put it back in the boats. Mr Bouguinville after a great deal of trouble made them understand that we were coming here merely to get wood and water and, having given them 18 piastres, I mean 18 small stones, and shown them 18 times the journey of the sun and indicated that then we would leave. They gave 9 stones and made us understand that they wanted to allow only 9 days and that we could sleep ashore, right away we erected the tents. These people seem to be uneasy and afraid of us.

The king and a few chiefs stayed during the night in the tents with Mr de Bouguinville and his gentlemen. During the night there was a light land breeze, no rain. In the morning, they worked to send ashore the majority of our barrels to get them well rinsed and filled in order to rid them of the bad smell the water of Magelant has given them. We have in this place 3 or 4 rivers whose water is excellent coming down from the mountains. There is no lack of canoes alongside with several types of food such as coconuts, bananas and other fruit, a little of different types of rock fish, crayfish and large crabs that are excellent. They all ask for nails, that is what they want the most. One has to beware of them, they are thieves, they took several items through the gunports and the windows of the cabins. They climb up alongside and steal everything they can lay their hands on. In daytime the winds have been NE to ENE since yesterday little breeze fair weather, high temperatures.

From 8 to 9 April.

We sent 20 men ashore who show signs of scurvy. The king gave a large house shaped like a store for Mr de Bouguinville and we took down the tents. Everyone is in the store. All their houses are in the same style, they are 50 to 60 feet long, very well covered with latania leaves. From the roofing to ground level, it is open work with small bamboos or rushes. They have 5 or 6 rafters holding the roof which is 12 to 15 feet high. The king's house is built exactly like the others. They do not wish us to have anything to do with their wives, it seems that they are very rigid over this matter, but

---

[1] Meaning apparently that a hawser was tied to a large anchor and linked to a lighter anchor, which thus prevents the larger one from dragging.

[2] Caro's spelling is wildly erratic. This is his version of Bougainville.

as for the girls they are available for anyone and this is also done in front of everyone in the open air. Most of these Savages have their buttocks painted black and several marks of the same kind on their hands, their arms and other parts of their bodies that are fairly close to marks made in France on arms with powder but here, I think it is done with the sap of some trees that they make these stains which I believe to be signs of distinction among these savages. During the night the wind freshened good E to NE breeze, the weather squally, the sea very rough over the reefs. Along the shore one can land only with difficulty, there is a small bar along the shore, the boats have to anchor off shore and send a towline ashore. During the night the Savages stole several items from the house in spite of the sentries, it is certain that if we do not provide a better guard they will steal everything they find and what they will be able to snatch. The king is the first and greatest of the thieves.

9 to 10 April.

We are working hard to obtain our supplies so as to leave, this roadstead is not too safe, yesterday at 8 p.m. Mr Duclaux came to tell Mr Giraudais on behalf of Mr de Bouguinville that we had to hurry and that he expected to be out of here within 8 days and feared some treason on the part of the Savages who had sent their wives and children into the mountain. During the night they did not stop throwing stones at our people who were guarding our barrels and on the house where everyone is lodged, they are devilish thieves. We do not want to fire on them but I think it will be necessary to do so. They are not to be feared because with a musket one would make them run a long way. We are working hard to collect our water. Mr de Bouguinville gave orders to fire on the Savages this evening if they do what they did the previous evening.

Note: our Savage boarder is still on board. He sleeps and eats there, he is as great a scoundrel as the other Savages.

10 to 11 April

During the day, the winds blew from NE to E light breeze, fairly fine weather and during the night the winds came from the land. Yesterday at 10 or 11 a.m. a Savage was killed by a pistol shot. We do not know who is responsible whether it is men from the *Boudeuse* or our people, however our men who are ashore carry no firearms or other weapons. During the night the Savages were very quiet, they did not throw any stones. We have bought about 80 hens since our arrival, they are not common. As for pigs, we find them in great abundance but lots of small ones.

Note: our Savage boarder left yesterday morning. We are quite pleased with this. The Savages no longer want nails, they want axes and scissors but we have none.

At 5.30 a.m., the *Boudeuse* drifted onto us on the port side which fortunately caused us no other damage than a broken studdingsail boom. We let out our cable on the SE, her cable was cut 2 or 3 fathoms from its fastening. We are in a very poor anchorage, the ground is full of corals, we must make haste to leave.

12 to 13 April.

Yesterday we placed in irons in the quarters four soldiers from the *Boudeuse* who had killed 3 Savages with their bayonets. All the Savages, men, women and children fled weeping into the mountains, all the canoes left at a signal the Savages ashore were making to them.

The Prince of Naseaux who was ashore came to warn Mr de Bouguinville who was on board his vessel and went ashore. Shortly after, we landed the chaplain. Mr de Bouguinville wanted to draw lots with the 4 soldiers to hang one of them, but as it was almost night, Mr de Bouguinville had them taken to the ship and placed in irons. All the women and children were retiring weeping and to everyone from the two vessels they met on the way they said *tailliot* and *mateaux*, which means 'friends' and 'they kill us'.

All this did not prevent two muskets in the night and the cauldron from being stolen from the surgeon's quarters and several other items. We fired several musket shots, but no one was caught. At 7 a.m., the two cables of the *Boudeuse* were once again cut. She was holding only by one cablet. We had to give two kedge anchors and our remaining cablets to moor the *Boudeuse* which only had her large anchor left. They could save only one anchor (the SE one) and have lost the NW one. We sent our boat to take soundings and see whether they might find a pass in the NW between the reefs and the land. At 11 o'clock our boat returned, they had found a passage that is much larger than the one we came in by and a very good bottom of sand and mud and a very good anchorage between the two islands that are on the reefs and the shore and a good distance from each other and not like here where we could not drift without making straight for the land and in the waves.

In the morning the Prince of Nasaux went to get the king and all his Savages and he made peace. They all came back each carrying a palm in sign of peace.

13 to 14 April.
We have finished everything and worked on avoiding the *Boudeuse* which is very badly moored. We veered our SE cable and heaved in our NW one.[1] We laid out a kedge anchor in the N, we took on all our people. At 6 p.m. everything was on board. During the night, the weather was fairly good, light land breeze. At 8 o'clock Mr de Bouguinville asked us if we were ready to weigh anchor. At 3 p.m. [sic] we replied that we would be ready at midday but could not get our SE anchor over which he was moored. At 10.30, Mr de Bouguinville again asked whether we could weigh anchor. We told him we could on condition that we left him our anchor and the NW cable which we would cut. As it was a spliced cable that was at the hawsehole, he sent us his longboat with a cablet, we secured it to our cable which we cut and left him the cable and anchor, then we raised our other anchor and left our longboat to raise our kedge anchor.

We sailed between the reefs and the land where we found the finest channel in the world with no dangers whatever going quite close along the reefs that are very easily seen, the winds ESE light breeze, fair weather, smooth sea. We had our boat ahead of us with Mr Leroy, the officer who had gone to survey the pass. We went through an area where we found another fine anchorage that looked similar where there must never be any movement of the sea being quite sheltered by the reefs and two little islands that are close to each other which are small sandy islands with a few

---

[1] I.e. to move the ship towards the NW anchor.

trees on them. There are 3 of them on these reefs, one other with the two I have just mentioned. We made approximately a good league between these reefs and the land steering N to N¼NW. At midday we had passed the end of the reefs; between this point and the land there is a pass of a good league and a ½ and we found ourselves in the open sea...

Saturday 16 April.

... Mr Lavary told us that since our departure from the island of Cythera, Mr de Bouguinville had drawn up and taken possession of the said island in the name of France and that he had named it New Cythera, that he had accordingly drawn up a document, that he had placed it in a well corked and sealed bottle and by night had buried it in the great hut that we had occupied as well as an oaken plank or post which is also buried with an inscription on the said plank in large block letters.[1]

Note: he has in his frigate our Savage who was 3 to 4 days on board our ship whom the king of this locality asked him to take and recommended him to Mr de Bouguinville as the brother of another king from another place. The king took him to the *Boudeuse* with 3 or 4 canoes loaded with a gift of bananas and coconuts. Mr de Bouguinville received him and promised to the local king to take all possible care of him. Mr de Bouguinville has named him Louis de Cythère.

Off the Samoan Islands

2 to 3 May.

... at 6 a.m. we sighted land bearing NW5°W to WNW5°N distance 15 to 16 leagues that gave the appearance of a very high land ... I am led to believe that the island we can see is the island of the Beautiful Nation [l'isle de la Belle Nation] or some other new island; as for the islands that are in our latitude as observed at midday, the nearest is the island of Hore [l'isle de Hore] 400 leagues away; as for Cocos Island [l'isle Cocos], we are 300 leagues away and that island is near 15° of latitude on the chart.[2] It appears more likely that these islands are the island of the Beautiful Nation than any other. It may be that it is badly shown in the chart by 45 to 50 minutes and even one degree, which would make me according to my dead reckoning only 3° 26′ too far W, which could well mean that since we left the strait we would have that difference or that it is, as I have already said, an island that has not been seen by navigators who passed a short distance from where we are further N or 15 to 20 leagues further S.

---

[1] This passage is somewhat garbled: Bougainville drew up and buried the Act of Possession before his departure. Caro obviously means that Layary told the officers of the *Étoile* after that ship had sailed about the commander's actions. Étienne Taillemite further points out: 'This comment is somewhat strange. Caro, like all the officers from both ships, signed the Act of Possession of New Cythera, he was therefore aware of its existence prior to the report from Lavary-Le Roy.' ET, *Bougainville*, II, p. 333.

[2] Caro's spelling may explain his 'isle de Hore'; it is probably 'l'isle d'or', 'the island of gold', a reference to the Solomon Islands and their reputed riches. On Robert de Vaugondy's map of the Pacific Ocean, drawn for De Brosses's *Histoire des navigations aux terres australes* published in 1756, an 'isle de la Belle Nation' appears east of the Solomons; it was Isla de la Gente Hermosa, a name given to Quiros's La Peregina, and is the island of Rakahanga. Cocos Island, also shown on Vaugondy's map, is Tafahi, discovered by Le Maire and Schouten in May 1616. The context of Caro's comments makes it clear that he was approaching the conclusion that Gente Hermosa was a substantial group of islands.

Wednesday 4 May.

 ... From 8 o'clock to midday we sailed NNW to NNE to round the E point of the large island that we ranged along a good half league from shore. This island seemed very safe and quite sheer, anchorages do not seem to be common around here. One would need to be very close to the land before finding ground. There is a great deal of smoke on the heights. We have seen several huts lower down quite close to the sea and some people walking along the seashore in the E part. We saw a very fine sandy cove but the sea appeared to be breaking all along the coast. I believe that a landing would not be too easy. Using the spyglass the island seemed to be very wooded on the low areas, numerous coconuts, banana trees and palm trees. The island is very high being visible from 18 to 20 leagues. It may be 7 or 8 leagues round, the other two islands are much smaller but almost as high. At 11.30, we saw several canoes along the shore. We continued to sail at the same distance from the land which seemed very safe without any dangers and very well inhabited and with the appearance of a very good island...[1]

4 to 5 May.

 ... At half past midday, a canoe came up in which there were 5 men. Shortly after, 3 or 4 others came up, we were sailing under the two topsails. We hove to for about ¼ of an hour, the *Boudeuse* was hove to ahead of us. These Savages are reddish, they have their buttocks painted like those of Cythera but not such handsome men. I do not believe they speak the same language, they do not even have the same tone of voice, they have long hair but as frizzy as those of Cythera, nor do they have a beard. They all seemed to be shaved, they seemed more mistrustful. When we gave them a rope to hold onto alongside, they took it only fearfully and did not remain alongside for long. They kept away from the vessel, constantly paddling to follow us. Their canoes seemed to me to be better finished than those of Cythera. They have small very well adjusted timbers, these canoes are all held up by an outrigger. Some also have a sail but their sails are quite different from those of Cythera, they are wide at the top and come to a point below, they are also made of straw, some had some coconuts, several some flying fish. We obtained 7 that were 10 to 12 inches long and 2 and a half inches thick that we bartered for nails, which they did not seem to know. They seemed quite surprised at seeing us and examined the vessel from top to bottom and seemed quite puzzled to see two canoes the size of the *Boudeuse* and ourselves.

Thursday 5 May at 8 a.m.

 We followed the coast a league and a half off, there did not appear to us to be a single place where one could anchor, the sea breaks a great deal along the shore. There is one small place where we saw numerous canoes coming from the land where the sea seemed to be calm. At 10 a.m., we sighted a second island bearing W¼NW approximately 10 to 12 leagues. We had several canoes by the vessel that kept a short distance away. They all had a sail except for 3 or 4 that had oars. They go very well and the Savages manoeuvre their canoes excellently. We are making close on two leagues an hour, when they filled their sails, they passed us as if we had been at anchor. There is always one man on the outrigger on the windward side, when it changes tack, he is

---

[1] The *Étoile* is sailing past the Manua Islands, part of the Samoan archipelago.

careful to go to the other side standing on a piece of wood that extends off the canoe and when the wind is astern, he goes for a while in the canoe. I counted up to 7 men in one canoe and in the other smaller ones 3 and 5 men. In one canoe there were two very ugly women and as reddish as the men. Judging from the large number of canoes we can see, this island must be well populated, it is also very wooded… Here are many islands we sight that, I believe, have not been seen by two or three navigators who have passed through these seas who make no mention of them in their accounts and the French map we use shows none of those we have seen so far in the latitudes we observe. Fortunately all these islands are very safe and very high, they are not like the first islands we sighted that in the night, [or] foggy weather, one could, in spite of all the precautions our condition necessitates, be lost on islands as low as those are, and how many did we manage to sail by in the night without seeing them and possibly passed quite close to them [judging] by the quantity of birds we have sometimes seen in flocks which is a sure indication of the nearness of land.[1]

Every day we mark the chart, we see no dangers nor does any memoir tell us that land was sighted in so many degrees of latitude. We have nothing to fear, we are 2 or 3 degrees further N and further S and that does not prevent us from seeing land in a quite different latitude and we are sailing fully confident that there is no risk in night time. As for me, I privately believe that we are making here quite a thorny voyage where we are incurring great risks not to mention the length of a crossing like ours, the sufferings and hardships we have to endure that continue to be very considerable and hard because one must take into account that we have been at sea since our departure from Montevidio, in 10 days it will be 6 months ago and those who will come after us, if any ever do, will still have great risks to run. This is enough on this subject and I am closing here the second part of my journal.

Off the New Hebrides

7 to 8 May

… Mr de Bouginville told us that on the island of Cythera some people from his frigate had received some pieces of gold from the Savages.

10 to 11 May

We are becoming used to our short stages and to the great heat we have been under for quite some time. At 6 a.m., we sighted land being two medium size islands of a good height. The most N island seemed to be the largest bearing from us W 9 or 10 leagues, the most S island WSW 5′W at the same distance…

21 to 22 May

… At 5.30 a.m. sighted land on the port side that seemed very high. We signalled this to the *Boudeuse*… This land seemed to consist of two islands the most N of which is much higher than the S one and steeper. We believe, according to Quiros's memoir, that this [is the] land he found in the [blank] and this shows it is that of the Holy Spirit and that he entered in a kind of gulf between two points after having covered 12 to 15 leagues, he found himself in a very good harbour and a fine bay where he was anchored in 6 fathoms of water sandy bottom where he spent more than a month. At

---

[1] This appears to be a reference to the atolls of the Tuamotus encountered early on.

9.30 being about 2 leagues from the most S point of the most N island, the *Boudeuse* altered course as we did, on the starboard tack, steered ENE good SE breeze until 10.30 when we bore away NW. I think that Mr de Bouguinville [believes] that we have nothing to do here, having all the memoirs and journals of all the navigators who have passed this way. We saw some high lands further in the bight formed by the two islands on the side of the most N island. From 10.30 to midday, we sailed NW to NW¼N, we hauled along the coast a good league off, we saw several fires on the mountains, there is no low land along the shore, everything is quite steep and seemed to me to contain no dangers. The sea breaks all along the coast…

22 to 23 May.

Good SE breeze, fine weather, calm sea, sailing under various sails, steered NNW to NW going along the coast a short distance off, ranging along the N island that continues to diminish, the N point is very low.

Note: off the N point of the said island, there is a very small flat island that is all rock not far to the said island, the sea breaks a great deal on it. A kind of breaker appears to stretch from the most N point of the island to this flat rock. We sighted over the N point of the said island that we have been ranging along since this morning a high land we believe to be another island. At 3.30 having rounded the N point and sailing along it a league off, we steered W then we found that this high land we can see is a much higher island than the two we have been seeing this morning and much more extensive. It is further S and further W. At 3.30, we sighted from the mast-head another island to the W¼NW and WNW, in a word land on all sides.…

At 6 p.m., we saw yet another island that bore W¼SW and WSW distance 15 leagues. At the same hour, spoke to the *Boudeuse* that told us we would tack here during the night and at daylight tomorrow, we would close in and go to examine the most S island, to see and try to find an anchorage to spend a few days…

At 6 a.m.,… continued S until [we were] 1½ leagues from this island then bore away to the SW. No sign of an anchorage or of the bottom except close inshore. We saw a canoe that examined us but from afar. At 8 o'clock, we were a short league from land, we could see several little sandy coves where the sea looked fairly good, the sand is blackish, the coast gave me the impression of a rampart hedged with rocks. We saw several inhabitants or Savages in one cove. Continued to range along a short league off until 8.30 when the *Boudeuse* hove to wind to starboard, facing NE, the drift is seaward.

The *Boudeuse* gave the signal to lower our boat, we armed it well on account of the Savages and it left to take its orders from the *Boudeuse*. At 9.45 the longboat and small boat from the *Boudeuse* and ours left the *Boudeuse* for the shore…

About the Islands of Pentecost[1]

23 to 24 May.

… At about 4 o'clock, the boats from the *Boudeuse* and ours left the shore and at 3 p.m. 5 canoes came up that approached us only within hailing distance. We gestured with our hats for them to come aboard, it was useless, they turned back and went back to land. At 4.30 our small boat came back having taken some wood they had cut

---

[1] An alternative translation of this sub-heading would be 'From the Islands of Pentecost'.

ashore to the *Boudeuse*. The boat brought a few coconuts from the land, they had found the Savages very mistrustful and all armed with arrows and stones. They were forced to fire a few musket shots wounding one very seriously … At 6 a.m … we see an island that seems considerable bearing WSW, very high, from the mast-head they can see the land from SSW to WNW distance 15 leagues. At the W and S point of the island where we landed, there is a pierced rock not far from the said point that from a distance looks like a ship in sail. We named the point of the said island Cape of Pillars [Cap des Pilières] …

26 to 27 May.

… At midday the *Boudeuse* signalled for us to lower our small boat … At 1.30 p.m., seeing a fine bay trending ENE approximately 6 to 7 leagues, we sent off the boat which we set out with the two boats from the *Boudeuse* to sound the entrance of this great bay which seems to be very deep and approximately one short league wide; at that time hove to wind to port facing SE¼S. We are a good half league from land. We sounded and let out 80 fathoms of line, no ground. In this bay in the NW part of this great gulf there are several very flat islands a short distance from the main island as wooded as the others. There is an island of a good height whose summit is quite flat like a table that seems very steep-sided. We can see numerous Savages along the coast and a few canoes. At 3 p.m. a canoe came up quite close to the vessel, there were 8 Savages in the canoe and they would not come any closer in spite of all the signs we could make, they were shouting a great deal and gesturing for us to go to them. After all their gestures, they turned back to make for the shore. Their canoes have outriggers.

Note: at 4.30, we heard several shots from blunderbusses and several muskets coming from our yawl that was near the land and firing on canoes that seemed to be close to her. They fired several volleys. The boats from the *Boudeuse* were about half way through the channel to the bay and had signalled with a flag at the top of the mast that it had found an anchorage. At 5 o'clock, the *Boudeuse* signalled to the boat to come back. They were about one good league from us on the side of the bay. The boats made for the *Boudeuse*, we remained hove to wind aport until 4 o'clock when we set off wind to starboard with the two topsails on short tacks to await the boats. The currents are bearing us towards the W. At 6.30, the boat came back. Mr Le Landois, the officer in the boat, told us that two canoes were approaching gradually with a thousands signs of friendship and displaying branches in sign of peace and always coming closer to the boat, it did not appear that they were intending to do any harm. They allowed them to approach confident [but] still on their guard, when they were a very short distance away, they shot two arrows, then they fired on the two canoes with the two blunderbusses which did no harm but with the muskets they killed 3 and wounded 3 or 4. They all threw themselves in the water and fled off to land. In one of the canoes there were 8 Savages and in the other 3. They each carried a bundle of arrows. They are all made of wood with nothing at the tip of the arrow. Our people fired at some Savages who were on the seashore but they do not know whether they killed or wounded any, but as for those in the canoe, they did see them fall. The boat was hoisted back in, then we made for the SSE to SE/4S under two topsails, good E breeze, fair weather, fair sea…

[At this point, Caro briefly speculates on whether they have in fact been sailing by Quiros's

Espíritu Santo, without drawing a conclusion. There are then several gaps in his Journal, with entries of a mostly navigational nature covering the period 4 to 7 June and 10 to 12 June.]

Off New Guinea

15 to 16 June.

Since the 12th of this month, we have had the most depressing weather in the world, rain, variable winds that worried us a good deal, we did not see land during those 4 days.

The winds E to ESE fresh breeze very fine weather, the sea fairly good, slight SE swell, keeping to a NNE to NE course starboard tack. At sunset, the finest weather in the world. We continued on the same tack throughout the night, under full sail. At 6 a.m., we sighted from the mast-head a large round island and five others smaller and lower. The large round island is visible from below bearing NE5°E 8 to 9 leagues. We signalled this to the *Boudeuse* which, when she saw it, went about and steered a SSE to SE¼S route, fresh breeze, fine weather.

After going about we sighted the mainland bearing NNW distance 15 to 16 leagues that seemed very high. We continued on a SSE to SE course until midday… At 8 a.m., we could no longer see land.

16 to 17 June.

The winds E to ESE good breeze, fair weather, fair sea. We kept to the S¼SE to SSE tack until 5.45 p.m. when we altered course, on the NE tack all sails set, at sunset we did not see any land, and continued on the NE tack until 1.30 a.m. when we changed course and sailed S to S¼SE good breeze, fair weather. At 5 a.m. altered course again to the NE tack. At 7 o'clock we sighted land from the mast-head bearing NNW to NW that appeared to be very high…

17 to 18 June.

…We are sailing here like blind men, we do not know where we are, no navigator has passed this way…

19 to 20 June.

… Our situation is a sad one, here we are bogged down to over our heads without knowing when we shall get out of it, without the slightest thing to give our sick, we have had no fresh meat of any kind for a long time. I would be a fool if I ever have to go round the world a second time.

24 to 25 June.

We have had the most sorrowful weather in the world since the 20th of this month and a bad sea. We are short of everything. We have seen no land since the 20th … On the 23d of this month, we spoke to the *Boudeuse*. Her entire crew is down to ¾ of water[1] a day and 13 ounces of bread. We are placing our crew on the same footing.

27 to 28 June.

… The *Boudeuse* is weary at seeing no land, she wants to see it as soon as possible and why not sail NE and NE¼N to get further E and more to windward? It would be better… At 6.30 a.m., we sighted an island bearing NW¼N 8 or 9 leagues of average

---

[1] I.e. three quarts.

height. At 7.30 saw from the mast-head another island N to N¼NE. At 8 o'clock sighted another island between N and E, the further N bore E5°N and the most S ESE in view…

New Britain

29 to 30 June.

The winds variables SE to E uneven and sometimes good breeze, fresh breeze and calm, sailing under full sail we steered NW¼W until 1 p.m. when we thought we could see a bay bearing from us NNE. We made for it to examine it and see if we could find a place to anchor and obtain wood and water. At 3 o'clock calm, we sounded and found 47 fathoms bottom of fine sand and small yellowish gravel. We were a good 3 leagues from the nearest land. At 4 I sighted the W point of the island to the N of us bearing NW distance 10 leagues, the most E point bearing ESE5°S 12 leagues. During the night there was a very light E to SE breeze, we tacked back and forth during the night with tacks of 2 hours under light canvas to stay during the night and find ourselves the next morning at the same distance from land. At 6 a.m. the *Boudeuse* signalled to lower our boat and send it to her. At 7.30 it left with the *Boudeuse*'s two boats to go to the shore to see if it could find an anchorage, at the same hour we sounded, we found 40 fathoms fine sand. At 6 a.m. I sighted the most E land of the most N island and the nearest bearing ESE, the most S and the most W from WSW to W5°N, the most S islands from S¼SE to SE¼S. We can see several islands from the topmasts from W to NW, in a word land everywhere. We continued to tack about and heave to and carried out various manoeuvres sometimes following the route taken by the boats we see along the shore.

At 9 o'clock saw several canoes along the shore, they gathered 10 or 12 of them that gradually came up close to us, there were a few that approached quite close to our vessel, they were making signs for us to go ashore and that we would find food and drink. These Savages are all blacks, their hair woolly like those of Guinea, fairly ugly faces. They have large canoes somewhat like those of the Malays. It goes very well by means of their oars. I counted up to 18 in the large canoes and 6 to 7 in the medium-size ones. They were all armed with bows and arrows. They were never prepared to step on board. They all have shells of pierced benie[1] on their foreheads, their arms and even their eyes. Their canoes have no outrigger. They each have a very well made straw shield which serves I believe to guard against arrows. I think they eat betel or some other herbs as their mouths were quite red. They stayed about an hour examining us then went back to the land … At midday we are at the same place as we were yesterday at midday, our boats are along the shore.

30 June to 1 July.

The wind ESE good breeze, fine sea, sailing under different sails, steering WNW, going along the coast like our boats, ranging along 2 to 3 leagues off. At 2 p.m., hove to wind to port facing SSW, the *Boudeuse* signalled to the boats to return. At 3 o'clock our boat arrived, it had not found any place where a ship could shelter. A short gunshot offshore, they found 40 fathoms of water, the ground in some places is very good

---

[1] The shell of the large clam tridacna was referred to in common parlance as a '*béni*' or '*bénitier*' as it was used in certain chapels and churches as a holy-water stoup.

and sometimes rock, the sea fairly rough along the coast like a kind of bar. They did not go ashore but followed it at a very short distance. They saw several Savages who did not seem to see our people. In all the small sandy coves, they saw a few huts and elsewhere they saw coconut and banana trees. We hoisted the boat back in. At 4 o'clock sailed WNW. At 5 p.m., we were in a considerable tidal race, in spite of the wind, the ship would not answer [her helm]. The look-out and several people who were aloft thought they could see breakers appearing from the W point of the island we were hauling along and going across to other islands further to the WSW. We believe it is only the strength of the tide that gives an appearance of breaking waves going against the wind. We spoke to the *Boudeuse* who told us they shared our belief and that we were going to tack about here all night and that tomorrow morning we would depart …

1 to 2 July.

… At 7 p.m. our boat arrived, we hoisted it in. The officer who was in it told us he had found the finest harbour in the world, that he carefully examined it and sounded everywhere with the boat and with the *Boudeuse*'s. This port is closed by the great island one leaves to starboard as one enters and by 3 or 4 small islands detached a short distance away from each other, very wooded. This port lies N¼NE and S¼SW and has a depth of about 1½ leagues. Its entrance is open from SSE to SW with the finest water in the world, one is sheltered from any bad weather. It can be compared to a basin, the slightest hawser could hold a vessel which can be secured to the trees of the small islands. The bottom is 19 to 7½ fathoms deep, grey sand where we would have anchored; in some places they found red sand and small shells. These small islands are covered with coconut trees. It seemed that this port abounds with fish and seems to have numerous sea turtles.

As the boats were sounding in this port at around 3.30 p.m., 11 canoes of savages came out from a kind of river that lies to starboard as you enter on the large island; there were in each canoe 14 to 20 men; in our two [boats] there were 15 men from the *Boudeuse* and 10 in ours. These canoes came up to our boats that were both ready to receive them. When the Savages found themselves close enough to attack our boats, they began to utter horrible screams and hurl stones and arrows. Our boats immediately fired their blunderbusses and muskets. At the first shot, they all fled and jumped into the sea from their canoes. Our boats seeing them all in disorder chased them and forced them to abandon their canoe that was about 30 feet from bow to stem. They killed 15 to 20 Savages and wounded many who escaped into the woods. Our boats had each taken one canoe from which the blacks had fled. They found in one of the canoes a man's jaw decorated with numerous shells that they believe to be a fetish. I think these Savages are cannibals. We found in their canoes some fish, crabs, coconuts and many small nets, their bows, arrows and spears are not dangerous. Our boats having concluded their fight, set out to leave the port. They saw at the entrance of a river 30 canoes that seemed to be well armed coming no doubt to help the 11 others, but seeing how the others had been received, they did not dare to revenge themselves and allowed our boats to leave quietly These Savages are not able to resist firearms. We regret this fine harbour.

2 to 3 July.

… We are getting away from our island and the harbour where we thought we would have been at ease to spend some time repairing our vessel that greatly needs it as well as everyone, but we must give up this thought....

3 to 4 July.

At 9 a.m. spoke to the *Boudeuse*. Mr de Bouguinville told us that he was going to sail W¼SW. Mr Giraudais asked him why we had not put into that harbour where the boats had been, he replied on account of the night and the currents and a shoal he had to starboard where he had taken soundings and found 7 fathoms rocky ground and that was a bank where there was not much water. We told him of the danger we had been in among the waves we had come so close to. He told us he had passed through the middle of the waves and had not seen ground below him, but surely he did not pass through the great waves we avoided. Surely there was not I believe enough water for him to pass seeing how the sea was breaking...

4 to 5 July.

... At 3 o'clock we saw 4 or 5 canoes 3 of which went up to the *Boudeuse*. They did not go alongside but went very close, they stayed about one hour close to the *Boudeuse*, then returned to the shore. At 5 p.m., we sighted 3 or 4 small islands to the NW that we took for vessels. We signalled this to the *Boudeuse* that replied; during the night we steered SSW light breeze and calm ... At 6.30 the *Boudeuse* gave the signal for us to lower our boat; she lowered her longboat that came alongside with Mr Duclaux, second officer of the frigate. We sent to his ship by his boat and ours 8 *tiersons* of brandy, 8 quarters of flour, 6 barrels of salt meat dried pork and beef and about two cords of wood. They have only 20 days of water left, they are short of everything. Mr Duclos told us that the canoes that had been so close to them had shot an arrow that landed on board. At 10 o'clock hoisted our boat back in and sailed SW¼S all sails set light SE to SSE breeze... From this view of the island, it seems at a quick glance that the currents are bearing us strongly to the NW so that we are passing it...

5 to 6 July.

... At 10 a.m. the *Boudeuse* hove to wind aport and lowered her boat and signalled for us to lower ours and ready it to go ashore. At 10.30, it left to go to the *Boudeuse* and then went with the *Boudeuse*'s boat to examine an area where there seems to be a bay formed by three islands, the two that are further W were fairly large and the one that forms the entrance with the point of the island is much smaller. We are approximately three short leagues from the closest land of this bay. We sailed close to the *Boudeuse* and hove to like her and remained thus until midday...

6 to 7 July.

... At 2 p.m., we sighted our two boats that appeared in the middle of the bay and flew a flag at the top of their mast which is the signal for having found an anchorage. The *Boudeuse* bore away and made for the two boats that were lying to N¼NE. We followed the same route as the *Boudeuse*, leaving her to go ahead to give her time to anchor. The boats were coming to meet us. At 3.30 we were between the small island that we left to port and the point of the large island to starboard. There is a good half league of channel between the two and it is quite steep to. One does not find bottom at 50 and 60 fathoms. Our boat has joined us, told us there was a good

216

anchorage where one was very sheltered but that there were 30 to 40 fathoms of water, bottom of sand and small gravel 2 cable lengths from shore in a small bay on the large island to starboard of us. We continued to follow the starboard coast that is the large island. All along the shore there are reefs where the sea breaks but that do not extend far out to sea, being off the point that forms the entrance to the harbour. We bore N to NE and up to E in order to make for the back of the harbour that may be 3 to 4 cables deep. The *Boudeuse* which we had lost sight of for a moment was at anchor, our boat was going ahead of us. At 4 p.m. we dropped anchor in 35 fathoms bottom of white sand, further into the harbour than the *Boudeuse*. After furling the sails, we lowered the longboat. We are waiting for the *Boudeuse* to moor so as to get our kedge anchor to moor ourselves,[1] having only that one anchor between us both. At 6 p.m., we laid out the kedge anchor at the back of the harbour on two cablets hauled quite taut for the night[2] …

It is a small cove of very white sand, the forest is so thick that it is not possible to enter without a great deal of trouble. There is no other long walk than along the seashore. There are three very pretty little streams or small rivers at the back of the harbour with very good water very easy to obtain. The *Boudeuse* took one and we another. We can see no indication that this place is inhabited, no Savages are visible or any appearance of hut and no other refreshments than wood and water.

7 to 8 July.

We lowered the topmasts, worked on cleaning the barrels to send them ashore to be filled. We caught on a line two to 3 *sardes* and a few *capitaines*.[3] There is a great deal of fish of various species in this harbour but only a few can be caught. We found a canoe belonging to the Savages in an inlet N of here, about 1 league that was high and dry near 2 or 3 rough Savage huts. Mr de Bouguinville going to see this bay in his boat found in a little bay a short distance further NW than the one where the canoe was an English inscription engraved on a small lead plaque that was half broken and rolled which we believe was tom by the Savages from the tree where it was nailed and it seems from the trees they have cut that they passed through here not more than 3 or 4 months ago and that it was the two vessels that went through the strait of Magelent before us.

One finds much rattan and canes in the woods. The rattan is excellent but the canes are not ripe and have a lot of knots close to each other. Many snakes of various sizes that are very noxious are found in the woods. We killed a large number. In the woods a short distance away there is a waterfall whose water drops from the mountain into 3 or 4 basins that produces the most attractive scene in the world. Since we have been here, it has always rained and been calm but outside there is a stormy gale.

---

[1] This is presumably the anchor which the *Boudeuse* had weighed for them on 13-14 April off New Cythera, which was needed back to enable the *Etoile* to moor with two anchors.

[2] This would mean that the bower anchor was to seaward and the kedge laid out by the longboat at the back of the bay on two hawsers heaved in to give a taut moor with the ship lying in the middle, riding to the seaward anchor when the wind was on shore and to the kedge when it was off shore.

[3] 'Probably the *Pelamys sarda*, a kind of tunny-fish of the scombridae family, that have black oblique bands on the sides. The *capitaine* (*Polynemus quadrifiliis*) is a kind of bony fish whose pectoral fins have filamentous points' (ET). The former is widely known as the bonito or striped tunny.

Sunday 10 July.

Always rain. They are working in the hold to ready some food to be sent to the *Boudeuse*. On this said day, died a sailor named Jean from St Malo. Continuation of rain.

[There are no entries between 10 and 22 July.]

Friday 22 July.

It has rained almost continually since we have been in this harbour. In the morning our boat went to the pass to see what the weather was like outside. They found S¼SE winds stormy gale, the seas high and always rain. We have finished our water and our wood. We are awaiting fine weather to get out.

At 11 this morning, we felt a movement in the ship that went on for a short minute approximately which had the same effect as one feels aft when they are dancing on the forecastle. We suspected that this movement could only be the result of an earthquake. We looked ashore, we merely saw the sea that had drawn back for an instant, all the reefs were uncovered, then recovered. The sea rose and fell several times in an hour, it went down suddenly by 4 good feet, it rose similarly. We had 3 young apprentices ashore who were filling water casks who felt the earthquake that seemed very powerful frightening two who wanted to return to the ship but the 3rd reassured them and having filled their barrels returned aboard. They told us what had happened to them.

25 to 26 July.

… This part of New Britain seems to be very populated, one can see a great deal of smoke along the shore, there seems to be quite a deal of cleared land but we can see nowhere for a vessel to anchor. The harbour whence we have come cannot be good as we saw no habitations or appearance of habitations…

25 to 26 July.

The wind variable from SE to S fresh breeze fine weather, fair sea sailing NW and NW¼W along the island of New Britain distance 3 to 4 leagues that so far has seemed very safe, sometimes some opening appears that could be a fine bay but when we approach it is only a very steep shore. At 1 p.m., we sighted an island bearing N¼NE 12 leagues, at 5.30 sighted another island to the NNW 14 leagues that seems small but very high. We are leaving all these islands to starboard and continue to range along the island of New Britain…[1]

27 to 28 July.

The winds have varied during the 24 hours, from NE, E, SE and S light breeze and calm, fine weather, fair sea. We steered NW¼W to WNW all sails set. At 6 p.m. the land was within the same compass point as in the morning. At 6 we were 3 good leagues from New Britain, one can see a great deal of smoke. During the night, there was a very light breeze…

---

[1] All these reference to New Britain are of course erroneous, the island in question being in fact New Ireland.

At 11.15 a.m. a canoe coming from New Britain in which were 9 Savages came up close to the vessel and was saying a great deal, that we could not understand. We made signs for them to approach showing them various items such as handkerchiefs and bonnets and other small items. They came quite close to the ship abaft the main channel[1] and stayed a long time examining us and discussing among themselves. Suddenly they all began to shout together, two or three Savages waving their kinds of spears that are merely pieces of wood very sharp at one end that cannot do much harm from a distance, the others began to throw stones at us, everyone was on deck looking at them. We were most surprised to see that 9 Savages risked throwing stones so close to the ship. We fired a shot at them causing one to jump from his canoe into the sea. We fired 4 or 5 more shots, most of them hid at the bottom of their canoe and the 3 boldest began to paddle like the devil to get away and returned to land. Their canoes are much the same as those we saw earlier…

28 to 29 July.

The winds have been variable E, SE and SSE light breeze and calms, fine weather, very fair sea, sailing W¼NW to WNW all sails set always along the coast of New Britain at a distance of 3 to 4 leagues which still gives the appearance of a very attractive island but no sign of any harbour or anchorage … These small islands that we have seen N of New Britain all look to be populated, we saw smoke in several places on the said islands. From midnight to dawn, we steered NW to NW¼W almost calm. We were again forced today to wage a little war. At 9 a.m., 6 canoes of which 5 had from 6 to 9 savage men in each and the sixth canoe was very small where there was only one savage. They approached gradually always displaying mistrust. The small one was coming ever closer to the vessel and was making much more noise than the others. They were raising their arms crosswise then lowered them and lifted them up again with much moving about. Mr Dampiere [sic] says these are signs of peace among the Savages of New Britain.[2]

They came up very close examining us and making gestures for us to go ashore. They remained more than an hour very close to the ship, then they let themselves drift astern and began to discuss all together.

Some time later, the small canoe in which there was only one Savage made for the shore possibly to fetch some reinforcements. Only a few arrows are visible in their canoes, in spite of that, like yesterday, we loaded 10 or 12 muskets and the boat's two blunderbusses which we placed on the quarterdeck handrail. They had all gathered about 200 paces away and were talking a great deal. Two canoes left the group and came up about halfway between the others and us and stopped to examine us. One of the Savages drew a sling he had around his neck and hurled a stone at us that passed above the ship, then a second one that fell on the forecastle without hurting anyone. We did not want to fire in order to let him come closer. We were making very little

---

[1] The channel or chain-whale of a ship is a broad thick plank projecting horizontally from the side abaft the mast to which the shrouds are secured. Each mast has its channels, the main channel being slightly abaft the main mast.

[2] Caro is probably referring to Dampier's comment: 'Their Signs of Friendship, are… often striking their Heads with their Hands'. Dampier, *A Voyage to New Holland*, ed. J. Spencer, p. 211.

headway, about half a league an hour. Shortly after, they sent us a third stone. Mr Giraudais fired a musket and at the same time we fired a blunderbuss and several muskets were fired.

Soon afterwards, simultaneously, they all threw themselves from their canoes into the sea. The furthest canoes fled without attempting to rescue the others. The Savages who were in the sea stayed hidden behind their canoes and were hauling the canoe off. When they were out of musket range, we saw them reimbark in their canoes but not in as good a condition as they were when they attacked us. In the canoe where there were 9 [islanders] and which was closest to us, 3 Savages were seen held up by the others and appeared by the glass to be quite wounded and 2 in the other canoe. They made for their islands. I believe that they will not come up so close to a vessel. The *Boudeuse* is half a league ahead of us. No canoe went up to her. They are only after the rearguard…

29 to 30 July.

Light E to ESE breeze, fine weather, fair sea, sailing under full sail NW to WNW along the island of New Britain still at a distance of 3 to 4 and sometimes 5 leagues… During the night very light breeze, fine weather, we are continuing our NW to WNW course under full sail. We notice from the sight of land that the currents are contrary and bearing a little to the ESE. At daybreak, we sighted several canoes coming from the island of New Britain. We were about 3 short leagues away. They came very close to the ship, about twenty and several others went alongside the *Boudeuse* a short distance ahead of us. One or two canoes being quite close to us threw us 5 or 6 bananas and some yams. We gave them some pieces of red cloth. They circled the ship several times. We gestured to them not to come close and to leave, which they would not do. They were all very well armed with arrows, slings and numerous stones in the bottom of their canoes. We readied a few muskets and we fired on several canoes with a couple of blunderbusses. They were not slow in throwing themselves into the sea from their canoes and the furthest ones fled. When they jump in the water, they take care to shelter behind their canoe and to swim away pulling it away from the ship.

They all returned to the land. Those who were by the *Boudeuse* also left, we believe that the *Boudeuse* also fired a few shots on the canoes alongside her. Most of these Savages had a long beard all painted white as was also part of their bodies. They were 6 to 8 in each canoe, their hair is painted in various colours…

30 to 31 July.

The winds remained variable from ESE to SSE almost calm, fine weather, fair sea, steering NW¼W to WNW. At 5 p.m., we saw 4 or 5 canoes that were examining the *Boudeuse* but from a good distance… During the night almost calm, we can hardly steer, at dawn quite calm, fine weather… At 8.30 a.m., we were becalmed and 2 leagues from the land nearest to the island of New Britain, several canoes came out from the said island and came to examine us from quite close. At 9 o'clock we had 20 around the vessel and [they] stayed astern of the vessel. There were also 3 small catamarans where there was only one Savage on each and 3 or 4 small canoes with one man in them. Those approached much closer and sometimes came up to touch the ship. In the large canoes, there were 6 and 8 and up to 9 Savages in each canoe. Another 7 or 8 others were coming and they were paddling with all their might and those who were close were gesturing for them to come quickly to take part in the capture. They seemed quite determined to attack us. They shouted from time to time. We made signs several times for them to leave, which they were unwilling to do, on the contrary they were ever coming nearer. Their canoes seemed well supplied with arrows, slings and heaps of stones. We fired a volley from 8 to 10 muskets and the two blunderbusses. The fight was much fiercer and more bloody than the other three. One small canoe they abandoned remained quite near the vessel, one was sunk. We saw 2 or 3 Savages who were rescued by the catamarans. They all fled as fast as they could and went to their islands. None of the canoes went to the *Boudeuse* which lay a short distance ahead of us. The small canoe in which there was only one man displayed much more aggressiveness than the others and came up much closer than the rest. In addition to their arrows and the sling, they had in their canoes clubs and a kind of wooden sword. We left the small canoe. A light breeze rose up from the ESE, we sailed NW¼W all sails set. At 11 a.m., a canoe came up from the small island to the NE. These Savages struck us as more friendly, they came alongside and traded a coconut and some other small items and then returned to their island without making any demonstration. There were 6 in this canoe, they made sign indicating that they were at war with those of New Britain. The island bearing NE from us is the most N of the 4 islands we saw N and E of New Britain. When we sighted it, it seemed to us that there was only one island but today that we have gone a little past it, 3 small islands appear a short distance from each other the most E one being the largest and highest...[1]

... We can see no more islands N or E of New Britain. I do believe that we are nearing the N end of New Britain. Dampiere shows the most N point in 2° 20 to 30 minutes of latitude. From the mast-head we can see no other point extending further N and W than the one bearing N¼NW3°N at midday 10 leagues. The latitude that Dampiere indicates [for] the most N point corresponds within a few minutes to our latitude observed at midday.

This island of New Britain seemed to us to be a very fine island. It is much higher in the S and E than in the N and W. It seemed very well populated judging from the number of canoes and fires we have seen since our departure from Port Praslin [Pralin] but we have not seen a place suitable for anchoring a vessel, it seems quite steep and safe with no dangers.

31 July to 1 August.

...At 3 p.m. a canoe came alongside that was from the small NE [island] which stayed some time tied up alongside and then returned to their island. They do not seem by far to be as nasty as the other Savages of New Britain...

...At 10 a.m., we sighted five canoes coming from the island of New Britain to look at us but from a great distance then returned to their islands...

[1] This small archipelago is known as the Tabar Islands. The French are now coming up to New Hanover.

APPENDIX 3

The Journal of Vivez

There are two records of the voyage compiled by the Étoile's surgeon François Vivez. The first, held in the archives of the Société de Géographie at Rochefort, was written by him shortly after his return to France; it consists of 52 pages and is not strictly a journal but a narrative containing a number of navigational details which he probably obtained from one of the Étoile's officers. Who this collaborator may have been is open to speculation. It was assumed by Martin-Allanic that it was Jean-Louis Caro, the ship's first officer,[1] but this view was not shared by Taillemite who suggested that another officer, La Fontaine, who is known to have been a friend of Vivez during the voyage, was a more likely source of information.[2] However, Vivez had been sailing since the age of seven, when he joined the Formidable *with his father who was that ship's senior surgeon, and was probably much more knowledgeable about navigation than the average ship's surgeon.*

A second manuscript is held at the Municipal Library of Versailles under the reference In-4° 126; it consists of 71 pages, ten of which are blank. In an introductory paragraph, Vivez states that he wrote a 'memoir' during the voyage, but that he decided not to try to get it published following the publication of Bougainville's own Voyage *'which leaves nothing to be desired'. Vivez was back in France at the end of April 1769, but his health was often indifferent and his first concern obviously would have been to return to his family after the strain of the long voyage. Probably towards the end of 1769 he began to compose this first narrative from the notes he had taken during the voyage, and it is likely that this work was finished late in 1770, by which time it was well known that Bougainville's journal was soon to be published. When this appeared, Vivez set his own narrative aside.*

Vivez sailed in the Tourterelle *in 1772 as senior surgeon and it was probably upon his return, and after learning of Commerson's death, that he took up his journal again and wrote what is now known as the Versailles manuscript, which he presumably intended to be a longer and more 'literary' work suitable for eventual publication. This was probably completed (although not re-read and corrected) in early 1774, the year in which Vivez took up another appointment, this time in the* Bricole.

The following extracts are taken from the Versailles manuscript. The Rochefort manuscript, however, has helped to determine the correct meaning of some of Vivez's more convoluted or careless second version. The extracts have been selected on the basis of their general interest and their relevance to the islands discovered or sighted and to the various islanders encountered on the way. Thus, the earlier part of the Journal, on Vivez's expedition in the Guerrier *in 1763, his later*

---

[1] Martin-Allanic, *Bougainville*, I, p. 762n.
[2] Taillemite, *Bougainville*, I, p. 128.

activities in Cayenne and his return to France in the Garonne, *followed by his account of Bougainville's attempted colonization of the Falkland Islands, have been omitted. Vivez then outlines the voyage from France to South America, with a lengthy description of the ceremonies surrounding the Crossing of the Line, which he ends with the observation, 'I am going to conclude at this point a narrative which, although intended to entertain my reader, runs the risk of becoming soporific'. A number of pages are then devoted to descriptions of life in Montevideo, Rio de Janeiro and Buenos Aires. His comments on navigation through the Straits of Magellan echo what was being written by Bougainville and others; they are followed by just a few lines reporting the crossing of the south-east Pacific Ocean, from 22 January to 22 March 1768.*

Through the Tuamotus

On the twenty-second of March [1768], fifty-fourth day after our departure from the strait, having sailed west throughout the night, at two o'clock we bore away by one point of the compass and at dawn we sighted on our quarter, distance one small half-league, five small drowned islets over which we would certainly have sailed had we not by chance altered course after midnight.[1] Continuing our route along the same course, we saw another land on the port side approximately ten leagues from the islets we passed so close to. Being abreast of it, we saw some men running down to the water's edge, raising their arms to make signals. They stayed for a moment on the shore and went into the woods. Then appeared some twenty other savages seemingly dressed in some kind of mats and a kind of square white hat, carrying flat sticks they used to make signals, [they] also returned into the forest except for two who remained on the shore beating the sea with their pieces of wood. They too left soon after and lit some fires. This island is probably one good league round, one can see some fine greenery and a land thick with trees, either coconut or palm trees judging by their appearance. We saw through the spyglass a large crowd and some huts that seemed to be built in a square, and some kinds of tall pyramids of stone, which led me to give it the name of *isle Garnies* until it receives another name.[2]

The shore appeared to us to consist of very white sand, we saw several reefs and, the winds being unfavourable, we sailed on. In the evening the weather became overcast with storm and rain. The next morning we had to heave to with the forestaysail, the weather being so black and so bad that from the stem we could see people at the bow only during lightning flashes. We fired several gunshots to stay with the *Boudeuse*. At six o'clock the weather having cleared, we sighted an island situated approximately thirty leagues west of the small one I have just mentioned.

Lakes [sic] Island

It is a very low land half of which is drowned, wholly wooded, decorated all round a lake by reefs that close it in like a basin and prevent a landing. We saw two boats with sails seemingly of medium size, a moment later another canoe where there

---

[1] In the Rochefort manuscript, Vivez credits this change of course less to chance than to his captain's instinct ('as if Mr Lagiraudais had smelt land'). There may be some deliberate dramatization in the rewriting of this incident in the Versailles manuscript, for in the earlier text the distance was given as four leagues, an estimate confirmed by Bougainville in his journal.

[2] 'The Well Appointed'. Bougainville will call it Lancers Island [Isle des Lanciers]. It is Akiaki.

were six or seven Indians sailing all along the shore inside the lake (whence it gets its name).[1] We saw several Savages on land who came to sit on the sand and seemed to me to be reddish brown. We saw there a considerable bay but unreachable on account of the number of reefs and breakers. Yesterday's island appeared to me to be more attractive than this one although the lake that surrounds it gives them a great deal of fish.

Stormy Archipelago[2]

Sailing the next day at fourteen leagues, the weather constantly stormy and wet, we found at a short distance seven fairly considerable islands one of which offered an extent of some twenty leagues but extremely low, partly drowned and wooded. These seven islands and others we have seen in this area make up the Stormy Archipelago. We passed a fair distance away from them so that we could not distinguish the inhabitants. As the winds and the currents were bearing towards them, and these drowned islands tend to have shoals off them, we only approached them at a good distance. To avoid the next islands of this type, we bore much further south, on which parallel we covered approximately three hundred leagues in spite of the contrary winds and the storm without finding any land but only a single fish they call the devil which has the flat shape of the ray, presenting a surface of four feet in every direction, which has on its head some very long black horns shaped like wad-extractors, the mouth very split, a knob on the back covered with prickles, the skin hard and rough, whose flesh is, they say, poisonous.[3]

Boudeuse Peak and Landfall on Cythera [Tahiti]

The second of April at ten in the morning, the weather being fairly fine but windy, we saw a very high round island that we named Boudeuse Peak [Pic de la Boudeuse] from the name of the frigate in which Mr de Bougainville was sailing, we were about ten leagues from it at the time. Steered towards it but shortly after the weather having cleared, we saw a much more extensive land on the port side towards which we set our course, intending to seek an anchorage there, our needs becoming pressing. The rain and other bad weather we had passed through during the last few days had brought about many cases of sickness, namely twenty clear cases of scurvy and the rest of the crew weakened and spiritless, having lived for four months on nothing but salt meat, a bottle of stinking and rotting water, brandy rationed, one meal only with wine, the biscuit beginning to go bad, the refreshments for the sick very scarce and our food at the officers' mess not much more appealing, if one recalls that on our departure from Montevideo a single squall had destroyed all our cattle and part of our poultry.

We had head winds that first day and the next which we spent tacking to get closer to land where we saw several fires and at eleven o'clock a canoe with a sail well off the point of the island. At about two o'clock a canoe came alongside in which were two Savages, which was followed by another in which there were one man and

---

[1] This island is Harp Island [Isle de la Harpe] or Hao. The 'lake' is the lagoon.

[2] Vivez, whose spelling and grammar are somewhat erratic, regards 'archipelago' as a femine noun, and writes 'Archipelle orageuse'. He will do so again later with 'Archipelle dangereuse'.

[3] *The Manta,* sometimes included in the category of devilfish, is a large ray. The anterior ends of the pectoral fins are free and project forward, which are used to scoop up small fish: these are Vivez's 'horns'. The tail, long and slender, bears a strong barbed spine connected to a poison gland.

two children, seemingly of a mulatto colouring, with black frizzy hair, their teeth remarkable for their whiteness, vigorous in appearance, of a good height, most wearing a loin cloth of a material made from the beaten bark of a tree. Their children were almost white. All the slaves were naked and had for all clothing merely a small belt made of coconut fibre. The canoes may have been a foot wide and fifteen to twenty feet long, having an outrigger to stop them capsizing. This land seemed to form two separate islands the headlands of which overlap. We went into the bay to see if there was a passage or anchorage. We saw there a large number of canoes and several huts tastefully built. Progressing into that bay, we also saw that a low land forming a superb bight joined the two alleged islands. In the evening, there were fifteen to twenty canoes around the ships and a large gathering ashore. We gave various items to those in the canoes in exchange for local fruits consisting of coconuts, bananas and a kind of bitterish apple, in addition to a few sucking pigs. We saw them with no type of defensive weapon, but on the contrary all holding palms in their hands, which gave us the hope that we would be well received.

There came on board from two canoes tied together at the side a Savage dressed as I have mentioned already, bearing his palm in sign of peace, accompanied in his canoe by a girl or woman of sixteen to seventeen years of age of very pleasing build wearing a loin cloth shaped like a sugar loaf on her head, one around her waist and the rest quite naked and white, one could say better than in Europe or at least equal for that age. At this charming sight, we soon were wishing for an early stop, our imagination worked a great deal from that time to know whether this beauty belonged to the country. How could such charming people be so far from Europe and how is it that in this island they are so white whereas all we had seen in the other islands since the time of our departure were different. This double canoe was rowed by ten slaves. We were then sailing at two leagues an hour, these people were rowing with all their strength. As soon as they were within range, we threw them a mooring rope, the chief jumped on the bow of the canoes, placed one foot on each and, displaying incredible strength to hold on to the rope, which we greatly feared might break or cause him to be thrown into the sea on account of the speed of the vessel at the time, which is why we told him to let go of the rope, which merely gave him added strength. Finally, after about a quarter of an hour of struggle, he succeeded in reaching the ship, catching the chain-stays and climbing on board holding his palm branch. He immediately sent back his canoe and presenting [sic] his palm to Mr Lafontaine who was the tallest of the ship's officers, who introduced him to the Captain to whom he gave his branch saying *tayeau*.[1] He told us by gestures that he had food and drink on land and invited us to go ashore. He asked us by gestures if we came from the direction of the sun and whether we were its children. We replied in the affirmative by a sign because their jargon was like no other tongue.[2] Not having been able to reach the back of the bight

---

[1] This is Vivez's spelling of the Polynesian or Maohi word for 'friend', usually *taio;* there was, of course, as yet no written form of the language to guide him or his fellows.

[2] This early conversation is not recorded in the Rochefort MS. It is possible that the chief was merely enquiring whether the French belonged to the same group as the Europeans who had preceded them, namely Wallis in the *Dolphin,* and who like them had been seen to arrive from the part of the ocean where the sun rose.

before nightfall, we decided to veer off to sea. Mr de La Giraudais, the Captain, suggested to the savage that he go home, that we were going far away and finally after repeating that we were going to the sun, he gestured that he did not care, that he wanted to follow us. He then very nimbly climbed into the lower rigging and after he had spoken a while to some thirty canoes that were surrounding us, they disappeared and there was then no way of sending him back. Mr La Giraudais who did not know what Mr de Bougainville's intentions were was not anxious to bring along this Savage and would have liked him to leave but he had made this impossible by the departure of the canoes. He then returned to Mr Lafontaine to whom he was making numerous caresses, embraced him, and gave him one of the loincloths he was wearing in which he wrapped him. Mr Lafontaine in return gave him a shirt, a large pair of trousers, a vest which they had a great deal of trouble to make fit because of his burliness. When he was decorated with this half-dress, he wanted and asked for a hat, which was supplied to him. One will easily understand that he provided us with much amusement before he was clothed by the manner he went about it. He was shown a mirror in which he did not seem surprised to see himself but he showed great surprise at seeing someone else behind him. It was evening, the time for prayer came, the bell was rung, at which sound he looked up above him in every direction, during the prayer he knelt like everyone else and was very well behaved, at the end when we shouted God Save the King he joined in. He then amused himself by going to mess with the sailors, sitting on the floor by them and eating a spoonful of soup alternately from different mess tins. Everything attracted his curiosity. He asked what the guns were for, when it was demonstrated to him that they were used for killing, he went pale and lost his countenance for some time.

When our suppertime arrived, he sat at table like us, he was watching all our movements to imitate us. He sniffed everything we gave him before tasting it. We gave him some water which he seemed to drink with pleasure and eagerness saying *vaye,* that is to say water. He refused the wine and the stew. As soon as he noticed that the ship's boys were there to serve, he made sure that he received the same attention and changed often.[1] What he seemed to find best was dried jam. He drank very frequently which caused us to fear that they had no good water among them. The signs we made to him left us uncertain whether he was telling us to bring some or that there was some.

After supper, we fired some rockets that he watched with great astonishment. We made him a bed on which he refused to stay, preferring to be below the deck on some sails. Throughout the night he kept on going up to see from the stars where we were going, having spent it tacking. In the morning our boat went along the coast to try to find a place where we might shelter but not having found any, we spent this next night like the previous one, impatient at not being able to land in a country that we had already seen such attractive samples of and afraid that Mr de Bougainville might give up.

---

[1] The Rochefort MS uses the same wording, so it is not clear whether he meant that the Tahitian changed places to get the ship's boys to move his plate or, as is more likely, requested different types of food and asked for his drinking cup to be refilled several times in order to enjoy being thus attended to.

The next day seventh of April, our yawl was given the same mission as on the previous day. She signalled an anchorage, we sailed towards it, passed between two heads of a reef and dropped anchor between them and the land where we moored permanently.

Hidden Story

To enable the reader to reflect on the stay in Cythera [Chythère] where we are staying, I shall digress from my journal for a story that he certainly does not expect. The reader may possibly be surprised that I did not tell him from the beginning about an adventure that began with us, but I thought it would make the story more interesting if I collected it in the same section. I hope that my motive will serve as an excuse and here are the facts.

A naturalist going round the world to deepen and increase the knowledge and productions of Nature, presumably wishing to have some new experience in this region, for this purpose took on board as his servant a girl in disguise supposedly from Burgundy,[1] changing names according to circumstances, although this has no real bearing on the matter. Leaving Europe in the bad weather we sailed through, she was very much affected by seasickness, as was her master, which removed the opportunity of complaining unless it was night time as the mutual attachment she had for her master made her fear or hope that he might have some weakness during sleeping time, [and] made her put up with the weariness of spending the nights in his cabin so as to be within reach to assist him. The special care she took of her master did not seem natural for a male servant, with the result that this quiet period of enjoyment went quickly by for the two people. After the first month, the peaceful rest they were enjoying was interrupted by a little murmuring arising from the crew about, they said, the presence of a disguised girl on board. Without hesitation, eyes turned towards our little man. Everything in him indicated a feminine man,[2] small of stature, short and plump, wide-hipped, shoulders in keeping, a prominent chest, a small round head, a freckled complexion, a gentle and clear voice, a marked dexterity and a gentleness of movement that could only belong to that gender completed the portrait of a girl who was fairly ugly and unattractive.

The leaders pretended to be unaware of this situation for a long time but the rumour having become too widespread, they advised the master that it was not proper to have his servant sleeping in his cabin, that it was causing some scandal and when he replied that this was casting aspersions on his sex, it was pointed out to him that it was therefore all the more unsatisfactory that he had not carried out this separation of his own volition. She therefore had to seek an asylum in the ordinary quarters in a hammock under the quarterdeck with the other servants. From the earliest days, these polite neighbours driven by curiosity wanted to visit their new guest, she was cruel enough to decline their offerings and to complain. Consequently they were punished and our so-called servant, to prove that our suspicions were ill-founded, assured us that in no way did she belong to the feminine sex but in fact belonged to

---

[1] Towards the end of his account Vivez states that she was believed to come from Picardy.

[2] Vivez uses the Spanish term *hombre* for 'man'. He wrote this account of the Jeanne Baret episode in a rather bantering and slightly salacious style.

the one from which the Mighty Overlord selects the guardians of his seraglio. After this scene, our man did his best to appear as he had claimed to, as much through heavy work or by the banter of his conversation. He worked like a black. During our period of call at the River Plate, she went to collect plants in the plain, in the mountains two or three leagues away carrying a musket, a game-bag, food supplies and paper for the plants that were always [...]. Scandalous gossip claimed that she suffered at Buenos Aires from an acute illness brought about by the care she gave her master to relieve him from the weaknesses he might have had during the nights when she watched over him. In the Strait of Magellan, these exertions doubled, spending entire days in the forest with snow, rain and ice to seek plants or along the seashore for shells. I do believe that she found herself repaid for this labour in the excursions by the rest she took in the plantings that her master was able to make when he found a soil suitable for a halt, as long as the harshness of the cold did not prevent it. Whatever may be the truth of it, it remains for me to say in her praise that she generally surprised everyone by the work she did.

The suspicions one had harboured lessened daily for lack of evidence, we were so accustomed to see her that opinions were divided when we reached Cythera [Cithère], but they soon revived. The Savage we had on board named Boutavéry, whose story I have told, having gone down into the ship's main cabin where all the crew had run to see him, saw our suspect in the crowd to whom he at once gestured from the bench where he was sitting, making proposals that were unequivocal shouting *Ayenene* which means girl in the local tongue. Since in the crowd she happened to be next to our armourer named Labare who had very effeminate features, they tapped him on the shoulder showing him and asking whether it was he, but with all the ardour and vivacity he could gather, he showed it was the other object, who lost her countenance and turning round went away with lowered head, but after this our Savage refused to eat and took no notice of anything we said to him. Nothing more was needed to confirm to the entire crew the nature of his sex and to convince the reader that her master looked disconcerted.

The next day, she went ashore to botanize with her master. As soon as they went down (for her master went down first), all the Savages tugged at her in one direction or another with a thousand shouts of *Ayenne* and already one determined individual was carrying her off within sight of her master as a starving wolf carries off his prey in the presence of the shepherd. It required an officer whom chance had brought there to drive off this multitude with his sword and frighten off the runner who let go. She was put back in the boat right away and sent back to the ship. After this incident, there was no question of going ashore. Boutavéri came fairly regularly every day to pay court in a manner that was very embarrassing for her and her master but amusing to us. Her master who was afraid of the Savage did not leave his side and overwhelmed him with small gifts. I do not know how chance led the Savage to say to him *Taratatance* that is to say Is she married? And as soon as he had made a sign in the affirmative, saying *maou*[1] without knowing what it meant, the Savage appeared to abandon his pursuing. He nevertheless continued to take a great pleasure in getting

---

[1] The word Vivez used, which changed Ahutoru's attitude towards Jeanne Baret, is *mahu*, a Tahitian transvestite. See Oliver, *Ancient Tahitian society*, pp. 371-4.

combed, powdered and dressed by him (or by her as it is time to agree with this [term]), which she did in a becoming manner. This game lasted until the departure from Cythera when Boutavéry left her regretfully to transfer to the *Boudeuse*, as I shall mention later. She remained on our ship where it can be appreciated there remained no doubt in anyone's mind after what had happened, but since there was no physical evidence, she dealt with the accusation by issuing some challenge to the servants who promised her an examination, which occurred at the next place of call in spite of the two loaded pistols she always carried by way of precaution and which she took good care to show them to impress them. Going ashore, one unhappy day when I do not know what had happened to the pistols, after having gone botanizing, her master left her ashore to look for shells and the servants who were there drying the washing took advantage of the moment and found in her the concha veneris, the precious shell they had been seeking for so long. This examination greatly mortified her but she became more at her ease, no longer compelled to restrain herself or to stuff herself with cloths. She finished the voyage very pleasantly, having suitors on all sides who did not lessen her fidelity towards her master. She ended up marrying a King's blacksmith at the Isle de France where I left them and have learned since that she was running a very happy household. Her master Monsieur de Commerson has since died in the same place without being able to give to the public his observations on the products seen on our voyage, which is a great loss for botany and natural history.

I preferred to conclude the story of Miss Baré supposedly from Picardy, although it took me away from my route, so as not to leave the reader in uncertainty and I shall now take up again my journal with the description of the island of Cythera.

This land, the extent of which we do not know, having seen only an area of about thirty leagues, is of an average mountain height, appearing broken everywhere from afar, the higher parts seem to form two islands on account of a low land of approximately three leagues distance and at the end of the bay I have mentioned. The south of this island seemed to me very precipitous and almost deserted although we saw some fires. I believe it to be inhabited only by deserters from the other one. From the beginning of the lowland, which is the most populated part and where the first chief of the island resides, up to the most northerly land, everything is inhabited and from stream to stream there is a new chief subordinate to the first. All along the shore of this northern land there is a low land edged by a sandy shore and some reefs a little offshore that make up a barrier for fishing of some twenty to twenty-five paces. A stream of good fresh water comes down from the mountains, the best we have drunk on the voyage and the one that kept the best. All the edge of this low land is covered with breadfruit trees, coconut trees, banana and other trees. Their houses are large and very well made, the uprights in herringbone pattern and mortised, six or eight make up a roof in cul-de-lampe,[1] the rest is small bamboo and the whole roof is made of artistically arranged latania leaves. The house of the king or first chief is in no way distinct from the others except that it is a little cleaner. The inhabitants are numerous

---

[1] A support, often ornamented, of an inverted conical shape.

here of several shades from mulatto to quite white who are I believe merely black-ened by the sun, all have black frizzy hair but none woolly. Most of the women as white as I had considered when I looked at the first who came alongside the vessel, having observed that three-quarters of those who were dark were elderly or vassals. The only occupation of the white women was to work under the huts seeing to the seines, mats and loincloths and to attend to the food.

They are divided into districts and from place to place they have a chief, each subordinate to the one of the low land. These chiefs are distinguishable by their white loin cloth as are the wives[1] and in addition both these are marked in blue or black with different patterns on various parts of the body according to their grade. They have a little hammer tipped with mother-of-pearl shaped like the quill with five to seven teeth which is used to draw lines on music-writing paper, and with that they strike the part, thigh or arm etc, where they want to make their drawings until they draw blood and the drawing is done. They have a little brush of pig hair they dip into paint and smear the area that had been pricked that never rubs out. All the other slaves go about naked with the exception of a belt covering their nudities. In addition the men let their beards grow which are greatly venerated among them.

As for their religion, I thought I noticed that they worshipped the sun before which they prostrated themselves when it rose and when it set all the more because, if one recalls, I have already stated that Boutavéri had asked us whether we came from the sun. Every evening upon its setting, they retire to their huts to adore some fetishes. They are broken shells, cut and sewn together in the shape of a bird with a long tail which they placed on large round platforms made of woven branches turning on a pivot. They also have some of these fetishes in all their boats. In addition they have small drums two feet long and six inches in diameter made of thick rushes over which they stretch dried pig's bladders, as well as some transverse flutes also made of small bamboo which they play with the nose, one hole underneath for the thumb and four holes above for the other fingers. The men play these two instruments at the same time.

They do not bury their dead. They have a religious leader differently dressed wearing a kind of sun perpendicularly on his head and a kind of collar over two feet wide all decorated with feathers and shark's teeth, who carries out the funeral cere-monies uttering horrible cries during which some of them have two quite large and thick oyster shells attached to a thread which they beat together producing a sound that lasts throughout the ceremony,[2] then they expose [the body] on a kind of stretcher where they leave it and come every day to pray and when it begins to rot, they carry it to a fire they make at the foot of the mountains where they reduce it to ash. Those who having committed some crime deserve a punishment are kept for

---

[1] There is inevitably an ambiguity in Vivez's and other observers' comments on island society because the French term *femme* means 'woman' as well as 'wife'. Nor was it possible for the French to suspect the degree of complexity of Tahitian society or the difference between a district or a local grouping living under a chief who was ruler of one or more villages. Although he refers to a 'king', Vivez also mentions senior chiefs and local chiefs, confirmation that the French realized that the island had a hierarchy of sorts.

[2] Banks described the costume as 'most Fantastical tho not unbecoming', *Journal*, I, p. 288. A pencil drawing, probably by Herman Spöring, of a chief mourner is held at the British Museum, Add. MS 23921.32. It shows that Vivez's description is accurate with the headdress of sunrays and the large collar.

times of storm. When they want to make it stop, they plead to the sun and sacrifice these unfortunates with blows of their spears and the bodies of those is not kept aside, as soon as the storm stops, they throw them in the sea.

The entire population is fully policed. Each chief cows all the slaves with a look. They have no personal property, everything is held in common. The chiefs give all they have to their subjects. They received our gifts with one hand and handed them out with the other. They cheated us and stole a thousand things from us but I believe it was only curiosity that made them do it. Their skilfulness taught them to filch with their feet as well as with their hands in spite of every precaution and care we might take during the night both in our hut and in the surrounding area. Finally in the full light of day by means of a fishbone attached to the end of a branch, they dragged the sheet from under one of our officers who was in his bed with a pleasant companion. They have tools made of a blackish pebble to work on wood and other small delicate instruments made of shell, all cutting very well and they have another stone to sharpen them. Their hooks for fishing which is very plentiful are also made with mother-of-pearl. They also fashion seines out of coconut fibres, but better made than ours.

Their food consists of a paste they make with bananas and a type of doughy breadfruit that comes from a tree shaped like a small pineapple which paste is used by them as bread, a great deal of poultry, pigs, fish also very abundant, the purslane and a kind of cress that we found most useful for our scurvy cases. Their soil seems very good. Mr de Bougainville sowed seeds of all kinds of vegetables and most were beginning to germinate when we left.

The [female] sex is generally very pretty and of a good lineage as I have already mentioned. The leading chiefs are the only ones allowed to have several wives, none is allowed to meet anyone but their husbands, as for the girls they meet whoever they please until they become pregnant, then she attaches herself to the most suitable among the workers who laboured at her vine. The king and the principal chiefs seem to have the right to choose and the women seem to pride themselves on pretending to. The men who are slaves seemed to me not to marry. I do not know whether they prevent the increase of the feminine sex, in spite of the reluctance I feel in imputing this crime to them, but it appeared to us that there were fifty men to one woman. The climate which is very hot in this region maintains the inhabitants in a kind of languor and is apparently harmful to them from the time of their birth, they being very poorly endowed by real Nature which has granted the fair sex all the favours likely to fill their days with happiness; there is sufficient proof of the structure of their men in that these women took the French for envoys from Vulcan and Venus without realizing themselves that they were granting a similar benefit. This charming sex, lacking all scruple, ignorant of every antiquated notion, came alongside the vessel to show us little signs of friendship by way of demonstration, displaying in order to persuade us all the attractions that Nature had bestowed on them to make us happy. As soon as we landed, they gave us half their clothing displaying every sign of passion and leaving us only with regret, and all the discomfort we felt, we who were not on our guard against this lack of scruples and the preconceptions of our climes, that we were unable to express our vulcanianism in public, because the crowd did not leave us. The

Frenchmen whose faces looked effeminate were constantly being tormented by the Savages who followed them everywhere and would seize them if they did not produce the proofs of their gender. They owed this to the servant in disguise whose story I have told, which led the Savages to think that they could find others. The women seemed to derive considerable pleasure from seeing us bathe when they were unable to see us in another way, which happened to us on a daily basis. Those who were to be pitied were the wives because the jealousy that reigns here as it does everywhere forced them to take steps to hide from their husbands who seem to me to be particularly inclined to refuse to share their most cherished belongings with a group of passing strangers from the other end of the world who, by giving a moment of pleasure to their loved partners, would make her realize how weak is the one she enjoys in her home. She bemoans her fate, envies the fate of all her fellow women in the belief that it was shared in proportion to the length of time she has spent with that foreigner, in the end sorrow gnaws at her, torments her by her inability to clear up this uncertainty.

The delights we tasted in this island, the beauty and the willingness of the fair sex made us give it the name of New Cythera. I would be more willing to endorse this choice, if the roses we had picked there did not have a stem lined with thorns of the type we wrongly believe we owe to Christopher Columbus, who surely never imagined that this part of the globe existed any more than the other countries as the origin of this sickness.[1] It is certain that it was not [brought] by the French to this country where it seems to have been prevalent for a long time, I am not speaking from hearsay but I saw two women who gave me sufficient proof and that I shall describe to the reader who would like to know. I had more from the report of my second surgeon, in addition to Boutavéri the savage we had on board who, on the third day he was there, gave us new personal evidence by which he made us understand that people were affected by it in his island but that their surgeons cured it with plants. A dozen victims in our ship and about twenty on board the frigate did not seem likely to have resulted from a first infection in the space of some eight days. I believe that one can agree with me that in all hot tropical countries where polygamy is allowed, this sickness is easily caught.

Let us now return to our journal and see what special events took place.

As soon as we had dropped anchor about two cables from land, Boutavéri got himself ashore in one of the canoes surrounding us. When he arrived a crowd of approximately three hundred savages gathered on the shore, clasped Boutavéri (dressed in the French style) in their arms and bore him in triumph on their shoulders under a row of trees; in the shade of their foliage, they sat around him forming a considerable circle, he talked to them for an hour and they rose all shouting for joy.

Mr de Bougainville, several of his officers and of ours, went to pay a visit to the leading chief who received them very well, made them sit on mats, and offered them

---

[1] These passages are convoluted and confused, with little punctuation and words omitted. The Rochefort version is more straightforward: 'This sickness that people claim came from Naples and that other nations blame on the French was certainly not brought here by them'. A comparison of the two MSS makes it evident that Vivez was endeavouring to rewrite his original version in what he believed was a more literary style, but that he did not have time to check what he had now written.

everything he had in the way of refreshments and some of his wives to Mr de Bougainville and the Prince de Narceaux [sic]. During this quiet interval, one of these gentlemen being at the edge of the forest fired a pistol at a bird, at the sound of which all the assembly was frightened to the point where it was very difficult to reassure them. In all this disturbance the prince lost a costly loaded pistol that may have been taken from him when they carried him on their shoulders as they always did to cross the streams. He complained to the chief and it was returned that very afternoon in the same condition as when he had lost it.

One can easily imagine that this scene cannot have given too good an opinion of us to the savages whom we saw carrying no defensive weapons, nevertheless in the afternoon, Mr de Bougainville had his soldiers landed with tents to camp ashore. When everything had been brought down, the local chiefs made us understand by gestures that we could do whatever we liked during the day but not stay overnight, by themselves picking up the effects that had been brought to return them to the long-boats. Mr de Bougainville after much talk and demonstrations, made them understand that we were coming for wood and water, and having given them eighteen stones, he showed them eighteen times the journey of the sun and that then we would leave. They gave us nine stones and made us understand that they wanted to allow us only nine days, finally after much discussion on both sides, they granted the fifteen days and that we could stay to sleep. We at once set up the tents, the king with several chiefs spent the first night there out of fear with Mr de Bougainville. The following days we emptied our barrels that we had landed taking care to rinse them carefully to remove the smell left by the stinking water from the Strait of Magellan. We landed thirty-five cases of scurvy who lived in a hut.

As we found the vessels to be badly moored, especially the *Boudeuse* which found itself close to the reef, we were forced to make haste and to re-embark our barrels as and when water was being collected.

The French my dear [com]patriots however much they pretend to be civilized, could not pass anywhere without degrading the nation, and so they were not slow in causing the savages their benefactors te feel the regrets that they should have foreseen at our presence. On the fourth day after our arrival, one of theirs fell victim by a shot to the brutality of one of the frigate's soldiers who was heated by wine and bed by debauch or greed into the mountains where, over some unknown argument, he fired a shot at him that went through his chest. This event did not actually lead to very much trouble but we saw them to be more reserved towards us and much more timid. On the day following this vile occasion, the *Boudeuse* which was moored among the reefs, had her cables cut, drifted towards us and caused us some damage. We were held only by the cable that was dragging us forward and we dropped a second anchor to balance this. The *Boudeuse* lost a second one, and had there been the slightest wind, would herself have been lost. In such an accident I wonder in such an eventuality what justice those who would have fallen into their power after starting to murder them could expect from a people who had shown them the utmost confidence.

In the evening, again we caught and put in irons four soldiers from the *Boudeuse* who had killed three savages with their bayonets, which correctly caused a dreadful tumult. All the Savages, men, women and children all fled from their houses running weeping into the mountains, taking away even the bodies of their dead in canoes which all fled to the open sea and those that were coming from the sea towards the shore also went back as the result of signals made by the Savages who were still on land. From all sides we heard only one shout *Tayo matao,* or 'friend kills'. Mr de Bougainville wanted to draw lots among the criminals to have one shot but as the sun had set and the preliminaries would have taken too much time, they were sent on board. People came to tell us that the Savages were armed and gathered in the forest. We took up arms several officers, in the moonlight and with the Prince, and went to the place that had been indicated where we found everything peaceful, the Savages collected in a group admittedly and the king alarmed. To all the Savages we met on the way we shouted in their tongue friends and they all replied in a tone of voice that expressed sorrow, friends you say and yet you kill. The prince and ourselves made various gifts to all those we met, that could not calm their understandable fear. We brought the king with us and several chiefs who spent the night under our tent.

Mr de Bougainville assured him that he did not want to do them any harm and that the criminals would be punished the following morning. Throughout that night we were on the look-out but that did not prevent the Savages from stealing from us two muskets, some swords and many other items from the very side of our people without their noticing it. As soon as it was daylight, Mr de Bougainville had the criminals landed and was going to carry out justice when all the savage chiefs gathered together and, moved by the fear of death evident in these unfortunates' countenance, out of a sheer humanity they would not have found in us, far from asking for revenge for their innocent blood that had been spilt, they all got together to beg mercy for them with a determined perseverance, which Mr de Bougainville granted after resisting for quite a time and if I may be allowed to express an opinion should not have done it. He told them that if they wanted them to be pardoned, he was asking them as an assurance of peace to bring back the Savages, women and others, who had fled, and that they wished them no harm. The king immediately sent slaves running [with a message] and we saw coming down from every part of the hills all these Savages laden with gifts in the forms of local foodstuffs which they gave in profusion to Mr de Bougainville without expressing the least bitterness towards us.

We learnt from these murderers that their crimes had been occasioned by a pig they had wanted to take by force and they lacked authority to do so, all the more in that for a flint button one could obtain a hen, for a nail or two a sucking pig, and for a hatchet when prices were at their highest one large pig. There was a quantity of nails that had been specially distributed to the crew for bartering. On the thirteenth of April at seven in the morning, the two cables of the *Boudeuse* having again been cut, she was holding only by a small anchor, her stern being very close to land. We lent her some new anchors and ropes that we were beginning to be short of, which brought us close to the point where we would have none, finding no other way of avoiding the danger, we raised camp on shore with such diligence that at six o'clock in the evening nearly all our belongings were on board. The next morning, Mr de Bougainville intending to leave and not being ready, asked Mr Lagiraudais against all his officers' hopes, if he could leave in the morning to ensure saving one of the King's vessels. Mr de Lagiraudais, who saw how precious time was, speedily carried out the finest manoeuvre possible. After readying the masts and the necessary manoeuvres, he cut the cable of the large anchor and left the longboat to raise it, he let the other one go which he gave to the frigate to hold her and we sailed with the greatest diligence, not going through the pass by which we had entered but following the coast going along the channel between it and the reef. We saw one league from where we had left a superb roadstead large enough to take sixty warships, but as the *Boudeuse* had lost seven anchors, that we had left some of them behind to secure the *Boudeuse,* we did not want to risk anchoring there. Once outside, we sent the yawl to guide the *Boudeuse* and spent the night tacking with a light wind and some rain. On the night of the fourteenth to the fifteenth, Mr de Bougainville had buried in the hut where we had camped an inscription taking possession of this island in the name of the King of France. As soon as the Savages saw us leave, they came in their canoes loaded with presents up to the *Boudeuse* and the leading chief gave the captain several gifts of local craftwork, begging him with tears in his eyes to take Boutavéry on board his ship and take him to our King. This was a very moving moment, in their canoes alongside the ship the women were weeping over our departure and could not hide the sorrow this caused them. They were constantly asking when we would return and we saw sadness give way to serenity when we gave them some hope of a return. Finally, we had to go but, unable to raise the anchors in spite of efforts at the capstan, Mr de Bougainville had recourse to a fairly strange stratagem. He opened two portholes opposite the capstan on the starboard side and had all the others closed. In front of the portholes there were three canoe-loads of women to whom he threw a few pearls and signalled to them to display themselves with all their charms, to which demand they bowed willingly as always so that our men working the capstan having seen them, driven by agreeable curiosity, pushed with all their might at the capstan in order to pass in front of the open port, which produced the expected result, the anchor that had been so stubborn was raised immediately.[1]

Off the Samoan Archipelago

After sailing for a fortnight, on the fourth of the month of May, we passed by three islands that we named the Three Cousins [Trois Cousines] on account of their similarity [to each other], very high, visible fifteen to eighteen leagues off.[2] They seemed to us to be very wooded, the seashore lined with huts with coconut and banana trees, seemingly however less populated than Cythera . Five or six rowing canoes came alongside, larger than those of Cythera, very elongated at the stem which is almost always in the water, and others coming back from fishing from whom we bought some flying fish eleven to thirteen inches in length.[3] The Savages here are redder than the Cytherans and Boutavéry understood nothing of their language. But he insisted that we kill them. They all have long hair, extremely fine white teeth, [are]

---

[1] This is a variant of the story that appears in Bougainville's printed *Voyage,* p. 190, but does not appear in his journal. Bougainville places this incident at the arrival of the *Boudeuse*; Vivez who was in the *Étoile* places it at the departure.

[2] In his Rochefort version, Vivez calls them the Three Sisters [Trois Sœurs]. They are the Manua Islands, consisting of Tau, Olosega and Ofu.

[3] There are a number of species of so-called flying fish, which belong to the *Exocoetidae* family.

all marked with black with drawings on various parts of the body, but for most of them it is only a panel that covers a whole thigh, an arm or part of the trunk. Their only clothing is a belt made of coconut leaves to cover their nakedness, they have no beard and do not appear to be as strong as the Cytherans. We did not see them with any weapons.

The next day, we coasted along another island to which we gave the name of Navigators [Navigateurs],[1] warranted by a dozen canoes with large sails that had an outrigger across it with its stay trapped to the top of the mast. A child placed on this outrigger gave the canoe the required balance according to whether he was closer or further from the end of the outrigger. Their ropes were made of coconut fibre and their sails of mats made of tree bark as at Cythera. They sailed with astonishing skill and speed, making twice the wake of the ship[2] which was making two leagues an hour, continually crossing and recrossing the bow so that we were afraid of crushing them at any moment.

They are apparently the same people as we saw the day before. They invited us to go to their place, seemed to be very good people, having no defensive weapons. They swim like fish, the ship making two leagues as I have mentioned, one of our men having thrown them a red jacket and a large pair of trousers that fell in the sea and were soon left behind the vessel, one of the Indians jumped into the water and went to fetch the clothes, we signalled to the canoe that was still following us to go and rejoin their comrade who was four leagues from land but they gestured to him to go ashore, which he did by swimming on his back. He made a bundle of the clothing, tied it around his body and went off to land. We were all the more surprised in that sharks were surrounding us on all sides. They have a strange way of catching them. All the Savages from one canoe hide, one beats the water with his oar and the one furthest back offer some bait at the end of a stick which he slips into a large noose of rope and as soon as the shark bites the bait, they drag his head forward into the slip-knot on which they pull and then they all together row the canoe until the shark drowns.

We then sighted an island off which we sailed that the rain and fog prevented us from seeing at our ease.

Off the New Hebrides [Vanuatu]

On Sunday twenty-second of May, we sighted a land in the shape of two islands ending in a low point. The *Boudeuse* which at the time was close to the land endeavouring to find the Bay of the Holy Spirit [Baye du Saint-Esprit], did not, as we did, see the land continuing towards the south, and seeing that it looked only like two separate islands ran before the wind. We had to follow her along the coast to discover a large round island remarkable for its height and fairly like the one named Boudeuse Peak in the land of Cythera and we gave this one the name of Etoille [sic] Peak [Pic de l'Etoille]. After that one we discovered several others that we counted up to ten, large and several very high and mountainous. We gave to all these islands the name of

---

[1] The island is Tutuila. 'Navigateurs' was the name the French applied to the whole archipelago.
[2] Going twice as fast.

Bourbon Archipelago [Archipelle de Bourbon][1] where we were surrounded as in a bay where we spent the whole night on various tacks to maintain our position. The next morning, we followed the coast as far as the furthest island without finding any appearance of an anchorage close to land. We saw three canoes with Savages who examined us from afar, not displaying the same confidence as those of Cythera. They stayed a short half-league away; there were four in one and six or seven in the others. The coast looked to me like a rampart with rocks very wooded everywhere which gives a green appearance to the shore. We sent the armed band towards the land to seek an anchorage quite close to the shore. We saw three other canoes surveying us from a distance like the earlier ones without coming near us. Our boat not having found any place suitable for anchoring, it was decided to go ashore to obtain wood for the frigate which was short of it and endeavour to get some fruits for our sick who were becoming numerous. When we arrived, we saw numbers of Savages coming down to the shore gesturing for us to go back to sea and pointing their arrows. They are mulatto, darkish red, of medium height, with ugly faces, frizzy hair, painted red white and black like the old monkeys of our colonies. They are quite naked apart from a banana leaf that is just enough to cover half their nudity. They also have blackened teeth covered with tartar, and bracelets of bone or fish teeth, most had a very large shell at the top of their forehead, the two sides of their heads shaved leaving only a finger length of hair growing in the middle from forehead to nape, all this to show the white shell they wear at the forelock. Some had very long bones five or six inches wide around their arms above the elbow placed on the inner side of the arm and tied with dyed ribbons, this was to make them stronger. For earrings [they had] turtle shells the size of a shoe sole, a piece of shell they place like spectacles on the cartilage of the nose, which almost blocks up the two nostrils and then rises up high towards the forehead. The reader will easily perceive that the taste for fashion has reached as far as these regions and that here it is most elaborate.

We also saw four or five women of whom two were fairly passable but the others abominably ugly, three had a breadfruit leaf to cover themselves and the others naked as in Nature unfortunately it was the ugly ones.

They all have their bodies covered with scabs and leprosy, which earned them the name of Lepers Island [Isle au Lépreux].

When we landed, the Savages moved away from us. One of them offered from a distance a small branch from a shrub as a sign of peace and the others always kept their arrows ready to be released and as we went forward retreated. About one hundred and fifty came along always out of the woods of whom a few finally approached us to whom we made a few gifts of handkerchiefs and some items of hardware that seemed to put them at ease. We entered the woods up to a hundred and fifty or two hundred paces from the seashore. There we wanted to cut down some coconut trees. They gestured for us not to. We cut down other trees they call *mapou* and some breadfruit trees they let us cut. They did not seem to be surprised at our presence and only tried to rob us. Their weapons were arrows most of them tipped with a fish bone on four

---

[1] Once more, sailing in the *Étoile* and not party to the senior officers' information, Vivez gives a wrong name: Bourbon was originally used for the Society Islands. Bougainville named Vanuatu the Great Cyclades [Les Grandes Cyclades].

sides, others with a piece of very hard wood eight to ten inches long with an infinity of teeth turned towards the handle, which are very dangerous all the more because we suspected them of being poisoned, the rest of the arrow is made of a reed. They have clubs of a very fine wood highly polished three to four feet in length, kinds of swords of the same timber and adzes like those of Cythera the cutting edge of which is made of bone. It is apparent that the Savages make war against each other, when on land, we saw them adjust their arrows at the slightest sound coming from the forest and remain on their guard on the side whence the sound came.

They have kinds of palisades made of a brownish stone with rushes or pikes to hold up the stones that make the shoulder-high walls. There are some in every path, we saw several they made in the forest very well aligned and several others going across these first ones.

A Small War

Having advanced a little into their ground, they gestured to us to go with them. A moment later they threw us a hail of stones. We fired in return two shots, then heard shouts from all sides. Mr de Bougainville ordered everyone back into the boats, the Savages still advancing towards us, aiming at us with their arrows intending to come within range of shooting at us. Fully convinced of their intent, we did not give them time for it. At the first demonstration, we paid them the courtesy of firing once more at them. One received a hit in the buttocks and left pleased with his gift together with his fellows, and left room for another, having approached from the other side, whom we entertained similarly with two discharges of all we had in the way of musketry. They fled at once leaving four dead on the spot. Moving away from the shore towards the ships, we saw the other Savages endeavouring to find where our arrows were. We heard them beat a kind of drum, this was presumably their call to arms.

We only found in this place, apart from the wood we cut, a few coconuts the size of a fist filled with brackish water, small bananas as scabby as the inhabitants of the country, also a few of these kinds of Cythera apples and a new fruit like a red fig-shaped cherry, in addition some cashews and some pigeons with grey feathers and red legs.[1] We kept the few local fruits we had for a few serious scurvy cases. I wanted to try to send one ashore for the land air, but he suffered such effects that we very quickly had to turn the boat back from the shore.

While the boat was away, several canoes came along that did not come near, making a thousand cries and gestures. We hoisted our boats back in and left for want of an anchorage. Throughout the night the Savages lit numerous fires on land. It [the night] went by almost without wind and in the morning we discovered some islands a continuation of the previous ones to which we gave the name of the Cape of Pillars [Cap des Pilliers] on account of its resemblance in their structure and position with the one I have already mentioned that forms the exit to the Strait of Magellan and we were hopeful that this one too would rid us of a passage where we could find no

---

[1] The Pacific pigeon, *Ducula pacifica pacifica*, is a fairly large grey and dark blue-green bird with bright red legs.

resources. In daylight all the fires seemed to be on the mountainous parts where presumably the Savages had retired, fearing a second landing. Their fears were groundless as in truth we had found nothing attractive in their country to make us want to.

I am convinced that all the vast land that surrounds the island I have just mentioned is not as unappealing as this, and that the Savages here are refugees from the others.

The next day twenty-sixth May at sunrise, we sighted three low islands well wooded approximately twenty leagues north of the former. We coasted along them during the morning, saw very few people, a few canoes with sails and oars that kept within speaking range. There were six or seven men in each canoe where all they had were their weapons and a few coconuts. They were inviting us to go ashore. In the afternoon the part we coasted along seemed more populated and cultivated, finding a bay that promised a good anchorage. We lowered the boats in which we went with our weapons. As we approached the land, five or six canoes came up towards us. The Savages are the same as those of Lepers Island although their land seems infinitely more attractive. They made signs for us to go ashore where we saw about sixty of them armed with bows and arrows. We circled the bay without coming close to shore for fear of grounding. One of the canoes in which there were three savages, two of them were rowing while the one in front, bending down for in case we saw him, shot an arrow at us that fell quite close to our boats. We made for it and the same man having shot a second arrow that came at the same distance as the previous one, we raised our oars to observe them. Those on land took up some branches with which they beckoned us. The Savage from the canoe who had shot two arrows had picked up his oar, and the one behind bending his bow to shoot, we fired three or four shots at them. One was wounded in the head. They all jumped into the water and went to join those on land who were still gesturing for us to go to them. We found other Savages along the forest who were waiting for us to be within range in order to shoot, but quite sure of their intent, we forestalled them with two discharges of blunderbuss and several musket shots. We saw three dead in addition to those who may have been wounded. Two canoes coming from seaward made for us and within half a musket range, they bent their bows and we opposed their plans as we had done with the others by musket fire. Like the others they jumped into the water and swam ashore, not waiting for any more. There were several in the woods on whom we fired who, as soon as they saw our shots, lowered their heads and as one can imagine the others could be misled by this. They also came as before as soon as we were out of range to look for our arrows and see how they were made. We rejoined the *Boudeuse's* boat which was a short league from shore and each went back to our ship....

[Vivez then gives brief details on the ship's route and on a visit to the Étoile *made by Bougainville on 28 May to discuss the shortage of food and obtain supplies for his frigate.]*

South of New Guinea

We continued sailing north, sighted several reefs, passed through considerable tidal races where there was an infinity of whole trees, branches, banana trees, small turtles, a great many fish which we could only look at, the water extremely muddy until the tenth of the month of June when we saw a substantial land mass to starboard and several small islets along this great land that is very high. In the evening, we saw it extending ahead of us with an infinity of islands.

We spent a fortnight tacking among these islands and breakers with the winds constantly blowing from where we had entered and preventing us from leaving, all the more because these lands appeared to continue further as we advanced into the wind. I would concede that it was imprudent to make such long tacks in this unknown part filled with shoals but our needy situation compelled us to. We had to succeed or perish, the moment was as pressing as it was uncertain whatever the position. The sea was dreadful throughout the time we remained in these parts, the roll greatly tiring the ship which was making water increasingly day by day and forced us to pump every hour, the continuous rain, all our rigging, ropes, sails, etc, breaking time and again, the mizzen mast considerably damaged, fortunately we had some preventer-stays[1] or rein-forcing stays fore and after to shore up the masts. By way of [poor] compensation and relief for the great weariness of our crew, to make up for the poor quality of the food, we had to reduce them to twelve ounces of bread, two ounces of vegetables and three quarts of water, as our sick were increasing considerably, their number including twenty scurvy cases with nothing to give them but water and bad rice and even then in small quantities. Our frightening state of uncertainty about our ability to get out from this abominable place where our only hope was in the tidal currents that seemed to run in our favour. During our stay in this gulf we caught two big-ears, a kind of tunny fish,[2] and two sharks. The reader can rightly imagine that we did not reject the latter in spite of our prejudice and that on the contrary we found them delicious, and often yearned for some although without result. We saw no inhabitants on these islands, [but] they light numerous fires on all of them. It would be useless for me to describe all these islands to the reader, all the innumerable reefs and breakers we saw during this painful stay as I believe no one would want to come here for a change of air, anyhow if there are a few I will give them some more detailed journal entries.

All these islands we have rounded make up in my view the southeastern extremity of New Guinea. The first land we sighted and which was the most northerly certainly had the appearance of an extensive land and situated in nine degrees forty minutes south and one hundred and forty-three degrees west. I estimated from the currents and other indications that there could be a passage to go between New Guinea and Carpentaria [Carpenterie].[3] We named this part Gulf of New Guinea [Golphe de la Nouvelle Guinée].

The twenty-sixth of June, twenty-first of our stay in this gulf, we finishing sailing past these lands. Seeing none on any side apart from the one we had just left, we set our course for the North-North-East. In the afternoon we lowered our small

---

[1] Preventer-stays (also 'preventer-shrouds', 'preventer-braces') are auxiliary ropes used to strengthen or replace the main ropes.

[2] As Vivez surmises, this is a kind of tuna, probably the large Pacific *thunnus*.

[3] Thus Vivez speculates on the possible existence of Torres Strait, which was still in doubt at the time. Carteret recognized that Bougainville came close to solving the riddle 'when he fell into the entrance of a Streights, which from its appearance might have lead him too far out of the way of the Mollucca Islands:' Wallis, *Carteret's Voyage*, I, p. 272. It fell to James Cook to settle the matter once and for all.

canoe and sent four *tierçons* of brandy to the frigate. Her crew had not had a drinks ration for three days. Like us they were famished. To ease our misfortunes, we found in the hold two barrels of flour and one of rice that had been eaten by rats leaving nothing,[1] although we have not stopped setting traps for them for a month and the sailors make their favourite dish out of them when they do not prefer selling them to us for twelve *sols* which is the fee imposed, but unfortunately they do not catch many of them.

Dangerous Archipelago

On the twenty-eighth in the morning, we sighted a very low island which became multiplied as we progressed. In the afternoon we saw among these some that were fairly high. We urgently needed to find a way through these islands (which are innumerable and grow like pumpkins) to go up north and seek an exit from this famous Western Sea that we have been sailing through for so long and enter the Molucca Sea [Mer de Molluc] where we hope to be assisted by the Dutch. The next morning, we saw reefs extending from one island to the next where we wanted to pass through that were about three leagues off. The calm that caught hold of us forced us to let ourselves go with the current and we were on the point of lowering our boats to haul the ship against the current, for fear of being dragged onto the rocks.

The next day, we sent a boat from both vessels to seek an anchorage or a pass. During their absence, ten to twelve canoes appeared on the coast and came up gradually around the ship. They were shouting a great deal, making signs with their oars that we should go ashore, that there was drink, food and rest available. There were twelve to eighteen men in each canoe, they have no outrigger, they are at least six feet high fore and aft, twenty to thirty feet long, two to three wide. The Savages are fine blacks, of a good height; the inside of their mouth is quite red and their teeth black which is caused by the betel they continually chew. They had boar's and fish teeth forming bracelets up to above the elbow, collars in three and four rows that seemed to be made of shiny pearls, earrings of mother-of-pearl as big as a hand. They had bows and arrows of a good size made of a black wood tipped with fishbone, a red ribbon at the other end, some spears, some shields well made of rattan and straw of different colours proof against any arrows, a few even against bullets. It appears likely that they would have shown us little quarter if we had fallen in their power.

The betel these Savages use, as in all the Molucca islands and Dutch colonies, is a plant from India whose leaf is similar to that of the currant-bush but with an extremely strong taste. This plant cannot be used without something to relieve it; the Indians chew it constantly with areca nuts and shell lime.[2] The areca nut is the fruit of a kind of palm tree from the East Indies the size of a large nut enclosing a stone the size of a nutmeg, it is a tart and bitter substance.

The winds becoming a little strong, the canoes left and our boats returned not having found any place where a ship might shelter, they did not go ashore but went

---

[1] Obviously, Vivez is writing this tongue-in-cheek, one of several occasions when he resorts to sarcasm or some form of heavy-handed humour.

[2] The betel *(Piper betle)* belong to the *Piperaceae* or pepper family. The areca nut is wrapped in a betel leaf, and a little quicklime is usually added to it.

very close, they saw several savages who did not seem surprised at seeing them. They saw numerous huts built of banana and coconut trees, finding no anchorage or passage, we decided to coast along these islands. Shortly afterwards we found ourselves in a tidal race strong enough to prevent the ship from making way although there was a strong gale, in which we remained almost three quarters of an hour, the sea becomes very agitated with rain and fog.

Further danger

On the first of July we found ourselves in the same tidal current as on the previous day unable to steer. We lowered the small boat to search for a refuge in a place that seemed very likely to contain one, being formed by three fairly large islands. We made for the entrance of the cove where our boats had signalled an anchorage. We saw an infinity of breakers extending across the channel. We thought this might be an effect of a tidal race, but the currents bearing us towards it bore us within earshot of a terrible roaring caused by the breakers which were rising almost to mast height and, arising in this way, made even the boldest tremble. We closed all the ports, but we were on the verge of being ground down as in a razor-sharpening mill. We did not dare bear away for fear of meeting some dangers on the side where the small islands were. In these circumstances which were bringing many thoughts into our minds, we saw ourselves being gradually carried towards the tail end of these breakers which we had to pass, fearing that our end was near, although the sea was much less stormy there than in the places we had sailed through. During these events we saw the bottom quite clearly below us. We thought for a long time that it was the currents swirling below us. We set all sails to get out of this cursed spot. Gradually we made away from the highest breakers, after which we made a new attempt to reach the anchorage that had been indicated, but having found new dangers, we had to give up, hoist back our boats and sail west to get away from all this land and although we were a distance away, we could still hear the roaring of the reefs. Our boat reported that the place where they had taken soundings was a superb harbour one league deep, horseshoe-shaped. The Savages came to attack them numbering twelve canoes with fourteen or twenty men in each well armed in the savage style. They fired several volleys at them, some of whom were hit, the others fled as best they could. The end of this skirmish took place in the middle of the bay where they captured two canoes. About forty then came to attack them that soon turned back when they saw the two canoes. These canoes were all planked and decorated with shells of the type of white *coris* called the egg[1] and also with bones of human jaws and arms.

On the second at dawn, we sighted several very high islands which, while increasing still further the number of our discoveries, did not provide us with the passage we wanted. All the maps of the travellers who claim to have sailed in these parts are quite false in respect of the position of the land. Meanwhile our wretchedness was increasing daily. Our men had lost their strength and we were short of everything. Need made us carry much canvas by day and by night sometimes more than we were in a condition to carry considering the condition of the ship and of the crew. All this

---

[1] The cowrie *(Cypraea),* a small marine snail common in the tropical waters of the Pacific and Indian oceans. The shell becomes ovoid as the animal matures.

latest land was called the Dangerous Archipelago [Archipelle Dangereuse].[1] The next day we saw, to the north of the new land which we had approached, several canoes leave and come alongside the *Boudeuse* where they stayed about an hour, then left. Mr Duclos, second-in-command of our companion ship, who came aboard the next day, told us that the canoes I have just mentioned were all armed with arrows, that one of them having come alongside, they gave them quantities of old clothing, that they then rowed away some sixty feet, and shot an arrow at them that landed on board. We sent on that day to the *Boudeuse* eight *tierçons* of brandy and eight *coiros*[2] of flour, and about two cords of wood. They are short of everything as we are. They had only enough water for twenty days. Once these items had been sent, we hoisted our boats back in and sailed to the extent the breeze allowed us.

New Britain

On the sixth [July] very early, we again sent our boats to seek an anchorage in the landmass we had discovered the previous day. As soon as they had signalled that they had found one, we made for a bay, leaving to port a very steep small island and as we advanced, we saw within a point to starboard the place that was intended for our anchorage and to port a large channel formed by a circular island separated from the main land. We dropped anchor in this starboard bay quite close to land sheltered by two islets at the entrance.

Need had made us want a place of call for a long time but we could not expect to find a less satisfactory one. Let the reader recall the state we were in and visualize us surrounded by a high and very steep shore, lacking any kind of inhabitant or animal, in a word a land filled only by an extraordinarily dense forest and with no fruit and no type of fish. Our only benefits were water and wood. Several streams of excellent drinking water come down from the mountains, a cascade right at the top formed by Nature making a marvellous picture, the water as it rises from its source springing up in the air in a jet twenty feet high, this area is surrounded by three basins shaped like holy-water stoops dug into the rock and which one might say were placed by the hands of an artist, twenty feet below and in steps on the seaward side is a large basin with several holes at the bottom which hurl water twenty feet away where it falls into another basin, finally from one tier to another and from one fall to the next, it reaches the sea by a short channel, having been well tossed up on the way. This cascade makes a considerable noise, one climbs up on each side by a kind of ramp, fairly difficult of access.

We found in a small sandy cove half a league from our anchorage an abandoned native canoe, and an English inscription in lead on the seashore that we judged to be from the English who preceded us in Galant Bay [Baye Galante] in the strait of Magellan. It is surprising that we kept to the same parallel and that we met exactly where they had anchored. We busied ourselves obtaining the necessary wood and water, repairing our rigging and masts, and we shared what food we had left with the

---

[1] The term had previously been used in the Tuamotus. The ships have sailed to the Solomons and are now reaching New Britain and New Ireland.

[2] *Cueros*, leather bags used to carry flour.

Boudeuse. The man Tonnelier, a sailor, died of a relapse of an inflammation of the lower abdomen. This is the first man we have lost. He was buried on this island.[1]

On the fifteenth we had an eclipse of the sun that lasted more than an hour, its immersion at eleven thirty and its conclusion at three minutes past midday. It was observed by Mr Verron, astronomer, which, by settling the correct longitude of the place, confirmed our view that we were in New Britain [Nouvelle-Bretagne] ....

Finally, on the twenty-fourth, the rain being very heavy and the wind almost calm, the frigate raised anchor towed by the boats of the two vessels and when she was outside the boats returned to fetch us and get us out in the moonlight by the same entrance as we had arrived. Throughout the night we stayed a good distance from land awaiting our longboats that came the next morning. They had gone to get our anchors, we hoisted them in and sailed off getting away from this coast of New Britain, always with the most abundant rain, and consequently we named the bay where we had entered Choiseul Bay [Baye Chouseulle] and the place of our anchorage Rain Harbour [Port à la Pluye].

The twenty-eighth, with very little wind, we continued to follow the same coast, we sighted in the offing, distance four or five leagues, several small round islands following the trend of the large land.[2] A canoe came up alongside containing nine Savages, all blacks, fairly handsome men and quite naked, gesturing to us to go ashore. They all stood up and began to shout together to endeavour to frighten us, at the same time, they threw about fifteen fairly large stones, still shouting. We were not expecting this attack, not believing them to be so bold, but a couple of musket being in a state of readiness, we fired about fifteen shots at them. Seven of them jumped into the water swimming to push their boats stem first, when they felt themselves out of range of our fire, they reimbarked, being still within range, we fired a few more shots and, looking with a glass, we saw one who was taking up a handful of water to put out the burning sensation in his back one of our bullets had made.

The next day, six canoes came up larger than any of those we had seen so far, with no outrigger, fairly well finished, adorned front and back with a carving of a fish design and painted. They contained thirty-two men in all. They are woolly and black, most having their wool dyed red, white and other colours, their mouths red and their teeth black, ruined by betel, a kind of red ribbon threaded through the nose, others appearing to come from the mouth, earrings and pieces of shell hanging over their stomach. Most had a piece of mother-of-pearl around their arms above the elbow, long feathers both spread over their heads and attached to their arms. Their weapons were spears, most of which had a piece of human arm-bone at the handle, clubs, types of wooden swords, picks, javelins, bows, arrows and slings around the forehead or hung tightly.[3] They remained around the ship until eleven thirty. One climbed behind

---

[1] This is another instance of the unchecked nature of the Versailles manuscript. The dead man was one Thomas, a barrel maker (a *tonnelier*).

[2] The *Étoile* is now sailing in a northerly direction off the coast of New Ireland. Vivez mentions the succession of islands the expedition will see on their starboard side: the Feni, Tanga, Lihir and Tabar groups.

[3] The text, literally, has 'or that they hold hanging in slavery'. The Rochefort MS has simply 'that girdles the head'.

the windows of the main cabin, wanting to take a basket hanging there, we made signs for him to go away but as we would not obey we had to resort to a few blows of rattan cane, he jumped into the sea and returned to his canoe. Shortly afterwards, they sent a small canoe to the shore in which there was only one man as a light frigate to seek reinforcements and when they believed themselves to be strong enough, they began a fight, which was quite brief, by throwing stones at us with their slings the first of which fell in the foretopsail from two hundred feet away, and upon the third one which was the agreed moment, we gave them a volley from four blunderbusses and some twenty muskets. They nearly all jumped into the sea and most abandoned their canoes. When they considered themselves far enough from the ship, those who could clambered back and returned to shore.

The next day thirtieth of the same month, with a light breeze, still following the same course, at sunrise, a quantity of canoes came towards us holding eight to ten men each, all dressed in the same style as on the previous days. We traded in a friendly fashion some pieces of cloth that seemed to please them for figs, bananas and cooked potatoes, which gave us the hope that we might part good friends, but they soon showed us that this was not their intention. When we felt that their visit had lasted long enough, we gestured for them to leave because their number was increasing considerably. They felt this insult, moved away from the ship and, uttering great shouts and clapping their hands to warn all those who surrounded the vessel they began to ready some thirty slings and swing them around with the most insolent expressions. We did not give them time to use them, two blunderbuss shots followed by several musket shots, there must have been quite a number of killed and wounded in this affair. When we fired, they always bent their heads and showed their backside.[1] Others jumped in the water and swam to push their canoes away from the vessel. All their dressing up became useless in consequence and if they had done it in order to capture us, they were mistaken especially as we were quite sure that our misfortunes would not have moved them and that they felt no other compassion for us beyond eating us like the cannibals they are which they have certainly done since then with Mr Marion.[2]

Three hours after this battle, four other canoes came alongside the frigate where, after bartering a number of items, they shot an arrow that landed at the feet of the Prince of Nasseau. They then did what they could to come to us, but as we were sailing too fast, they could not catch us up. This wind did not last long and in the evening we saw another three canoes looking at us. When night was falling, we amused them as well as the frigate with flying rockets, fear seized them and they vanished. We had had this attention throughout the previous evening. It was always our turn to engage in these affrays, we had some red weathervanes on our masts that delighted the islanders so much that none went to the *Boudeuse,* which situation had lasted since Cythera.

---

[1] The display of buttocks was a standard form of challenging or displaying contempt towards an enemy.

[2] The Rochefort MS merely says, 'we were in no way willing to serve as food for these cannibals.' The reference to Marion confirms the later date of composition of the Versailles MS: Marion Dufresne, leader of the *Mascarin and Castries* expedition, was killed and eaten by New Zealand Maoris on 12 June 1772. See Duyker, *An Officer of the Blue*; Kelly, *Marion Dufresne at the Bay of Islands.*

At sunrise, some twenty canoes came as on the previous day, they surrounded the becalmed ship, these were gesturing to ten others that were on their way to come quickly and seemed better disposed than ever towards a combat. We also saw coming from the land three kinds of catamaran[1] which are entire trees tied and threaded with branches so that they can neither capsize nor sink. We made signs for them to go away but on the contrary, believing that they had us already in their control, they surrounded the vessel with greater attention, making a great deal of noise and speechifying, especially the jugglers.[2] Finally as they did not want to go and [we did not want to] let them increase in number, we fired several volleys of our usual light musketry that forced them to abandon several of their canoes alongside the ship and two were sunk.

This latest battle was much more bloody than the previous ones and lasted close to half an hour, there were many wounded, we saw three or four dead thrown onto the catamarans. They stared at each other and seemed very surprised, either at the effects of the gunpowder or the whistling of the bullets, and so after the second shots as always they lowered their heads, showed their backsides or jumped in the water.

In the afternoon a canoe of Savages came up from the small islands I have already mentioned which are scattered along this coast and are five or six leagues distant from it. They appeared most affable, absolutely similar to the others except that they were circumcised. All the top of their heads was painted blue and the sides red. They came alongside the vessel, we traded one coconut, their arrows, small bells the size of a thumb made of little cylindrical shells, and collars of small *burgeau*.[3] We gave them in exchange much more than they expected. They made us understand that they were at war with the people of New Britain with whom we had daily dealings. Shortly after their departure, a second one came from the same district with whom we dealt in a similar manner.

All this part of New Britain from Rain Harbour is very attractive, flat and well populated, however we saw nowhere to drop anchor. In the first days of August, we were again visited by a few canoes from New Britain who found themselves in too small a number to wage war against us. We finally sailed past these lands we had been hauling along for several days and which end in a low drowned point. We were again greeted by a selection of rain, calms and storm.

Thousand Islands

Ten leagues away, we again saw several drowned islands of a circumference of about one league through which the winds forced us to pass, in order to, from there, fall in with a multitude of others, which led us to name this place the Thousand Islands [Milles Isles], all flat and mostly drowned, among which we stayed until the eleventh of the month. As these islands multiplied as we made what little progress the constant storms and rain allowed us, we left these low islands to sight three others extremely high with hillocks higher than any on the coast of Brazil or in the strait of

---

[1] This provides a good example of Vivez's highly erratic spelling. He refers to these craft as *'cartié maron'* in the Versailles MS and as *'quartié maron'* in the Rochefort MS – which translate as 'brown quarter'.

[2] I.e. the leading warriors who were issuing challenges, as against the rowers.

[3] The *Turbo marmoratus* is a gastropod found in Indo–Malayan waters whose shell is widely used for decorative work in Malaya and Japan.

Magelant. We coasted along them slowly, the weather unchanged.[1] They were followed by six or seven others and several islets. We also passed through several considerable tidal races of a yellowish colour carrying entire pieces of wood and a thousand spoiled things from the shore. The sea which was very rough in all this area, caused us much trouble, our leak being always the same and a piece ten to twelve feet long fell off our sheathing.

We sent for the frigate's chaplain to administer [final rites] to two moribund scurvy cases who we estimated would not last the next day, being unable to give them the slightest food other than rice water.[2] These two were additional to twenty other declared cases and the remainder of the crew very ill.

The frigate is in no better condition than we are, we have continued to sail past new islands one after another, details of which would require a volume which would make me exceed the limits I have set myself. We were gradually approaching countries that had been visited by a few travellers and we had to get out of this immense Western Ocean in which we had been sailing since the strait of Magelan to enter into the Molucca Sea where we hoped to find help among the Dutch. Time was pressing because the storms we had encountered for some time were announcing the approach of a season when the winds are constantly unfavourable, which made time very precious to us in addition to the extreme necessity in which we were placed.

Departure from the Western Sea

We therefore endeavoured to get out on the twentieth of August by the strait of Salvati [Détroit de Salvati] but the contrary winds that had already reversed in this region and the constant presence of dangers made us abandon this plan. After sailing towards the open sea, all having our wretched dinner, we saw suddenly a rocky bottom twenty or thirty feet below us which took only a moment to pass, but if it had had teeth long enough to bite us as we passed, it would have been quite fatal after having avoided all the dangers we had come upon. On this shelf we captured a horse-mackerel,[3] a deep-water fish caught on a single line we had out, we showed it no quarter and out of politeness invited it to dine with us unceremoniously with its bones.

[The subsequent entries deal with the passage through Dampier Strait and the voyage to the Dutch establishments.]

---

[1] The *Étoile* has now sailed away from New Ireland, past the St Matthias Group and is sailing through the Admiralty Islands. The name Thousand Islands did not originate on the *Étoile*; it was used because the French thought these islands might have been those so described by Jacob Roggeveen in July 1722.

[2] Martin-Allanic writes of brandy *(eau-de-vie)*: *Bougainville*, I, p. 777. However, both MSS state clearly that what the men were given was *eau de riz*.

[3] 'Probably a *scombresox saurus*, a very long bony fish' (ET).

APPENDIX 4

The Journal of Fesche

Charles-Félix-Pierre Fesche, a volunteer on the Boudeuse, *compiled a substantial journal totalling 1,015 pages bound in three volumes, which is held at the Musée d'Histoire naturelle, Paris, under the reference MS 1896–1898. It begins with the departure from Paimbeuf, upriver from the port of St Nazaire, and ends with the stay at the Isle de France. It was written up during Fesche's stay in Mauritius, although it follows the standard pattern of a shipboard journal. However, it is not simply a fair copy of Fesche's original journal and there is considerable evidence that he collaborated with Saint-Germain in rewriting his primary text, and furthermore that he asked Commerson to read and comment on it. The issue of the true authorship of this manuscript has been the subject of speculation for many years.[1] It seems clear that Fesche and Saint-Germain collaborated up to the point when the expedition sailed from Tahiti. However, Saint-Germain's own manuscript cannot be found and until it is, the part played by each man in the compilation of this journal cannot be elucidated.*

In the following extracts, annotations clearly made by Commerson are shown by way of footnotes.

Fesche begins with the departure from Paimbeuf in November 1766. Almost all of the entries until the Boudeuse's *arrival at the River Plate in February 1767 are of a navigational nature. He then devotes several pages to comments on the life, flora and fauna of the countryside. Navigational details on the voyage to Rio de Janeiro follow, with some details on the town itself, the same procedure being followed with the move to Buenos Aires and the voyage to and through the Straits of Magellan. He has only a little to say about the voyage towards the Tuamotu Islands and, as will be seen from the following extracts, he is sparse on the islands themselves, commenting: 'I am not aiming to give a description of these people because, if we anchor, I shall detail the circumstances and story of the place, not having yet seen enough of it to report on it'. This was his general practice, and when he reaches Tahiti he continues with navigational details and little more than a mention of the Act of Possession. He then, however, compiles a quite comprehensive report on Tahiti, its products and inhabitants.*

Through the Tuamotus
Monday 14 [March] to Tuesday 15.

Variable East to NE winds light breeze, almost calm; all sails set, the studding sails rigged; at 3 p.m. lowered a boat to take a harness–maker soldier to the *Étoille*[2] to make

---

[1] See Dorsenne, *La Nouvelle-Cythère*, p. xxiii; Taillemite, *Bougainville,* I, pp. 124-7; Martin-Allanic, *Bougainville,* I, p. 452.

[2] Fesche regularly spells the name of the second ship with a double 'l'.

leather breast-plates in case of Savages in the islands we are going to; the iron ones provided by the King are rather too heavy to be worn in hot countries; on board our ship all our gentlemen have one.

Thursday 17 to Friday 18

The winds variable from South to East light breeze, fair weather, the sea smooth, all sails set. At 3 p.m., the *Étoille's* boat came alongside to bring some papers for Mr de Bouguainville.[1]

Monday 21 to Tuesday 22.

The winds variable from ENE to East fresh breeze, fine weather, smooth sea, all sails set. At 6 p.m., lowered the studdingsails and the topsails, clewed up courses to wait for the *Étoille* which is very far behind us. At 11 p.m., being abeam of us, let out all sail after speaking with her and told her that at midnight we would alter course, that is to say that, since we are sailing W¼NW, at midnight we would steer W. At 1 a.m., reduced canvas to await the *Étoille* which we had lost from sight, at 6 a.m. the *Étoille* signaled the presence of land consisting of 4 islands bearing from us SEE5°E [sic] distance 7 to 8 leagues. At 6 a.m., while the *Étoille* was signaling land, we saw a 5th island straight ahead of us bearing W, distance 5 leagues. At 9 à.m. we signaled the *Étoille* to take soundings, which she did and lowered 200 fathoms of line without finding bottom, being one and ½ league from land. At 9 a.m. we were N and S of the W point, distance ¾ of a league. At 9 o'clock, as Mr de Bouguainville had not seen any place suitable to land or safe for boats, he set sail; we immediately saw 3 men coming out of the woods and at first we believed them to be Europeans, Mr de Bouguainville right away steered close to the wind to take these Europeans off if such they were. The first 3 men went to hide in the woods and ¼ of an hour later we saw 12 or 15 of them; our gentlemen saw through their spyglass that they were carrying long sticks,[2] I have no doubt now that they were Savages. We set sail at once. Once we were well away, the Savages lit fires at several places on land. The *Étoille* spoke to us at 11 o'clock and told us they had seen a suitable place for a landing, 30 to 40 men and a dozen huts and at the East point a canoe ashore with a sail. The 4 islands and the 5th lie NE and SW, distance 12 to 14 leagues.[3] The island is very attractive, seemed to me to be lying East and W and may be 2 leagues in length, is well wooded everywhere, we saw palm trees, coconut trees, lemon trees, I am quite certain that there are plenty of refreshments to be taken from this island, it being inhabited by Savages. The sea is very unsuitable for a landing, especially the two points where the sea breaks with incredible strength. The entire shore is covered with sand. Its latitude is certain, as for its longitude it is only estimated.

Tuesday 22 to Wednesday 23.

Variable E to SE winds light breeze, the weather stormy, thunder, lightning, rain. At 4 a.m. we clewed up the topsails during a squall. At 6 a.m. saw land bearing

---

[1] This is Fesche's standard spelling of the commander's name. This same spelling appears in Saint-Germain's text, but with another and more frequent variant: 'Bouquainville'.

[2] The island is Akiaki; the 'long sticks' mentioned by Fesche were wooden spears, which led to Bougainville naming the island Lancers Island.

[3] 'Latitude of the first 4 islands 18°49′, longitude: 141°28′, latitude of the second 18°30′, longitude: 141°52′' (Note by Commerson).

NE¼N, distance 3 to 4 leagues, then we saw it stretching as far as the NW. At 6.30 a.m. we hove to for an hour to wait for the *Étoille* and at the same time let the weather ease until 7.30 a.m. when we hoisted up the topsails, stormy gale then calm until 7.45, when the wind veered to SE. Then we sailed close-hauled towards the NE and making for the land. At 8 o'clock changed to the other tack. At 8 a.m. we sighted the East point bearing NE¼E distance 4 leagues, the W point NW¼S [sic] distance 6 leagues and we went about to get to windward of it all, then follow the land, which we did. At 10 a.m., we saw that one league from land there was a bar that was inaccessible to boats; we immediately sailed along the coast, continuing to find the same bar. The said land is all drowned and detached in islet[s] distant ¼ league from each other, between these islets there are low lands, the islets are well wooded; the said land seems to extend quite a way because I have seen at least 20 leagues of land and the further one goes the more one discovers. The bar goes on all along at approximately the same distance; on the other side of the bar, one sees about one league of sea as calm as a pond. At midday sighted the E point bearing NE¼E, the most W islet NW. At midday saw fires lit by the Savages; saw also 5 or 6 canoes both with sails and with oars, one with a sail which is following the coast inside the bar and goes far better than the others. With the spyglass one can make out from the ship their rigging and 8 men in it.

Wednesday 23 to Thursday 24.

The winds variable from East to ENE light breeze, fair weather, fair sea, all sails set. At ½ past midday sounded and let out 100 fathoms of line without finding bottom. At 4 p.m. we sighted the East point of the tongue of land bearing SE¼E distance 3 leagues, the most westerly land NW distance 4 leagues, saw the bar all round the coast and the canoe with a sail apparently coming out of the said tongue to come alongside us; consequently we changed course from time to time remaining during the night approximately so as to sight land in daylight. At 6 p.m. we sailed SSE, at 3 o'clock altered back sailing N¼NW. When we were facing the South, we fired 4 flying rockets to try to astonish the Savages if they see them, the *Étoille* did likewise. At 5 a.m. let out all the reefs in the topsails that we had taken in for the evening and let out all sails to make for the land being rather far from land. At 6 a.m. we sighted the most W land bearing N¼NE distance 4 to 5 leagues. At 7 a.m. we saw that the said bar dominated the whole of the coast and that there was no way to land, the said land being drowned on all sides; consequently we bore away to W¼NW and continued on our way. As we bore away we sighted the most W land of the said island NNE of the compass. At 8 o'clock sighted the most N land bearing NNE5°E, distance 4 to 5 leagues. At 9.30 sighted the W end of the tongue of land bearing ENE5°E, distance 6 leagues.[1]

Thursday 24 to Friday 25.

The winds variable from S to NE through East light breeze, the sea fair, all sails set, squally and rainy weather; at 5 p.m. saw land from the topmasts, at sunset we sighted the most N bearing WNW3°W distance 6 to 7 leagues, the most S SW¼S2°S

---

[1] 'In the margin of the other manuscript was written: 'my bearings give me for the island's latitude: 18°12′, longitude: 143°13′.' (Note by Commerson). This position is that of Harp Island, present-day Hao, in latitude 18°10′ to 18°30′ and longitude 143°14′ east of Paris.

distance 6 leagues; at 6.30 we took in all the reefs in the topsails and sailed close-hauled. At 8 p.m. light airs, let out one reef in each topsail; at 10 p.m. light airs, let out all the topsail reefs. At 10 p.m. went about running SE. At 10.30 p.m. we spoke to the *Étoille,* they told us that by their reckoning the said land extended from the NW to South of the compass. At 2 a.m. altered course. At 5.15 sighted the same land from W to SW¼S, the weather squally and rainy. At 8 o'clock sighted the same land the most North bearing W¼SW5°S distance 3 leagues and the most South SSW distance 5 leagues. At 9 a.m. we sailed towards NW until midday when we changed to W, the land running on to the NW, I believe this land may be 15 to 20 leagues long.

Friday 25 to Saturday 26.

The winds variable from NE to ENE light breeze, fine weather, fair sea, all sails set. At half-past midday saw another land bearing SW distance 6 leagues seemingly running towards the SE; at one 15 saw another land to the NW¼W seemingly running to the NW. At 6 p.m. sighted the land the most NW of the land seen at 1.15 bearing NW¼W to the NW5°N distance 7 to 8 leagues. At 6.30 p.m. furled the studdingsails and topgallant sail, at 10 p.m. took in the main sail, at 11 o'clock the foresail. At one in the morning broached to wind to starboard and signaled to the *Étoille* to do the same. At 5 a.m. filled the sails. At 5.30 saw another land to the South distance 4 to 5 leagues seeming to trend SW. At 7.30 a.m. sighted some breakers to the SSE that appear to end the SW point of the said land distance 3 leagues. At 8 a.m. we sighted the NE point of the said land bearing SE5°S distance 6 leagues. At 9.30 we saw the said land of the breakers bearing ESE5°S distance 6 leagues. At 9.30 saw another land to the NW¼N distance 6 to 7 leagues looking like 3 small islets detached from each other very low. At midday we sighted the said land looking like islets bearing the one N¼NW, the 2 others N¼NW5°W distance 6 leagues, the port one, that is to say the one where we saw breakers, bearing ESE distance 8 to 9 leagues.

Saturday 26 to Sunday 27.

The winds variable from NE to SE light breeze, fair weather, fair sea, all sails set. At 4 p.m. the *Étoille* signaled the presence of land. At 7 p.m. we spoke to the *Étoille* who told us that they had seen land to the SW distance 10 to 12 leagues. At 2 p.m. hove to wind to port, at 5 a.m. filled the sails and at 6.30 signaled to the *Étoille* to alter course.[1]

The Call at Tahiti

Friday 1 to Saturday 2 [April]

.... At 7 a.m. went about, the weather squally and rainy; at 10 o'clock sighted land bearing NNE5°E, at 11 saw another land very high to windward bearing W¼NW. At midday sighted the highest mountain of the high land bearing W¼NW and the starboard bluff called Boudeuse Peak [Pic de la Boudeuse] bearing NE¼E distance 4 to 5 leagues.[2]

---

[1] The two ships are now leaving what Bougainville called 'this dangerous labyrinth of islands' and which he will name the Dangerous Archipelago, a name they would retain for many years.

[2] 'Bluff' is rather modest a description for Mehetia which rises to 1,427 ft (435 m). This is the island Bougainville named the Boudoir. The expedition has now come into sight of both Mehetia and Tahiti, Fesche's 'another land very high'.

Saturday 2 to Sunday 3.

The winds variable from NW to ENE, light breeze the weather squally and rainy. At 6 p.m. sighted the South point of the high land bearing W2°N distance 10 to 12 leagues, Boudeuse Peak ENE4°N distance 4 leagues; calm from 2 until 6 p.m. when a light breeze came up. At 11.30 with a fresh breeze, we double-reefed the top-sails and clewed up the mainsail. At 2 a.m. altered course on the SE tack, at 8 a.m. the weather foggy and rainy. At midday the South point of the high land bore W2°N, the most N bearing WNW3°W and the most N point of a low land that one sees continuing to the high land NW°W5°W distance 5 leagues.

Sunday 3 to Monday 4.

Variable NW to NNE wind, light breeze, the weather squally and rainy. At 3 p.m. altered course. At 6 p.m. bearing of the South point SW¼W3°S, the middle of the land W distance 4 to 5 leagues. At 5.30 changed tack, at 7 o'clock changed again, at 10 o'clock changed tack, at 11 o'clock changed again, spent the night on short tacks to stay within a short distance until daylight. At 8 a.m. sighted the point of the most South island bearing SW¼S3°S, the point of the island the most Northerly W, the back of the bay WSW. At 8.30 altered tack, at 9.30 changed again following the most advantageous tacks to reach the middle of the land where we thought there was a bay and consequently an anchorage. At midday bearing of the North point W¼NW, the middle of the bay WSW, the South point SSW5°W, at 10 a.m. we saw a boat leave the South point wind astern, I think it wishes to come up to us.

Monday 4 to Tuesday 5.

The winds variable from the W to N light breeze, fair weather; we entered this sort of bay. In the afternoon about 100 Indian canoes came up alongside. These Indians are affable, handsome and well built, the canoes have an outrigger and are fairly similar to those of the Malays, I believe they come from China or thereabouts. The men have a long beard similar to a Capuchin friar's.[1] I do not claim to give a description here of these people because, if we anchor, I shall detail the circumstances and story of the place, not having yet seen enough of it to report on it. These Indians brought us several bunches of bananas and a sucking pig as a gift, presenting to us some branches of trees a signal used by these people signifying peace. At once Mr de Bouguainville gave them 2 dozen woollen caps and several knives, they were very pleased and left us shortly after; there are 5 or 6 men in each canoe, they are made with good planks stitched together but very narrow and consequently very unstable and so they have an outrigger. We continued to sail into this bay but could never find bottom, so we rapidly altered course in order to tack throughout the night and in the morning sent our boats to take soundings in search of an anchorage.

At 5.30 p.m. after altering course, bearing of the N point of the said island NW¼N4°N, a low land at the N point NNW4°W, the South point SE¼S4°E; at 10

---

[1] The Capuchins arose out of reforms of the Franciscan Order and spread to France from 1573. The emphasis was on Francis of Assissi's vow of poverty and the Capuchins are often referred to as mendicant friars. Although their simple form of dress was prescribed by their rules, there was no particular requirement as far as their beard and hair were concerned. What Fesche implies here is that the beards he saw were uncut or untrimmed.

o'clock altered course again, tacked all night. At 4 a.m., as we were bearing towards the land, a number of canoes came alongside to barter bananas. It is apparently not the first time they have traded as they give nothing for nothing. They display very good faith.

At 4 a.m. sighted a large very high land further N than the one we saw earlier, its low point bore W 5°S. the most S point of the island SE¼S 4°E, a low land East of the said N land W 5°S. At midday bearing of the most N point of the island by the low point E of it W 5°S, the most S point SE 4°E, the N point of the bay SE, an islet a little offshore to the W of the said point SE 5°E.

Tuesday 5 to Wednesday 6.

The winds variable from W to NW light breeze. At midday being one league from land, we lowered the boat to take soundings and the *Étoille* did likewise to seek an anchorage; the *Étoille*'s signaled that they had found a fairly good ground in 12 to 15 fathoms of water, being abreast the cascade distant by ⅓ league from land and the reefs; a little further NW our boat found 9 to 10 fathoms sandy bottom, distance from land ½ a league NE of the said waterfall and after having run 3 or 4 frigate lengths to the E, found 20 to 25 fathoms bottom of rock and coral. We bore away at once ⅓ of a league to leeward of the waterfall; sounded then and found 25 to 30 fathoms bottom of coral and shells, our boat following us along the shore ½ a league from land and ¼ of a league from the ship found 4 fathoms of water rocky ground, the N point of the bay bearing ESE 5°S, the low point N of the island WNW, it was then 3.30. Mr de Bouguainville seeing there was no good anchorage, called back his boats and at 4 o'clock we filled the sails. At 5 o'clock hoisted the boats back and at 5.30 tacked and sailed on the port tack.

Bearings at 6 p.m.: the most N point W 5°S, the low point at the N of the island W¼SW, the head of the first bay S¼SE, the most S point of the island SE¼S. Calm all night and tacked variously until 8 a.m. when we saw that this large land was separated from the other and, seeing at 8 a.m. that there was no anchorage this way along the coast, Mr de Bouguainville bore away to seek an anchorage in the bay we had seen the previous day.[1]

Bearings taken when we made for the bay: the N point of the island WSW 5°W, the S one of the same island WSW 2°S, the low point of the large island WSW 5°S, the N point of the bay SSW, the S point of the island SSE 5°E. At midday bearing of the most N point of the bay WNW 3°W, the true point of the N of the bay S 3°E, the most S point of the island SE 2°E.

Wednesday 6 to Thursday 7.

The winds variable NW¼N to NNW, fresh breeze, various courses until one o'clock when we sent the boats ashore to take soundings and the one from the *Étoille* likewise. At one thirty the *Étoille*'s boat sent a signal that it had found a good anchorage, we bore away at once and came to anchor at 2 p.m. in 34 fathoms of water

---

[1] Having sailed along the Isthmus of Taravao separating the Taiarapu Peninsula from the larger and more mountainous Tahiti-nui, the French are veering north and will put in at the first likely anchorage, Hitiaa, which however was far less suitable as a sheltering place for large vessels than Point Venus where James Cook will arrive a year later.

bottom of blue and grey sand, shells and gravel and moored SE and NW in the following position after warping two cables further North because there were two reefs a short distance astern of us. The *Étoille* did likewise and our boat that had been almost to the back of the bay returned.

Bearings of the anchorage the N point of the bay NW¼N3°N, the most N islet N¼NW2°W, the middle one N¼NW5°N, the nearest and the most Southerly N¼NE4°E, the point of the starboard reef as you enter ENE, the south point of the bay SE3°E, the point of the port reef as you enter SE4°S, the back of the bay S2°E, the watering place WNW.

During the night the winds came from ENE to East light breeze, almost calm. We lowered our topmasts and lower yards and continued the work of mooring.

[Fesche's entries for the next nine days are mostly of a navigational nature, including details of the difficult anchorage and the loss of the anchor. Then follow a substantial report on Tahiti and a vocabulary of the Tahitian language. There are striking similarities between this section and Remarques et descriptions de l'isle de Cythère *by Saint-Germain.[1]]*

Comments and Description of the island of New Cythera

On 2 April 1768 we discovered an island at 10 a.m. bearing NE¼N of the compass; this island is called Boudeuse Peak, is isolated and is like a true sugarloaf, very high and of small circumference and very well wooded, its latitude is 17° 52′ observed and its longitude according to my dead reckoning 150° 29′. An hour later, we discovered another one very high to the W¼NW5°N and tacked variously until the 6 April when we dropped anchor at 5 p.m. The said Bay of Cythera [Baye de Cythère] where we anchored is in latitude 17° 33′ observed and longitude 151° 42′ W of the Paris meridian, and may be 4 to 5 leagues deep. Its N coast runs approximately SSE and NNW and its S one NE and SW, the latter is not as long as the other, the opening between the two is about equivalent to the depth, the N one is lined by reefs that allow a passage N of 3 islets near it, see the new pass discovered by the *Étoille's* boat indicated above. As soon as we had dropped anchor, the frigate was surrounded by canoes in several of which there were women, and, following encouraging gestures by a few Frenchmen, one of the island women comes [sic] on board accompanied by an old man and several of her compatriots. She was tall, well built and had a complexion of a whiteness most Spanish women would not disavow. Several Frenchmen, [who were] gourmets and to whom an enforced fast of several months had given a ravenous appetite, look, admire, touch. Soon the veil that hid the charms which a regrettable modesty no doubt requires to be hidden, this veil I say, is soon lifted, more promptly it is true by the Indian divinity herself than by them, she was following the customs of her country, a practice that alas the corruption of our ways has destroyed among us. What brush could depict the marvels we discover when that troublesome veil happily falls, a retreat reserved for Love alone, none other could rest within it, an enchanting grove planted no doubt by that god himself. We fall into a state of ecstasy, a lively and gentle warmth spreads over our senses, we burn, but decency, that monster which so

[1] See De la Roncière, 'Le Routier inédit d'un compagnon de Bougainville', pp. 226ff.

often fights against the will of men, comes and opposes our vehement desires and makes us plead in vain to the god who presides over pleasure to render us invisible for a moment or merely to fascinate for an instant the eyes of all those present. This new Venus, after a long wait, seeing that neither the invitations of her fellow citizens and especially of her elderly mentors, nor the eagerness she was herself displaying to offer a sacrifice to Venus with one of us whichever he might be, could make us cross the boundaries of decency and of the prejudices built up for us, a sentiment she may have interpreted to our disadvantage, left us with an offended look and ran away in her canoe.[1] This one occasion was sufficient to give a poor impression of the gallantry and the boiling ardour so widely identified among Frenchmen, if our stay in this island that we have named New Cythera on account of its inhabitants' customs had not given us the opportunity of fully erasing the poor opinion they must have had of us. In the evening all the Indians were dismissed, only 5 remained on board, who dined with us. After supper we fired a few flying rockets, which at first caused them a fairly considerable fear which then changed into admiration. The very day after we anchored, Mr de Bouguainville went ashore accompanied by several officers; they were received by the chief who accompanied them everywhere with a thousand demonstrations of friendship, returning from their walk, he led them to his home and had a collation brought to them consisting of all the country's fruits and some dried fish. Some time before getting back on board, an officer noticed that one of his pistols was missing, they told the chief and endeavoured to make him understand that it was a dangerous weapon and could kill. The chief immediately flew into a rage, ill-treated many of the Indians and had stripped several whom he no doubt suspected, examined them from tip to toe, but not finding the pistol, he went into his home, and brought back a mat of woven reed, which he presented to the officer who had suffered the loss and which he forced him to accept, making him understand that he would do everything in his power to find it for him. The next day we went ashore, the chief brought back the pistol lost on the previous day and received gifts in exchange.

I shall outline facts that will appear to many to be falsehoods, but those who know me can be sure that what I shall report as having seen is absolutely correct. As for what I know only from hearsay, that is up to them to decide. I have heard accounts of many happenings that I myself do not accept as truthful and that is why I do not mock the opinions of others.

The way I was received, as was everyone else, was fairly unusual. There were three of us, we go off with the intent of taking a walk escorted by a group of islanders, we arrived at a hut where we are welcomed by the master of the house, he firstly

---

[1] This episode inspired Bougainville to write his famous passage evoking the image of Aphrodite/Venus, traditionally a goddess rising from the foam (*aphros*) at the Greek island of Cythera, and the judgment of Paris (who was brought up by shepherds, hence 'the Phrygian shepherd'), which was essentially a literary allusion aimed at his classically-educated readers and does not correspond to any of the accounts given in the shipboard journals, including his own: 'In spite of all our precautions, one young woman came aboard onto the poop, and stood by one of the hatches above the capstan. This hatch was opened to give some air to those who were working. The young girl negligently allowed her loincloth to fall to the ground, and appeared to all eyes such as Venus showed herself to the Phrygian shepherd. She had the Goddess's celestial form … and never was a capstan heaved with such speed:' Bougainville, *Voyage,* p. 190.

shows us his possessions, making us understand that he was waiting for his wives who were due to arrive shortly. We go together, he shows us the tree the bark of which is used to make the loincloths they wear as their clothing and tell us the names of all that country's fruits. After some time spent strolling, we returned to his home where we found his wife and young girl aged 12 or 13. We are made to sit, they bring us coconuts and bananas, we are invited to eat, we conform to their wishes. We then see each one of them pick up a green branch and sit in a circle around us, one of those present took a flute from which he drew pleasant soft sounds and they brought a mat that they laid out on the open space and on which the young girl sat down. All the Indians' gestures made us clearly understand what this was about, however this practice being so contrary to those established for us[1] and wanting to be sure of it, one of us goes up to the offered victim, makes her the gift of an artificial pearl that he attaches to her ear, and ventures a kiss, which was well returned. A bold hand led by love slips down to two new-born apples rivals of each other and worthy like those of Helen to serve as models for cups that would be incomparable for their beauty and the attraction of their shape. The hand soon slipped and by a fortunate effect of chance, fell on charms still hidden by one of their cloths, it was promptly removed by the girl herself whom we then saw dressed as Eve was before her sin. She did more, she stretched out on the mat, struck the chest of the aggressor, making him understand that she was giving herself to him and drew aside those two obstacles that defend the entrance to that temple where so many men make a daily sacrifice. The summons was very appealing and the athlete caressing her was too skilled in the art of fencing not to take her right away had not the presence of the surrounding 50 Indians, through the effect of our prejudices, put the brake on his fierce desires, but however great the ardour that drives you, it is very difficult to overcome so quickly the ideas with which you have been brought up. The corruption of our morals has made us discover evil in an act where these people rightly find nothing but good. It is only someone who is doing or thinks he is doing evil who fears the light. We hide in order to carry out such a natural action, they do it in public and often. Several Frenchmen, less susceptible to delicacy, found it easier, that same day, to shrug off these prejudices.

After some time spent in that hut, our eyes finally weary of looking and touching, we withdrew, the residents quite displeased at seeing us so reluctant to share the spoils and even telling us so. We walked to the place that had been chosen to set up a camp and a hospital, the Indian chief showed displeasure at the erection of tents and the placing of sentries and he told us by signs to return to the ships, his indications were not ambiguous for he picked up several of the soldiers' sacks and carried them himself into the boat, making it clear to Mr de Bouguainville that this was to be done with all the others. Our intention was not to use force to achieve our ends, mildness being the only method we wanted to use. Mr de Bouguainville picked up eighteen pebbles which he presented to him, he made him understand fairly easily that he wished to remain on their territories only for that number of days by showing the

---

[1] Fesche on several occasions uses the expression *'pour nous'*, 'for us' and not 'by us' or 'among us', to underline the view that European – or more correctly Christian – morality had led to the establishment of a code of behaviour which was then imposed on members of society. This view was shared by a number of supporters of the Rousseauist philosophers, notably on this expedition by Commerson.

sun which he made rise and set. The chief gathered several others, they held a council together where it was decided that they would cut back by a half; the meeting over, he came up to our commander and, rejecting half the pebbles he had given him, showed that he was allowing us only nine sunrises and nine sunsets; Mr de Bouguainville persisted, making them touch and see by filling some barrels with water and showing them also that we needed wood, finally, by a compromise, they granted 15 at the request of those in the council and mostly this king's women who seemed to be very well disposed towards us. They granted us a fairly large house, in which we set up tents for the sick and the soldiers on guard. On the first night and the following one, the king of the district we were in had supper with the officers who were ashore and slept in the same hut. He had placed sentries of his nation around the hut, either to keep watch over us, or to prevent the islanders his subjects from disturbing us. I shall give here the description of the country, the manners of its inhabitants, their practices and customs.

The inhabitants of the island of Cythera seem to consist of two different peoples, and this is what leads me to believe it, it is the enormous difference in their colours, some are whiter than quadroons and mixed breeds, the others have the colour of the least white mulattoes; the former are nearly all of a height and build that is infinitely superior to the average Frenchman, being 5 foot 6 inches to 6 feet in height, the latter who are more numerous are commonly of a height of 5 feet 3 or 4 inches. They are generally well built, have a great head of black hair, and long although slightly frizzy, some let it float in the wind, others wear it tied up on their head and slightly to one side as do the Chinese. They rub it as they do their body with an oil they obtain from coconuts which gives them a smell very unpleasant to us but pleasant to them. They wear their beards very long, particularly people of status. They pull out all their body hair, legs and armpits. They wear as clothing the bark of a tree[1] which I think I have seen in Santo Domingo [St-Domingue][2] and with which they make fairly attractive cuffs, imitating lace. These cloths are extraordinarily long and wide, they have them in different colours, some are dyed a red as attractive as scarlet with the juice of a leaf they boil and squeeze, others are brown, others are speckled and dyed with colours into which gum is worked and which make a kind of oilcloth, yet others are yellow and others are white. Their red is very fine when it is fresh and in no way inferior to scarlet, it deteriorates a great deal when it is used. They also have ponchos made with a cloth of reed with Greek-style designs, they have also mats, all the men wear genital pouches and are very modest. This is the way they prepare these cloths: they remove the bark from the tree and take off the outer layer with pieces of shell. They place the remainder on pieces of squared timber of 5 or 6 inches on each

---

[1] Polynesian tapa cloth is made from the bark of the paper mulberry, *Broussonetia papyrus,* related to the mulberry genus. As its name indicates, it is used in the East to manufacture paper: 'the Otahite cloth … as is used in the Islands, *Moruspapyrifera* Linn, the same plant as is used by the Chinese to make paper', Banks, *Journal,* I, p. 444. See also *The Discovery of Tahiti,* ed. Hugh Carrington, p. 210n.

[2] This is one of the indications that Fesche was not the sole author of this manuscript: he had not been to Santo Domingo at the time he was writing this. However, neither had Saint-Germain, nor Commerson. The possibility, raised by Étienne Taillemite, is that Bournand, who was an officer in the *Boudeuse* and had previously sailed to the West Indies, supplied this information which somehow Fesche appropriated. See Taillemite, *Bougainville,* I, p. 127n.

side and beat them with other pieces of wood and tie them together in this way until they have the length and width they intend. These cloths form several layers, all these cloths are so artistically made that we were unsure for a long time whether or not they had looms.

Most of the women are fairly white, tall and well built, like the men their buttocks are painted black with wreaths of flowers around. To do this they use an instrument made up of an extremely thin piece of shell, toothed at the end like a comb and attached to the end of a small stick half a foot long. They dip the piece of serrated shell into the colour, and place it on the skin which they pierce by striking the handle with another stick held in the other hand. The applied colour enters the holes and remains there permanently. This operation is painful, the skin swells at once and remains in that condition for several hours. I believe that this instrument is used only for the wreaths of flowers and drawings of men, and the other characters they engrave on their legs, thighs, arms and hands, they simply use brushes for the entirety of the buttocks.[1]

The women have the same style of dress as the men, they wrap their bodies with one of their cloths from the midriff to mid-thigh, another serves them as a small cloak, a few have braided black hair which they wear as headbands, like all the men they have pierced ears, flowers take the place of earrings, but they have been replaced by the pearls and other glass trinkets we have given them.

Marriage.

Polygyny is allowable among them, many have several wives, the chiefs especially. Their marriages are, I believe, made in public. I make this supposition on the basis of what happened to possibly two-thirds of the Frenchmen, the fathers and mothers who brought their girls, presented them to the one who pleased them, and urged them to consummate the task of marriage with them. The girl struck the chest of the one to whom she was being offered, uttered a few words that expressed, from the meaning we have attributed to them, the surrender she was making of herself, lay down on the ground and removed her clothing. Several made a fuss when it came to the point, however they allowed themselves to be persuaded. During the operation the islanders themselves, always present in large numbers, made a circle around them, holding a green branch, sometimes they threw one of their cloths over the actor, as in Cythera they covered the happy lovers with greenery. If one of them happened to have a flute, he would play it, others accompanied him singing couplets dedicated to pleasure.[2] Once the operation was over, the girl would cry, but would easily recover her composure and make a thousand caresses to her new spouse as well as to all those who had been witnesses.

There is some evidence that these are the same ceremonies as are used in their weddings; there may be some other formalities required, I believe this all the more readily because an officer from the *Étoille* to whom a young Indian girl had offered

---

[1] In spite of inevitable generalizations, this description is fairly accurate. Tattoo designs were applied with a special chisel, called a *ta,* fitted with a shell, the wing of a bird or in some cases a human bone; the tip was dipped in pigment, placed on the skin and tapped with a light wooden mallet. A description of the tattooing of a young girl is given by Banks, *Journal*, I, p. 336.

[2] The instrument was a nasal flute, called *vivo*, made of a bamboo stick some 30 cm long and 35 cm in diameter, in which four or five holes had been made. See Teuira Henry, *Ancient Tahiti*, p. 276.

herself, [but who was] not favourably disposed, a Cytheran, the same one who joined us on board to follow us in our travels, took the girl and showed him how he should act. If there were no other formalities than those for a marriage, he would not have acted in this way. Moreover all they did for us can only be viewed as honours they wished to pay to strangers.

Married women are a model of faithfulness to their husbands, we saw several even ask them for permission to disrobe at our request, which they grant readily, they also grant the right to touch and sometimes hand them over totally, but it seems that they punish with death men caught in the act of adultery without the husband's permission. The P[rince] of N[assau] one day wanted to caress one of the wives of a chief who was no doubt jealous. Possibly she was his favourite, the king our friend stopped him immediately with a slight show of anger shouting at him that she was married and made several gestures that showed it was likely they killed those who placed themselves in the situation I have mentioned. As for those who are unmarried or widowed, they are free and prostitute themselves with whoever takes their fancy, and so one can appreciate the kind of life most of the French led in this fortunate island; they gave themselves to us at first without asking for any reward, simply eager to give us some pleasure, but soon self-interest became their guide, they insisted on presents.

Religion.

I do not know of any religious practices among these people.[1] In a few huts I saw some statues of men and women. To find out if these were indeed idols they adored, we prostrated ourselves before them, they began to laugh; we gave them kicks and other signs of contempt, same laughter on their part. These statues are placed on a wooden column, carved and bored through. They use them as ornaments for their canoes, I think that is the only use they put them to. If they had some divinities to whom they paid homage, it would no doubt be to pleasure which concerns them ceaselessly, believing themselves to be happy only when they offer a sacrifice to it or share in the sacrifice of others. They have nevertheless some priests, but whose only function is to take care of the dead.

When one of them expires, after having rubbed him over with coconut oil, they carry him to places reserved for this purpose together with all his vestments. This place is surrounded by a balustrade and at the back there is a hut. The dead person is raised on a kind of scaffold and his clothes deposited in a corner of the house. They leave him thus exposed for a lengthy period; we saw several who remained there throughout our stay and who had already been affected by corruption before our arrival. They do not leave them in the same posture, they lie at one time on their stomach, at others on their back, there is a man in the enclosure to guard it. They were unwilling to let us in but we approached the hut as close as we wished. I questioned

---

[1] Early visitors all found it difficult to discover religious practices, seeking as they were a theological and liturgical structure that could be understood in Christian or in Graeco-Roman terms. After saying 'We have not yet seen the least traces of religion among these people, maybe they are intirely without it', Joseph Banks went on to write: 'Religion has been in ages, is still in all Countreys Cloak'd in mysteries unexplicable to human understanding ... What I do know however I shall here write down wishing that inconsistencies may not appear to the eye of the candid reader as absurdities' and gave a lengthy and fair interpretation of what he had observed: *Journal*, I, pp. 277, 369ff.

this man, I asked him by signs and as best I could if they buried them or burned them; he understood me perfectly and made me quite clearly understand that they buried them, himself digging the ground, and showing me that the dead man would be taken there. Several of our gentlemen heard a guard say of a body *emoe,* which, in their tongue, signifies sleeping, which would lead me to believe that they consider death to be a sleep but in any case they must find this sleep quite long because I do not believe resurrection to be more frequent among them than it is among us. From time to time, the priests, after dressing up in their adornments which, like those of the chiefs, consist in a very high round bonnet, decorated by various feathers, fish teeth and large mother-of-pearl shells, the body wrapped in a prodigious number of cloths, go to the dead man, accompanied by his women and parents, cover him with the cloths he wore in his lifetime, rub him with oil while uttering words we could not understand and blow between his 2 legs one of which they hold in each hand. During this ceremony, the women, no doubt out of propriety, weep abundantly, but several of the French who happened to be present at their ceremony easily caused this to be followed by most immoderate laughter through the signs and propositions they were making to the prettiest of them, propositions that were accepted. Let one draw from this whatever conclusions one wishes. I was forgetting to state that, outside the precinct where the body was laid, there was a space of 7 to 8 feet long by 4 to 5 feet wide covered with stones. At one of the extremities one could see two wooden poles two feet from each other dug into the soil and rising about 4 feet strengthened with cross-pieces making a kind of ladder and decorated with reddish fruit that grow on small trees along the shore and whose leaves are rather like rushes. These fruit, 3 inches long and one wide on each side, which I have not seen them put to any use and which grow in bunches, were arranged over each other, facing them was a bundle of reeds tied in its middle with a garland of reed; one could see all around some coconuts and other fruit. What is the use of all this?

Ideas of the country, its products, manufactures and happiness of its inhabitants

A superb plain stretches along the shore, not very wide admittedly, the mountains coming down to put an end to it ½ a league away, it is full of huts built with rushes and covered with the leaves of these same shrubs, but much more artistically than the thatched cottages of our French peasants. These huts are large, a few lengths of timber at intervals support the weight of the roofing. Everything they own in cloths, baskets etc. hang from these pillars. A few have doors made of cane, others a balustrade about 2 feet high that closes them off. The entire plain watered by a thousand streams of various sizes, not far from each other, is covered with fruit trees intended to provide these people's food. These trees are coconut, banana trees of various types, two of which are unknown in our colonies, one producing fruit of an astonishing size, the other fairly large but reddish yellow, the flesh being green and with a fairly unpleasant taste. The tree we name bread tree, a tree new to us, whose fairly long leaves are extremely serrated and whose fruit is rather like our bullock's heart custard-apples,[1] they cook it under embers. They have also some custard-apples and a kind of fruit of a bitterish taste shaped like a hog-plum but infinitely larger,

---

[1] The common custard apple, *Annona reticulata,* was common in the West Indies and known in France. Its name in Tahiti was *nono.* The banana with green flesh and 'a fairly unpleasant taste' is the

with a stone covered by numerous very tender filaments and containing 3 or 4 pips. Several claim that it is what they call mangrove-fruit at the Isle de France. The natives call it *erii*, the tree that bears them is as high as our open-air pear trees.[1] The soil provides them with yams and sweet potatoes, I have seen some *taiove* and indigo, several told me they had seen some manioc, I did not find any, but if they have it, they make no use of it. They also have some wild sugarcane. The hills near the plain are not difficult to reach; they are not forested, several are cultivated but they plant only root crops; in the plains one sees only fruit trees and other kinds of wood, exceedingly soft, with which they build their canoes.[2] When it is cut a kind of milky sap exudes with which they make a kind of pitch which they use.

The country is fertile in pigs, one sees them in almost every hut, they have also numerous hens and roosters but I believe that those natives [who own them] live in distant homes in the mountains, we seldom saw two in the same hut and nevertheless on the last 4 or 5 days we bought 400 of them between the two ships ¾ of them roosters.

Fishing provides them with yet another means of sustenance, they have hooks of various shapes, some ordinary, other imitations of flying fish, all made of mother-of-pearl. To work them, they use coral with which they file the shell until it reaches the desired shape. They also have a kind of very large swivel-hook made of wood, with a tooth at the end and intended no doubt to catch very large fish. They also have nets made like ours, rushes take the place of cork and stones that of lead. They have as well some very long concave baskets with lengthwise at each end a plank 6 inches wide. They presumably place them between the reefs and the fish enter them. They eat it smoked or more frequently raw, there are several species, we saw very few. They also have shells but the eagerness with which many people [among us] ran after them prevented me, as an ordinary being lacking all the items they prized, to obtain all the varieties, moreover the continual trouble we were in also presented some obstacles. These people have minds that are very disposed towards theft, they are the cleverest scoundrels I know. They resorted to several stratagems to rob us. Do they steal from each other or was it their admiration for our belongings that caused them to steal from us as they did? Here are several tricks they played on us: I saw one who had come on board with 3 or 4 of their cloths that covered him from the shoulders to the foot slip under his clothes a spade 4 feet long and tipped with iron. He was about to get into his canoe when, suspicious of him because he had already been caught out in several other thefts, we wished to examine him, how great was our surprise when we saw this tool. Others, aware of the special esteem we had for women, brought several very pretty ones on board who offered themselves to the first come. An elderly man held in special respect by them as far as we could tell led three of them into Mr de Bouguainville's room and urged him most pressingly to enjoy their favours. Mr de Bouguainville resisted but it was still impossible for him not to be distracted to the point where, while they were there, an achromatic glass was stolen from him and

---

plantain banana of which there are two varieties in Tahiti, the *meia* or coastal plantain, and the mountain variety or *fei*; it requires cooking prior to eating.

[1] Fesche in his vocabulary gives '*evii*, a kind of apple', in which case this is the *hevi*, or Tahitian apple, *Spondia dulcis* park.

[2] The *parau*, often made from the soft hibiscus *H. tiliaceus*.

probably several other items, but he only noticed the glass, a little late it is true as the women and the old man had already got back into their boat and were halfway to shore. Mr de Bouguainville immediately despatched a boat to chase this crafty filcher who, suspecting the reason for the boatmen working their oars with such ardour towards his canoe and seeing that he could not avoid being caught up, resigned himself, came back to meet them and gave up the stolen item. What trick could be craftier than that one, bringing women so that he could rob more easily while they were amusing their host? An officer from the *Étoille* was seated ashore, his sword under his arm, an islander passes by, jumps on the sword and runs away, the officer runs after him, the Cytheran draws the sword from its scabbard, still running and throws this latter item at him, the office believes that it is the sword itself he was going to receive on his face, bends over to let it pass, then turns round to pick it up. Meantime, the other runs on and is still running. They carried away handkerchiefs and shirts from next to the people who were guarding them, bayonets from soldiers and a thousand other items of every kind, but the most surprising theft is the one they carried out during the night of Tuesday 12 to Wednesday 13 April and that I shall mention under that date.[1]

If happiness consists in the abundance of all things necessary to life, in living in a superb land with the finest climate (a land that produces everything with no cultivation to speak of), in enjoying the best of health, in breathing the purest and most salubrious air, in leading a simple, soft, quiet life, free from all passions, even from jealousy, although surrounded by charming women, if these women can themselves even form part of that happiness, then I say that there is not in the world a happier nation than the one whose home is New Cythera. When I say that these Indians lead a free life, it is not because they have no chiefs whose subjects they are. They have some as I have already stated, but these chiefs are possibly chosen by themselves. It is anyhow indispensable for people exposed to warfare as these are to elect men who can lead them and maintain a certain discipline.

In a small area, there are several [chiefs], they all have people to serve them. Are they slaves? I do not believe it, unless they are prisoners taken from enemies, maybe they are descendants of families that, less industrious and more lazy, neglected to set up plantations and who, rather than die of hunger, have attached themselves to the wealthier ones to be fed by them on condition that they will render all the services they can, such as going fishing, working the canoes, etc. These chiefs have also men armed with a kind of halberd who accompany them and whom we have often seen carrying them in various places to prevent the crowd from coming near. The only signs of authority they have given is to arrange the return of many items stolen from the French, to ill-treat those who had stolen them or whom they suspected (although they themselves did not neglect the opportunity of taking our belongings, they repeatedly urged us to kill those we might catch red-handed), to drive away with sticks the other Indians, when they saw there were too many of them and they could

---

[1] There is no reference to a theft under that date in Fesche's journal. However, Saint-Germain gives details of several thefts, including the loss of two muskets and a cauldron, without giving precise dates for these events.

upset us, and on a single occasion one of them had someone beaten who had disobeyed them. As far as we could make out, all the chiefs appeared very united among themselves, there are some however who placed guards on the boundaries of their land to prevent us from entering, but it was enough to show one's teeth, they at once became gentle as lambs and would themselves act as guides and drive away those who were coming too close to us.

An unbroken peace is the only thing that could make their happiness more perfect, but from everything I have seen, I think I can declare that these people know war. Why did all those who came alongside in their canoes bring branches of greenery, a signal of peace among all nations? What use would be all the weapons they have, consisting of bows usually made with a white wood with a veined inner section, reed arrows two feet long tipped with a piece of extremely hard singed wood, kinds of halberds 6½ feet in length also made of extremely hard wood, slings and pikes approximately 10 feet in length that they throw with extreme skill? Whence would come the wounds that afflict many of them, and wounds made by all these different weapons; all this indicates clearly that these people sometimes have disputes, possibly among themselves, possibly also with the inhabitants of some neighbouring islands.[1]

I am quite sure that they sometimes sail over long distances, for otherwise of what use would be the immense 60-foot canoes, carefully sheltered under especially made hangars? These canoes are joined together in pairs by crosspieces to make them steadier, they are very narrow, their stern is extraordinarily high. At the bow they have a plank of some 12 feet, 6 of which make up a deck for the canoe to keep a fair amount of water out when they pitch and the other 6 overhang. At the stern they have a castle shaped like a canopy held up by carved columns; this castle is, I think, portable and can be placed where they like, it is no doubt the place where chiefs and women stay. Single ones have outriggers that take the place of the second one. Their canoes are built of an extremely soft wood and made up of several pieces that they tie together by means of ropes passed through holes bored with the inner part of a spiral shell; to work the wood they use a kind of adze made of an extremely hard black stone that I believe to be black jasper, they make it very sharp. They have some very large ones, we tried them out on some oak, they give it a great polish. They do their caulking with the outer husk of the coconut and the milky gum that exudes from the bark of the tree they build their canoes with serves as pitch. I do not know whether those of the first size have only one sail, the medium ones had only one, made of mats, placed centrally, their masts have supports placed on a wooden crosspiece, their sails are fixed and can neither be hauled down nor clewed up, the internal part is straight, as a lower boltrope they have a fairly large bent stick, the side one is a pliable stick that assumes the shape of the sail. They use very well made oars. All the islanders generally are very good swimmers, we saw many whose canoes had filled up or sunk following the loss of their outrigger, hold up for more than an hour in the water to repair them and bail them, the others pay not the slightest attention to them, unless there are women in these canoes, then all those that are nearby come up and the men jump eagerly into the water and bring them up into the one that is nearest.

---

[1] The hard wood in question is the *toa,* ironwood.

The 5th or 6th day after our arrival, several people on a walk found a dead man by whom were some weeping women rubbing him with coconut oil. They approach, the women explain that he had been killed with firearms by imitating the noise, they examined the wound and saw that the shot had been fired point blank, even the powder having entered the wound all around; the next day Mr de Bouguainville sent the senior surgeon to examine this wound and verify the fact. The surgeon confirmed it. Mr de Bouguainville made without result all the enquiries needed to discover the author of this act, determined to punish it with the utmost severity. It was clear that the unfortunate man had been killed by someone who was careless or a professional scoundrel, if he had had a good reason to do this he would have reported it to the officer in charge ashore or to Mr de Bouguainville. The natives were indignant at this murder and began to move and to remove all their possessions from their house The same evening, 4 soldiers went out of the camp in spite of having been expressly forbidden to, carrying their bayonets, and went on for ½ a league. There, they wanted to buy a pig for two nails; upon the owner's refusal, they took the pig forcibly as they themselves admitted and ill-treated the Indian. He may have wanted to defend himself, his compatriots came to his assistance, the soldiers killed or wounded several. Meanwhile the Prince of Nassau arrived at the place where these tragic events had occurred. All the women and the men in tears came up to him, kissing his hands and chest, begging for his help and explaining the misfortune that had just occurred. A sailor from the *Étoille* who was passing nearby, found an Indian being carried away having sustained two bayonet wounds. Mr de Nassau ran at once after these soldiers and caught up with them. He hit several with the flat edge of his sword and brought them back to camp. Mr de Bouguainville was immediately advised of this unfortunate incident, he went ashore and had the soldiers placed in irons at the door of the shed, firmly determined to shoot several, if he found convincing proof. He even sent the chaplain ashore so that the execution could take place right away to make an example and at the same time to show this people that the entire nation was not implicated in this murder. All the Indians had already left the area for ½ a league from the place we were occupying. Fortunately for the soldiers there were no proofs and only a simple instruction not to leave the camp with weapons but no order published. The commander sent them back to the ship. The chief of the place where we were returned in the evening bearing a banana tree, sign of peace, offered it to the Prince of Nassau and kissed his chest weeping abundantly and saying: *taio, taio, mate,* which means in their language: you are our friends and yet you kill us. He then withdrew without tarrying any further. This general departure made us fear that these people might be rebelling against us and might attempt some undertaking. We doubled the guard, placed a group to the right of the camp, on the bank of the stream, and another on the left. I was sent ashore for a few nights. Around 9 p.m. we saw several lights coming down along the river, which caused us to remain on alert for some time. At 7 p.m. an Indian had come covered with their cloth and foliage, he jumped around for a long time,[1] calling us from time to time *taio maté,* shouting like a devil. We led him into the hangar where we gave him some food. He held in his hand a fairly thick bamboo, he

---

[1] Fesche uses the word '*jongla*', juggled; he is describing a war dance.

had by him the bar of justice which had served for the soldiers,[1] and without losing his composure, he had already slid two-thirds of it in his bamboo when it was noticed, he was sent off at 10 o'clock and led back beyond the guards.

At midnight, the Indians stole from within the very tent of the post situated on the watercourse, while it was occupied by an officer, a volunteer and some sailors, all men from the *Étoille,* two muskets and a fairly large cauldron without the sentries noticing it. In all likelihood they were asleep as well as those in the tent. The two muskets were between two people as was the cauldron. This theft is incomprehensible, they can only have taken them away by uprooting a few tent pegs and pulling them out underneath it, but how did they manage to steal these objects placed between two people without making the slightest noise?; a cauldron is not stolen like a snuffbox, still it may be the first post where such an adventure has occurred.

A sailor came at once into the shed to ask for a lantern to see whether he might not by chance find them nearby; as he came back, his musket in one hand, the lantern in the other, he saw leaving some grass close by two or three Indians who had been crouching there. He immediately fired and at the same moment five or six other shots were fired but which did not hit anyone. We searched around but we found nothing. The rest of the night passed fairly quietly. The next morning very early, the Prince of Nassau, myself and three others, left to attempt a reconciliation. When we had taken a few steps, it began to rain, the bad weather discouraged us. I nevertheless urged these gentlemen to go as far as the king's house to see if he had left anyone there and if the dead man who had been exposed fairly close to it had also been taken away. We continue on our way. We in fact found that the dead man had been moved away and that several people who served the king were guarding his house. At first they did not dare let us approach them too closely, however they were reassured by our demonstrations of friendship, they took a great deal of trouble to keep us from going further but seeing that we took no notice of anything they might be saying to us, they helped us to cross a river where there was a fair amount of water by carrying us on their backs, they even escorted us and led us to the place where the chief our friend had retired.

On the way, all the women came up to take our hands, kissing them as they did our chests, their faces wet with tears. We had more than two hundred islanders following us. From the furthest spot he could see us, the chief came to meet us a branch of greenery in his hand; he offered it to Mr de Nassau and embraced him as he did with all of us his arms around us and weeping. He immediately drove away all those who were surrounding us. These various spectacles brought tears to our eyes. Were these Savages? Certainly not, on the contrary we were the ones who had behaved like barbarians and they acted like gentle, humane and well-regulated people. We murdered them and they did only good to us. The chief led us into a nearby hut and had bananas and coconuts brought to us which we ate and drank. After a while, we rose and urged him to come back with us, which he did. We arrived at the camp with a good supply of hens and pigs we had bought on the way. Mr de Bouguainville who

---

[1] The 'bar' in France as in England is an area where the prisoner stands 'at the bar', with his defender close by. This area is usually enclosed and has some form of handrail. Fesche refers here to some metal bar that symbolized this area and which the islanders attempted to steal.

happened to be on land made gifts to all the chiefs who were soon gathered around such as axes, nails, scissors, generally all types of iron that they prize greatly, for all the bartering we have done for hens and pigs etc was only with nails, axes and artificial pearls. The chief our friend sent for all his women who came and wept with joy all the morning. He also had enough bananas brought for one ship. They had shown us on the way the place where the bayonets had been used which they made very clear. The rest of the day went by in the usual way.

The next day they compelled several people to fire muskets and pistols. We killed a pig in front of them, we pierced one of their canoes from side to side, which caused them a deal of surprise but not as much as when they saw two flying parakeets brought down with a single shot. The same day, the muskets that were stacked around the pile of arms fell through the carelessness of some people who had come too close. One fired in falling, fortunately no one was wounded. This shot frightened the Indians who rushed out in a crowd and knocked over the quadrant belonging to Mr Verron, the pilot sent by the Court as observer, which was near the entrance. It was seriously put out of true. In the evening, Mr de Bouguainville, after sending back all the Savages, had a hole dug in the ground and placed in it a plank on which he had engraved the extract from the Act of Possession of the island of New Cythera in the name of His Christian Majesty written as above.[1] In addition he placed the names of the people making up the staff and chief warrant officers in a corked and sealed bottle. The same morning the *Étoille* had sailed through a pass discovered between the reefs and the shore, see above. On the morning of the 15th, we raised our anchor.

The chief our most zealous supporter came on board to make his farewells and embraced us, we gave him some presents, he urged us to return, which we promised to do, he made us a gift of bananas and a pig.

He had brought with him another of his islander friends, a very intelligent man, and invited us to take him with us, which we agreed to do with a promise however at his request to bring him back. This same Indian, when we arrived, had gone on board the *Étoille* and sent back his canoe and stayed on board for the 3 days we spent tacking within sight of the island before dropping anchor. He was at once provided with clothes but I am firmly of the belief that this poor wretch will long regret his foolish action, as I consider his return to his home to be impossible; lucky if the sorrow of being at sea for a lengthy period does not deprive him of the temporary pleasure he will feel upon seeing France. The main motive that makes him act in this way is his desire to be married for a while with some white women. That is all I can say on New Cythera.[2]

[Following this section and until 23 May, Fesche's journal consists largely of navigational information.]

Through the New Hebrides
Sunday 22 to Monday 23 May.
    … At 7 a.m. we sailed SW and sighted the NE point of the island bearing E5′S

---

[1] The full text was included in Saint-Germain's *Remarques*.
[2] Following this entry, Fesche appends a vocabulary of 183 Tahitian words.

distance 1½ leagues after realizing that this was only islands, and so I was wrong yesterday thinking that this was the land of the Holy Spirit [St-Esprit] because they are only islands, well wooded, with few inhabitants judging from appearances for we saw no huts along the shore only in the forest, and from time to time a little smoke. At 7.30 a.m. we saw a few canoes belonging to the Savages very far away watching us go by. Their canoes have an outrigger from what I could see with the spyglass, and are oar-propelled, I do not know whether they have knowledge of sails.

Bearings at 8 a.m.: the most N land of the first island bearing NNE4°N distance 7 to 8 leagues, the NE point of the most S island ENE3°N, the SW point SW distance 4 leagues.

At 8.15 being a short league from land, we took soundings and let out 40 fathoms without finding ground. At 8.30 a.m. we hove to starboard to the wind to lower the boats and we signaled the *Étoille* to send her boat. At 9 all the boats were gone with the *Étoille's*, being well armed to fetch a little wood and collect some refreshments if they found any. There are 2 officers with 4 volunteers and 20 soldiers. When they are back I shall know more both about the products of the soil and about the country's inhabitants. At 10.30 changed tack and tacked several times from 10 to midday waiting for our boats that are to come back at night.

Midday bearings: the most N land of the first island NNE distance 8 leagues, the NE point of the most S E5°N distance 2 leagues, the most SW point of the same island SW¼S distance 5 to 6 leagues.

Monday 23 to Tuesday 24 May.

The winds boxed the compass, light breeze, light airs, fair weather, fair sea. At one after midday we lowered the yawl and Mr de Bouguainville with 3 officers went ashore to see where our boats were, if there was much wood, and finally to take a walk. We stayed hove to until 5 p.m. At 4.30 the boats came back, we hoisted them in right away and set the topsails and the foresail. Mr de Bouguainville said the country was very beautiful, well wooded with all kinds of trees and that there was a great deal of fruit which he had loaded on the boats such as coconuts, bananas etc. in quantities. He told us that the people who inhabit this country are a very ugly type of man, short of stature, full of sores and leprous, that he had been forced to fight them although they had put up with a great deal from them; firstly when the 2 boats were about to land, 3 or 400 Savages came down on the plain wanting to prevent them from landing but this did not stop the officer [who] made everyone land and immediately disposed the soldiers in such a way that they could fire at the first alarm they might cause and right away began to cut down some wood. During this time, they all came to see the wood being cut, they numbered 3 or 4 hundred looking at us and when one of them shouted out, they all went back into the forest and a quarter of an hour later came back in the same manner all holding a handful of arrows in one hand and a bow in the other and often they even had an arrow at the bow, ready to fire and making clear that it went right through one's body. When they came back to watch us, they were all in the same situation pointing the arrow even close enough to touch us, so you can see what a trouble it was to have a day like this. Note that their arrow is very dangerous, first of all it is made of a piece of reed, then of a length of iron wood and at the tip there is a very pointed and serrated fishbone 8 to 10 inches long. It is true that not all are serrated but made of a very smooth and sharp bone. I am sure that they can do a great deal of harm. Once the wood was cut, we got them to carry the logs to the boats in exchange for some lengths of red cloth. They were never willing to barter their bows at any price, making us understand that they were at war against each other; as for their arrows, they gave some in exchange for some handkerchiefs, but very few. They also have a kind of club that they were never willing to hand over, clear evidence that they are fighters; they are all naked, men and women, and wear only a testicle pouch made of quite well woven reed. They have no huts, they live in the woods and their hutments consist merely of two forks port and starboard and a top in the middle over which they place many banana leaves, making the cover. As for canoes, they have some but they were so far off we could not see clearly how they were made. That is all we have seen of these people. At about 5 p.m., as Mr de Bouguainville was getting every-one and everything that was ashore back in the boats, the Savages at once came to the level ground and waited until there was no one left on land and as soon as they saw us rowing out, they threw a hail of fairly sizeable stones at all those who were in the boats. We endured this without a word but a little later, as the boats were continuing to go out, one of them angry at not having caught anyone, came forward until he was in the water up to his loins to fire an arrow at someone in the jolly-boat but he had to pay for his impudence for an officer who was in the yawl, seeing his manoeuvres, fired at him just as he was about to let go of the arrow, which caused him to lose much of its strength and it hit no one; he was seen to come out of the water right away, running a little and fell on the shore, no doubt dying. The other boats immediately fired a volley. I cannot say whether many were killed in this general discharge but the first one was killed and certainly several others. That is all I know about these leprous Savages.

At 6 p.m. sighted the N point of Aurora Island [l'isle de l'Aurore] bearing NNE3°E distance 9 to 10 leagues. The NE point of Lepers Island [l'isle aux Lépreux] E¼NE1°N distance 3 to 4 leagues, the SW point of the same island SW¼S2°S distance 4 leagues. At the same hour saw land to the SW distance 12 to 15 leagues. At one in the morning took in the foresail.

Bearings at 8 a.m.: the N point of Aurora Island NE¼N4°E, the NE point of Lepers Island ENE5°E. At midday sighted the SW point of Lepers Island bearing SE¼S3°E distance 2½ leagues, the most NE point in view ENE3°N, the most N land of Aurora Island NE¼N.

    .....

Thursday 26 to Friday 27 May.

The winds variable from NNE to SE light breeze, fair weather. At 2 p.m. we lowered the boats to send them to take soundings at the back of what looked like a bay to seek an anchorage as well as the *Étoille*'s boat, forbidding them to set foot on land. The officer of the *Étoille*'s boat being faster went off more quickly than the others, being almost ashore, he fired all his artillery at some wretched Savages who gave no appearance of wishing us any harm, for as we were sounding, 2 canoes came within hailing distance to urge us by their signals to go ashore, but being formally forbidden to land, we continued to sound and found a fairly good anchorage but off an open shoreline with no appearance of a bay, all this is nothing but islands. In the evening the boats came back and the officer in charge of the boats gave his report to Mr de Bouguainville that the *Étoille's* boat had fired a number of shots at the Savages. The officer from the *Étoille's* boat explained that, being not far from land, two canoes had shot some arrows of which he showed two, that he had been forced to fire at them to drive them off and believed he had killed three or four.

Bearings at 6 p.m.: the most E point of the NE land ESE4°E distance 7 to 8 leagues, the most W point of the same land W5°S distance 6 leagues, the low lands at the back of the bay of the same land NNE4°E, the W point of the S island SSE2°S, the E point of the same island SE3°E. Tacked several times during the night.

Bearings at 8 a.m.: the most E lands of the NE lands E¼NE, the most W point of the same land W¼NW4°N, the middle of the bay NE, the most W point of the S island SE¼S3°S, the E point E¼SE3°E.

Midday bearings: the most NW land to NW¼N4°N distance 8 to 9 leagues, the most E lands E¼NE.

Mr de Bouguainville has named these islands the Bourbon Archipelago [Archipel de Bourbon].

Saturday 28 to Sunday 29 May.

The winds variable from SE to SSW fresh breeze, fair weather, fine sea, all sails set. At midday lowered the boat and Mr de Bouguainville went on board the *Étoille* and had supplies brought over such as wood, brandy and beans. At 3.30 he returned, the boat was hoisted back in and we remained hove to 1¼ hours, then we filled the sails and let draw with all sails. Bearings at 6 p.m.: the most NW land NNE, the most E bearing E. ...

[The entries for the period from 29 May to 1 July are incomplete and contain mostly navigational details.]

The Solomons, New Britain and New Ireland

Friday 1 to Saturday 2 July.

The winds variable SE to E light breeze, almost calm, at 1 p.m. we lowered the boat to go and survey an anchorage together with the *Étoille's*; we sounded at 1.15 and found a depth of 20 fathoms bottom of coral. At 2 p.m. saw the bottom under us, whereupon we sounded and found 10 fathoms coral and tacked variously until 3 o'clock when we were at the same point as at midday according to our bearings. At 3.45 the boats signaled a good anchorage, we then made for the bay where the boats had been sounding so that we passed over the ground we had seen at 2 o'clock, from then on we sounded at intervals and found 7 to 9 fathoms of water coral ground and on our starboard side there did not appear to be more than 3 fathoms, but the closer we got to shore the more the depth increased, which led me to believe that it was only a shelf, the tide being very strong and against us as I have already stated. Then Mr de Bouguainville changed his plan of going to anchor in this bay and called back the boats and at 6 p.m. we recorded the following bearings: the N point of the bay NW°N distance 1 to 1½ leagues, the S point of the W island SW distance 5 to 6 leagues, the middle of a small island on the port side W3°S distance about 6 leagues, the S point closest to us of the N land SE°E. At 6 p.m. the boats returned having struggled to catch up with the ships and told us that at 3 o'clock 9 or 10 canoes quite large and well armed rowing strongly and shouting like the devil came towards them in the bay where they had found an excellent anchorage, but as the *Étoille's* boat was quite close to them, it was told to stay on its guard and to watch for attack from port while [we] watched to starboard; once the canoes were within range of their arrows, two left the group rowing between the land and the boats to go around and apparently trap us, but we did not give them time to carry out this fine manoeuvre because since they believed themselves to be within arrow range, all those that were ahead of us kept their bows bent ready apparently to let fly as soon as the canoes would be behind us ready to do the same. Mr de Nassau said it was necessary to shoot first which he did immediately and after him there was a general discharge of the musketry made up of 2 volunteers, 7 soldiers, the officer and the crew of the boat, the *Étoille's* boat did the same for its part. The Savages were not frightened of this right away, they took up a shield made of rushes and artistically so and shot arrows at us that did not reach half-way, then we fired a second volley that made them lose heart because they began to move away from us and try to escape but the officer ordered the men to row towards them to endeavour to board and capture them. Meanwhile the musketry was continuing although more defensively, but they were faster than us and we could never catch them, they fled but we returned towards 5 or 6 others who had withdrawn into the forest where there were still 4 or 5 blacks dragging their canoes along the shore, being in the water up to their knees shielded by their canoes, but when we arrived we fired such a volley that we made them all abandon their canoes and thus we captured one as did the *Étoille's* boat, the which were laden with coconuts, arrows, spears, betel and areca, the ordinary sustenance of the Indians. After this clash the commanding officer sent the *Étoille's* boat to us in accordance with his instructions and we stayed in the channel while we watched the frigate manoeuvre in order to anchor. During this time Mr de Nassau wanted to go ashore with our prize, but as it lacked stability he capsized and almost drowned and we lost everything that was in it. As for the *Étoille's* [prize], the tide was so strong that the boat, struggling to reach the ship, was forced to let it go after removing from it an item that was strange to Europeans, it was a man's head that they had eaten, proof that they were coming towards us with a clear intention but they did not know our strength and they certainly regretted it as I believe that from the two boats the number of killed or wounded may reach 25 to 30 men out of their approximate number of 50. As for us we suffered no harm. The best anchorage in the said bay is on the port side in 10 fathoms, the middle island bearing NNE, the starboard point S5°W and the port one SSW5°S. The said bay seemed to me to be a bight 2 to 2½ leagues deep, the entrance is little more than ⅓ of a league wide but at the anchorage it seems to be ½ a league; the entire port side consists only of a peninsula with some nice sandy coves and signs of the presence of turtles, as in the said bay we missed one, one can assume it is not the only one, also I saw a great many fish and these sandy coves would be useful for the seine. All these islands have coconut trees, there are also numerous parrots and other birds, which would make it a good place in case of necessity, sheltered from every wind except the south to SW which is very infrequent. We signaled the boat to return to the ship as mentioned above, the commander not wishing to anchor here, we took Mr de Nassau who had gone ashore and we were back on board at 6 p.m.

Bearings at midday: the most W point of the land we coasted along SE2°S distance 7 leagues, another point of the same land and more ENE bore ESE2°S distance 8 to 9 leagues, the most S point of the W land S2°E distance 9 leagues, the most S small islands in view S5°E distance 7 to 8 leagues, the N point of a pass on the W land SSW5°S distance 9 leagues, the S point S°SW distance 10 leagues, the most N land WNW3°W distance 14 to 15 leagues.

Monday 4 to Tuesday 5 July.

The winds variable from ESE to S light breeze, almost calm, all sails set. At 4 p.m. 3 canoes with Indians came up not daring to approach us too much but gradually we gained their confidence by throwing over the side (from which they stood a good space away) small pieces of red cloth and presenting a red flag, this is what made them decide to come alongside; when they were alongside, they gestured for us to give them the flag but we did not agree, we gave them a knife which they called *bouca bouca* and in exchange[1] they gave us a coconut, we also bartered with them some pieces of cloth, handkerchiefs and a thousand other things for bows and arrows; a quarter of an hour later they left very angry with someone who had wanted to barter a suit for some arrows and a bow, when the individual had the bow and the arrow, he did not hand over his clothing, he gave him a loincloth that is to say a length of very feeble material we obtained at the island of Cythera. When the Indians saw this, they fled and when they were at some distance from the ship, they fired an arrow on the person who had cheated them and who was still outside the vessel, but fortunately the arrow did not reach him. When they left they gestured that they would return, that they were going to fetch some coconuts, however they did not return. These Indians are of a very fine black colour, well built, of a good height, some are pock-marked, they have very red mouths, they say this is on account of the betel they eat; as for their canoes it [sic] is well made and very light; I believe them to be of light planks, the inside is all made of reeds; there are 4 or 5 men in each, [they] do not take much in the way of food with them as we saw only one coconut in all.

Bearings taken at 6 p.m.: the most S point of the SE land in view SE, the most NW point of the low land SE5°S distance 3½ leagues, the most S point of the same land S4°E distance 5 leagues. Sighted land at the same time bearing NW to NW°W.

Bearings taken at midday: the most S land SE°S4°E distance 9 to 10 leagues, the most N ESE2°S distance 9 leagues, the small island N to NNE3°N distance 7 to 8 leagues.

Bearings taken at 8 a.m.: the N point of the low land ESE5°S distance 6 leagues, the S point of the said island SE5°E distance 5 to 6 leagues, the S point of the island we had seen between Cape Laverdie and the low land SE distance 8 leagues, the land bearing yesterday evening NW to NW°W bore at 8 a.m. N°NW to NNW in view.

---

[1] The word Fesche uses here is 'in revenge', but it can be regarded as an error in the copying. The repeated use of the word *bouka* led the French to name the island Bouka, a name it has retained as Buka. Saint-Germain understood the word to be the local name of the coconut (*Routier inédit,* ed. La Ron-cière, p. 246).

At 8 a.m. lowered the longboat to go on board the *Étoille* to fetch food supplies such as flour, brandy, etc. At 10 o'clock hoisted the boat back in.

Tuesday 5 to Wednesday 6 July.

The winds variable SE to SSW fresh breeze, fair weather, fair sea, all sails set. At midday sighted two islands one bearing N distance 10 leagues which we lost sight of.

Bearings at 5.30: the middle of the other island seen at midday N5°W distance 12 leagues, the E point of the great land NW°N3°W distance 10 leagues, the closest land of the same island NW°W distance 7 to 8 leagues, the W point of the same land W2°S. At 11.30 p.m. altered course, at 3 o'clock altered again; at 6 a.m. sighted two islands further W than those of yesterday.

Bearings at 8 a.m.: the SW point of the most W island SW°W4°W distance 10 to 11 leagues, the NE point of the same W°SW3°S distance 10 leagues, the SW point of the middle island W5°S distance 10 leagues, the NE point of the same island NW°W2°W distance 9 leagues, the W point of the most E island NW°N distance 8 leagues, the closest land to us of the same island N°NW distance 5 to 6 leagues, the most E point of the same island N°NE1°N distance 8 to 9 leagues.

At 9 a.m. we hove to await the *Étoille* to lower her boat as well as ours in order to seek an anchorage.

Wednesday 6 to Thursday 7 July.

The winds SE fresh breeze. fair weather, we tacked variously to wait for our boats. At 2 p.m. our boats signaled a good anchorage, we then bore away and set course to go and anchor in a bay we have called Port Praslin where we dropped anchor at 3 p.m. in 33 fathoms of water bottom of fine sand with the following landmarks: the S point of the small island at the entrance that we left to port W°SW, the N point of the same island W°SW3°W, the starboard point as you enter the bay W°SW1°S, the portside one W°NW, the S point or outer point of the island that lies in the middle of the bay NW°W3°W, the N one NW5°N, the back of the small bay where we have anchored SE°E, the point of the reef that lies off the N head of this bay N2′W, the latter point N, and we moored SE and NW. During the night, quite calm and rainy.

Estimated distances in leagues covered by the ship during the years 1766, 1767, 1768

From France to Montevideo	2478 leagues
From Montevideo to the Malouine Island	695⅓
From Malouine Island to Rio janeiro	761⅓
From Rio janeiro to Montevideo	461
From Montevideo to Cabo Virgenes [Cap des Vierges]	635⅓
From Cabo Virgenes to Cabo Deseado [Cap Désiré]	118
From Cabo Deseado to the island of New Cythera	1896
From the island of New Cythera to Praslin Island	2102
Plus for the goings and comings from Montevideo to the Maldonados, Riogenero ditto	140
Total	9287 leagues

Days at anchor in Port Aquarius [Port Verseau] in the Eastern part of New Britain

Thursday 7 to Friday 8 July 1768.

The winds SE light breeze, rainy weather. We underran our cables with our longboat to see if they had not been damaged by the rocks, then we sent people ashore to cut wood and fill the barrel with water. Our gentlemen went walking ashore to examine the land and found a port there where the English had collected wood and water not long ago, judging from the cutting about a year and a half to two years ago.[1] We found an inscription engraved on a lead plaque nailed to a large tree with small nails 12 to 15 feet up.[2]

Friday 8 to Saturday 9.

The winds S, the weather wet. At 6 a.m. we raised our anchors to check the cables and we dropped them soon after. Every morning some men are taken ashore to get wood and others to fill the barrels together with a corporal and 4 fusiliers to protect the workers from the Savages if any should come, having found during the morning hidden in the woods a canoe belonging to the Savages, proof that they come here sometimes.

Saturday 9 to Sunday 10.

The winds variable, fine weather all afternoon, rain during the night, during the day we took the workers and the soldiers ashore as usual each with a tent to shelter when it rains, which happens very frequently.

Sunday 10 to Monday 11.

The winds variable SSE to SSW light breeze, fine weather. At 6 a.m. secured the sheet anchor that was in the hold to the cathead. At 6 p.m. a man from the *Étoille* died, our chaplain went to bury him ashore close the river where the *Étoille* was getting its water.

Monday 11 to Tuesday 12.

The winds SE, fair weather. At 8 p.m. we put out a kedge anchor as we seemed to have dragged on our anchor. A man who was fishing with the seine has been bitten by a fish or a snake; on his arrival on board we had to see to him right away and made a small incision in the side where he had been bitten and [he] suffered considerably. He showed how the animal was made and I saw from his indications that it was a snake, it was of the thickness of a thumb, about an arm long, with a white mark on its head, I have myself killed several of this same type ashore that people were sure was a snake. Firstly it had bitten him in the thumb, he caught it in his hand and threw it far from him then the animal came back and bit him in the side.

At dawn we worked to relay our moorings and raised the port anchor that was foul and we lowered it as soon as we had raised it and completed our remooring. The man who was bitten is much better.

Tuesday 12 to Wednesday 13.

The winds variable S to SW light breeze, fair weather. At 6 a.m. we erected a

---

[1] Carteret's visit was in fact in September 1767, i.e. less than a year earlier.

[2] This is at variance with Bougainville's statement that the plaque was found in the sand and broken up, probably by the islanders who tore it down. It is evident from this that Fesche was not part of the shore party on this occasion.

machine to support our mizzenmast. At 8 a.m. we raised it because its base was all worn and it was split, we cut off approximately 15 feet and remade the foot to fit in the keelson and woolded it. There was this morning an eclipse of the sun which started at 1.30 a.m. and finished at half-past midday.

Wednesday 13 to Thursday 14.

The winds SE light breeze, fair weather, we continued to collect wood and water.

Thursday 14 to Friday 15.

The winds SE light breeze, the weather rainy throughout the night, still collecting wood and water.

Friday 15 to Saturday 16.

The winds SSE light breeze, the weather overcast and wet; we finished collecting our wood and water.

Saturday 16 to Sunday 17.

The winds SE light breeze, the weather rainy throughout the twenty-four hours.

Sunday 17 to Monday 18.

The winds ESE stormy gale, the weather rainy with intervals.

Monday 18 to Tuesday 19.

The winds variable from SE to SSE fresh breeze, the weather overcast and rainy, waiting for the *Étoille* to be ready to sail, our longboat getting water for her daily.

Tuesday 19 to Wednesday 20.

The winds variable from SE to East, the weather rainy throughout the twenty-four hours.

Wednesday 20 to Thursday 21.

The winds SE, continual rain all day.

Thursday 21 to Friday 22.

The winds S fresh breeze, the *Étoille* being ready to sail, Mr de Bouguainville sent the *Étoille*'s boat to see what the wind was outside, the officer reported that outside the winds were S, stormy gale and that the sea was very wild, at 10.15 a.m. there was an earthquake.

Mr de Bouguainville went at 5 p.m. to get an oak plank buried on which was engraved the following inscription for taking possession of the island: [In the year seven hundred and sixty eight, the 12 July, we Louis Ant. de Bougainville, infantry colonel, captain of King's ships, commanding the frigate *Boudeuse* and the *Étoile*, in the name of and by order of His Most Christian Majesty, under the ministry of Mr de Choiseul, duc de Praslin, we have taken possession of these islands: in witness whereof, we have left the present inscription corresponding to the act of possession that we are taking to France].[1]

---

[1] The wording is left blank in the Fesche MS, but appears thus towards the end of Saint-Germain's journal.

Friday 22 to Saturday 23.

The winds SE, rainy weather, our boat went outside the pass to see how the winds were and reported that outside they were S; the *Étoille's* boat went to see if it was possible to pass between the island and the land and leave from where to all appearances the English went out, in the evening the officer reported that the passage was very good.

Saturday 23 to Sunday 24.

Rainy weather; at 4 a.m. the wind quite calm, we began to unmoor and sent a hawser ashore to hold our stern and act as a spring to port and for starboard we had the kedge anchor. Just as we were right above the outer anchor, the breeze that was previously ENE changed to SW to WSW and continued thus until midday.

Note: Tuesday 12 to Wednesday 13 ditto there was an eclipse as indicated above and Mr Véron, astronomer sent by the Court, determined the longitude of the said bay at 149° 44′ 15″ of the Paris eastern meridian and observed its latitude at 4° 49′ 27″ South.

According to my dead reckoning, I do not appear to differ very much since I am in 152° 28′ east, there is only a difference of 2° 44′, which makes 54⅔ leagues, so I shall not change the longitude until I reach a place where I can take a sure point of departure on my chart.

Sunday 24 to Monday 25.

The winds having continued SW until 3 p.m. when it became quite calm; at 4.30 we were working the hawser and the cablet in order to anchor but as we were finishing, the breeze came up from the NE, we at once hauled with the capstan and when we had taken it up we slipped the stern hawser which we at once hauled aboard and passed the cablet to the *Étoille's* longboat which will go and raise it and at 5.30 p.m. we sailed and steered W°NW to SW°S from departure until 6 p.m.…. The *Étoille* also sailed at 7.30 p.m. and at 9.30 [to] 10 was abeam.

[The following few entries are mostly of a navigational nature, as are the brief entries covering the period 1 to 8, 9 to 14 and 20 to 24 August.]

New Ireland to New Guinea

Thursday 28 to Friday 29 July.

The winds variable from NNW to SE passing through E, light breeze almost calm, all sails high. Bearings taken at 6 p.m.: the NW point of Great Britain [sic] in view W¼NW2°W, the SE one SE2°S in sight, a small islet to the N of Dubouchage Island [isle Dubouchage] NNE4°E distance 9 leagues, the most N point of Dubouchage Island NE¼N3°E, the most S point of the same island E¼NE, the middle of the small island to the N of Susannet Island NW4°N distance 8 to 9 leagues, a small islet to the NE of the latter island NW¼N3°W distance 9 leagues, the NE point of Susannet Island [isle Susannet] NW3°N, the SW one NW3°W. The weather stormy with lightning throughout the night…

… At 8 a.m. two canoes with Indians came up which were never willing to come closer than musket range, they are blacks, quite handsome and well built, they have canoes with outriggers with very well made reed sculpture at the bow and stem. At 10 a.m. they left and as they went they threw us a stone with a sling which did not reach the vessel but was not far from it; if they had thrown a second one they would have paid for their insolence as there were 6 fusiliers ready to take aim at them and as well 4 guns loaded with grapeshot. We heard 4 or 5 shots on the *Étoille* which they were firing at canoes that were alongside, they must have thrown some stones at them. In Dampierre's voyage, he says that he had seen a number of them and that being at anchor in a bay, he was impelled to name it Slingers Bay [Baye des Frondeurs][1] on account of the number of people he saw using slings. The weapons these Indians have are the slings and the spear, as for arrows I have not seen them with any. The tides have once more carried us to the SE and so I shall again correct my longitude and my run.

Friday 29 to Saturday 30 July.

The winds variable from NNE to SE through E light breeze almost calm, all sails set. Bearings taken at 6 p.m.: the SE point of the island of New Britain SE4°S, the NW point of the same island in view W, the S point of Dubouchage Island E¼NE2°E, the N point of the same NE¼E, the middle of a small island further N NE3°E, the SW point of Susannet Island NW3°N, the NE point of the same NW¼N3°N, a small island to the N of Susannet Island NNW4°N…

… At 6 a.m. several Indian canoes came alongside being in number ten to twelve canoes in which were five or six blacks, they were not as unsociable as usual as they came right away alongside our ship intending to barter potatoes for knives and pieces of red cloth, they showed very bad faith in their exchanges for they took away several knives and other items they were being given in exchange for arrows and spears, but when they had the items of trade they asked to be first given they did not hand over the items they had shown. They have their hair and beard dyed, they all have around their arms some bones of animals I do not know, they have some glassy items through their noses that hang down even to their mouths, they also have them on their ears. Their canoes are built as I mentioned yesterday. At 7.30 a.m. they all left having heard some shots fired at canoes that were alongside the *Étoille,* apparently they had had stones thrown at them. I think they might have wounded a few because they fired about 12 shots. At 10 o'clock 2 more canoes came up that hooked a length of cloth without giving anything and then went to the *Étoille* apparently to try to snatch something else, possibly musket shots which will not be lacking if they threaten to throw stones, but the winds freshening, they were unable to catch up the *Étoille* and went back to shore quite upset at being unable to steal anything from us.

Saturday 30 to Sunday 31 July.

The winds variable from SE to S light breeze almost calm, all sails set. Bearings taken at 6 p.m.: the most NW land of New Britain W¼NW distance 7 to 8 leagues, the most SE SE5°E, the most SE point of Dubouchage Island E3°S distance 12

---

[1] Fesche's comment is correct. Sailing south along the coast of New Ireland, Dampier endeavoured to go ashore, 'Which when the Natives in their Proes perceived, they began to fling Stones at us as fast as they could, being provided with Engines for that purpose; (wherefore I named this place Slinger's Bay)': Spencer (ed.), *A Voyage to New Holland,* p. 209.

leagues, the most NW E3°N, the most SE point of Susannet Island NE¼E4°E distance 4½ leagues, the most NW N2′W distance 4½ leagues …

… At 7.30 p.m. two canoes came halfway to examine us, night failing, Mr de Bouguainville had three flying rockets fired to try to astonish the Indians. We saw during the night several fires along the coast. At 8 a.m. there were 12 to 15 canoes all together around the *Étoille* and none was prepared to come to us. At about 8.30 to 9 o'clock we heard some thirty shots being fired and we saw all the canoes leaving except one in which all the Indians had been killed or had swum away to the other canoes. The Savages must have thrown stones at them to bring about such a fusillade. The number of killed is uncertain but I am sure there were some as I saw a canoe drifting about and since it was calm the canoe remained for nearly an hour close by the vessel. At 10 o'clock as the wind was freshening a little I lost sight of the said canoe, I believe the Indians will soon come to collect it if they can find out about it.

Bearings at midday: the SE point of Susannet Island E4°S distance 7 leagues, the NW one NE¼N2°E distance 7 to 8 leagues, the most E land of New Britain in sight SE¼E2°S distance 6 to 7 leagues, the W one W¼NW3°W distance 8 leagues.

…

Monday 8 to Tuesday 9 August.

… all these lands are only an archipelago of very low and almost drowned islands, all covered with coconut trees, we saw some inhabitants with the spyglass, and some very fine huts along the seashore of the first two seen at midday from which at 5 o'clock we were N and S…

Sunday 14 to Monday 15 August.

The winds variable from NNW to SSE through N and E, light breeze almost calm, all sails set, the weather stormy during the night accompanied by much lightning and very violent thunder. During the night I saw Saint-Elmo's fire over our weathercocks, it stayed for about ¼ of an hour.[1] Bearings taken at 6 p.m.: the most W land of the main land SW distance 15 leagues, the most W of the most W island S3°W distance 2½ leagues, the most E point of the same island S¼SE2°E distance 2½ leagues, the middle islet SSE4°S distance 2½ leagues, the middle of the islet furthest E SSE1°S distance 3 leagues. At 6.30 a.m. saw land as far as the WSW … There is a very strong tide in this area, I am not sure where it flows, every day we see quantities of trees passing alongside.

Monday 15 to Tuesday 16 August.

The winds having boxed the compass, almost calm throughout the afternoon, with a fierce tide. I noticed during the afternoon that the tides were taking us considerably towards the N since in the evening we hardly saw the land which we had seen at midday 3 or 4 leagues away and we were sailing W. The sea was extraordinarily changed and was as white as it could be in any river. We took soundings on several occasions and ran out 200 fathoms of line without ever finding ground having a lead

---

[1] A discharge of atmospheric electricity which appears on the masts during a storm. It was regarded by Mediterranean sailors as the visible sign of the protection of their patron saint, Saint Erasmus or Sant' Ermo.

of 80 at the end of the sounding line. During the night the currents bore us so much to the N that we could not see land at dawn the next day.

Tuesday 16 to Wednesday 17 August.

The winds variable from SSW to NW through W light breeze, almost calm, all sails set, at midday we bore away to join the *Étoille* which is 2½ leagues to leeward. At 5.30 we spoke to her and [the captain] told us that over these recent days when there was such a tide, he had not been able to steer his ship, there being so little wind he had to give in to the tide which was much stronger than the wind. Mr de Bouguainville asked Mr de La Giraudais why they had fired so many shots at the New Britain canoes, he replied that it was the excessive number of stones thrown at him both by hand and with slings that had forced him to fire at them. At 5.30 p.m. we altered course, the weather rainy and stormy from 2 to 3 a.m., at 2 o'clock took in the mainsail and the topsails during a squall accompanied by thunder and lightning, the lightning flashes were so bright that one could not open one's eyes. At 2.30 I think the thunder fell very close to the ship according to everyone's opinion, we were running under bare poles, our pumps sealed and the hatches closed. At 3 a.m. everything calm, set the topsails, the weather rainy.

Wednesday 17 to Thursday 18 August.

The winds variable SSW to W light breeze almost calm, at 3.30 p.m., we saw land bearing from S¼SW 5°30′W to SW 5°30′ distance 16 to 18 leagues. At 9 a.m. we saw an island bearing from us SW¼W. At midday the same islet bearing SW¼W 1′W distance 8 to 9 leagues. The Prince of Nassau, tired of eating salt meat, had a dog killed (which he had obtained by barter from the Savages in the strait of Magellan) which he had served at table, all the gentlemen ate some and found it excellent, it is the last quadruped left on board with the exception of the rats that are eaten every day; we have eaten dogs, cats, rats, the leather from the yards, etc. . . .

Thursday 18 to Friday 19 August.

The winds variable from SE up to W through S, light breeze, fair weather, smooth sea, all sails set. At 5 p.m. sighted the said island of yesterday midday bearing SW 2°W distant 8 leagues. At 10 p.m. changed course. At 6 a.m. sighted the NE point of the said island from aloft bearing SW¼S 2°S, the SW point SW¼S 3°W distant 10 to 11 leagues. At 10 a.m. altered course.

Bearings at midday: the most W land in view SW distant 13 to 14 leagues, the most E bearing SSW same distance. The currents have carried us a considerable distance to the N or NNW, I am not sure, but according to my midday altitude reading I have reached the equinoctial line and by my reckoning am 11 leagues S of it; all the other sailors reach roughly the same conclusion.

. . . .

Wednesday 24 to Thursday 25. Feast of St Louis, King of France.

The winds variable SSE to SSW, good breeze, fair weather, calm sea, all sails high. At 2 p.m. saw some inhabitants going about and a few canoes with 3 huts, as we hauled along the coast we saw only those. . . .

Note: it is usual in all His Majesty's ships to give a double ration to the crew and provide an entertainment, as for us, far from giving out a double ration, we cannot give them their ordinary ration and [they have] to eat some wretched bacon and rotten beef.[1]

[The remaining entries, dealing with the navigation through the Dutch East Indies and across the Indian Ocean to Mauritius, contain little of relevance to the crossing of the Pacific and its aftermath. But the following brief extract from Fesche's journal entry of 29 to 30 August sheds an interesting light on relations between Bougainville and La Giraudais at that point in time:]

Monday 29 to Tuesday 30 August.

They [the men of the *Étoile*] have killed a turtle weighting about 140 pounds with a musket shot, which resulted in a good meal. Mr de La Giraudais invited Mr de Bougainville to come over and eat his share at supper but he refused. I think Mr de La Giraudais acted very badly in not sending over his share to the commander, for a number of reasons, in the first place because he owes him everything, and secondly because he is his commanding officer. If such a thing had happened on our ship, we certainly would have divided it into two shares.

[1] The Feast of St Louis was the equivalent of France's national day under the Old Régime.

APPENDIX 5

The Journal of Nassau–Siegen

There are three copies of the Prince of Nassau's 'Voyage de la frigate La Boudeuse, *et de la flûte* L'Étoile, *au Paraguay et sur les Côtes de la Californie'. The first, probably written up during the last stages of the voyage from notes taken at various times, is held at the Archives Nationales, Paris, collection de la Maison du Roi, ref o¹ 569 7 No. 28. In writing this account, Nassau had based himself on a 'Journal de navigation' or log, which has not been preserved. A second manuscript, based on the first version of his 'Voyage' and carried forward to the end of the voyage, is held at Affaires Étrangères, Mémoires et documents de France 2115, folios 128–75. The third, MS C, essentially a copy of this second version, is held at the Archives Nationales, under the reference Marine BB4:1001 No. 2. These last 'evidently represent a more finished version, intended to be read, possibly to be published. Certain passages from Version A were omitted, no doubt on account of their licentious nature.'[1]*

Nassau was a passenger, not a naval officer, with very little technical knowledge of navigation. This gives his account a more personal, less technical flavour. Occasionally, however, he does include brief details of the latitude and longitude.

The first fourteen pages provide an account of his depature from Paris with Bougainville in October 1766, the crossing to South America, and his time in the Falkland Islands, in Rio de Janeiro and in Buenos Aires. Approximately four pages deal with the passage through the Straits of Magellan and his meetings with the Patagonians and Fuegians. The following cover the Pacific stage of the voyage and have been taken from his original text, which is rather more complete and informative.

Through the Tuamotus

Having entered the South Sea, the proximity of Cape Horn still caused us to encounter fairly bad weather. But the SE trade winds caught us in the 32nd degree of latitude. We began by seeking Davis Land, but as fruitlessly as the Dutch admiral Rochevin did in 1721.[2] We were slowed by the Westerlies and by calms and began to be greatly affected by such a long crossing. The prodigious number of rats that had multiplied in our hold were eating up everything. We were short of water and everyone was reduced to a very meagre ration. We desalinated seawater with Mr Poissonier's cucurbit. The water produced was very good. Lacking fresh food, we had several sailors attacked by scurvy. Finally on 22 March we sighted land. It formed four small islets that seemed almost to be touching. We passed too far off to be able to say

---

[1] Taillemite, *Bougainville*, I, p. 134.
[2] On Davis Land and Roggeveen's search, see Introduction, 'The Voyage'.

anything about them. They are in 18°48′ of southern latitude and 138°25′ of longitude west of Paris. We saw another island on the same day. We sailed fairly close to it, it is raised, round, it has a diameter of approximately one league, the seashore is sandy, the rest of the land is covered with fine trees of various species. This island, although very small, is inhabited by Indians who came out from under the trees to watch our ships pass by. They seemed to be armed with long pikes which they brandished as they looked at us. It is in 18°30′ of southern latitude and 139°43′ of longitude west of Paris. The next day we saw another island formed by two tongues of narrow land separated by a lake on which some large canoes were manoeuvring. We also saw several men on the shore. These lands are very low, surrounded by a reef, one sees here and there small fairly pleasant clumps of trees. We sailed along the S coast. The E point of this coast is only in latitude 18°26′ and longitude 140°15′.

On the 24th we sighted another land the most N point of which is in latitude 18°1′and longitude 141°58′. The most S point is in 18°27′ and longitude 141°45′, the NW point in 17°22′ and longitude 142°25′.

On the 26th another coast trending SW, the most East point of the north coast is in 17°56′ of latitude and 142°51′ of longitude. The same day, we saw two islets in 17°26′ of latitude and 143°14′ of longitude. We did not see those lands [close] enough to be able to say anything about them. They are very low and wooded.

The Call at Tahiti

On 2 April, we saw a peak to leeward, we named it after the *Boudeuse*. It lies in 17°56′ of latitude and 148°23′ of longitude. We sighted at the same time a very high and large land on the windward side, which decided us to make for it. Coming up on the 3rd to the N coast of the island, we saw a canoe which, without coming close, came to examine our vessels and returned with several others to meet us. The Indians, unarmed, were carrying banana trees branches which they offered to us. One Indian, unable to catch up with our frigate that was sailing too fast, went alongside the *Etoille*[1] and jumped onto her with unbelievable eagerness, sent back his canoe and then made it clear that he wished to remain with the strangers. Without delay he provided a number of refreshments. It did not take us long to realize that this very intelligent man was very curious about all products of our industry. We spent two days seeking an anchorage. We had hardly reached it when one of these islanders, full of the confidence that innocence inspires, climbed on board with his wife. He asked for our friendship, the woman showing most willingly all the perfection of a fine body, offered everything she had that could gain the heart of the newcomers, but the King's regulation, which no doubt had not foreseen such a circumstance, prevented us from responding to her civility on board.[2] She left quite displeased. When we landed, we found 3 to 400 people along the shore, among which was the chief who led us to his home.

The country is as beautiful as it could be, forests, fertile vales, streams and gardens make up a charming setting in which the inhabitants have located their houses. One is troubled by none of those insects that are so troublesome in hot countries. These

---

[1] Although Nassau-Siegen's spelling and syntax are generally superior to those of his fellow compilers, with the exception of Bougainville, he consistently spells *Étoile* with a double 'l'.

[2] This is one of the passages omitted from the later version, which was intended for publication.

people belong to a superb race, the men are usually 5 feet six inches to six feet tall. We even saw one who was 6 feet 4 inches. They all are handsome and well built, they have very long hair which they rub with coconut oil and have long beards. The women, also of a good height, have fine large eyes, pretty teeth, European traits, a soft skin, Nature was pleased to grant them perfect bodies. These Indians are copper-coloured. The women are whiter. There seems to be another much browner race among these people. The men and the women paint their backsides in a way that cannot be erased and is not unpleasant. These consist of two large regular marks decorated with a very well made drawing. The men dress in a piece of cloth they wrap round their bodies and tie gracefully around their waist. Often they wrap their bodies in another length of cloth. Their hair is tied on their heads in the manner of the Chinese. The women, covered from the waist to the knee, usually have another cloth over their bodies. Their clothes have neither cut nor style, they are only drapings according to the taste of the person who dresses.

When we arrived at the chief's, we visited his house which, like those of the other inhabitants, was very large and built of reeds in such a way as to let fresh air enter. They are very skilfully covered with leaves. Inside, it was very clean, but the only furniture one saw were the utensils needed for the necessities of life. Two wooden statues, one representing a man, the other a woman. The chief and his family sat down with us on the grass, the retinue remained standing. Servants brought fruit which we were given to eat. After the meal, he decorated us with some local clothing and accompanied us to our boats. On the way back we met two men who forced us to stop to listen to them. One was singing and the other accompanied him on a kind of flute that has only three holes and that the Indian played with his nose. This music which their tendency for pleasure leads them to make up, was monotonous but gentle, it expressed the gentle affections of the soul. This is how this courteous nation carried out the duties of hospitality towards us. I went back to shore the next day with Mr d'Oraison. As soon as I reached the shore, a group of Indians came to offer me a woman. I had no wish to be disobliging towards these good people but even less to appear publicly in the way they required. I withdrew politely with gifts and explaining my keenness to call on their chief. When I arrived at his home, they served us fruit, then the women offered me a young girl. The Indians surrounded me and each one was eager to share with his eyes in the pleasure I was about to enjoy. The young girl was very pretty but European preconceptions require more mystery. An Indian used very singular means to further excite my desires. Happy nation that does not yet know the odious names of shame and scandal. If wise people carry out these ceremonies in association with the planting of seeds, why should the reproduction of the finest species of things ever created not also be a public festival? These peoples, being given over to pleasure, multiply them and vary them as much as they can, allowing themselves several wives who are strictly restricted to conjugal duties; however their husbands permit them to take their pleasure with other men, but such a liberty taken without their permission would cost the life both of the woman and of the man who had been thus favoured. The girls their own mistresses dispose of their charms as they wish.

Mr de Bougainville sent a detachment of soldiers ashore who establish a camp, then the leaders of the country gathered and made representations to the chief who came to tell us that we had to return on board. We made him understand that, needing wood and water and food, it was essential for us to sleep on land but that we would leave in 18 days, then a second council proposed nine days and ended up granting us the time we were asking for.

Wanting to spend the length of our stay ashore, I had a tent [erected] where the chief spent the night with me. Mr de Bougainville slept on shore in case of an attack but the night was very peaceful. The chief asked him to show him some rockets like those we had let off the day before on board. Mr de Bougainville had a few fired that so frightened him that he dreamt about them all night and, to gain our friendship, each time he woke he sent one of his guards to bring us some gifts. At daybreak one of his wives arrived whom he made sleep with us. These Indians offered us women as being the objects they most cherished, undeniably these well deserved this distinction. They each in turn used all their charms to please us. Here is one example. I was strolling in a charming place, carpets of greenery, pleasant groves, the gentle murmur of streams inspired love in this delicious spot. I was caught there by the rain. I sheltered in a small house where I found six of the prettiest girls in the locality. They welcomed me with all the gentleness this charming sex can display. Each one removed her clothing, an adornment which is bothersome for pleasure and, spreading all their charms, showed me in detail the gracefulness and contours of the most perfect bodies. They also removed my clothing. The whiteness of a European body delighted them. They hastened to see whether I was made like the locals and pleasure quickened this research. Many were the kisses, many the tender caresses I received! Throughout this scene, an Indian was playing a tender tune on his flute. A crowd of others had lined up around the house, solely preoccupied with the spectacle.[1] We were living amidst this gentle nation like allies and friends. The chief, the leading men constantly made us gifts and we had numerous opportunities to barter with the people for the food we greatly needed. Inexpensive iron tools formed the basis of the trade. They helped us in our work. Our only complaint concerned the thefts which were made so skilfully that all our precautions could not prevent them. One of our people killed an Indian with a pistol shot. We did not know who the murderer was, nor consequently the reasons that had led him to commit such a barbarous act. The inhabitants were so frightened by it that they left their homes and carried away their belongings into the mountains where they went. We calmed their fears with some difficulties, but they were renewed shortly after by an event of a similar nature.

I lived on good terms with the Indians whom I liked greatly. Appreciative of the small gifts I made them, they displayed a real affection towards me. Strolling along the seashore one day, I found myself suddenly surrounded by tearful men and women who threw themselves in my arms while others, kneeling, clutched my knees. I could not at first guess the reason for their distress and I endeavoured to tell them we were their friends. Alas, they replied, these friends are killing us and they made me understand that three residents had been killed with bayonets. I endeavoured without success to alleviate their pain, each one was leaving his home and loading his belongings

---

[1] Not surprisingly, Nassau also omitted this section from the later version.

in canoes. I wanted to put what they were taking away back on land. What did I find? A body which these unfortunates feared we would not sufficiently respect. Deeply moved by the spectacle I had witnessed, I went straight away to find Mr de Bougainville to whom I told everything. He went ashore and following the suspicions that fell on four soldiers of being responsible for this trouble, they were immediately placed in irons in the presence of the Indians. The chief once more came to the camp, he was carrying a banana tree palm which he offered to me, then embraced me with tears in his eyes and left without delay. Mr de Bougainville doubled the guards ashore for fear of an attack. Alas, either these Indians did not know about revenge or the fear of our firearms, the mere sound of which made them lie down on the ground, dashed their spirits. The night was very peaceful, but the least of our fears was to have to face an attack from a crowd of Indians. A coral bottom on which we were anchored had three times cut our cables and three mooring ropes. For 36 hours we saw our two ships on the verge of perishing on the coast. What would have become of us 6000 leagues from our fatherland in an island unknown in Europe, among people who were angry against us! What a situation! The next morning at dawn, I went with four people to fetch the chief. On the way I met one of his brothers who made me understand that his house was deserted and that he had left with all his family for a place a league and a half from there. I begged him to take me there. Several inhabitants wanted to oppose my going into the country but finally I found the chief and his wives with a crowd of people. The chief came to meet me with a banana tree palm in his hand, and embracing me he offered it to me. I made every demonstration of peace and friendship towards him. The women threw themselves weeping into my arms but in the end I succeeded in reassuring them. I brought the chief with five or six hundred Indians back to our camp, all carrying fruit for us. Mr de Bougainville having seen me from his ship returning with this multitude of Indians, came ashore with presents for the chief and the leaders. Peace being restored, these good people came in greater numbers than ever and thefts began again with renewed vigour. Mr Bougainville, without the Indians knowing, had an oak plank buried in the camp on which was inscribed the taking of possession of the island in the King's name together with a bottle in which we sealed a copy of the inscription written on paper with the names of the ships, and the senior and warrant officers.

The next day, as we were leaving, the chief came on board with his wives who were weeping at our departure. They urged us to return to see them. The chief gave us the same man who had given himself on the very first day to the *Etoille* [sic] on condition we brought him back. This Indian, who is one of the leading citizens of this island, corresponds in no way to what we have said about these people in relation to size and face but he has for all that all the intelligence possible, which makes up well for the absence of the beauty that Nature has bestowed so well on his compatriots. The island appeared to us to be excessively populated and we found only one imperfect man in the multitude we saw. Each locality lives under a chief who has great authority over the people. He issues orders as he wishes. His wives and relatives are respected. Differences of rank are very marked. We saw only weak traces of religion among these people and their practices seem to indicate that they have no other gods than love and pleasure to which they constantly offer sacrifices and towards which all

inhabitants' houses which are round and less numerous than at Cythera. Having seen a few canoes coming towards us, we waited for them hoping they were bringing a few refreshments. The first to come, the Indians seeming reluctant to come alongside, Mr de Bougainville had a boat lowered and wanted to go to see what they had in the canoe. But realising his intention they fled with great shouts. They released a white bird they had tied up and threw a few coconuts into the sea, hoping to gain time while we were picking them up. Mr Bougainville gave up his pursuit, realizing that they would jump into the sea rather than surrender. Others came who apparently had not seen the first ones being chased. They bartered a few cloths. We saw they had a cock which they were unwilling to trade, this leads one to think that they are rare in this island. We noticed they had a white bird in each canoe,[1] we also saw some pieces of shell, which might make one hope that turtles might be found. These islanders are tall and well built, with no beard. They paint their rear in blue up to the middle of the thigh. Our Cytheran could not make himself understood by them and their two languages seemed to be unrelated. Their canoes are very well made, decked in the fore part and stern with an outrigger like in Cythera, they have a triangular sail made of mats.

On 5 May, having sighted a large land, we made for it and hauled along part of the South coast the most Westerly point of which lies in latitude 14°21′30″ and longitude 173°49′. From this point, the coast runs for 5 leagues towards the SE¼E and then 7 leagues NE up to the most E place. It is large, attractive, a range of high mountains covered in trees and some fairly attractive plains, which made a very pleasant landscape. One presumes we would have found an anchorage. Several canoes came out built like those we had seen the previous day, but these Indians seemed shyer than those of the previous islands. They did not dare approach. The same day we saw another island the South coast of which trends NNE and SSW from latitude 14°7′ and longitude 174°10′ to latitude 14°13′ and longitude 174°35′. A thick cloud and the distance prevented us from seeing the end of this coast.[2]

On the 11th we saw an island that extended North and South from latitude 14°21′ and longitude 179°27′ to latitude 14°16′ and longitude 179°30′. We firstly thought that it was separated and formed two but the next day we realized that there was only one, which made us believe that what we had taken for a strait is a wide bay covered in forests in which we saw several fires which proves it is inhabited.

The discovery of this series of islands which we saw from 14°11′ of latitude and 169°54′ of longitude west of Paris to 14°26′ of latitude and 179°30′ of longitude west of Paris strikes me as very useful for the search for these famous islands of Solomon [Isles Salomon] so rich in gold and precious stones. The Abbé Prévôt tells us that

---

[1] The brevity of this remark does not make identification easy, but it may have been a tern, possibly the Black-naped *tern, Sterna sumatrana*, known locally as the *gogosina*, or less probably a Sooty tern, *Sternafuscata, or Gogo uli*, although these two, although white, have black markings. They are prized by local fishermen because they hover over shoals of small fish and then dive for them. A local legend tells of Chief Manu who kept a tern in a cage in his canoe: 'when Manu saw his bird getting restless … [he] knew the tiny mafua fish were swarming and that the bonitos would be plentiful:' Corey & Shirley Muse, *The Birds and Birdlore of Samoa*, p. 37.

[2] Having sailed past the small Manua group, the ships are making for the larger islands of the Samoan archipelago, Tutuila, Upolu and, though probably too far north, Savaii.

Mindonna first saw the Solitary Island [Isle Solitaire] in 10°40' of latitude and 157° west of Paris and nine days later with a favourable wind he discovered the island of the Holy Cross [Isle Ste-Croix], one of the Solomons.[1] He covered in the nine days a good 260 leagues in a westerly direction. Therefore he reached the 170° degree of longitude west of Paris. The Spanish charts situate these islands 170° west of Paris. Mr de Lisle[2] gives latitude 11° for Santa Cruz [Ste-Croix]. I agree with him in this but not with his longitude which he shows as being 172° east of Paris. Mendoce with whom Mindana had apparently previously made the same voyage to the Solomon Islands places them between the 7th and the 12th degrees of latitude but not so far to the west and so I am holding on to the position between 170° and 180° of longitude that seems to be the most credible, all the more in that Gallego, a companion of Mendana, places them in the same location.[3] It is moreover stated in Mendoce's account that in those waters, the Spanish travelled along 150 leagues of coast as far as 18° of latitude and looted an Indian town where they found grains of gold hanging as ornaments in the houses. In truth, I cannot persuade myself that the Spanish exactly followed a coast of 150 leagues without seeing several islands and probably the very ones we have seen since they went down to 18° of latitude. There is every reason to expect to find today in the area through which we have travelled the same riches as they found in the past, and so this chain we saw between 170° and 180° west of Paris and everything that lies NW of it up to the Solomon Islands seems to me to warrant careful investigation.[4]

Off the New Hebrides

On 22 May we discovered to the East two forested islands and we saw numerous fires on them. Their coastlines lie N°NW[5] and S°SE. They are of similar lengths and separated by a channel approximately two leagues wide. We named the more southerly one Pentecost [Pentecôtte] and the second one Aurora [Aurore]. The most southerly part of Pentecost is in 15°50' of latitude and 165°42' of longitude East of Paris and the most Northerly point of Aurora is in 14°50' of latitude and 165°27' of longitude. They are at the most two leagues wide. After rounding the North of the latter, we saw the west coast of a third extending approximately 4 leagues ENE and

---

[1] The Abbé Prévost was the author of *Histoire générale des voyages,* a large compilation published between 1746 and 1770. Alvaro de Mendaña y Neyra, sailing in 1595, discovered 'a low round island in 10°40' S, which they named Isla Solataria and which is identifiable with Niulakita in the Ellice Islands:' Jack-Hinton, *The Search for the Islands of Solomon,* p. 116. The island is now known as Nurakita. Santa Cruz was discovered by Mendaña on 7 September; it belongs, not to the Solomons but to the New Hebrides archipelago.

[2] Guillaume Delisle (1675-1726), Royal Geographer, published a total of 134 charts and maps.

[3] Alvarez de Mendoza is credited in Antonio de Herrera's *Descripción de las Yndias Occidentales* of 1601 with 'the discovery of the *Insulae Salomonis* which... he attributes to "Alvarus de Mendoza"' (Jack-Hinton, *The Search for the Islands of Solomon,* p. 101). Garcia Hurtado de Mendoza was Viceroy of Peru at the time Mendaña's voyage was being planned; his biography, published in 1613, was a work much consulted by eighteenth-century French geographers. Hernan Gallego was the Chief Pilot on Mendaña's first expedition of 1657-8 when the Solomon Islands were first discovered. See Dunmore, *Who's Who,* pp. 113-14, 179-80

[4] This paragraph is a good illustration of the confusion and myths that surrounded the Solomons at this time.

[5] Corrected to N¼NE in the later MS.

WSW and 8 leagues N°NE and S°SW. We named it Lepers Island [Isles des Lépreux]. Its most Easterly part [is] in latitude 14°17′ and longitude 165°37′ and its most Southerly part in latitude 15°30′ and longitude 165°5′ East of Paris, the same day a small round island [hove into sight] situated in latitude 14°54′ and longitude 165°32′. We named it Etoille Peak [Pic de l'Etoille].

The next day 23rd, sailing along the western coast of Lepers Island, we saw numbers of fires and a few gatherings of men. 4 or 5 fairly small sailless canoes seemed to come out to examine our vessels but without approaching us. This behaviour did not indicate a trustful nation! As we were in great need of wood, Mr Bougainville sent three well-armed boats ashore to cut some. I wanted to be part of the first landings and not miss the opportunity of examining a country where we would be staying only a few hours. I landed on the island. As soon as we had left our canoes, two groups of Indians came forward with their bows bent and ready to shoot and stopped two hundred paces from us, then I walked alone towards these inhabitants to prevail upon them by my advance. After demonstrations of peace on my part, three of them advanced three paces, responded to my approaches and touched me on the hand and soon we were surrounded by more than five hundred. The officer commanding the detachment had some wood cut without opposition on the part of the Indians who nevertheless kept us still surrounded in their midst, their bows at the ready. We made them a few gifts, they picked some fruit and peacefully let our sailors pick some. They also bartered a few arrows but were not willing to trade their bows, their wooden swords and the clubs a number of them carried. It was in this island that we saw for the first time a new race of men different [both in respect of their features] and by their size from those we had found until then.[1] The inhabitants of Lepers Island are black but of a mixed complexion, ugly and ill-proportioned, almost all covered with ulcers and leprosy. They have no clothing, they cover their waist with tree leaves or a piece of crimson-painted fine matting. The women carry their children on their backs in a scarf. These blacks are armed with serrated arrows that we thought were poisoned, wooden swords and clubs the thick end of which is carved into a half-ball. All this is of very fine wood and well made. We did not see any of their houses, I saw them climb coconut trees with greater facility than a monkey displays. While we were ashore, a group of about three hundred men from the neighbouring settlement came to attack those we were with, but the latter whose numbers had gone on increasing, split their forces, part remained with us and the others went off to chase the enemies. Mr de Bougainville landed just as we were finishing loading the wood and the fruit. The blacks, until we were getting in the boats, had been satisfied with keeping us among them, but we were hardly in our boats when they sent us a hail of stones and a few arrows. We fired a few shots at them, which frightened them a great deal, as they saw several of their number falling down wounded.

From 22 May to the 27th we also discovered a continuation of this coast almost North and South from latitude 14°23′ and longitude 165°5′ to latitude 16°16′ and longitude 165°30′, that is to say almost 40 leagues of land. Mr de Bougainville said he

---

[1] The French have passed from Polynesia to Melanesia; the physical and cultural differences between the two groups have now become apparent.

had found the lands he was seeking. After leaving Lepers Island, we placed ourselves in latitude 14°40', the position that Quiros allocates to a great bay in the land of the Holy Spirit [St-Esprit] which he discovered in 1606 and we found a passage in the West in 14°47'. At first we took this pass that runs East and West for the entrance to Quiros's bay which he says is North and South. I saw only two islands South of this passage that seemed to have any length. The northern lands seemed much more extensive. One sees some very high mountains fairly forward in the country and some fine plains. The coast is superb, covered in fruit trees. Some bare land as found in mining country appears in several places on the mountains. These lands are very populated, the inhabitants are woolly [haired] Blacks, we saw numbers of them on the shore of the pass and several in canoes that did not dare approach our ships. This is no doubt the same race as at Lepers, a neighbouring island I have already mentioned. That is what we could see of these lands from our vessels. Quiros states in his narrative that in latitude 14°30' and longitude 187° west, which gives 167° east of Paris, he saw a long high coastline with land West and SE of him, that he landed on the W one where he saw white, black and mulatto men, that he then went to anchor on the SE land in a great bay in 14°40' of latitude. The countryside he adds is superb, one can find there tasty fruit, aloes, ebony, nutmegs, stones, marcasite, silver, game, poultry and heavy beasts. Quiros's longitude differs by a mere 2° from the one we observed on these shores. In their charts Messrs de Lisle and Buache did not vary by much from this when locating the lands of the Holy Spirit. We sailed about 15° East and as much West of these lands which we found, always keeping to the latitude provided by Quiros without finding any other and of all the ancient discoveries they are the best indicated in the author's journal.

After the pass, we saw an appearance of a bay in the Northern part, we wanted to anchor there and to this end we sent the boats to take soundings. I went in the small boat to see this from closer, suspecting that the inhabitants might attack them. The *Etoille's* boat having neared the land, fired several volleys at the blacks who were on the shore and at canoes that were going back to land after having come very close to our boats to show us fruit and make us understand that we had to fetch it on land, without making the slightest hostile gesture, but the *Etoille's* officer said he had to fire because the blacks intended to shoot arrows at him. Several of these unfortunates were killed. We found a good anchorage in 40 to 25 fathoms one league from land but closer and with a lesser depth one finds rocks. These anchorages are in no way similar to those of the bay Quiros mentions and cannot be the same. Quiros, landing on the land of the Holy Spirit, saw land to the W and SE, the coast was therefore trending approximately SW and NE. The part we saw which is the Eastern one trends approximately NS. The Spanish navigator was probably in another part of this great island. The coast, continuing the bay where there is an anchorage, seemed to trend NW. The next day 28th, we left these lands, sailing West without changing latitude...

South of New Guinea

During the night of 4 to 5 June, we were very close to becoming stranded on a small very low lying sandy island situated in 15°47' of latitude and 145°50' of longitude. 2 days later, we found two very extensive banks of rocks in 15°17' of latitude and 143°35′ of longitude. Such repeated dangers made us decide to sail North and NE. This was the only discovery we had made since Quiros's land, although we had always kept to the same latitude, so the geographers who placed these lands beyond the meridian where we found them made a mistake. I believe nevertheless that if we had been a little further west when we made for the North, we would have found other lands and that it would have been advantageous to have been able to follow that coast as far as Dampierre's passage.

Sailing through these totally unknown seas, we found ourselves on the tenth in the morning in a great gulf formed by lands, some bearing West in latitude 10°31′ and longitude 144°2′, the others to the North in latitude 10°15′ and longitude 144°30′ and still others in latitude 10°31′ and longitude 145°16′. These lands are very high and covered with attractive forests. A charming scent came from them. We saw one canoe and many fires, which led us to believe it was highly populated.

Continually finding land to the SE of this gulf and having a head wind with a frightful sea, we were very concerned that we might not be able to round them for a long time and at the same time we had to seek a place of call that would supply refreshments. We were short of wood and water, our crews were worn out by the constant strains they had been subject to since they had left the River Plate, scurvy was beginning to spread among them, we were forced to cut down their bread rations and reduce them to 11 ounces per 24 hours and the water to ¾ of a bottle. As for us, long reduced to the same salt meat as the crew's that was rotting, we ate rats with pleasure.

Finally on the 25th finding ourselves closer to land to our windward, we rounded a cape to which we gave the true name of Deliverance [Délivrance]. It lies in latitude 11°24′ and longitude 147°49′.

On 28 and 29 June, we saw several islands 4 to 5 leagues in length each, lying NNE and SSW, the middle of the central island lies in latitude 7°22′ and longitude 147°5′ and the other in latitude 7°13′ and longitude 151°18′, a third situated between 8°11′ of latitude and 151° of longitude. I saw a fourth whose Northern part rose up to latitude 7°40′ and longitude 152°29′. Its S part was hidden by the previous island. I saw yet another island in 7°40′ of latitude and 152°35′ of longitude, its Northern part fog-covered could not be seen. It is not always possible at sea to sight an entire length of coasts, that is the work of several voyages.

During the first days of July, we sighted a large island whose South coast goes from latitude 7°21′ and longitude 152°48′ up to latitude 6°42′ and longitude [blank]. It then turns East as far as 6°43′ of latitude and 152°27′ of longitude.[1] All these islands make up a sinuous strait rendered difficult by the currents and tidal races found there, and the shoals one finds there also make them very dangerous.

On the [blank], seeing our great need of water and refreshments, we sent the boats to the South coast of the large island to seek an anchorage but it was in vain. The coast seemed fairly populated although covered with a very dense forest down to the edge of the sea. Some blacks came with their canoes around our boats and several came several times very close to our ships. The next day we wanted to examine a port we saw off the NW point of the same island. I went ashore. The port seemed very

---

[1] Choiseul Island, one of the Solomons.

fine, we found several sandy grounds from 10 to 20 fathoms from the entrance as far as a small island behind which it goes in by more than another league. I saw numbers of coconut trees and many other refreshing fruit that we had so much yearned for. We were still examining the anchorage when the inhabitants, who had remained hidden until then, came out with 11 canoes from a river behind some reefs at the entrance to the port. There were about 150 to 160 armed men. Their numbers, against 25 to 30 men, and our enclosed position between the land and them, inspired too much confidence among them. They advanced in two lines, rowing with all their might, within a musket range from us, they broke and surrounded us uttering their war cries and making many menacing gestures with their bows they held ready to shoot, but when we saw them taking aim at us, we fired a volley from our muskets and our blunderbusses, which I think affected the arrows they shot at us. Certainly none hit us in spite of the great number they shot. The blacks, having learnt the effect of our noisy weapons which killed them after piercing their shields, took flight. We took two canoes from them which they abandoned. We found the weapons of the dead men, bows, arrows and spears tipped with fish bones, shields made of reeds, much fruit with the head of a man that had been grilled and probably eaten, a grim omen for those who might have fallen into their hands. We were taking the two canoes to our ships when 10 others that were also coming to attack us, seeing the result of the first affray, withdrew. These blacks paint their woolly hair red or white, as for the rest they are naked and have only some leaves to cover their waist. Their canoes without outriggers are very large and can hold 25 men. The bow and the stern rise by 11 or 12 feet, decorated with pearl shell and figurines representing one or two animals or a man's head.

The night and the very strong currents that were taking our ships onto a rocky shoal made us abandon the plan we had of anchoring in this fine harbour and the next day, seeing no land to the North, we found ourselves out of this pass that I advise all travellers to avoid.

Off New Britain

After following the lands that lie NNW and SSE as far as Cape Laverdy situated in latitude 5°38′ and longitude 153°7′, we sailed along an island that is only 7 leagues from it,[1] it is low lying and wooded and approximately 15 leagues round. It seemed to be very populated. Three canoes with inhabitants, blacks like those who had attacked us a few days earlier, came close to the ships, we threw them a few gifts that they picked up and bartered against a few coconuts they called *bouca* as well as some arrows. One of them with whom I had bartered a piece of cloth against some arrows gave me one. The piece of cloth I had thrown towards him having been picked up by blacks from another canoe, he bent his bow and aimed at me. His comrades having prevented him from letting it go, they left and urged us to go ashore and as soon as they were a good distance away, the displeased black shot his arrow at me.

On the 7th, we sighted a vast coast stretching out towards the South as far as latitude 5°11′ and longitude 146°48′. At first it seemed to consist of 3 islands. We found a harbour on it in latitude 4°48′ and longitude 140°44′ where we dropped anchor

---

[1] Amended in the later text to two leagues. The island is Buka.

and it is only on leaving this harbour that we saw it was New Britain [Nouvelle Bretagne].[1] This is a fine and large harbour, an island at the entrance creates two passes, vessels are well sheltered. We dropped anchor the same day in 30 fathoms muddy ground. We found in this harbour neither fruit nor resources other than a little fish. The country is very unhealthy, it rained during almost the entire length of our stay. The land is nothing but mountains and rocks covered with a fine forest and with these reeds one uses to make canes down to the seashore. Several of our people were attacked by scurvy there and the condition of those who had it already merely worsened, but we considered ourselves very fortunate to have all the wood and water we wanted. To be frank, in such a situation, one is satisfied with very little. No inhabitant made an appearance. We merely found two huts and two abandoned canoes on the shore and places where people had eaten shells and boars the remains of which were still visible. Undeniably the inhabitants of the interior go there to fish and hunt especially as we saw boars in the woods. We found near one of these huts a piece of lead broken and rolled up where English words and half-words were written. This is the form of this inscription.[2] One and a half leagues from there we saw a tree carved in espalier shape where three nails still remained and two lengths of rope that had held the inscription. Several trees had been cut with European tools. I do not doubt that English ships had anchored fairly recently in the same place, two or three months before us and that they were the same as those whose traces we had found in the strait of Magellan. These vessels had sailed from Europe at least five months after the French ones, and had gained at least two months on them because of the regrettable delays the *Etoille* had caused us. With so few resources in this port that I believe to be the worst in the South Sea, we hastened to leave it. I spent my time looking for shells, there are some very fine ones, but the search is fairly dangerous on account of the large number of venomous insects and snakes there are on the reefs. One of our sailors was stung and was cured only after enduring frightful pains and convulsions…

On the 21st at around 11 a.m., the effects of an earthquake were very distinctly [felt] on board our ships. The duration of the movement was of 2 to 3 minutes, it was also very clearly felt on land. The sea rose very high at the time of the quake but fell right away. Dampierre says it is a common happening in the country.

The sailors discovered on board the *Etoille* a girl disguised in men's clothes who worked as a servant to Mr Commerçon. Without suspecting the naturalist of having taken her on for such a tiring voyage, I like to allocate to her alone all the credit for such a bold undertaking, forsaking the peaceful occupations of her sex, she had dared to face the strains, the dangers and all the happenings that morally one can expect in such a navigation. The adventure can, I believe, be included in the history of famous girls.[3]

After taking possession of the island by leaving an inscription as evidence, we left this port on 24 July, well supplied with wood and water.[4]

---

[1] The island is in fact New Ireland. It southernmost extremity, Cape St George, lies in 4°52′.
[2] No text or sketch of the plaque is given.
[3] This paragraph is out of place: Jeanne Baret's gender had been discovered much earlier. Nassau omitted it altogether from his later version.
[4] 'Nassau's first version ends here. The text that follows has been taken from MS C' (ET).

On 2 August, we rounded the NW point of New Britain which is situated in [blank] of S latitude and [blank] of longitude. The most E part of this island is in [blank] of latitude and [blank] of longitude. I have given the position of the port where we anchored and of the most S place we saw, therefore the coast we followed with the one S of Dampier Passage and the W part that Europeans have not yet surveyed, forms a fairly vast country. The coast we followed offers a magnificent scene intersected with mountains and plains mostly cleared and everywhere well inhabited. These people hastened to come in their canoes from their settlements towards our vessels and made signs inviting us to go shore. From among the great number of canoes that was continually surrounding us, one man only hurled a stone at us with his sling, from too far to hurt anyone. This did not decide us to fire at them, but soon we saw from afar, without guessing the reason, our consort start a running fire of muskets on some thirty canoes that were around her. These unfortunates, attacked by superior weapons, fled with great losses. The next day, a larger number of canoes from another locality approached this same vessel which again fired on them. We were too far away from the store ship for Mr de Bougainville to put a stop to this carnage and it made us impatient to find out why part of the blacks had been so peaceful towards us while others were waging war against our second vessel, but a few days later, I discovered that the people of the *Etoille* had begun to cheat the blacks when buying fruit off them, a dispute followed and the commanding officer inhumanely had them fired on. The cruelty affected the next day even those who, coming from a different locality, knew nothing of the previous day's happenings. They told me of having seen on the second day, during the continuing shooting directed at defenceless wretches, one youth swimming towards an elderly man, no doubt his father, who, having been wounded, was drowning, and being killed just as he was saving him. Will people who call themselves civilized always be more cruel than those we call savages?

The inhabitants of New Britain are black and have wool on their heads and chins that they paint red or white, with no other kind of dress than a few leaves and bracelets of pearl shells; they displayed strong and well built bodies. Their canoes, about 30 feet in length, carried up to 18 men. They were decorated fore and aft with roughly carved figures. We saw several small islands on the E coast of New Britain that seemed very populated. We sighted on the 9th between [blank] of latitude and [blank] of longitude a chain of low islands, all well wooded and populated. I even saw on the smallest one which was hardly more than half a league in circumference, several men and a few houses on the seashore. This whole area is crowded with people although they do not have as much in the way of agricultural resources as we have in Europe. One night our store ship was close to running aground on these lands where there are strong currents that should always be feared by navigators who are unaware of them.

The 11th of the same month brought us a view of New Guinea [Nouvelle Guinée] in latitude [blank] and longitude [blank]. We followed it as far as a point that forms the most N part of this country situated in [blank]. It may be easily identified through two small islands that lie very close to it, on one of which shortages and hunger made us look without success for some succour. After this point, the coast trends SW where we could see it stretching as far as [blank]. The first discoverers by sea were better navigators than writers, those who published their narratives were more authors than seamen or geographers and the work of both groups was totally useless during this voyage. The position of Cape Mabo which other voyagers earlier identified as leading from the South Sea to the Moluccas [Moluques] being very uncertain, we decided to leave at this point the land of New Guinea and bear W where we sailed after numerous difficulties among many islands…

[The remainder of Nassau's journal deals with the voyage through the Dutch East Indies and, even more briefly, with the return voyage from the Isle de France to Brest. He then concludes:]

The success of the voyage, in spite of the usual obstacles one comes across in such an expedition and in spite of the special obstacles caused by the storeship *Étoile*, is entirely to the credit of Mr de Bougainville who, coming from a career in the land-based forces in which he had distinguished himself, has just completed the greatest undertaking one could expect from a sailor accustomed to taking command of men. He was able to keep together his crews in the midst of the greatest hardships with the kind of determined zeal that alone can overcome these difficulties. For their part, the men easily accepted the leadership of their commander and, whereas in England they have concealed from the ordinary soldiers the places [to be explored] and the dangers they could expect on a voyage around the world, and whereas their Government believed that these men, patriots though they were, could only be kept at their task if their pay was doubled, the French on the contrary travelled everywhere with confidence and put up with exhaustion and food shortages out of a sense of duty.

We shall in time be able to to assess the gains which the sciences and the kingdom can expect from this voyage. The Navy will benefit from a shining example, and the nation, which is well disposed towards extraordinary matters, will be encouraged to direct its mind and its thoughts towards the sea. The spirit of discovery is reviving in Europe. The various books spreading among the reading public are causing individuals to turn their attention to it.

These new undertakings will bring about a memorable epoch in the world, and we shall at last achieve a complete knowledge of the globe. The new Columbuses and Cortezes have also new fields of glory to explore, but this is the century of humanity, and we must at least express the hope that Europeans intend to become acquainted with their brothers in the southern hemisphere for no other reason than to teach them the truth and make them happy.

APPENDIX 6

The Commerson Documents

Philibert Commerson did not live to complete the detailed narrative he had meant to write, but he left a large number of notes and drafts. He died in Mauritius in March 1773 and, according to Taillemite, 'the collections and papers he had collected were brought to France in the Victoire, commander de Joannis, who arrived at Lorient in May 1774. The consignment consisted of 34 crates which were forwarded through Le Havre and Rouen and handed over to Jussieu at the Royal Gardens. The latter was to sort out and publish Cornmerson's work, but this task was never carried out and the papers kept at the Musée d'Histoire naturelle library have remained in that condition.'[1] Commerson had apparently begun, with his usual enthusiasm, to write a shipboard journal, but as was so often the case with him his enthusiasm waned as other interests caught his fancy, and that manuscript petered out after 34 pages, which did not take the narrative even to the time of the Étoile's arrival in Montevideo in April 1767.

The 'Mémoires pour servir à l'histoire du voyage fait autour du monde par les vaisseaux du Roi la Boudeuse et l'Étoile pendant les années 1866–1768 pour être rédigé par nous Philibert Commerson' foreshadows the writing of a major account of the expedition, which he did not live to complete or even begin. The 'Mémoires' consist of five substantial notebooks and reveal a close cooperation between Commerson and Pierre Duclos-Guyot, the young volunteer sailing in the Étoile and the son of the Boudeuse's second-in-command. The first two are in fact by Duclos-Guyot, with annotations and appendices by Commerson, two others consist of notes taken by Commerson, complemented by additions by Duclos-Guyot, and the fifth is the work of an anonymous author with a note by Commerson.

The 'Mémoires' as such add little to our knowledge of the route and the navigation of the Étoile. More interesting are the 500 or so notes or apostils known to have been written by Commerson himself. A number of these have been lost or mislaid, but they relate largely to the voyage across the Atlantic, the stay in South America and the earlier part of the navigation through the Straits of Magellan. The following extracts are from MS 2214 in the Bibliothèque du Musée, and are taken from a section entitled 'Continuation of the Marginal Notes'

The call at Tahiti
Description of the island of New Cythera where we anchored on 6 April 1768 at approximately 3 p.m.
Men.

The Savages of this island are all whites. There seem to be no blacks among

---

[1] Taillemite, *Bougainville,* I, pp. 134-5.

them, whom they appear to loathe.[1] They are mostly naked as the back of my hand excepting the heads of families and others who cover themselves with a kind of alb or poncho of beaten cloth made from the second bark of a tree of the pine tree type as far as the fruit is concerned but with different leaves, that is to say narrow like reeds and a foot or a foot and a half long. The sun and water turn the Savages who are almost most all handsome men a very deep brown. The tallest (of whom we measured a few) are 6 foot 4 to 5 inches tall, the others of a fine ordinary height, black hair, with a bearded chin like the Capuchins, black eyebrows, eyes level, small closely spaced teeth, white as enamel, in a word very handsome faces and those of the older men very respectable. They are very strong and very skilled especially at thieving, everything being good for them. Their buttocks are painted in a bluish colour from the waist to mid-thigh. Their natural parts concealed by a piece of cloth that goes round their buttocks, passes below the thighs and ends finally in a reef-knot they tie with both ends. They can all swim very well, the proof is that if we threw something at them in the sea, they sped swimming after it although the ship was making 4 to 5 knots and, being unable to return to their canoes or our ship, they would swim ashore which was sometimes more than a league away. If some of their canoes capsized, as often happened in coming too close to the ship, this simply caused the spectators and the actors to laugh and the latter soon righted it. That is what we saw.

Women and girls.

Stand comparison with the finest European brunettes except that they are less white. They have large eyes, blue or black and level, black eyebrows, a coquettish and seductive glance but bold, a small mouth, small closely spaced teeth, fine black hair usually tied above the back of the head, a fine bosom, nice plump hands and even finer arms, in a word all their body is exquisitely proportioned with the exception of the feet which are a little too big and the legs too heavy. These women are very skilled, they made small items that are most beautiful for this country, i.e. small hats for themselves, mats or blankets to lie on with designs on them and small bags to store all their little utensils, all made of reeds or straws, admirable small tresses of hair and tree bark and finally a kind of muslin of beaten bark to cover their bosom and their sex. Their buttocks are also painted like the men down to mid-thigh, however they do not all cover themselves as at least 3 quarters have their bosom bare and sometimes the rest as well. It seemed to us that as soon as a girl is married or at least has agreed to stay attached to one man, she is faithful to him and does not concern herself with relations with others, the death penalty being so they say imposed against adultery. On the contrary girls as long as they remain so are very free to do whatever they like, jealousy being unknown in the celibate state.

The women are as good swimmers as the men. We saw a canoe capsize alongside our ship with 3 men, 2 women and one child at the breast. As soon as she found herself in the water, the mother took the child underneath the armpits to hold its head above

---

[1] This comment would seem to be based on Ahutoru's reaction to the Melanesians he saw later on the voyage and possibly on his attitude towards blacks encountered on Mauritius. A later comment suggests that the sun and seawater had a tanning effect on the Tahitians' otherwise white skin.

water and swam approximately the ship's length to reach another canoe without worrying in the least, let alone the other women who began to laugh at the adventure.

Food.

These Indians live on pork, chicken, birds, fish, shells and fruits of various kinds such as sweet potatoes, yams, etc… They cook a fruit the size of melons which they call *aourou* and is of the coniferous species, it seems, as it has a rind that is incised or marked out in diamonds and compartments (this fruit, according to Mr Poivre, is called *rima* in the Philippines, the travellers' breadfruit),[1] with others of different types in a hole they dig in the ground near their huts and around and at the bottom of which they place some kinds of pebbles. Then they place the fruit on top and over them some pieces of wood placed cross-shaped on top of each other, after which pebbles are added on top and all around until raised a foot approximately above the ground, then they set fire to it. Once the fruit is cooked, they knead it and place it as a paste in banana leaves apparently to preserve it or to keep it fresher. They eat it at all their meals, and store it as a provision in all their canoes when going out to sea. They often gave some to our sailors but they could not adapt themselves to it.

They have no other liquor or drink than water from streams and coconuts. That is about all their food.

Laws and customs.

They seemed to be ruled by a senior chief whom they respect more than fear and then by heads of families, and this from one river to another which apparently forms the boundary to each little state. Peace and unity seem to reign unbroken among them. Each family has its own hut separate from the others, however we saw some very large ones that appeared to be used for gatherings possibly to carry out work in common or share meals. They sleep on the floor on mats of rushes and most have as a pillow for the night at the same time as a seat during the day a small bench made in this shape and all in one piece. The women place themselves next to their husbands with their children and others lie down pell-mell except no doubt the men and the wives who sleep separately.

Ceremonies.

Their dead are firstly exposed in a kind of chapel on a kind of mortuary display bed, then on a small table or grill situated in the place which is warmed the most by the sun's rays, the whole is surrounded by stakes.[2] The area in front of a hut [is where] they place themselves to carry out their ceremonies,[3] which are dancing and jumping

---

[1] Commerson's description is correct. The breadfruit tree *(Artocarpus altilis)* produces a fruit that is roughly the size of a small melon with a rough, somewhat alveolate rind. Its Tahitian name is *uru*. Commerson, above and again in a short appended vocabulary, gives *aourou*, possibly a mis-hearing of *e uru*. Pierre Poivre (1719-86) became in 1766 the *Intendant* or civilian head of the colony of Isle de France; he was a noted botanist and endeavoured to promote a French spice trade. See on him Ly Tio-Fane, *Mauritius and the Spice Trade*; Malleret, *Pierre Poivre*

[2] Presumably a reference to the *fare-tupapau*, the mortuary hut which Banks called 'the house of corruption': Beaglehole, *The Endeavour Journal of Joseph Banks*, I, p. 377.

[3] Some of Commerson's sentences are incomplete or unclear, which makes one believe that he did not have time to revise his notes. He is presumably referring here to the sacred or ceremonial area in front of the entrance to a hut, known as a *marae*.

around the dead man with bowings of the head towards him as a mark of respect and from time to time, that is to say daily, they turn him in another direction, rubbing him morning and evening with coconut oil while repeating the above-mentioned ceremonies until he is dry. What makes us believe this is that we saw one who was dry, hanging by the neck with his hands tied and crossed over the stomach and with a large gorget made with shark's teeth and feathers of poultry or birds.

N.B. I do not know how the one who reported this fact found this out. As for me, I believe these good people consider death simply as a long sleep, they treat them for several days as though they were merely asleep and it seems in the hope they will wake up, even stationing men around them with fly-whisks to keep insects away and providing food to use when they might wake. When finally the bodies have quite rotten away (for we saw some on the stands which were very black), it seems that they end up burning them as I saw several holes filled with ashes around these mortuary enclosures and these ashes seemed to be the remains of their pyres.

In all the confusion caused by the murders, they never forget to remove their dead, so great is the respect they bear them that we saw one in particular being taken away in a canoe in spite of the unbearable stench it gave out. One does not know what they did with it because in spite of the restoration of peace, the dead did not reappear.

Fertility of the island and its productions.

There are in this island numerous coconut trees, banana trees and *heris*[1] and other unknown fruit, yams and sweet potatoes, birds of various types, hens, pigs of the wild boar species and dogs, a lot of water game such as ducks, teals, herons, snipe, in the woods numerous turtle doves and wood-pigeons not to mention the parakeets that are etc...[2] This soil [once] cultivated would accept all kinds of cereals as in France. There is no lack of sugar cane or of the kind of arum they call *longe* in the Isle de France.[3] Mr de Bougainville made them plough a small square in which he planted wheat, barley, maize, haricot beans, beans, lentils, etc ... lemon and orange seed (which were absent in this island) with many other domestic seeds such as cabbage, salad, rape, lettuce, chervil, parsley, etc....

We showed them how bread was made with wheat by showing this wheat which we crushed in front of them, from the flour we made some dough which we cooked in the oven after which we made them taste the bread which they found excellent, finally explaining that what had been sown would give them this same bread and the other vegetables we served at table. In case these crops did not succeed the first time, we gave them a second set of all the seeds to renew them. To these gifts we added a very precious one which was a pair of turkeys although no more remained on board, making them understand it was for breeding and not for killing. We forgot to give them any pigeons... Mr de La Giraudais etc... [sic]

The entire top of this island possibly 24 to 25 leagues along (I would be more

---

[1] In his vocabulary, Commerson gives *ari* for coconut. The Tahitian word is in fact *haari*.

[2] Another instance of Commerson's incomplete sentences.

[3] The word should be 'songe': it is the taro, *Colocasia esculenta* of the order Araceae or arums. Details on its uses in the islands of the Indian Ocean will be found in Gurib-Fakim, Guého and Bissoondoyal, *Plantes médicinales de Maurice*, I, 1995. I am grateful to Dr Ly Tio-Fane for this information.

inclined to say thirty) and approximately eighty leagues or possibly a hundred around, is entirely forested and rises in fairly high mountains but making the most pleasing scenery that could be seen, there being no bare or sterile patches. We did not ascertain the diameter of this island because we stayed only 8 days and we were busy getting water which is perfect and keeps as well as could be expected. One can nevertheless assume that this island is large and is possibly only part of a considerable archipelago.

Barter with the Indians.

We did not feel that any ships came to this island before us or even that they trade with other Indians, at least we found nothing among them that indicated this apart from the dogs, the pigs and the hens we found.[1] (It will be seen that we [now] think quite differently and why we do.)

These people do not know iron (this is incorrect, they were the first to ask us for some), nails or mirrors or anything we first showed them. At first they refused bread and biscuits however etc … they valued them greatly after they discovered them. For one or two 3-inch nails on could obtain I hen or a pig, ditto for a mirror of 6 *liards*,[2] ditto for crystal beads, beads, rosary beads, sleeve buttons of lead or pewter, artificial pearls and any other shiny item to be placed around the neck, on the arm or ears. Towards the end they were willing to barter only for large nails and agricultural and carpentry tools but above all for axes which had amazed them by the ease with which one could cut trees down with them and in so little time. They hated knives and scissors because they considered them harmful; as for their loincloths, they cut them very skilfully with a reed or bamboo blade.

Tools and weapons.

Are all of a grey and black stone close to touchstone and shaped to cut and polish even wood fairly cleanly.[3] They are shaped like axes, adzes and scissors for cutting wood. They have nothing else to build their huts, canoes, etc. which are admirably well made.

Their weapons are nothing except the spear, sling, bow and arrows. These arrows do not look dangerous for they are made simply out of reeds and are tipped only by a small piece of hard wood an inch long but not sharp. They throw the spear so skilfully that from twenty to 30 paces they seldom fail to penetrate the trunk of a banana tree. We did not see them using their slings.

Our muskets and even more our guns inspired the greatest fear in them. When

---

[1] Trade is motivated by the desire to obtain by some form of barter items that one does not already possess. Dogs, pigs and hens had long been available in Polynesia. The dog *(kuri)* was brought to New Zealand when the Polynesians first colonized the country; pigs and hens were so well known from olden days that their origin figures in Polynesian tales, such as the legend of Matuapua'a who gave birth to the pig in Borabora, and Tumoana-arifa and his wife who gave birth to the hen in Raiatea. See Henri, *Tahiti aux temps anciens*, pp. 392–3.

[2] *The liard* was a copper coin of low value. The common expression 'not to have a *liard*' is the French equivalent of the English 'not to have a bean'.

[3] The stone was often a black dolerite or basaltic rock obtained from the island of Maupiti.

we fired, they imagined we all wanted to kill them.[1] Furthermore being very timid, they trembled and fled when they saw the slightest gesture of menace. If they were called back or one did nothing likely to frighten them, one found oneself surrounded by 2 or 300 but it was not difficult to make them move aside, a gesture with the hand was enough, let alone a dog one might set on them. I saw a small ship's boy of some three and a half feet make himself feared by them, with a switch in his hand, he made them flee in their hundreds before him and there is no instance of their ever raising a hand against any of us....

[Their] language very clear-toned has a great deal of affinity with Italian and consequently very musical, and so all Tahitians are natural musicians and express poetically and in song everything that affects them either pleasant or painful. Our prince Poutaveri broke out into impromptus at any time.

As for instruments, they have mainly the flute shaped approximately like our transverse flutes but one would not imagine the organ they play it with. They play it with the nose, to one nostril of which they place the mouthpiece blocking the other with the thumb, the other fingers being used to modulate their tones but their very simple music allows only two and a half, nevertheless it is not unpleasant.

The two large valves of a pearl oyster arranged more or less in the manner of our castanets are also used and [used as] instruments for amusement and to trick or as a rattle to call and warn each other, etc...

Off the New Hebrides

The boats that returned from Lepers Island [Isle des Lépreux] being back reported that the Savages of the bay where they went to take soundings were very ill built, small and almost all leprous; they are almost black, frizzy-haired like the negroes, the tallest do not seem to exceed 5 feet in height (I assume it is smallpox that has caused the retrogression of this species). They are moreover very unpleasant and perfidious. It was not possible to barter for any refreshments with them and when they saw everyone in the boats returning to the ships they began to shoot arrows without however being able to hit anyone. We fired a few shots by way of reprisals from which 2 Savages were killed and 2 wounded.

On Friday 26 May we hove to, ... Mr de Bougainville having given the signal to lower the boat to go and sound in a bay and seek a suitable anchorage between various small islands. When the boats were near the shore of the larger land, a number of Savages gathered who seemed to be making many gestures of friendship either from the land, or from their canoes. They held some banana leaves and gestured to go ashore where one would find drink and food. The mistrust they showed, being unwilling to come close to us in spite of their great number, caused a similar feeling among us, and not without reason for in the middle of all these demonstrations of friendship they shot some arrows 2 or 3 times and from various canoes at our boat

---

[1] This feeling was no doubt inspired by their experience with Wallis the previous year, e.g.: 'applyed to the Great Guns and gave them a few rounds and Grape shot...we let them come within about three or four hundred Yards of the ship, then fired a three pounder loaded with seventy Musquet Balls among the thickest of them,...and to add the more to the terror they were in we fired two round shot amongst them when they were about a mile from the ship': Carrington, *The Discovery of Tahiti*, pp. 154, 156.

which however wounded no one, their arrows being tipped with a fairly heavy thorn of hard wood and consequently not very suitable to send far. We then fired back with two blunderbusses loaded with ball and several musket shots which soon cleared two of the nearest canoes, the others having fled. Four or five of these Savages were killed and at least as many wounded, the latter having jumped into the water to flee with the others. A strange thing was that, when they wanted to shoot, they turned the prow towards us and took aim concealed behind the former which more or less covered them as best it could by its movements. These Savages are frizzy-haired and black but not leprous like the previous ones. Mr de Bougainville having heard firing, fired a gun as a signal for the boats to return.

Note that only the *Etoile's* boat was attacked, the *Boudeuse's* having always kept to seaward to take sounding and being in no hurry to come to the rescue. It is true that Mr Landais who was in command of the boat, had neglected to go on board the *Boudeuse* and even to go to the officer in charge of the boat to receive his orders hurrying instead to be the first ashore. The Prince and the *Boudeuse's* officer having tried to reprove him, he had the impudence to reply with such insolence even to the Prince that I heard it said (from the Prince) that he was on the point of shooting him but that fortunately good sense prevailed and he merely expressed his contempt to him. A little while later, the commander having come on board, Mr Landais was reprimanded for his two infractions, but it is [like] rain falling on tinplate, which merely rusts it more.

We captured a few weapons from the Savages but although I was present and had shared the risks equally with the others, I could not obtain anything, Landais took everything.

This was the same day that Mr Vivés poisoned the wound on my leg[1] that was almost cured and caused the return of a sore that lasted another two or three months, and in spite of that I insisted on going ashore, that is to say to take part in the above action although we did not set foot on land.

Towards New Britain

Description of the Savages who live on the islands seen on the 29–30 June 1768.

At about 10 a.m. on 29 June, 9 canoes came near the ship in which were some 70 to 80 men with ugly features, their hair dyed white and red as well as their faces and several others parts of their body, their teeth black and yellowish. They wore a kind of ribbon of red rushes attached in some cases to the nose, in others to the ears, the arms and armpits, a kind of bracelet the size of a hand at the middle of the arm, that is to say above the elbow. These bracelets are made with a very white material of gold or ivory or perhaps the operculum of a very […] shell common in these seas (and at Batavia), the others in braids or in plaited reeds or rushes; others wore shell plates the size of a hand attached to their forehead and finally others had 3 or 4 small animal teeth (boar's?) attached to their nostrils in the shape of a moustache. We first of all threw a few lengths of red bunting which seemed to please them, then a corked bottle in

---

[1] In a journal entry of 18-19 April, Duclos-Guyot reports that Commerson went over to the *Boudeuse to* attend to Bougainville who had been ill for several days. In the margin, Commerson wrote, 'I damaged my leg going across and thanks to Mr Vivès'.

which were chaplets and blue, white and green necklaces, but either they did not see this bottle being thrown (it was of white glass) or they did not deign to fetch it. These Savages made numerous gestures inviting us to land and that we would find everything to eat such as coconuts, bananas, sweet potatoes and yams that they showed us; we made them understand that this was our plan, without however trusting all these appearances of friendships too unreasonably, this seemed to satisfy them. At the same moment, the weather changed with signs of a heavy storm, which made them leave and we contented ourselves with sailing along the coast. We have reason to believe they are cannibals on account of the human jaws we found in the decorations of their canoes.

New Britain

From the 1st to Saturday 2 July 1768.

Mr de Bougainville gave the signal at 11 a.m. to lower the boats to survey off the W point of the N land a bay that seemed very suitable for an anchorage on condition the bottom was good. We sent Mr Donat. At around 2 p.m., the boats from the two ships had reached it and went quite far into this bay to sound the bottom and examine the country so that the islets that line the main land soon hid them from sight. This made us feel optimistic about there being an anchorage. At around 3 p.m., we heard the muskets and blunderbusses they had taken being fired at eleven canoes full of Savages who had gathered to surround them and attack them in this manner, with success had they not been on their guard. These Savages seemed so sure of their success that they celebrated in advance by waving their spears and readying their slings and arrows. When the Prince of Nassau, Mr D'Oraison and Mr Donat, officers in the boats, saw that it was time to forestall them and that they were within good range of their arms, they fired at them and killed approximately a dozen, the others jumped in the water and abandoned their canoes two of which we took. We found in them bows, quantities of arrows and spears and shields with various fruit such as coconuts, yams, manioc flour, etc.... but we could not bring the canoes aboard, the sea being choppy and they were filling with water. One should further point out that on leaving the bay, the officers saw some thirty canoes pass through the mouth of a small river not far away, coming more to reinforce their comrades than to rescue them but when they saw what had happened, they went back.

Shortly after, that is to say around 4 o'clock, we saw our boats going along the shore with a white flag at their mast, the agreed signal of a good anchorage, which led the *Boudeuse* to advise us to sail there, that she was going to make for the boats, sailing NE and NE°N, starboard tack close-hauled. Manoeuvring thus, we came upon a tidal race, so-called, or to be more precise a shoal where there were only 5 or 6 fathoms of water over which the sea broke heavily and towards which the tide was bearing us, so that we were not surprised when we saw the bottom as clearly as if there had been only 12 to 15 feet of water. In spite of this they said, and Mr de La Giraudais maintained this more stubbornly that anyone, that it was a shoal of fish. It was only because of the repeated shouts of the men on the forecastle who could see much more clearly and on the evidence of one officer who was coming from the stern windows that they finally decided to throw the sounding lead and found some said 4 or 5 fathoms, the others 5 or 6 because the current was such that the lead was dragging and one could not be precise. Whichever it was, we came out of this danger, the greatest of all those we have run throughout this navigation, more by good luck than good manoeuvre. The bottom consisting of reefs and corals (white in slabs) and the terrible strength of the waves would have put us in the utmost danger had we struck it but once. We closed all the starboard gun ports in time, as well as those of the gunners' storeroom because the waves broke as on the shore. The manoeuvre we carried out when we came to was to alter course and make to windward to avoid the waves, then we furled all the smaller sails until the strength of the rollers had passed, after which we bore away and steered NNE and finally found ourselves in a sea as calm as in a pond, with 7 to 8 fathoms of water. The winds were then SE to SSE light breeze and fair weather. And so we escaped with nothing more than a good fright.

You should have seen these renowned seafarers in this dangerous situation go pale, at a loss and not knowing which way to turn. The captain was the first to lose his head and one can truthfully say that it was the forecastle that saved the quarter-deck, that is to say for those who are not sailors, that the tail ruled the head, if there was a head. As for me, I found the spectacle altogether so singular, I was so engrossed in sorting out who was right about the so-called shoal of fish (and one must agree that the sheets of white coral that passed so quickly under the keel were somewhat like one) that the peril had gone before I realized its extent, but even if I had known it I make bold to say that I am not susceptible to the degree of fear I saw appear on the faces of the La G[iraudais], the Garaux, the Houris, etc …[1] The sailor[s] seemed to notice the danger only in order to deal with it with as much swiftness as courage. The winds were at the time SE to SSE fortunately light breeze and fair weather.

At 6 p.m. the *Boudeuse* having been able to judge before us the suitability of the bay, Mr de Bougainville decided that it was not wise to make for it and sailed away, to everyone's deep regret for we were short of everything and mainly water. We followed him and the boats were recalled. These boats sounding ahead of us were still signalling from time to time 6 to 8 fathoms of water bottom of coral and sand.

At 6.30 the boats had returned and we immediately hoisted them in. Then we sailed NW and NW°W all sails set to regain the open sea because we were only some ¾ of a league from land and feared the calm and the tide.

The officers from the boats reported that the place they had surveyed formed a very fine bay with a very good harbour where one would find 13 to 19 fathoms of water bottom of red sand and shells with small pebbles, that one would be sheltered from every wind as in a good basin (excepting the SE wind veering to SW passing by S) within reach of a river and wood which was available in abundance, coconut and banana trees, yams and other fruit and fish, cuttlefish and turtles etc. so that it was the most suitable place for us in our circumstances, being short of everything, and so everyone was grumbling in no uncertain tones about the leader but it was already too late to regret it.

At half-past one in the afternoon of Wednesday 6 July 1768, the boats that had

---

[1] The targets of Commerson's somewhat acerbic comments are: La Giraudais, captain, probably Caro, the first officer, and Ourry, a volunteer taken on at the Falkland Islands.

been sounding in the bay came towards the two ships with their white flag at the mast, signal of a good anchorage, we filled the sails at once and bore towards them steering N°NW and NNE to reach the said anchorage with a variable SSE to SSW light breeze.

At 2 o'clock being abreast the point that forms the entrance to this bay, we noticed a great bay to starboard which made us alter to starboard and steer E°SE and ESE to reach it. We arrived in it and dropped anchor at 3 p.m. in 30 fathoms bottom of very fine grey sand with a few worn shells and we moored ESE and WNW two cables from land, then we sent the boats to shore to prepare the place where we were to obtain wood and water. The *Boudeuse* did likewise. There are several small rivers in this bay and one is sheltered as in a basin. The trees reach down to the water's edge. We found no inhabitants, but only a handsome canoe abandoned with its spears,[1] the Indians having presumably fled when they saw us arriving, frightened no doubt by hostilities committed by the English of whom we found quite recent traces as I shall state elsewhere.[2]

[1] The word in the text is 'sagayes', but this is possible an error for 'pagayes', i.e. paddles.
[2] 'If any additional notes were written, they have not survived' (ET).

BIBLIOGRAPHY

Works relating to Bougainville and his voyage

Aragon, L. A. C. d', *Un Paladin au XVIIIe siècle: le Prince Charles de Nassau-Siegen*, Paris, 1893.

Arnaud, E. et al., *Colloque Commerson*, Centre universitaire de la Réunion, St Denis, 1973.

Boissel, T., *Bougainville ou l'homme de l'univers*, Paris, 1991.

Bougainville, L. A. de, *Voyage autour du monde par la frégate du Roi* La Boudeuse *et la flûte* L'Etoile *en 1766–1769*, Paris, 1771; second edition in 2 vols, 1772; English translation by J.R. Forster, London, 1772.

——, *Voyage autour du monde*, edited with introduction and notes by M. Bideaux and S. Faessel, Paris, 2001.

Cap, P. A., *Philibert Commerson, naturaliste voyageur*, Paris, 1861.

Carré, A., 'L'Expédition de Bougainville et l'hygiène navale de son temps', *Journal de la Société des océanistes*, 24 (Dec. 1968), pp. 73-5.

Cazaux, Y., *Dans le sillage de Bougainville et de La Pérouse*, Paris, 1995.

Chevrier, R., *Bougainville: voyage en Océanie*, Paris, 1946.

Denoix, L., 'Les Bateaux du voyage de Bougainville', *Journal de la Société des océanistes*, 24 (Dec. 1968), pp. 55-58.

Diderot, D., *Supplément au voyage de Bougainville* (ed. G. Chinard), Paris, 1935.

Dixmerie, N. Bricaire de la, *Le Sauvage de Taïti aux Français*, Paris, 1770. New edition with preface and notes by Grand, A. R., Matteï, J. F., and Guiart, J., Papeete, 1989.

Dorsenne, J., *La Vie de Bougainville*, Paris, 1930.

——, see Fesche below.

Dowling, J. K., 'Bougainville and Cook' in W. Veit, ed., *Captain James Cook: Image and Impact*, I, Melbourne, 1972, pp. 25-42.

Dunmore, J., *Monsieur Baret: First Woman around the World 1766–68*, Auckland, 2002.

Dussourd, H., *Jeanne Baret (1740–1816), première femme autour du monde*, Moulins, 1987.

Duyker, E., 'Josselin and Alexandre Le Corre: Early French Voyagers to Van Diemen's Land and New Holland', *Explorations*, 13, December 1992, pp. 9-13.

Elliott-Scott, G. F. (ed.), *The Life of Philibert Commerson D.M., Naturaliste du Roi: An Old-World Story of French Travel and Science in the Days of Linnaeus, by S. Pasfield Oliver*, London, 1909.

Espitalier-Noël, R., 'Jeanne Baret 1740-1807', in *Dictionnaire de biographie mauricienne*, Port-Louis, 1993, pp. 1475-6.

Fesche, C. F. P., *La Nouvelle Cythère (Tahiti): journal de navigation inédit* (ed. J. Dorsenne), Paris, 1929.

Gerbe, J., 'La Famille bressane de Philibert Commerson le découvreur du bougainvillier: Essai généalogique', *Regain, recherches et études généalogiques de l'Ain*, n.d., pp. 42-7.

Hammond, L. Davis (ed.), *News from Cythera, A Report of Bougainville's Voyage 1766–1769*, Minneapolis, 1970.

Jacquier, H., 'Jeanne Baret, la première femme autour du monde', *Bulletin de la Société des études océaniennes*, 12, No. 141, 1962, pp. 150-56.

Jimack, P., *Diderot's* Supplément au Voyage de Bougainville: *a critical guide*, London, 1988.

Kerallain, R. de, *Les Français au Canada: la jeunesse de Bougainville et la Guerre de Sept Ans*, Paris, 1896.

Kimbrough, Mary, *Louis-Antoine de Bougainville 1729–1811: A Study in French Naval History and Politics*, Lewiston, 1990.

La Roncière, C. G. M. B. de, *Bougainville*. Paris, 1942.

——, 'Le Premier Voyage français autour du monde', *Revue hebdomadaire*, Sept. 1907, pp. 22-36.

——, 'Routier inédit d'un compagnon de Bougainville: L.A. de Saint-Germain, écrivain de la *Boudeuse*', *La Géographie*, XXXV, Mar. 1921, pp. 217-50.

Lefranc, G., *Bougainville et ses compagnons*, Paris, 1929.

Martin, J. E., 'Essai sur Bougainville navigateur: la genèse de sa carrière maritime', *La Géographie*, Nov-Dec 1929, pp. 321-45.

Martin-Allanic, J. E., *Bougainville, navigateur et les découvertes de son temps*, 2 vols, Paris, 1964.

Monnier, J. et al., *Philibert Commerson, le découvreur du bougainvillier*, Châtillon-sur-Chalaronne, 1993.

Montessus de Ballore, F. B. de, *Martyrologe et biographie de Commerson*, Chalons-sur-Sâone, 1889.

Olivier, D., 'Studies in the Anthropology of Bougainville's Solomon Islands', *Papers of the Peabody Museum, Archaeology and Ethnography*, Harvard, XXIX, pp. 1-4.

Orian, A., *La Vie et l'oeuvre de Philibert Commerson des Humbers* [sic], Port-Louis, Mauritius, 1973.

Pascal, M., *Essai historique sur la vie et les travaux de Bougainville, suivi de la relation de son voyage autour du monde*, Marseilles 1831.

Role, A., *Vie aventureuse d'un savant: Philibert Commerson, martyr de la botanique 1727–1773*, Saint-Denis, Reunion, 1973.

Ross, M., *Bougainville*. London, 1978.

Roy, P. G., 'M. de Bougainville aux îles Malouines', *Bulletin des recherches historiques de Québec*, September, 1831.

Saint-Germain, L. S. de, see La Roncière, above.

Sotheby & Co., *Catalogue of the Highly-Important Papers of Louis de Bougainville F.R.S. (1729–1811)*, London, 1957.

Taillemite, E., *Bougainville et ses compagnons autour du monde 1766–1769*, 2 vols, Paris, 1977.

——, 'François Vivez: un chirurgien rocquefortais autour du monde', *Actes, découvertes, colonisation*, Colloque CORAIL, Noumea, 1993, pp. 239-58.

——, 'Le Lieutenant Caro et sa relation inédite du séjour de Bougainville à Tahiti', *Journal de la Société des océanistes*, XVIII, Dec. 1962, pp. 11-19.

Thiéry, M., *Bougainville, Soldier and Sailor*, London, 1932.

Touchard, M.C., *Les Voyages de Bougainville*, Paris and Papeete, 1974.

Vinson, E., *Célébrités créoles: Philibert Commerson*, Saint-Denis, Reunion, 1861.

Waggaman, B., *Le Voyage autour du monde de Bougainville: droit et imaginaire*, Nancy, 1992.

Other Works

Aman, J., *Les Officiers bleus dans la marine française au XVIIIe siècle*, Geneva, 1976.

Amherst, W. A. T., and Thomson, B. (eds), *The Discovery of the Solomon Islands by Alvaro de Mendaña in 1568*, 2 vols, Hakluyt Society, London, 1901.

Anson, see Somerville, also Walter, below.

Auget, P. and Moorgat F., 'Études dentaires dans les établissement français d'Océanie', *Acutalités odontostomalogiques*, 37, Paris, 1957, pp. 111-25.

Bachaumont, L. P. de, *Mémoires secrets de Bachaumont* (ed. P. L. Jacob), Paris, 1859. Original ed. Paris, 1780.

Banks, see Beaglehole, below.

Baré, J.-F., *Le Malentendu Pacifique, des premières rencontres entre Polynésiens et Anglais, et ce qui s'ensuivit avec les Français jusqu'à nos jours*, Paris, 1985.

Beaglehole, J. C. (ed.), *The Endeavour Journal of Joseph Banks 1768–1771*, 2 vols, Sydney, 1962.

——, *The Exploration of the Pacific*, 3rd ed., London, 1966.

——, *The Journals of Captain James Cook on his Voyages of Discovery*, vol. 1: *The Voyage of the* Endeavour *1768–71*, Hakluyt Society, Cambridge, 1955.

——, *The Life of Captain James Cook*, London, 1974.

Bellin, J. N. *L'Hydrographie française ou recueil des cartes dressées au Dépost des plans de la Marine pour le service des vaisseaux du Roy*, Paris, 1756.

——, *Le Neptune français*, Paris, 1753.

——, *Petit Atlas maritime*, 5 vols, Paris, 1764.

——, *Recueil des cartes réduites contenant les parties connues du globe*, Paris, 1748.

Bériot, A., 'Fin du XVIIIe siècle: les grands découvreurs', in C. Benoit et al., *Sillages polynésiens*, Paris, 1885, pp. 37-56.

——, *Grands voiliers autour du monde: les voyages scientifiques 1760–1850*, Paris, 1962.

Bougainville, H. Y. P. P. de, *Journal de la navigation autour du globe de* La Thétis *et* L'Espérance *pendant les années 1824, 1825 et 1826*, 2 vols, Paris, 1837.

Brasseaux, C., *Scattered to the Wind 1755–1809*, Lafayette, 1991.

Broc, N., *La Géographie des philosophes: géographes et voyageurs français au XVIIIe siècle*, Paris, 1974.

Brosses, C. de, *Histoire des navigations aux terres australes*, 2 vols, Paris, 1756.

Bruijn, J. R. et al., *Dutch-Asiatic Shipping in the Sevententh and Eighteenth Centuries*, The Hague, 1987.

Bulkeley, see Smith, below.

Burney, J., *A Chronological History of the Voyages and Discoveries in the South Sea or Pacifick Ocean*, 5 vols, London, 1803-17.

Byron, see Gallagher, below.

Callander, J. *Terra Australis Cognita, or Voyages to the Terra Australis or Southern Hemisphere during the sixteenth, seventeenth and eighteenth centuries*, 5 vols, London, 1766-8.

CampbelI, E. C., 'Savage Noble and Ignoble: The Preconceptions of Early European voyagers in Polynesia', *Pacific Studies*, 4, 1980, pp. 45-59.

Carrington, H. (ed.), *The Discovery of Tahiti: A Journal of the Second Voyage of H.M.S. Dolphin...written by George Robertson*, Hakluyt Society, London, 1948.

Carteret, see Wallis, below.

Clarke, G. F., *The Expulsion of the Acadians*, Fredericton, 1955.

Corney, B. G. (ed.) *The Quest and Occupation of Tahiti by the Emissaries of Spain during the Years 1772–1776*, 3 vols, Hakluyt Society, London, 1913-19.

Dahlgren, E.W., *Les Relations commerciales et maritimes entre la France et les côtes de l'océan Pacifique (commencement du XVIIIe siècle)*, Paris, 1909.

——, 'Voyages français à destination de la mer du Sud avant Bougainville (1695-1740)', *Nouvelles archives des missions scientifiques*, XIV, 1907, pp. 423-568.

Dalrymple, A., *An Account of the Discoveries made in the South Pacifick Ocean previous to 1764*, London, 1767.

——, *An Historical Collection of Several Voyages and Discoveries in the South Pacific Ocean*, 2 vols, London, 1770-71.

Dampier, W., *A Collection of Voyages*, 4 vols, London, 1729.

——, *A New Voyage round the World,* London, 1697.

——, see also Spencer, below.

Daubigny, E., *Choiseul et la France d'outre-mer après le Traité de Paris*, Paris, 1892.

Delille, J., 'Les Jardins' in *Oeuvres de J. Delille*, Paris, 1834, Chant II, p. 18.

Des Cars, Duc, *Mémoires*, Paris, 1893.

Devèze, M., *L'Europe et le monde à la fin du XVIIIe siècle*, Paris, 1970.

Dorsenne, J. *La Nouvelle-Cythère*, Paris, 1929.

Duchet, M., *Anthropologie et histoire au Siècle des lumières*, Paris, 1971.

Dunmore, J., 'Dream and Reality: French Voyages and their Vision of Australia', in I. Donaldson (ed.), *Australia and the European Imagination*, Canberra, 1982, pp. 109-21.

——, *French Explorers in the Pacific*, 2 vols, Oxford, 1966, 1969.

——, 'L'Imaginaire et le réel: le mythe du Bon Sauvage de Bougainville à Marion du Fresne', in Mollat, M., and Taillemite, E. (eds), *L'Importance de l'exploration maritime au Siècle des lumières*, Paris, 1982, pp. 161-8.

——, 'Rivalités franco-anglaises dans le Pacifique 1700-1800', in *Le Colloque d'Akaroa*, Waikanae, 1991, pp. 88-93.

——, 'Rousseau's Noble Savage: A New Zealand Case History', in W. Veit, ed., *Captain James Cook: Image and Impact*, Melbourne, 1979, pp. 160-72.

——, *The Expedition of the* St Jean-Baptiste *to the Pacific 1769–70*, Hakluyt Society, London, 1981.

——, 'The Explorer and the Philosopher: Diderot's *Supplément au voyage de Bougainville* and Giraudoux's *Supplément au voyage de Cook*' in W. Veit (ed.), *Captain James Cook: Image and Impact*, I, Melbourne, 1972, pp. 54-66.

——, *The Journal of Jean-François de Galaup de la Pérouse 1785–1786*, 2 vols, Hakluyt Society, London, 1994.

——, *Visions & Realities: France in the Pacific 1695-1995*, Waikanae, 1997.

——, *Who's Who in Pacific Navigation*, Honolulu and Melbourne, 1994.

Duviols, J. P., *L'Amérique espagnole vue et rêvée: les livres de voyages de Christophe Colomb à Bougainville*, Paris, 2000.

Duyker, E. , *An Officer of the Blue: Marc-Joseph Marion Dufresne, South Sea Explorer, 1724–1772,* Melbourne, 1994.

——, see also under Le°Maire below.

Eisler, W., *The Furthest Shore: Images of Terra Australis from the Middle Ages to Captain Cook*, Cambridge, 1995.

Elliot-Joyce, L. E. (ed.), *A New Voyage and Description of the Isthmus of Panama by Lionel Wafer, surgeon*, Hakluyt Society, Oxford, 1934.

Fairchild, H. N., *The Noble Savage: a Study in Romantic Naturalism*, New York, 1928.

Faivre, J. P., *L'Expansion francaise dans le Pacifique de 1800 à 1840*, Paris, 1953.

Faessel, S., 'Le Mythe de Tahiti, ou comment une réalité devient une fiction', *Comité de Documentation historique de la marine,* Vincennes, 1995, pp. 211-32.

Fendon, E. N., *Early Tahiti as the Explorers Saw It 1767–1797,* Tucson, 1981.

Ferrando, R. (ed.), *Pedro Fernando de Quiros: Descubrimento de la regiones australes*, Madrid, 1986.

Fleurieu, C. P. de, *Découvertes des François en 1768 et 1769 dans le sud-est de la Nouvelle-Guinée et reconnaissance postérieure des mêmes terres par des navigateurs anglois qui leur ont imposé de nouveaux noms*, Paris, 1790.

Freycinet, L. C. D. de, *Voyage autour du monde exécuté sur les corvettes de S.M. L'Uranie et La Physicienne en 1817–1820*, 5 vols, Paris, 1827-39.

Frézier, A. F., *A Voyage to the South Sea and along the coast of Chile and Peru in the years 1712, 1713 and 1714*, London, 1717.

Froger, F., *Relation d'un voyage fait en 1695…par une escadre de vaisseaux du Roy, commandée par M. de Gennes*, Paris, 1698.

Gallagher, R. E. (ed.), *Byron's Journal of his Circumnavigation 1764–1766*, Hakluyt Society, London, 1964.

Gautier, J. M. 'Apogée et déclin du mirage tahitien en Angleterre et en France (1766-1802)', *Journal de la Société des Océanistes*, 6, Dec. 1951, pp. 270-73.

——, 'Tahiti dans la littérature française à la fin du XVIIIe siècle: quelques ouvrages oubliés', *Journal de la Société des Océanistes*, 3, 1947, pp. 46-56.

Giraud, Y., 'De l'exploration à l'Utopie: notes sur la formation du mythe de Tahiti', *French Studies*, 1977, XXXI, pp. 26-41.

Goebel, J. , *The Struggle for the Falkland Islands*, New Haven, 1927.

Gonnard, R., *La Légende du Bon Sauvage*, Paris, 1946.

Gough, B. M., *The Falkland Islands, Malvinas: The Contest for Empire in the South Atlantic*, London, 1992.

Graaf, N. de, *Voyages aux Indes orientales et d'autres lieux de l'Asie*, Amsterdam, 1719.

Gunb-Fakim, Guého, Bissoondoyal, *Plantes médicinales de Maurice*, Port-Louis, 1995.

Hanke El Ghomri, G., *Tahiti in der Reiseberichterstattung und in der literarischen Utopien Frankreichs gene end des 18 jahrhunderts*, Munich, 1991.

Harlow, V. T., *The Founding of the Second British Empire 1763–1793*, 2 vols, London, 1952-64.

Hawkesworth, J., *An Account of the Voyages undertaken by Order of His present Majesty for making Discoveries in the Southern Hemisphere, and successively performed by Commodore Byron, Captain Wallis, Captain Carteret and Captain Cook*, 3 vols, London, 1773.

Henry, T., *Tahiti aux temps anciens*, Paris, 1968, transl. of *Ancient Tahiti*, Honolulu, 1928.

Heyerdahl, T., *American Indians in the Pacific*, London, 1952.

Horner, F., *Looking for La Pérouse: D'Entrecasteaux in Australia and the South Pacific 1792–1793*, Melbourne, 1995.

Howe, K. R., 'The Fate of the "Savage" in Pacific Historiography', *New Zealand Journal of History*, 11, 1977, pp. 137-54.

Hume, H., 'Seamen and Savants: The Interaction between Naval Officers and Civilian Scientists on the Pacific Voyages of Exploration and Scientific Research', University of Latrobe thesis, 1998.

Jack-Hinton, C, *In Search of the Solomon Islands 1567–1838*, Oxford, 1969.

Jacquemont, S., 'Le Mythe du Pacifique dans la littérature', in Rousseau, M. (ed.), *L'Art océanien*, Paris, 1951.

Jacquier, H., 'Le mirage et l'exotisme tahitiens dans la littérature', *Bulletin de la Société des études océaniennes*, 7, Nos 72-4, 1944-5, pp. 3-7, 50-76, 91-114.

Kelly, C., *La Austrialia del Espíritu Santo*, Hakluyt Society, 2 vols, Cambridge, 1966.

Kelly, L. G., *Marion Dufresne at the Bay of Islands*, Wellington, 1951.

La Barbée, Linyer de, M., *Le Chevalier de Ternay*, Paris, 1972.

La Borde, J. B. de, *Histoire abrégée de la mer du Sud*, 3 vols, Paris, 1791.

——, *Mémoire sur la prétendue découverte faite en 1788 par des Anglois*, Paris, 1790.

La Caille, N. L., *Caoelum australe astelliferum*, Paris, 1763.

Lalande, J. J. de, *Connoissance des mouvemens célestes pour l'année 1767*, Paris, 1745.

Langdon, R., *The Lost Caravel*, Sydney, 1975.

Le Maire, Jacob, *Mirror of the Australian Navigation*, Introduction by E. Duyker, Sydney, 2000.

Levi, A., *Guide to French Literature: Beginning to 1789*, Detroit, 1994.

Levron, J., *Choiseul: un sceptique au pouvoir*, Paris, 1976.

Ly Tio-Fane, M., *Mauritius and the Spice Trade*, 2 vols, Port Louis, 1958-70.

Magellan, see Stanley of Alderley, below.

Malleret, L., *Pierre Poivre*, Paris, 1974.

Mannevillette, J. B. N. d'Après de, *Le Neptune oriental ou Routier général des côtes des Indes orientales et de la Chine*, Paris, 1745.

Margueron, D., *Tahiti dans toute sa littérature*, Paris, 1989.

Markham, C. (ed.), *The Voyages of Pedro Fernandez de Quiros, 1565 to 1606*, 2 vols, Hakluyt Society, London, 1904.

Mayr, E., *Birds of the Southwest Pacific*, New York, 1945.

McCormick, E. H., *Omai: Pacific Envoy*, Auckland, 1977.

McEwen, C et al., *Patagonia*, London, 1997.

Meek, R. L. *Social Science and the Ignoble Savage*, Cambridge, 1976.

Mendaña, see Amherst and Thomson, above.

Métra, F., *Correspondance secrète politique et littéraire*, Paris, 1787.

Mollat, M. and Taillemite, E., *L'Importance de l'exploration maritime au Siècle des lumières (A Propos du voyage de Bougainville)*, Paris, 1982.

Montbard, Mme de, *Lettres tahitiennes*, Paris, 1786.

Moutemont, A., *Voyages autour du monde par Bougainville, Cook, Marion, Lapérouse…* Paris, 1853.

Mulert, F. E., *De Reis van Mr Jacob Roggeveen*, The Hague, 1911.

Muse, C. & S., *The Birds and Birdlore of Samoa*, Washington, 1982.

O'Reilly, P., *Bibliographie de Tahiti et de la Polynésie française*, Paris, 1967.

——, *Tahitiens, répertoire biographique de la Polynésie française*, Paris, 1975.

Oliver, D. L., *Ancient Tahitian Society*, 3 vols, Honolulu, 1974.

Parkinson, S., *A Journal of a Voyage to the South Seas in His Majesty's Ship* The Endeavour, London, 1773.

Pearson, W. H., 'European Intimidation and the Myth of Tahiti', *Journal of Pacific History*, IV, 1969, pp. 199-217.

——, *Rifled Sanctuaries: Some Views of the Pacific Islands in Western Literature*, Auckland, 1984.

Pernetty, A. J., *Histoire d'un voyage fait aux isles Malouines en 1763 et 1764, avec des observations sur le détroit de Magellan et sur les Patagons*, 2 vols, Paris, 1770.

——, *Journal historique d'un voyage aux isles Malouines fait en 1763 et 1764*, 2 vols, Berlin, 1769, Paris, 1770.

Pingré, A. G., *Mémoire sur le choix et l'état des lieux où le passage de Vénus du 3 juin 1769 pourra être observé*, Paris, 1767.

——, *Mémoire sur les découvertes faites dans la mer du Sud avant les derniers voyages des Anglois et des François autour du monde*, Paris, 1778.

Portlock, N., *A Voyage Round the World*, London, 1789.

Prévost d'Exiles, A. F., *Histoire générale des voyages*, 20 vols, Paris, 1746-70.

Pritchard, J., *Louis XV's Navy 1748–1762: A Study of Organization and Administration*, Kingston, 1987.

Quiros, see Ferrando; also Markham, above.

Rey-Lescure, P., *Abrégé d'histoire de Tahiti*, Papeete, 1970.

——, 'Les Premiers Européens à Taïti', *Bulletin de la Société des études océaniennes*, 1953.

Ricklefs, M. C., *A History of Modern Indonesia since c.1300*, London, 1991.

Robson, J., *Captain Cook's World*, Auckland, 2000.

Roggeveen, see Mulert above; also Sharp, below.

Rossel, E. P. E. de (ed.), *Voyage de Dentrecasteaux envoyé à la recherche de la Pérouse*, 2 vols, Paris, 1808.

Rousseau, J. J., *Discours sur l'origine et les fondements de l'inégalité parmi les hommes*, Amsterdam and Dresden, 1754.

Ségur-Dupeyron, M. P., *La France, l'Angleterre et l'Espagne après la Guerre de Sept Ans*, Paris, 1866.

Sharp, C. A.. *The Discovery of the Pacific Islands*, Oxford, 1960.

——, (ed.), *The Journal of Jacob Roggeveen*, Oxford, 1970.

——, *The Voyages of Abel Janszoon Tasman*, Oxford, 1968.

Six, G., *Dictionnaire biographique des généraux et amiraux de la Révolution et de l'Empire*, Paris, 1934.

Smith, A. D. H. (ed.), *A Voyage to the South Seas in His Majesty's Ship the* Wagner *in the years 1740–1741, by John Bulkeley and John Cummins.* London, 1927.

Smith, B., *European Vision and the South Pacific 1768–1850*, Oxford, 1960, Yale, 1984.

Smith, H. M., 'The Introduction of Venereal Disease into Tahiti: A Re-examination', *Journal of Pacific History*, 10, 1975, pp. 38–45.

Société de l'histoire de l'île Maurice, *Dictionnaire de biographie mauricienne*, Port-Louis, 1941–86.

Somerville, H. T., *Commodore Anson's Voyage into the South Seas and around the World*, London, 1934

Spate, O. H. K., *The Pacific since Magellan* : I *The Spanish Lake*, Canberra, 1979. II *Monopolists and Freebooters*, Canberra, 1983.

Spencer, J. (ed.), *A Voyage to New Holland: the English Voyage of Discovery to the South Seas in 1699 by William Dampier*, Gloucester, 1981.

Stanley of Alderley, *The First Voyage Round the World, by Magellan*, Hakluyt Society, Oxford, 1874.

Taitbout, *Essai sur l'île de Tahiti*, Avignon, 1779.

Toussaint, A., *A History of Mauritius*, London, 1977 (translation by W. E. F. Ward of the 1971 French edition).

Troudé, O., *Les Batailles navales de la France*, Paris, 1867.

Valentijn, F., *Oud en Niew Oost-Endien*, Dordrecht, 1724–6.

Vermeulen, T., 'The Dutch Entry into the East Indies', in *European Voyaging towards Australia* (Hardy, J., and Frost, A., eds), Canberra, 1990, pp. 35–43.

Vibart, E., *1767–1797: Tahiti, naissance d'un paradis au Siècle des lumières*, Brussels, 1987.

Wafer, see Elliot-Joyce, above.

Wallis, H. (ed.), *Carteret's Voyage round the World*, 2 vols, Hakluyt Society, London, 1965.

Walter, R., *A Voyage round the World in the years MDCCXL, I, II, III, IV, by George Anson Esq, compiled by Richard Walter*, London, 1748.

Williams, G., *The Great South Sea: English Voyages and Encounters 1570–1750*, London, 1997.

Wood, G. A., *The Discovery of Australia*, London, 1926.

Wroth, L. C., *The Early Cartography of the Pacific*, New York, 1944.

INDEXES

Ahutoru (Louis, Poutavéri), xlii, lviii-lx, lxiii, lxxii, lxxv; meeting with the French, 60, 203-4, 226-7, 282; character, 108, 121, 297n1, 301; and Baret, 97n2; voyage to France, xxxv, 72, 75, 86-7, 196, 208, 236, 267, 285; health, 88, 171n1; at Samoa, 82; at New Hebrides, 93, 236, at New Ireland, 121; and monkey, 159-60; in Dutch East Indies, 150, 171n1; return, xxix, lviii, lix

Albani, Francesco, 73

Anson, George, xxiv, lxxii, 26, 203

Arnould, Sophie, xxxiii

Banks, Sir Joseph, xxxiii, lxxi, lxxiiin1, 75n2, 82n3, 93n2, 203n2, 231n2, 260n1, 298n2

Baret, Jeanne, xxxii, xlviii, 228-30; life, xli-xlii, lxix, 179; and Bougainville, 97; in Tahiti, lxiv, 228-30; in New Ireland, 280, 293; *baretia*, xlii

Baudin, Nicolas, lxxvi

Bellin, Jacques-Nicolas, 3n4, 4-6, 36, 27-9, 50, 96, 101-2, 108

Boucher, François, 62, 73

Bougainville, Jean-Pierre de, xxv

Bougainville, Louis-Antoine de, life, xxiv-xxvii, lxxvii; character, lx, 87, 104; marriage, xxvi; early plans, xix-xx, xxv; Falklands colony, xx-xxiii, xxv, 224; earlier voyage to Magellan, 19; health, 22, 77; servants, 194-6, 198; *bougainvillea*, xli, li, lxxvi-lxxvii; — Bay, lxxvi, 18, 23; — Island, lxv, lxvii, lxxvi; Mt — , lx, lxxvi; — Peak, lxxvi, 134n1; — Reef, lxiv, lxxvi; — Strait, lxiii, lxxvi, 94, 113n2, 116n1, 139n2

Bougainville de Nerville, xlv, l, li, 194, 196

Bourdet, dentist, 28

Bournand, Alexandre de Lamotte-Laracé de, xlviii-xlix, 20, 113, 170, 183, 258; life, xxvii-xxviii, lxii; servant, 194; — Bay, 19; —Island, lxvii, 127

Bouvet de Lozier, xliv

Buache, geographer, lxvii, 290

Buet, François-Nicolas, xxxviii, li

Buffon, Georges-Louis, lxx, lxxiv

Byron, Commodore John, in Magellan, xxii, 11, 36n1; Pacific voyage, xxiv, lv, lxxii-lxxiii, 52n1, 145

Caro, Jean-Louis, ix, xli, lxiv, lxv, 171, 173, 223, 304; life, xxxiv-xxxv; journal, 199-222; at Tahiti, 202-8; off Samoa, 208-10; off New Hebrides, 210-12; off New Guinea, 213-14; at New Ireland, 214-22

Carteret, Philip, xxiv, liv, lv, lxvi, lxxiii, lxxiv, lxxvi, 116n2, 117n1, 145n5, 149n3, 172n2, 241n3; in Magellan, 11n2, 18n5, 26, 217; in New Ireland, lxvii, 118-19, 121n1, 124n5, 125n2, 217, 274n1; meets Bougainville, lxix

Chapelle, Nicolas, stowaway, 151n1, 152

Charnières, Charles-François de, xxxix, 46n1

Chenet, Jean, drowned, 38, 191

Choiseul-Praslin, duc de, xx, l, lxxi, lxxiii, 70, 275; duchesse de, lviii; — Island, lxv, lxxi, lxxvi, 113-16, 291. (See also Port Praslin)

Cicero, 73, 87

Claret de Fleurieu, lxvii, lxxviii.n3

Columbus, Christopher, 88, 233, 295

Commerson, Philibert, xxxviiin2, li-lii, lxxi, lxxvi-lxxvii, 77n1, 97, 125n2, 249, 257n1; life, xl-xli, 230, 293; brother, xxxii, xl, xli; character, xxx, xxxv-xxxvi, xl-xli, 304; journal, xxxviin3, 296-305; in Magellan, 12, 19; and Tahiti, lvi, lvii, lxxi-lxxiii, lxxv, 68; in New Ireland, lxvii, 119-20, 123, 304; at Mauritius, lxix, 179; — Island, lxviii. (See also Jeanne Baret)

Constantin, apprentice, xxxviii

Cook, James, xxiv, xxxiii, xlii, xliv, lv, lxii, lxiii-lxiv, lxx, lxxii-lxiii, lxxvi, 179n3, 241n3; and Australia, 18n5; and Polynesia, 48n2, 50n2; at Tahiti, lvi, lxi-lxii, 61n1,

66n3, 72n3, 254n1; and New Hebrides, 89n2

Couture, Denis, master, 138, 184

Dalrymple, Alexander, lxxxiii

Dampier, William, liv, lxiii, 101-2, 115, 125, 130, 163n1, 219, 222, 277, 293; — Strait, 113, 137n1, 139n1, 291, 294

D'Anville, Jean-Baptiste, cartographer, 165

D'Après de Mannevillette, cartographer, 164-5, 166-8, 173, 177

Davis, Edward, liii-lv, 41, 281. (See also Davis Land)

De Brosses, Charles, xix, xxiv, lxii, lxvi, lxx, lxxiii-lxxiv, 32, 95, 208n2

Delisle, Guillaume, cartographer, 288, 290

Denis, boatswain, 113-14

D'Entrecasteaux, Bruny, xxvi, lxxvi, 109n2, 147n3

De Gennes, Jean-Baptiste de, 18, 24, 30

Donat, Joseph, xxxv, 303

D'Oraison, Henri de Fulques, 283, 303; life, xxviii, 183; servant, 194, 196; — Island, lxvii, 127n1

Du Bouchage, Jean-Jacques-Pierre de Gratet, li, lxxv, 20, 25, 81, 121, 194; life, xxviii, lxix, 179, 183; — Island, lxvii, 127n1, 128-9, 296-7

Duclos-Guyot, Alexandre, xxx, 184. (See also, Guyot, Alexandre)

Duclos-Guyot, Nicolas-Pierre, xxv, xxix-xxx, lxxiv, 67-8, 194, 216, 244; life, xxvi-xxvii, 183; in Magellan, lxxiv, 18, 27; — Bay, 15; — Island, lxvii, 118

Duclos-Guyot, Pierre, lxix; life, xxx, xxxvi, 179, 184; shares journal with Commerson, xxxviin3, 296

Dudley, D.R., cartographer, lxvi

Dumont d'Urville, Jules S.C., 134n1

Fénelon, François de, 73

Fesche, Charles-Félix-Pierre, ln2, 38n1, 77n1; life, xxx-xxxi, 179, 183-4; at Tahiti, 60n2, 70n2; journal, 249-79

Frézier, Amédée-François, 32-3

Gallego, Herman, lxvi, 288

Guyot (aka Duclos-Guyot), Alexandre, voyage to Magellan, 10n1

Hawkesworth, John, lxi, lxxiii

Hercouet, René, xxxv

Herpin, Pierre-Joseph, 179, 185

Hervel, Joseph, 179, 198

Heylen, Peter, lxvi

Juan y Santacilla, Jorge, 37

Kerguelen-Tremarec, Yves de, xliv, lxxvi

Kerhué, Jacques-Marie de Crazemel de, 92, 194; life, xxviii-xix, 144n3, 184, 196; — Island, 129n1, 130

Klein, Jacob, 124

Labare, armourer, 229

La Caille, Nicolas-Louis, astronomer, 177

La Fontaine-Villaubrun, on Étoile, xxxv, 223, 226-7

La Giraudais, François Chesnard de, xxxv, xlvi, xlviii-xlix, l, 7-8, 18, 40, 34, 108, 138, 141, 216, 220, 224n1, 226-7, 235-6, 279-80, 299, 303-4; life, xxxiii-xxxiv; earlier voyage to Magellan, 9, 71-2

Lalande, Lefrançais de, xxxix, lxx-lxxi

Landais, Pierre, xli, 32, 212, 302; life, xxxv-xxxvi

La Pérouse, Jean-François de Galaup de, x, xxvi, lv, lxx, lxxvi, 71n2

Laporte, Louis-Claude, 44-5, 66n5, 67; life, xxxii, 184

Launay, Julien, 152, 187

Lavaisse, Jean-Baptiste, chaplain, 74, 184, 207, 248, 265, 274; life, xxxii, 179

Lavarye-Leroi, Pierre-Marie, 208; life, xxxvi

Le Clerc, Kop, 145, 150

Le Corre, Josselin, 196, 197; life, xxix, 183-4; — Island, 130

Le Maire, Jacob, lx, lxiii, 83, 84n1, 117n1, 127n1, 136n1, 208n2

Lemoyne de Montchevry, Jean-Robert-Suzanne, 113; life, xxxvi-xxxvii, 179, 196

Leroi, Isaac, pilot, 78, 193, 207

Loran, Jean, 120

Louis XV, xliv, lxv, lxxiii, 70, 124

Macbride, John, at Falklands, xxii-xxiii, l

Magellan, Ferdinand, lxiii, 36

Marcant (also Marcand), 32-3

Marion Dufresne, xxix, lviii, lix, lxxvi, 246

Maurelle, Francisco, 131n1

Mendaña, Alvaro de, lxiii, lxvi, 288

Merven, Joseph, port captain, 178

Michau, clerk, xxxvii-xxxviii

Molière, 96, 106

Montesquieu, Secondat de, 120n6

Narborough, John, 14n1, 24, 34-5, 37

Nassau-Siegen, Charles-Othon d'Orange et de, xlix, liii, lxv, 75n2, 194, 198, 271-2; life, xxxiii; in Tahiti, lvii, 60n2, 69, 207, 234-5, 260, 265-6, 284-6; at New Hebrides, lxiii, 288-9, 302; in New Ireland, lxvii, 293-4, 303; dispute with Landais, xxxv, 302; journal, 281-95

Oger, Charles, 179, 185; — Island, 134-5

Ouman (Houman), Hendrick, 146-7, 149-50

Oury (Ourry), Jean-François, pilot, 179, 185, 304

Pernetty, Dom, lxxii, 9n5

Pingré, Alexandre-Guy, astronomer, liv, lv, 148n1, 176-7

Poissonnier, Pierre-Isaac, 44n1, 281

Poivre, Pierre, xxxi, xxxix, xli, lxix, 172n3, 178, 183, 298

Pompadour, Marquise de, lxxi

Prévost, Antoine-Pierre, 24-5, 287

Quiros, Pedro Fernandez de, lv, lxii-lxiii, lxvi, lxxiv-lxxv, 199, 208n2, 290; in Polynesia, 47-8, 50n2, 51, 53-4, 200-202; at New Hebrides, 89n3, 95-6, 101, 210, 212

Riouffe, Alexandre-Joseph, xxxvii

Rogers, Woodes, 163

Roggeveen, Jacob, liv, lv, lxiii, lxxv, 52n1, 53-4, 76, 96, 248n1, 281

Romainville, Charles Routier de, xxxix; in Magellan, 19, 37; at Tahiti, 56; at Port Praslin, 122n2; remains in Mauritius, 179

Rossel, Elizabeth-Paul-Édouard, 147n3

Rousseau, Jean-Jacques, xi, lvi, lxxii, 12, 25, 107, 257n1

Saavedra, Alvaro de, 131n1

Saint-Germain, Louis-Antoine Starot de, xlvii, xl, lxv, 78n2, 249, 250n1, 255, 263n1, 272n1, 275n1; life, xxxi-xxxii, 179, 180n1, 184, 198

Samiento de Gamboa, Pedro, 9, 15n3

Schouten, Willem, (see also Le Maire), liii, 84n1, 127n1, 208n2; — Islands, 136n1

Shortland, John, lvii

Solander, Daniel Carl, lxxi, lxxiiin1

Surville, Jean-François-Marie de, x, liv, lv, lxvii, lxxvi

Suzannet, Jean-Baptiste-François de, life, xxxviii, xxix, 183; servant, 194; at Tahiti, 61; — Island, lxvii, 127n1, 128-9, 276-8

Tacitus, 28

Tasman, Abel, lxiii, 11n1

Terence, 96n5

Thomas, barrel maker, 245

Torres, Luiz, 96n1

Tostain, Jean, 175, 188

Tremergat, Louis de, 180

Ulloa, Antonio de, 37

Valentijn, Fr., 143

Vancouver, George, 18n5

Vaugondy, Robert de, lxvi, 208n2

Véron, Pierre-Antoine, xv; life, xxxix-xl, 196; instruments, 1n4, 2, 121; observations, lxvii, lxix, lxxv, 1-3, 6, 26, 57, 78, 85, 87, 101, 109, 161, 176-7; made in Magellan, 13, 17-20; in the eastern Pacific, 36, 38-9, 42-4, 46, 52, 54; at Tahiti, 61, 68, 70, 267; at Samoa, 81, 84; at the New Hebrides, 91; in New Ireland, lxvii, 120-22, 245, 276; at Buru, 151; at Mauritius 179; *Veronia*, xl

Virgil, 7n1, 24n1, 29, 32, 36, 63, 71, 102, 109

Vivez, François, surgeon, xli, l, lviiin1, lx, 52n1, 97n2; life, xxxviii, 223-4; at Tahiti, 60n2, 225-36; off Samoa, 236-7; off the New Hebrides, 237-40, 302; off New Guinea, 240-42, at New Ireland, 244-7; and Commerson, 302

Wafer, Lionel, liv-lv

Wallis, Samuel, xxiv, liv, lv, lxxiii, 26n5, 145n5; in Magellan, 11n2, 35n1; in Tahiti, lv-lvi, lx-lxi, lxxiv, 55n2, 60n1, 82n3, 204n1, 226n2, 301n1

Wood, John, in Magellan, 14

INDEX OF SHIPS

Afrikaanisch Galley, 76n1

Aigle, xx, xxii, xxvii, xxx, xxxi, xxxvi, xxxix, lxxii; earlier voyage to Magellan, liii, 10, 18n1, 19n2,

Alliance, USS, xxxvi

America, xxvii

Andromaque, xxviii

Anna, 26

Barbeau, xxix

Barbue, xxxv

Batchelor's Delight, liii–liv

Belle-Poule, xxvii

Bien-Aimé, xxvi

Boudeuse, xvi, xxiv–xx, xlii–xliii, xlviii; masting, 1n2; officers and men, xxvi–xxxiii; voyage, xlix–lxx

Bricole, xxxvii, 223

Capricieuse, xxviii

Castries, 246n2

Chézine, xxvii

Comte de Lamoignon, xx

Confiance, xxix

Côte d'Or, xxxvi

Coulisse, xxxviii

Diadème, xxxix

Dolphin, lvi, lx–lxi, 60n1, 145n5

Duc-de-Duras, xxxiv, 199

Duchess, 163n1

Duke, 163n1

Eendracht, 83 n4

Endeavour, xxxiii, xlii, lvi, lxiv, 179n3

Esmeralda, l, lii

Étoile, xxv, xxx, xxxi, xxxii, xxxiv, xlii–xliii, xlviii, 20; at Montevideo, xxxvii, li; officers and men, xxxiii–xlii; voyage, l–lxviii; earlier voyage to Magellan, 10, leak, 20

Étourdie, xxix

Fine, xxxiv

Flamand, xxx, xxxvi

Flore, xxxvii

Formidable, xxxviii, 223

Garonne, xxxv, xxxvii, xxxviii, 224

Gloire, xxxiv

Guerrier, xxvi, 223

Héros, xxxvi

Hoorn, 83n4

Inconstante, xxxvii

Jason, l

Languedoc, xxvi, xxx, xxxviii

Liebre, l

Lys, xxxiv

Magnifique, xxix

Malicieuse, xxxix

Marquis-de-Castries, xxix

Mascarin, lix, 246n2

Mortemart, xxx

Nuestra Señora de los Remedios, l

Osterley, xxvii

Patroite, xxxvi

Philippe d'Orléans, xxx

Roebuck, 101n2

Rossignol, xxix

Sainte-Barbe, 32n3

Saint-Esprit, xxxvi

Saint-Laurent, xxxiv

Sceptre, xxix, xxxix

Sphinx, xx, xxxiv, xxxv

Swallow, lxix, 172n2

Tamar, 145n5

Terpsichore, xxxi

Tourterelle, 223

Utile, 172

Victoire, xxvii, 296

Vigilant, xl

Villevault, xxxiv

Zéphir, xxxv

Acadians, xix,xxv, li
Admiralty Islands, lxviii
Ascension Island, lxix
Australia, lxiii, lxiv, lxxv-lxxvi, 100-101
Batavia, lxviii-lxix, 145-6, 149, 168, 172, 179, 197
Bonito, 41n2, 42, 78, 217n3
Boudeuse Island, 132
Breadfruit, 203, 230, 232, 238, 261, 298
Brest, xxvi, xxviii, xxix, xxxvi, xlviii, xlix, 186-7, 189-92, 194
Buenos Aires, xxx, l-li, 12, 26, 46n2, 224, 229, 249, 281
Buru, lviii, lxviii, 142n1, 143, 146-8, 151-3, 175, 196
Canada, war in, xix, xxiv-xxv, xxvii, xxix, xxxiv-xxxv, 22n2, (See also Acadians)
Candia (Heraklion), 107
Celebes (Sulawesi), 152-5, 160, 162
China, xliv-xlvi, xlix, lxxvi, 258
Choiseul Island, lxv, lxxvi
Coconuts, 48, 51, 54, 59, 82, 92, 115, 119, 131, 148, 160, 162, 201, 203, 205, 208-9, 212, 215, 221, 224, 230, 236, 238-9, 240, 243, 257, 268, 271-2, 278, 287, 289, 292, 298; fibre, 232, 237; for clothing, 226-7; oil, 63, 66, 73, 258, 260, 264-5, 283, 299; weapons, 200n1
Davis Land, liii-lv, 41, 76n1
Dollond, telescope, 121
Dutch East Indies (Indonesia), lxvii-lxix, lxxvi, 134n1, 140-73. (See also, Batavia, Buru, Java, Moluccas)
Dysentery, lxix, 171n1, 173, 175, 179
Easter Island, liv-lv
Espiritu Santo (Land of the Holy Spirit), xliv, xlv, lxii-lxiv, 210, 268, 290
Falkland Islands, xlii, 26n5, 273, 304n1; English at, xxii-xxiii, l-li; French settlement at and voyages to, xi, xx, xxii, xxv, xxvii, xxx, xxxiv-xxxvi, xxxix, xliv, xlv, xlix, l-li, lxxii-lxxiii, 224, 281; Spanish

at, l-li; similarity with Straits of Magellan, 12 ,17, 23
Futuna (Hoorn Islands) 86
Indonesia, see Dutch East Indies
Iron, at Tahiti, 82
Isle de France (also Île de France, Mauritius), xi, xxvii, xxix-xxxii, xxxv, xxxix, xli, xlvi, lix, lxix, lxxi, 143n2, 172n3, 173, 177-80, 197-8, 249, 296, 297n1, 298n1, 299n1
Java, 165-71
Jesuits, li-lii
Leprosy, lxiii, 91, 93, 238, 289
Louisiades, lxv, lxxvi, 105-11
Magellan, Straits of, xi, xxii, xxv, xxxixn1, xlv, lii-liii, lxxiv, lxxvi, 6-36, 107, 136, 224, 229 234, 239, 244, 249, 279, 281
Malouines, see Falkland Islands
Manta ray, 225
Mauritius, see Isle de France
Melanesians, lviii, lxv, lxxv, 114-15, 237-40, 242, 245, 268-9, 289-90, 292, 294, 297
Moluccas, xxxix, 139-42, 149, 158, 163n1, 165, 242, 295; search for spices, xxxv, xlix, lxviii, 172n3
Montevideo, xxxvi-xxxvii, xlix-lii, 1, 5, 5n1, 20, 45, 123, 196, 210, 224-5, 273, 296
Nantes, xliii, xlv, xlviii, xlix, xlii, 188, 190, 193-4
New Britain (see also New Ireland), lxv, lxvii, 113, 117-29, 141, 143n1, 150, 214, 244, 302-3
New Cythera (Tahiti), lvi-lxii
New Guinea, xliv-xlv, lxv-lxvi, lxviii, lxvi, 84, 96, 102-9, 133-9, 142, 165, 240-2, 294-5
New Hanover, 129n1, 221n1
New Hebrides (see also Espiritu Santo), xxxviin3, lxii-lxiv, lxvi, lxxiv-lxxvi, 88-97, 210-12, 237-40, 267-70, 288-90, 301-2; canoes, 209-10, 212, 268, 289; islanders, 92-3, 95, 209-12, 239-40, 268-9, 289-90, 301

New Ireland (see also New Britain),
  xxxixn1, lxvii-lxviii, lxxvi, 117-30,
  214-22, 244-7, 274-6, 292-4, 303-4;
  islanders, 127-9, 214-21, 245-6, 277-9
New Zealand, 11n1
Patagonians, lii-liii, 9-10, 30, 199, 279;
  meetings with Bougainville, 10-12, 21,
  26, 28-30; "giants", liii, lxxiv, 11; huts, 17;
  child's death, 29-30
Philippines, xliv, xlix, 298
Polynesians, xlii, lv-lxii, lxv, lxxv, 48, 50,
  115n2, 199-204, 224-7, 250, 253-67, 282,
  296-300; migrations, 48, 82n1, 200. (See
  also Tahiti, Samoa)
Port Praslin, 118n2, 122n2, 125n2, 245,
  273
Portuguese, li-lii
Psalms, 27
Rats, 41, 281; as food, 97, 242, 279, 291
Rio de Janeiro, xxxvii, xxxix, li, lxxvi, 196,
  224, 249, 273, 281
Rochefort, xxvii, xxxi, xxxiv, xxxvi,
  xxxvii-xxxviii, xli, xlv-xlvi, xlviii, xlix, l,
  97, 196
Saint-Servan, 184-91, 194-5, 197
Saint-Malo, xx, xxii, xxvi, xxvii, xxx, xxxvi,
  lxx, 184-91, 193-5, 197-8
Sala-y-Gomez Island, lv
Samoan Archipelago, lx, lxxv, 79-84, 86,
  208-10, 236-7, 286-8; canoes, 81, 83, 209,
  236-7
Savage (Noble Savage), xi, lxxii, 12, 28, 61,
  72-3, 121, 263
Scurvy, lxviii, 54-5, 87, 117, 120, 122, 124-5,
  133, 136, 139, 146, 147n3, 151, 158, 173,
  205, 225, 234, 239, 241, 248, 281, 293
Solomon Islands, lxiii, lxv-lxvii, lxxiv, lxxvi,
  110-17, 208n2, 270-3, 237-8; canoes, 112,
  272; islanders, 112, 114-15, 117, 272
Spain, concern over Falklands, xxii-xxiii,
  xxv, xxliv; and the Pacific, lxi, 37n4, 82n3;
  and New Hebrides, lxiii
Spices, 141, 144, 148-9, 155, 172, 298n1;
  cloves, 143, 145, 155; nutmeg, 155
Tahiti, xxxix.n1, xlii, lv-lxii, lxxiv, 55-75,
  225-36, 255-67, 272-3, 296-301; Baret at,
  288-30; canoes, 59, 81, 83, 209, 226-7,
  264, 286; disease at, lx-lxii, 72, 233;
  inhabitants, 59-64, 66-75, 82, 112, 226-36,
  253-67, 282-6, 296-301; myth of,
  lxxi-lxxii, lxxiv, 60n2, 72, 75, 263; report
  of gold, 210; women, 60, 63, 70, 73, 231-2,
  236, 255-7, 259, 282-3, 297-8
Terra Australis, liv, lxii-lxiii, lxvi
Tierra del Fuego, liii, 6-9, 15-16, 21-3, 27;
  inhabitants, lii, 9-10, 28
Torres Strait, 241n3
Tuamotu Archipelago, 48-54, 199-202, 210,
  224-5, 249-51, 281-2
Vanuatu, see New Hebrides, Espiritu Santo
Venereal Disease, lx-lxii, 73, 88-9, 233
Virgins, Cape of, 6-9
Water supply, 41, 44-7, 54, 62, 64, 67, 101-2,
  108, 117-18, 146, 204, 213, 227, 234, 281,
  291; in New Ireland, 216-17, 244, 275,
  305; at Tahiti, 205, 230, 258; in the
  Tuamotus, 201
Yaws, lxi

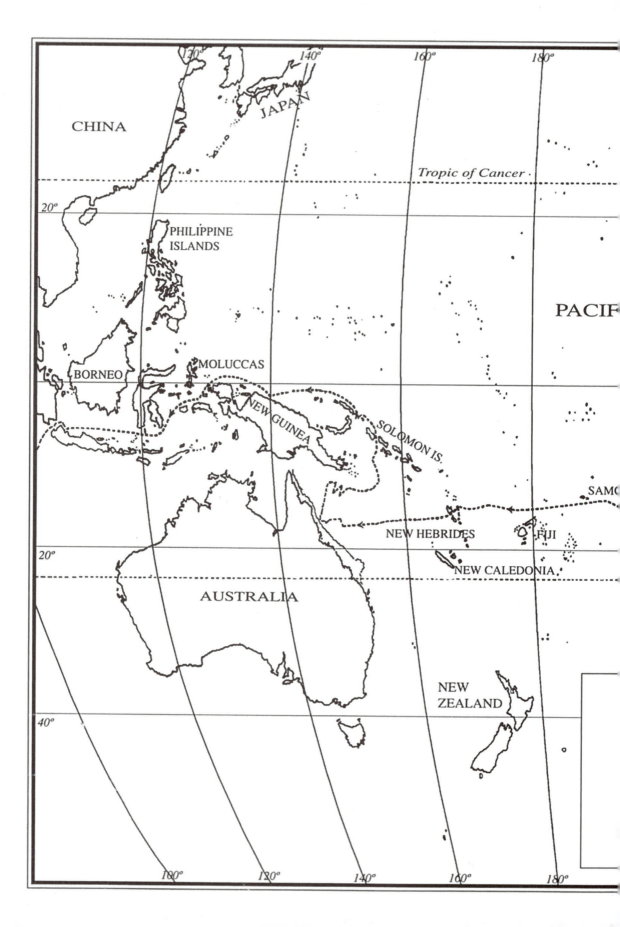